Novel Materials and Water Purification

Towards a Sustainable Future

Chemistry in the Environment Series

Editor-in-chief:
Dionysios D. Dionysiou, *University of Cincinnati, USA*

Series editors:
Rajasekhar Balasubramanian, *National University of Singapore, Singapore*
Triantafyllos Kaloudis, *Athens Water Supply and Sewerage Company (EYDAP S.A.), Greece*
Rafael Luque, *University of Cordoba, Spain*

Titles in the series:
1: Graphene-based 3D Macrostructures for Clean Energy and Environmental Applications
2: Metallurgical Slags: Environmental Geochemistry and Resource Potential
3: Functional Hybrid Nanomaterials for Environmental Remediation
4: Emerging Nanotechnologies for Water Treatment
5: Biological Treatment of Industrial Wastewater
6: Advances in Functional Separation Membranes
7: Persulfate-based Oxidation Processes in Environmental Remediation
8: Advanced Ozonation Processes for Water and Wastewater Treatment: Active Catalysts and Combined Technologies
9: Environmental Nanopollutants: Sources, Occurrence, Analysis and Fate
10: Photo- and Electrochemical Water Treatment: For the Removal of Contaminants of Emerging Concern
11: Ionic Liquids for Environmental Issues
12: Novel Materials and Water Purification: Towards a Sustainable Future

How to obtain future titles on publication:
A standing order plan is available for this series. A standing order will bring delivery of each new volume immediately on publication.

For further information please contact:
Book Sales Department, Royal Society of Chemistry, Thomas Graham House, Science Park, Milton Road, Cambridge, CB4 0WF, UK
Telephone: +44 (0)1223 420066, Fax: +44 (0)1223 420247
Email: booksales@rsc.org
Visit our website at books.rsc.org

Novel Materials and Water Purification
Towards a Sustainable Future

Edited by

Grigorios L. Kyriakopoulos
National Technical University of Athens, Greece
Email:gregkyr@chemeng.ntua.gr

and

Miltiadis G. Zamparas
Hellenic Open University, Greece
Email:mzamparas@upatras.gr

ROYAL SOCIETY
OF **CHEMISTRY**

Chemistry in the Environment Series No. 12

Print ISBN: 978-1-83767-049-9
PDF ISBN: 978-1-83767-166-3
EPUB ISBN: 978-1-83767-167-0
Print ISSN: 2516-2624
Electronic ISSN: 2516-2632

A catalogue record for this book is available from the British Library

The Royal Society of Chemistry is a charity, registered in England and Wales, Number 207890, and a company incorporated in England by Royal Charter (Registered No. RC000524), registered office: Burlington House, Piccadilly, London W1J 0BA, UK, Telephone: +44 (0) 20 7437 8656.

For further information see our website at www.rsc.org

Printed in the United Kingdom by CPI Group (UK) Ltd, Croydon, CR0 4YY, UK

Preface

The current international literature is concentrated mainly on dispersed technological approaches for water resource remediation, and adopts a localized way of demonstrating these technological applications. In addition, the current literature often fails to cover and represent the wide geographical range of the technologies and materials applied, both nationally and globally. In response to these research constraints this book "*Novel Materials and Water Purification*" provides a rigorous and systematic approach for the direct and indirect use of novel materials for water management and water purification. These approaches are not limited to aqueous outflows, but also extend to cover a variety of water analysis services, such as in urban, suburban, rural and isolated contexts. As the global water crisis motivates industry to look for alternative water supplies, novel materials present a significant potential for use and for the safer disposal of hazardous water sources. Moreover, the potential for re-using otherwise lost sources of waste, gray or untapped water for further use have emerged.

More specifically this book provides an in-depth review of current technologies and emerging applications, including: biomaterials, advanced nanomaterials, polymeric materials, green novel materials, magnetic absorbents and photocatalysts, nanocellulose, graphene-based materials and engineered macrostructures, quaternary ammonium anion exchangers prepared from sugarcane bagasse, titanium dioxide-based composites, as well as technological and procedural advances, including: synthesis of materials, membranes and desalination in water purification. The book thus provides an overview of niche treatment technologies for water source purification, and includes the following chapters.

Chapter 1: *Biomaterials for Water Purification: Dyes, Heavy Metals, and Pharmaceuticals*, by Vinayak Adimule, Nidhi Manhas, Shashanka Rajendrachari.

Chemistry in the Environment Series No. 12
Novel Materials and Water Purification: Towards a Sustainable Future
Edited by Grigorios L. Kyriakopoulos and Miltiadis G. Zamparas

Published by the Royal Society of Chemistry, www.rsc.org

Chapter 2: *Sustainable Synthesis of Green Novel Materials for Water Purification*, by Amal I. Hassan, Hosam M. Saleh.

Chapter 3: *MXene-based Separation Membranes for Water Purification and Desalination*, by Akbar Alia, Ayaz Ali Memon, Jun Yang, Javeed Akhtar, Khalid Hussain Thebo.

Chapter 4: *Advanced Nanomaterials for Removal of Emerging Organic Pollutants from Water*, by Adedapo O. Adeola, Patricia B. C. Forbes.

Chapter 5: *Magnetic Adsorbents/Photocatalysts for Water Purification: Progress and Challenges*, by M. G. Galloni, H. M. Adeel Sharif, A. Grainca, M. Rizwan Haider, R. Djellabi.

Chapter 6: *Nanocellulose-based Membranes for Water Purification: Multifunctional Nanocellulose Extraction, Characterization, Modification Strategies, and Current Release in Water Treatment and Environmental Remediation*, by Saly R. El Dakkony, Mahmoud F. Mubarak, Adel A.-H. Abdel-Rahman.

Chapter 7: *Graphene-based Materials for Water Remediation: Recent Advances on Pollutant Sorption, Photodegradation and Filtration*, by Bruno Henriques, Nicole Ferreira, Martina Kocijan, Milan Vukšić, Burak Ünlu, Serbülent Türk, Mahmut Özacar, Gil Gonçalves.

Chapter 8: *Responsive Polymeric Materials: Advances in Membrane-based Technologies for Water Treatment Processes*, by Perla Cruz-Tato, Alondra S. Rodríguez-Rolón, Jessika Pazol, Eduardo Nicolau.

Chapter 9: *Graphene-based Engineered Macrostructures for Water Purification*, by Chirag Batukbhai Godiya.

Chapter 10: *Textile Azo Dye Removal Using a Quaternary Ammonium Anion Exchanger Prepared from Sugarcane Bagasse*, by Natália Roberta Marques Tanure, Liliane Catone Soares, Luisa Cardoso Maia, Amália Luísa Pedrosa Xavier, Oscar Fernando Herrera Adarme, Guilherme Max Dias Ferreira, Eduardo Ribeiro de Azevedo, Luis Henrique Mendes da Silva, Leandro Vinícius Alves Gurgel.

Chapter 11: *Iron Oxide Nanomaterials for Water Purification*, by Ateeq Rahman, Likius Shipwiisho Daniel, Veikko Uahengo.

Chapter 12: *Factors Influencing TiO_2-based Composites for Water Decolorization: A Systematic Review*, by Likius Shipwiisho Daniel, Veikko Uahengo, Ateeq Rahman.

Chapter 13: *Enviro-friendly Nanomaterial Synthesis and Its Utilization for Water Purification*, by Huda Sharbini Kamaluddin, Katabathini Narasimharao.

Therefore, the book "*Novel Materials and Water Purification*" is an integrated and in-depth reference handbook in which diverse technologies and research approaches have been collected and systematically addressed in relation to water purification using innovative materials. By covering a wide geographical range and technological background for water management and water purification, as well as by demonstrating the multifaceted technological, operational and fabrication perspectives for water purification worldwide, this is a timely reference book of special interest for researchers of a wide spectrum of educational and scientific specializations, such as researchers and professionals in water science, water managers, materials scientists, professors, and university students undertaking research in the interdisciplinary scientific fields of environmental sustainability and technological competency.

The editors
Miltiadis G. Zamparas and Grigorios L. Kyriakopoulos

Contents

Chapter 1 Biomaterials for Water Purification: Dyes, Heavy Metals, and Pharmaceuticals 1
Vinayak Adimule, Nidhi Manhas and Shashanka Rajendrachari

1.1 Introduction 1
1.2 Characterization 2
1.2.1 Elemental Compositional Analysis 3
1.2.2 Scanning Electron Microscopy Investigation 3
1.2.3 FT-IR Spectroscopy Studies 3
1.2.4 X-ray Diffraction Investigation 4
1.3 Results and Discussion 5
1.3.1 Role of Zero-, One-, Two-, and Three-dimensional Nanomaterials for Polymeric Membrane Development 5
1.3.2 Role of Chitosan-based Nanomaterials and Developing Polymeric Membranes 8
1.3.3 Chitosan–Metal Oxide-based Polymeric Membranes 12
1.3.4 Role of Initial Dye Concentration 13
1.3.5 Performance Comparison of Various Types of Chitosan-based Polymers 14
1.3.6 Treatment of Water Contaminated With Various Pharmaceuticals 14
1.4 Conclusions 14
Authors Contribution 16
Data Availability 16
Funding 16

Chemistry in the Environment Series No. 12
Novel Materials and Water Purification: Towards a Sustainable Future
Edited by Grigorios L. Kyriakopoulos and Miltiadis G. Zamparas

Published by the Royal Society of Chemistry, www.rsc.org

Conflict of Interest 16
Acknowledgements 16
References 16

Chapter 2 Sustainable Synthesis of Green Novel Materials for Water Purification 24
Amal I. Hassan and Hosam M. Saleh

2.1 Introduction 24
2.2 Innovative Approaches Using Nanotechnology in the Remediation of Water 26
2.2.1 Photocatalysis: Transforming Water Treatment with Nanomaterials 26
2.2.2 Recent Advances in Water Purification Nanocomposites 28
2.2.3 ZnO: An Alternative Semiconductor for Water Treatment 30
2.2.4 Innovative Membrane Separation Techniques 32
2.2.5 Polymer Membranes for Enhanced Filtration 32
2.3 Green Materials for Bioremediation of Water 33
2.3.1 Bacteria 33
2.3.2 Protein Nanofibrils 35
2.3.3 Seaweed 36
2.3.4 Hydrogels and Aerogels 36
2.3.5 Carbonaceous Materials 37
2.3.6 Green Nanoparticles for Environmental Remediation 38
2.4 Future Prospects of Sustainable Materials for Water Purification 38
2.5 Conclusion 40
References 41

Chapter 3 MXene-based Separation Membranes for Water Purification and Desalination 49
Akbar Ali, Ayaz Ali Memon, Jun Yang, Javeed Akhtar and Khalid Hussain Thebo

3.1 Introduction 49
3.2 Methods for Preparation of MXene Membranes 51
3.3 Applications for Water Purification and Desalination 54
3.4 Concluding Remarks and Prospective Views 59
List of Abbreviations 60
Acknowledgements 61
References 61

Chapter 4 Advanced Nanomaterials for Removal of Emerging Organic Pollutants From Water 66
Adedapo O. Adeola and Patricia B.C. Forbes

4.1 Introduction 66
4.2 Graphene and Graphene-based Composites 68
4.3 Carbon Nanotubes and Composites 69
4.4 Concluding Remarks 71
References 75

Chapter 5 Magnetic Adsorbents/Photocatalysts for Water Purification: Progress and Challenges 78
M. G. Galloni, H. M. Adeel Sharif, A. Grainca, M. Rizwan Haider and R. Djellabi

5.1 Introduction 78
5.2 Synthesis of Magnetic Materials 81
5.2.1 Synthesis of Nanozerovalent Iron 81
5.2.2 Synthesis of Magnetite (Fe_3O_4) 83
5.2.3 Synthesis of Maghemite (γ-Fe_2O_3) 86
5.3 Catalytic Activation of $NaBH_4$ Towards Organic Pollutant Reduction 86
5.4 Fenton and Photo-Fenton of Magnetic Materials 88
5.5 Photocatalytic Oxidation of Pollutants 90
5.5.1 Photoactivity of Bare Magnetic Materials 90
5.5.2 Heterojunction-based Magnetic Photocatalysts 91
5.5.3 Role of Insulator 92
5.6 Conclusions 94
References 94

Chapter 6 Nanocellulose-based Membranes for Water Purification: Multifunctional Nanocellulose Extraction, Characterization, Modification Strategies, and Current Release in Water Treatment and Environmental Remediation 101
Saly R. El Dakkony, Mahmoud F. Mubarak and Adel A.-H. Abdel-Rahman

6.1 Introduction 102
6.2 Macro and Micro Sources of Cellulose 107
6.3 Types of Nanocellulose 107
6.3.1 Cellulose Nanocrystals 107
6.3.2 Cellulose Nanofibrils 108
6.3.3 Microfibrillated Cellulose 109
6.4 Modern Technologies for Preparing Nanocellulose Membranes 109

6.4.1 Bottom-up Methods 109
6.4.2 Top-down Methods 110
6.5 Advantages of Using Nanocellulose in Membranes 111
6.5.1 Antifouling Nanocellulose Membranes 112
6.5.2 Auto-self-healing Property of the Membrane 117
6.6 Conclusion 120
List of Abbreviations 120
Acknowledgements 121
References 121

Chapter 7 Graphene-based Materials for Water Remediation: Recent Advances on Pollutant Sorption, Photodegradation and Filtration 126
Bruno Henriques, Nicole Ferreira, Martina Kocijan, Milan Vukšić, Burak Ünlü, Serbülent Türk, Mahmut Özacar and Gil Gonçalves

7.1 Graphene and Its Derivatives 127
7.1.1 Preparation of Graphene and Its Derivatives 128
7.1.2 Reduction of Graphene Oxide 130
7.2 Graphene Composites for Photodegradation of Aqueous Organic Contaminants 132
7.2.1 Synthesis Methods of Graphene-based Composites 133
7.2.2 Graphene Binary Composites 134
7.2.3 Graphene Tertiary Composites 135
7.3 Graphene-based Composites for Sorption of Metal Ions 136
7.4 Graphene-based Membranes for Water Filtration 139
7.5 Separation Mechanism and Properties of Graphene-based Membranes 142
7.6 Conclusion and Future Perspectives 147
References 147

Chapter 8 Responsive Polymeric Materials: Advances in Membrane-based Technologies for Water Treatment Processes 155
Perla Cruz-Tato, Alondra S. Rodríguez-Rolón, Jessika Pazol and Eduardo Nicolau

8.1 Introduction 155
8.1.1 Conventional Membranes and Intrinsic Challenges 155
8.1.2 Responsive Materials—Next-generation Materials 156

8.2 Direct Stimulation of Polymeric Responsive Membranes 157
8.2.1 pH-responsive Membranes 157
8.2.2 Hygro-responsive Membranes 159
8.2.3 Ionic Strength-responsive Membranes 159
8.3 Indirect Stimulation of Responsive Membranes 160
8.3.1 Temperature-responsive Membranes 160
8.4 Field-responsive Polymeric Membranes 160
8.4.1 Electro-responsive Membranes 161
8.5 Catalytic Membranes 166
8.5.1 Photocatalytic Membranes 167
8.5.2 Biocatalytic Membranes 171
8.6 Challenges and Future Directions 174
References 176

Chapter 9 Graphene-based Engineered Macrostructures for Water Purification 183
Chirag Batukbhai Godiya

9.1 Introduction 183
9.2 Graphene-based Macrostructures and Synthetic Methods 185
9.2.1 Pristine Graphene 186
9.2.2 Graphene Oxide and Fabrication Methods 186
9.2.3 Reduced Graphene Oxide and Fabrication Methods 187
9.2.4 Chemically Modified Graphene and Fabrication Methods 188
9.2.5 Graphene-based 3D Nanostructures and Fabrication Methods 189
9.2.6 Polymer-based GBMs 190
9.2.7 Inorganic GBM Hybrids 191
9.2.8 Biomass-based GBMs 192
9.2.9 MOF-based GBMs 192
9.3 Adsorption of Organic/Inorganic Pollutants on GBMs 193
9.3.1 GBMs for the Uptake of Cationic Contaminants 193
9.3.2 GBMs for the Uptake of Anionic Contaminants 195
9.3.3 GBMs for the Removal of Non-ionic Pollutants 196
9.3.4 GBMs for the Removal of Adsorbates of Varied Water Solubilities 199
9.4 Conclusions and Perspectives 202
References 203

Chapter 10 Textile Azo Dye Removal Using a Quaternary Ammonium Anion Exchanger Prepared From Sugarcane Bagasse 208
Natália Roberta Marques Tanure, Liliane Catone Soares, Luisa Cardoso Maia, Megg Madonyk Cota Elias, Amália Luísa Pedrosa Xavier, Oscar Fernando Herrera Adarme, Guilherme Max Dias Ferreira, Eduardo Ribeiro de Azevedo, Luis Henrique Mendes da Silva and Leandro Vinícius Alves Gurgel

10.1 Introduction 209
10.2 Materials and Methods 210
10.2.1 Bioadsorbent and Dye 210
10.2.2 Synthesis of STEA 210
10.2.3 Characterization of SB and STEA 211
10.2.4 Adsorption Studies 211
10.2.5 Isothermal Titration Calorimetry 214
10.3 Results and Discussion 215
10.3.1 Chemical and Structural Characterization of Raw SB and STEA 215
10.3.2 Adsorption Studies 216
10.3.3 Enthalpy of RGY Adsorption on STEA 226
10.3.4 Adsorption Thermodynamics 228
10.4 Conclusions 231
Acknowledgements 231
References 231

Chapter 11 Iron Oxide Nanomaterials for Water Purification 234
Ateeq Rahman, Likius Shipwiisho Daniel and Veikko Uahengo

11.1 Introduction 234
11.2 Wastewater Treatment Conventional Methods 235
11.3 Nanotechnology 237
11.4 General Method for the Preparation of Iron Oxide NPs by Biosynthesis Route 238
11.4.1 Biosynthesis of Iron Oxide Nanoparticles 238
11.4.2 Removal of Contaminants from Wastewater Iron Oxide NPs 241
11.5 Characterization of Iron Oxide NPs 244
11.5.1 Characterization of Fe NPs by FTIR Spectroscopy 244
11.5.2 Characterization of Fe NPs by XRD 244
11.5.3 Characterization of Iron Oxide NPs by Electron Microscopy 245
11.5.4 Characterization of Fe NPs by Thermogravimetric Analysis 246

11.5.5 Characterization of Iron Oxide NPs by Nitrogen Adsorption/Desorption Isotherm 247
11.5.6 BET Surface Area of Iron Oxide NPs 248
11.6 Conclusions and Future Outlook 248
Conflict of Interest 249
Acknowledgements 249
References 249

Chapter 12 Factors Influencing TiO_2-based Composites for Water Decolorization: A Systematic Review 256
Likius Shipwiisho Daniel, Veikko Uahengo and Ateeq Rahman

12.1 Introduction 256
12.1.1 Dye As an Aquatic Pollutant 256
12.1.2 Advanced Oxidation Process As Solution to the Degradation of Dye 258
12.1.3 Fundamental Principles Governing Photocatalyst Performance 259
12.1.4 TiO_2-based Composite As a Promising Photocatalytic Material for Water Decolorization 262
12.1.5 Specific Research Questions 263
12.2 Study Methodology 264
12.2.1 Search Design: Eligibility Inclusion and Exclusion Criteria and Grouping 264
12.2.2 Search Strategy 265
12.2.3 Study Selection and Data Extraction 266
12.3 Results and Discussion 267
12.3.1 TiO_2-based Composites for Water Decolorization 267
12.3.2 Current Drawbacks Identified in the Literature 268
12.3.3 Factors Influencing Catalyst Design 269
12.3.4 Supportive Substrate As Factor Influencing Catalyst Design 270
12.3.5 Influence of Light Intensity and Irradiation Time 274
12.3.6 Influence of Initial Dye Concentration 277
12.3.7 Influence of Solution pH and the Water Matrix 278
12.3.8 Influence of Temperature 279
12.4 Challenges and Future Development 281
12.5 Conclusion 281
Acknowledgements 282
References 282

Chapter 13 Enviro-friendly Nanomaterial Synthesis and Its Utilization for Water Purification **298**
Huda Sharbini Kamaluddin and Katabathini Narasimharao

13.1 Introduction 298
13.2 Green Synthesis of Adsorbent and Photocatalyst Materials 301
13.2.1 Carbon-based Materials 302
13.2.2 Metals 307
13.2.3 Metal Oxides 316
13.3 Conclusion and Prospects 339
List of Abbreviations 340
References 341

Subject Index **353**

CHAPTER 1

Biomaterials for Water Purification: Dyes, Heavy Metals, and Pharmaceuticals

VINAYAK ADIMULE(0000-0002-2198-6601)[a*], NIDHI MANHAS(0000-0002-6049-2145)[b] AND SHASHANKA RAJENDRACHARI(0000-0002-6705-763X)[c*]

[a]Angadi Institute of Technology and Management (AITM), Savagaon Road, Belagavi-590009, Karnataka, India; [b]School of Sciences, Indira Gandhi National Open University, Maidan Garhi, New Delhi, India; [c]Department of Metallurgical and Materials Engineering, Bartin University, Turkey
*E-mail: adimulevinayak@yahoo.in, researcher@vinayakadimule.in, shashankaic@gmail.com

1.1 Introduction

Access to pure water is vital for the plant and animal kingdoms and is an emerging problem, as global demand increases. Many researchers and scientists are working to resolve the issue and to develop new materials with enhanced features for water purification.[1] Chitosan is a naturally occurring polysaccharide and possesses remarkable properties, which have attracted particular attention, such as wide availability, renewability, interesting structural features, low cost, and impressive performance in water treatment processes.[2] Chitosan is well known for its interesting properties, such as being a non-toxic biopolymer, biodegradable, and biocompatible, and it can be obtained commercially from shrimps.[3] The polysaccharide polymer

Chemistry in the Environment Series No. 12
Novel Materials and Water Purification: Towards a Sustainable Future
Edited by Grigorios L. Kyriakopoulos and Miltiadis G. Zamparas

Published by the Royal Society of Chemistry, www.rsc.org

exhibits unique functionalities and structural features and is widely used in the fields of agriculture, biomedicine, and in environmental protection.[4] Chitosan can be obtained from chitin by enzymatic and chemical methods and has also been used in biomedical applications. Chitosan compounds are highly soluble in water due to their polymeric structural features.[5] The adsorption mechanism offered by chitosan and chitosan-based polymeric membranes has been widely explored for the removal of pollutants.[6] Chitosan also possesses unique and attractive features that can be applied in the fields of agriculture,[7] water purification,[8] cosmetics,[9] and biomedicine.[10] It also possesses unique antibacterial properties and is a promising candidate for the removal of salts (desalination).[11] For example, Motshekga *et al.* synthesized a chitosan-supported Ag-ZnO nanocomposite with bentonite for the effective removal of disinfectants from water.[12] In another report, Kamal *et al.* synthesized a chitosan-ZnO nanocomposite coated onto microfibrillar cellulose and studied its antibacterial properties against *Escherichia coli*.[13] Certain biosorbents with smaller particles show high capacity in the removal of pollutants from water due to their large surface area, which increases the availability of binding sites present on the adsorbents.[14] Metals and heavy metals constitute a large portion of the pollutants found in waste water from industry. In particular, heavy metals cause environmental problems due to their high toxicity and non-biodegradable characteristics.[15] At certain concentration thresholds heavy metals can cause serious health problems for humans, with adverse effects on cardiovascular, renal, gastrointestinal, peripheral, and central nervous systems.[16] Cellular dysfunction occurs as a result of heavy metals combining with cellular components that contain essential elements, such as O, N, S *etc.*, which in turn leads to modifications of enzymes or proteins.[17] In aquatic bodies, elements such as Hg, Co, Cr, Al, Zn, Ni, Cu, Pb *etc.* are considered as priority elements to for removal from the water.[18] Readily available biowaste products are used to remove or reduce the concentration of these heavy metals in waste water.[19]

This chapter discusses the use of biodegradable polymers, *i.e.*, chitosan, for the effective removal of heavy metals, dyes, and pharmaceutical small organic compounds from waste water, as well as the use of biosorbents derived from biowaste and their mechanism of action in the removal process, the development of polymeric membranes with various sized nanomaterials, chitosan-based nanomaterials, chitosan-based nanocomposite membranes, and chitosan-based metal oxide membrane development. The polymeric membranes developed have been used to compare their performance with other polymeric membranes, for the treatment of water contaminated with heavy metals, dyes, and various pharmaceutical molecules.

1.2 Characterization

Characterization of the polymeric membranes was carried out using different techniques. A Malvern Panalytical instrument was used for C, H, and N elemental analysis. Functional groups were identified using Fourier-transform infrared (FT-IR) spectroscopy (Agilent) in the wavenumber range 400 to 4000 cm^{-1}. The morphology of the as-developed polymeric membranes was observed using a Jeol JSMIT800 instrument. The crystallite size, phase, and microstructure of the as-developed polymeric membranes were observed with scanning electron microscopy (SEM) using a Jeol JSMIT800 instrument.

1.2.1 Elemental Compositional Analysis

Elemental compositional analysis of the different chitosans (CTS) synthesized was performed with a CHN analyser. From the elemental compositional analysis, a significant decrease in C, H, N composition was observed. Table 1.1 summarizes the C, H, N composition of the various chitosan structures.

1.2.2 Scanning Electron Microscopy Investigation

SEM analysis was carried out in order to understand the change in morphology of the as-developed membranes. Figure 1.1(a and b) displays the change in morphology of chitosan and chitosan nanocomposites.

1.2.3 FT-IR Spectroscopy Studies

FT-IR was performed to understand the change in the functional groups of the modified and unmodified chitosan, and to provide reliable information for prediction of the structural changes taking place in the chitosan. Figure 1.2(a and b) shows unmodified and modified chitosan. The unmodified chitosan shows asymmetric and symmetric stretching of N–H groups appearing at 3400 cm^{-1} and 3250 cm^{-1}, respectively, while the primary amine appears at 1650–1580 cm^{-1} as a broader peak. The peaks in the region 1564 cm^{-1} to 1628 cm^{-1} relate to the bending vibration of the N–H peak. The modified chitosans, especially chitosan–glutaradehyde (CTS–GL), show a broad –OH peak in the range 3200 to 3400 cm^{-1}. For the modified chitosans the sharp appearance of the N–H bands disappeared due to the formation of

Table 1.1 Elemental analysis for chitosan and modified chitosan samples. Reproduced from ref. 20, https://doi.org/10.3390/ijms23042396, under the terms of the CC BY 4.0 license https://creativecommons.org/licenses/by/4.0/.

		Content (%)		
Entry	Sample	C	H	N
1	Chitosan (CTS)	40.27	7.91	6.22
2	Chitosan–glutaradehyde (CTS–GL)	43.01	7.23	5.81
3	Chitosan–epichorohydrin (CTS–EC)	37.18	6.17	5.61
4	Chitosan–diethylenetriamine (CTS–DET)	43.19	7.34	8.94
5	Chitosan–monochloroaceticacid (CTS–CAA)	42.54	6.98	7.38

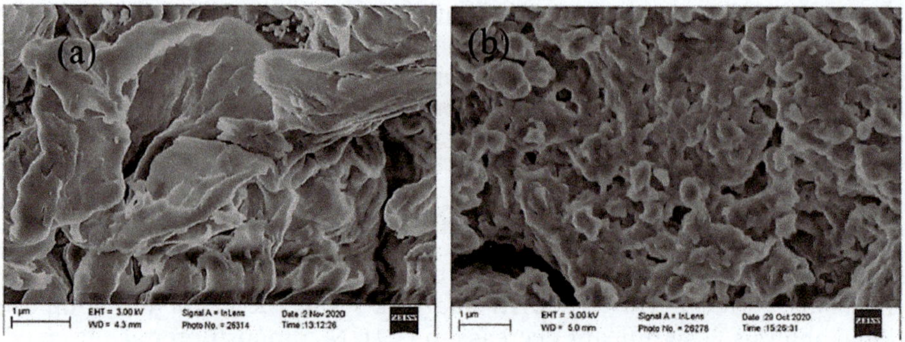

Figure 1.1 Scanning electron microscopy images for unmodified chitosan (a) and CTS–CAA (b). Reproduced from ref. 20, https://doi.org/10.3390/ijms23042396, under the terms of the CC BY 4.0 license https://creativecommons.org/licenses/by/4.0/.

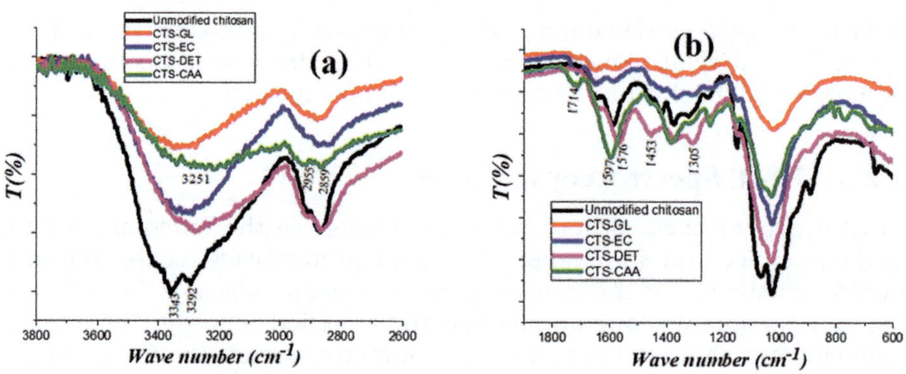

Figure 1.2 FT-IR spectra of unmodified chitosan, CTS–GL, CTS–EC, CTS–DET, and CTS–CAA in the wavenumber range 3800–2600 cm^{-1} (a) and in the wavenumber range 1900–600 cm^{-1} (b). Reproduced from ref. 20, https://doi.org/10.3390/ijms23042396, under the terms of the CC BY 4.0 license https://creativecommons.org/licenses/by/4.0/.

cross-linking, while additional bands appeared in the region 2859 cm^{-1} to 2955 cm^{-1} due to the presence of CH_2 groups.

1.2.4 X-ray Diffraction Investigation

To understand the phase and crystallinity of the unmodified and modified chitosan, X-ray diffraction (XRD) spectra were recorded, as displayed in Figure 1.3. The dominant peaks appearing at 2θ values of 10.22° and 19.77° correspond to the unmodified chitosan membrane, while the modified chitosan–CAA shows strong peaks at 2θ = 8.259° and 22.145°, respectively. Modification of the structure of the chitosan can be inferred from the shift in the position of the 1st peak to a lower 2θ, value while the second peak is

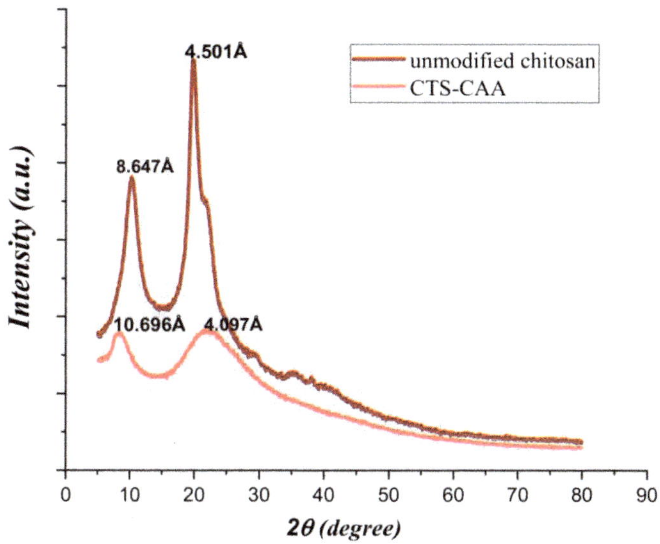

Figure 1.3 XRD diffractograms of unmodified chitosan and CTS–CAA. Reproduced from ref. 20, https://doi.org/10.3390/ijms23042396, under the terms of the CC BY 4.0 license https://creativecommons.org/licenses/by/4.0/.

shifted to a higher 2θ value. The peak broadening in the XRD spectrum shows a high degree of functionalization inside the polysaccharide membrane.

1.3 Results and Discussion

1.3.1 Role of Zero-, One-, Two-, and Three-dimensional Nanomaterials for Polymeric Membrane Development

Over the past decades different polymeric membranes have been developed and used efficiently in the purification of water or in water treatment processes. Certain conditions and features are generally needed to optimize the process of separation of contaminants and to improve the efficiency of the separation process.[21] Nanoparticles (NPs) embedded into the polymeric matrix improve certain parameters, such as stability, flexibility, and separation grade, and can be used for effective removal of contaminants from waste water.[22] There are different nanomaterial-based biosensors with various applications (Figure 1.4). Abedini *et al.* synthesized a new polymeric membrane using TiO_2-cellulose using a sonochemical technique and used it in the water treatment process.[23] The water flux greatly increased with increase in TiO_2 concentration up to 20 wt%. Chaloupka *et al.* reported antibacterial properties for Ag NPs dependent on their size, shape, and chemistry. Because of their chemoselective nature, Ag NPs are found to decrease the bacterial concentration and spread on the cell membrane.[24] Cao *et al.* synthesized a

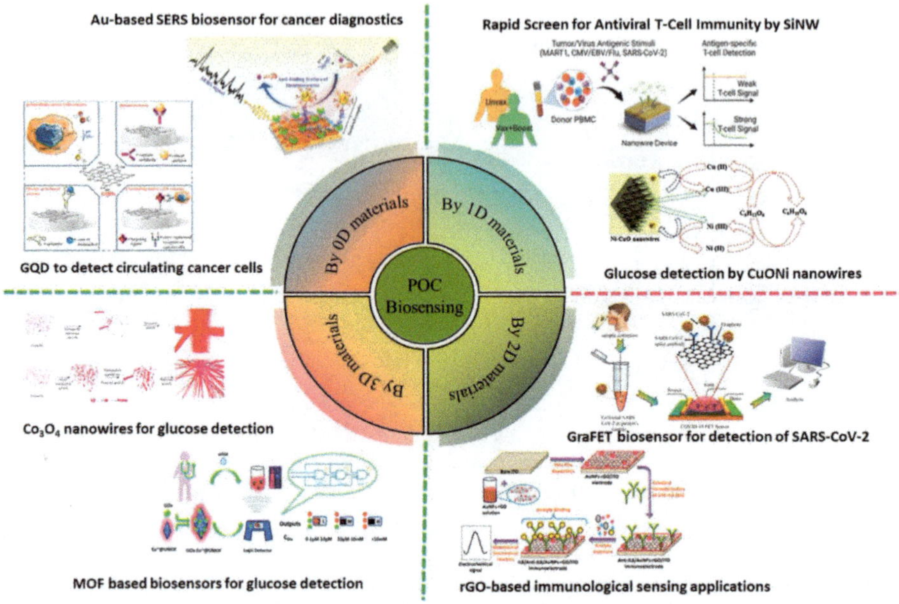

Figure 1.4 Illustration of zero-, one-, two-, and three-dimensional nanomaterial-based point-of-care biosensing devices. Reproduced from ref. 26 with permission from Elsevier, Copyright 2022.

new membrane by inserting Ag NPs and vitamin C as a reducing agent.[25] Haider *et al.* reported the antibacterial activity of the polyethersulfone membrane against *E. coli* upon incorporation of Ag NPs.

Cu NPs are robustly used in the synthesis of polymeric membranes. Cu NPs and their derivatives present antibacterial and antifungal properties against different microorganisms.[27] Xu *et al.* synthesized membranes with antibacterial properties for Cu NPs incorporated in chelated polyacrylonitrile membranes. On the other hand, ZnO NPs have attracted much attention due to their diverse antibacterial and antifungal properties and their ability to attract hydroxyl groups, with a larger exterior system compared to other inorganic NPs.[28] Similar results were obtained for Al_2O_3 NPs and they showed significant improvement in hydrophilicity and fouling properties.[29] The most common methods used in the synthesis of polymers are interfacial polymerization, surface deposition, and structure entrapment.[30] Zinadini *et al.* reported the synthesis of Fe_2O_3 NPs on the formation of polymeric membranes.[31] Niksefat *et al.* studied the morphology and performance of a sheet of Si NPs, and as the composition of Si increased the water permeability of the polymer and flux increment also increased.[32]

In recent years one-dimensional (1D) NPs have gained much interest due to their attractive and unique features, such as flexibility, porosity, *etc.*, and they are widely used in sensors for environmental monitoring, bioengineering, *etc.* One of these interesting materials is carbon nanotubes (CNTs), which display enhanced permeability, physiochemical, mechanical, and thermal properties without any decrease in their selectivity.[33] Celik *et al.* synthesized a CNT-functionalized polymeric membrane to increase permeability and decrease the fouling rate. The synthesized CNT-functionalized polysulfone polymeric membrane is widely used in the removal of heavy metals.[34] Shen *et al.* synthesized methyl methacrylate-reinforced CNT polymers and showed enhanced water permeability and increased flux, by 62%, compared with pristine composite membranes.[35] To exhibit high potential for water treatment processes, polymer membranes have been decorated with ZrO_2 and TiO_2 NPs.[36] Pan *et al.* reported the synthesis of thin-film composite membranes for the removal of various pollutants,[37] which also showed remarkable antibacterial properties against *E. coli* and *Staphylococcus aureus*.

In recent years, scientists and researchers have developed two-dimensional (2D) nanocomposite materials with superior properties, such as chemically active surface area, and superior transport properties, with respect to their shape, size, and structure[38] (Figure 1.5). When incorporated with polymeric membranes, the 2D nanocomposite materials showed superior contaminant removal from water.[39] 2D nanomaterials have also been used in the desalination process. Graphene is a 2D material that has been used in the desalination process due to its honeycomb-like structure.[40] The honeycomb structure facilitates 100% removal of salts from water, as well showing removal of 99% of organic contaminants when coated with PVDF.[41] Graphene oxide (GO) is produced from the oxidation of graphite and possesses more hydrophobic functional groups, such as $-NH_2$, $-COOH$, $-OH$, *etc.* Water purification can be enhanced by incorporation of GO with the polymeric structure.[42] Yang *et al.* reported the synthesis of GO in nanofiltration membranes, which were found to be effective in water and dye purification.[43] Xiaowei *et al.* reported the synthesis of chitosan dispersed with GO and this has been widely used in water purification systems.[44] Gao *et al.* synthesized di-vanadium pentoxide (V_2O_5) in the presence of hierarchal TiO_2 NPs for the effective removal of organic contaminants from water.[45]

Currently there are larger number of inorganic materials that are three-dimensional (3D) in structure, such as silicates, zeolites, *etc.* Zeolites, in particular, are used in gas and liquid separation, desalination, and in the formation of polymeric membrane reactors.[47] Zeolites show enhanced permeability, greater selectivity, high thermal and mechanical properties, *etc.* The presence of zeolites in the structure increases the hydrophilicity, diversifies the molecular structure, and provides greater stability, as described by Huang *et al.*[48] Certain nanocomposite membranes have shown greater chemical reactivity (catalysis) when fixed with CO_2. Ahmad *et al.* synthesized polymeric membranes with amounts of silica at different concentrations.[49] The newly formed silica showed greater permeability than the simple membrane and the results showed that antifouling properties were enhanced, which in turn increased the oil and water separation.[50] Nanoparticles showing fluorescence and sensing properties can be used to detect heavy metals

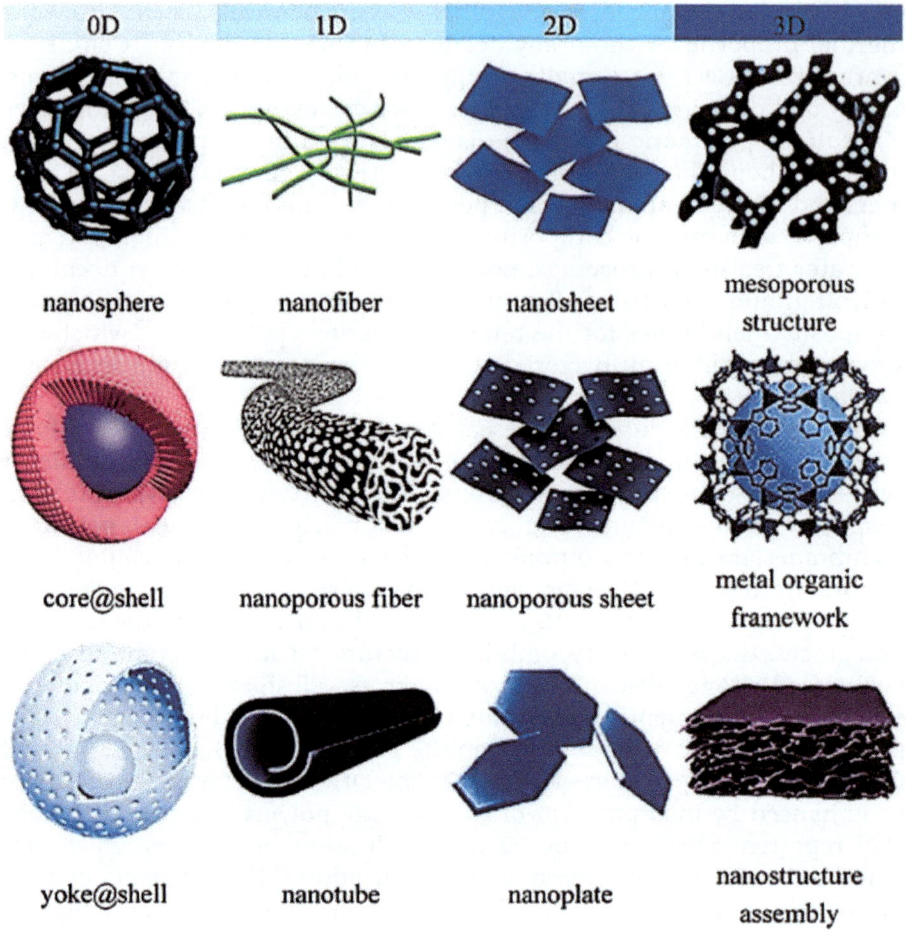

Figure 1.5 Different shapes and sizes of nanoparticles. Reproduced from ref. 46, https://doi.org/10.3390/membranes10100297, under the terms of the CC BY 4.0 license https://creativecommons.org/licenses/by/4.0/.

in water treatment applications.[51–55] There are many different 0D, 1D, and 2D nanoparticles used in water treatment, as shown in Table 1.2.

1.3.2 Role of Chitosan-based Nanomaterials and Developing Polymeric Membranes

The process of water purification involves removal of contaminants from waste water.[120] In recent years diversified and novel improved polymeric membranes have been developed for water purification.[121] Various membranes have been developed with biomaterials and other particles deposited on them for efficient removal of contaminants from water. However, the process of removal of large- and small-sized particles depends on the pore size

Table 1.2 Applications of nanoparticles in nanocomposite polymeric membranes for water treatment. Reproduced from ref. 56, https://doi.org/10.3390/ma14092091, under the terms of the CC BY 4.0 license https://creativecommons.org/licenses/by/4.0/[a]

Nanoparticle	Membrane type	Targeted application	Reference
ZnO	MF	Wastewater treatment, copper ions, and VOC elimination from water systems	57–59
	UF	Humic acid, salt, micelles elimination from solutions, treatment of waste waters	60–64
	NF	Water treatment, separation of Rhodamine B	65–67
	FO	Desalinization and wastewater treatment	68
	RO	Salt, bivalent ions, and bacterial retention models	69
$AgNO_3$	UF	Evaluation of antibacterial properties and salt elimination (Na_2SO_4)	70
	RO	Evaluating antibacterial properties and salts (NaCl)	71–73
	DCMD	DCMD desalination of sea water through silver nanoparticle deposition	74
	PRO/RO	Evaluating antifouling and antibacterial properties in water treatment membranes	75
Ag NPs	MF/UF	Evaluating antifouling properties in water treatment membranes—mixture model—BSA	76
	UF	Water purification, evaluating antifouling properties in water treatment membranes	77–81
	NF	Evaluating antibacterial properties in water treatment membranes	82
Bio-Ag_0	UF	Evaluating antifouling and antibacterial properties in water treatment membranes	83
	NF	Evaluating antibacterial properties and salt removal	84 and 85
Cu NPs	UF	Wastewater treatment and evaluating antifouling properties in membranes	86 and 87
	RO	Evaluating antibacterial properties in water treatment membranes and salt removal	88
TiO_2 NPs	MF	Evaluating antifouling properties using whey solution	89
	UF	Evaluating antifouling properties in water treatment membranes. Treatment of water systems	90–95
	FO	Evaluating salt rejection (NaCl)	96
	MF/MBR	Evaluating antifouling properties	97
TIP		Evaluating antifouling properties	98

(continued)

Table 1.2 (*Continued*)

Nanoparticle	Membrane type	Targeted application	Reference
CNTs	NF	Evaluating antifouling and salt rejection (NaCl, Na_2SO_4), filtration system applications	98–102
	UF	Water system treatment and biofouling application. Water treatment for applications	102–105
	MF	Effluent treatment through a membrane bioreactor	106
GO	MF	Effluent elimination with high dye content. Wastewater filtration	107–113
	NF	Soft water production. Water treatment	114 and 115
	RO	Salt removal (NaCl, $CaCl_2$ and Na_2SO_4)	116
	FO	Desalination processes for sea water	117
Graphene	UF	Wastewater treatment	118
	NF	Water purification	119

[a]UF, ultrafiltration; MF, microfiltration; NF, nanofiltration; FO, forward osmosis; RO, reverse osmosis; DCMD, direct contact membrane distillation; MBR, membrane bioreactor; TIP, titanium tetraisopropoxide; VOC, volatile organic compounds.

of the particles.[122] Many processes are used for the removal of contaminants from water, such as microfiltration, nanofiltration, ultrafiltration, *etc.* In the use of chitosan-like biomaterials, their physical character, adsorption ability, and selectivity determine the contaminant removal from waste water[123] (Figure 1.6).

For water treatment, changes in the structure of chitosan with hydrogels or beads,[124] nanoparticles,[125] membranes,[126] fibres,[127] powders,[128] *etc.* are needed. The increased adsorption capacity of the chitosan-supported membranes is due to their elemental composition, surface area, pore volume, size, *etc.*[129] Saifuddin *et al.* reported structural variation in chitosan when doped with Ag NPs and this was used for the removal of pesticides in agriculture. Magnetically activated C-doped chitosan can be used as an adsorbent for the removal of antibiotics and smaller pharmaceutical compounds, such as ciprofloxacin, amoxicillin, erythromycin, *etc.* Nadavala *et al.* synthesized a novel chitosan-based membrane with sodium alginate and calcium chloride, which is widely used in the removal of phenols.

Pollutants, such as inorganic oxides, and organic pollutants, such as dyes, drugs, pesticides, detergents, and phenols, play a vital role in the water purification process. For the removal of organic contaminants, the concentration, pH, and contact time influence the absorptivity for removal of the pollutant from waste water. Studies carried out by Bibi *et al.* summarized the adsorption capacity of modified chitosan membrane-based systems with polyvinyl alcohol (PVA), nanotubes, and silane, which were responsible for the removal of naphthalene.[153] Fan *et al.* reported the synthesis of magnetic graphene oxide-modified chitosan for the effective removal of methylene blue.[154] In

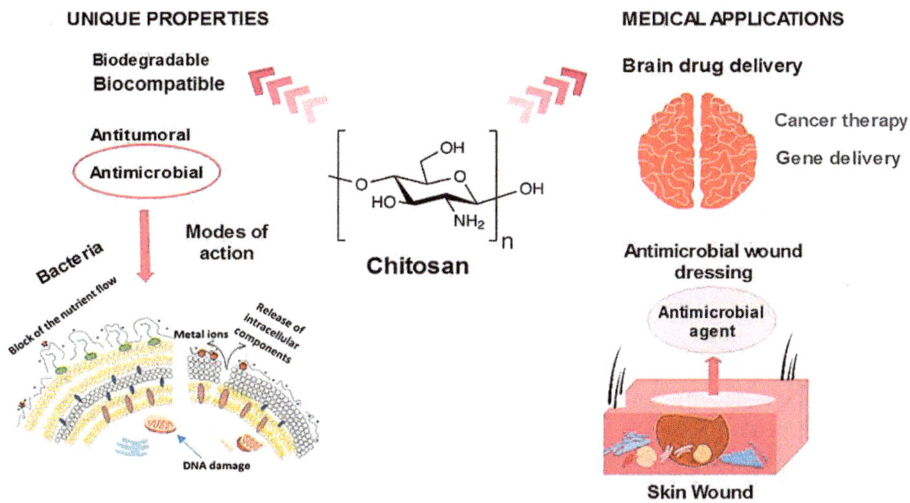

Figure 1.6 Schematic diagram of chitosan with different properties and biomedical applications. Reproduced from ref. 130, https://doi.org/10.3390/molecules24101960, under the terms of the CC BY 4.0 license https://creativecommons.org/licenses/by/4.0/.

their reports, the modified chitosan with magnetic graphene oxide was characterized by SEM, XRD, and FT-IR techniques. The change in the parameters for the modified chitosan showed higher adsorption of methylene blue when compared with unmodified chitosan. Chitosan with GO nanocomposites doped with Cu NPs were reported by Khan *et al.* for the effective removal of 4-nitrophenol compounds from water.[155] For the effective removal of polyethylene glycol (PEG) and PVA, chitosan supported by a silica membrane was used by Herman *et al.*[156] The study showed enhanced permeability, selectivity, and increased mechanical features for the membrane. The study also showed that higher amounts of PEG resulted in increased pore diameter and increased the flux rate. However, increase in PVA concentration decreased the pore diameter and flux rate. Rosdi *et al.* synthesized chitosan polymeric membrane with silica added for the effective removal of Pb(II) ions from waste water.[157] The composite membrane showed greater selectivity and effective removal of Pb(II) ions compared with the unmodified chitosan. An evacuation permeation process was carried out for the novel development of chitosan membranes containing zeolite, which was further loaded with glutaraldehyde, for the removal of trace metal ions, such as Cr, As, Pb, *etc.*[158] Carbon nanotubes were loaded with chitosan for the development of polymeric membranes and are effective for the removal of heavy metal ions.[159] A special membrane was also developed with chitosan-biochar (CBS). The as-prepared CBS composite has shown higher absorptivity, which can be attributed to the large pore size of the CBS, which is extensively used in the removal of heavy metals, antibiotics, and pharmaceuticals from waste water.[160] The diverse

Table 1.3 Diverse applications of chitosan with various compounds. Reproduced from ref. 56, https://doi.org/10.3390/ma14092091, under the terms of the CC BY 4.0 license https://creativecommons.org/licenses/by/4.0/.

Material type	Compound	Selected applications	References
0D	Ag NPs	Health and environment, textile industry	131–141
	Cu NPs	Biomedical, biofouling, water treatment	
	Zn NPs	Heavy metal removal, adsorbent in water applications	
	TiO_2 NPs	Water treatment, antifouling applications	
	Glycerol	Burn dressing applications	
	PEG	Wound dressing applications	
1D	CNTs	Wastewater treatment	142–144
	TiO_2 CNTs	Proton-exchange membranes	
	Attapulgite	Proton-exchange membranes for fuel cells	
	Cellulose/halloysite	Biomedical applications	
2D	GO	Adsorbent in water purification, food industry	145–152
	SiO_2@PVDF	Supercapacitor, membrane separation applications	
	MMT	Drug delivery systems, food packing applications	
	LAP@Ag NPs	Food packing applications	

applications of chitosan, the added compounds, and the material type are summarized in Table 1.3.

1.3.3 Chitosan–Metal Oxide-based Polymeric Membranes

Various metal oxides when combined with chitosan show significantly improved adsorption ability and are widely used in the removal of contaminants from waste water. Among these, ZnO, MgO, Fe_3O_4, *etc.* show a significant improvement because of their high semiconducting characteristics. Among the various metal oxides, ZnO–chitosan has shown remarkable properties due to its increased surface area and semiconducting properties[161] (Figure 1.7). The as-formed ZnO–chitosan hybrid materials are effective in the removal of organic contaminants from waste water, compared with unmodified chitosan. On doping with ZnO, the hybrid material showed an increase in the number of active centres, formation of a surface-strengthening effect, and improved surface corrosion resistance properties.

MgO NPs are used in water purification treatments, especially for removal of organic contaminants, due to their high chemical stability and non-toxicity.[163] MgO NPs were dispersed into a chitosan polysaccharide matrix for the effective removal of reactive blue dye in a very short contact time.[164] Additionally, Fe_3O_4 NPs also showed good adsorption characteristics, especially

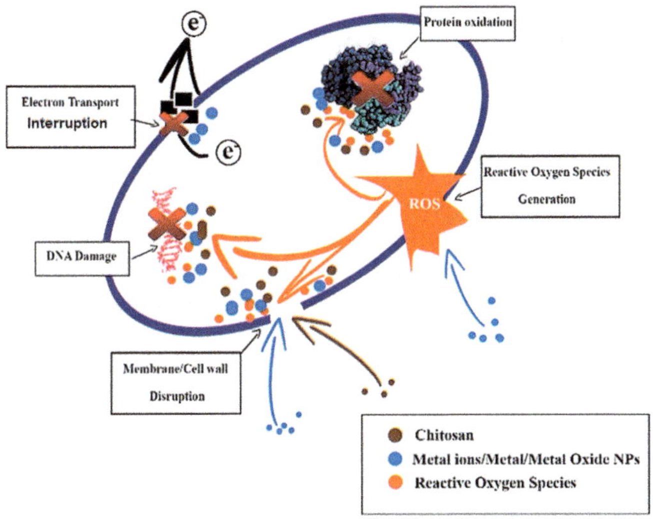

Figure 1.7 Functioning of chitosan-based metal oxides. Reproduced from ref. 162, https://doi.org/10.3390/polym9010021, under the terms of the CC BY 4.0 license https://creativecommons.org/licenses/by/4.0/.

Table 1.4 Metal ion adsorption and desorption on modified chitosan over three cycles. Reproduced from ref. 20, https://doi.org/10.3390/ijms23042396, under the terms of the CC BY 4.0 license https://creativecommons.org/licenses/by/4.0/.

Cycle	1		2		3	
Metal	Adsorption (%)	Desorption (%)	Adsorption (%)	Desorption (%)	Adsorption (%)	Desorption (%)
Copper	100	80.23	91.56	76.54	88.47	74.36
Zinc	100	84.21	93.42	79.89	89.75	78.89

when used in the magnetic separation process. Abu El-Soad *et al.*[20] prepared chitosan-functionalized recyclable biomaterial for the adsorption of Cu(II) and Zn(II) ions in aqueous media (Table 1.4).

Adimule *et al.* synthesized nanocomposites of rare earth metals and they exhibited enhanced binding sites with considerable increase in their specific surface area.[165–169] These can be further applied by researchers to modify chitosan and study their applications.

1.3.4 Role of Initial Dye Concentration

Biosorption isotherm models have been utilized to study the initial concentration of dye influencing the biosorption process. The initial concentration of the dye was studied in the range 50–400 mg L^{-1} at a constant time interval of 400 minutes. With increase in the concentration of the initial dye, the biosorption capacity of the polymer substantially increased. At the initial dye concentration on the polymer, more binding sites are available for biosynthesis of the dye molecule. Transformation from the liquid state to the solid state occurred with the initial biosorption process. Increased mass transfer of the dye molecule occurred with respect to initial dye concentration. Further biosorption capacity of the dye remained constant after some time, which can be attributed to the saturation state of the chitosan polymer, with no more binding places available.[170] Due to saturation of the binding sites, excess dye molecules are not adsorbed. Similar behaviour has been reported for chitosan–PVA, chitosan–MgO, and chitosan-activated C.[171]

1.3.5 Performance Comparison of Various Types of Chitosan-based Polymers

Chitosan performance for both anionic and cationic dyes is summarized in Table 1.5. An optimal pH range of 8–11 is required for the removal of cation dye using chitosan polymer, whereas a pH of 2–6.8 (acidic range) is optimal for the removal of the anionic dye. The biosorption process takes place in the temperature range 25 °C–45 °C. The process of biosorption is affected by contact time, initial pollutant concentration, solution temperature, *etc.* The dye removal efficiency depends upon the solution pH. The pH of the solution affects the charge of the polymer composites and this suggests that the dye removal process is chemisorption.[172] Chemisorption involves electronic interaction and exchange of the charges between the chitosan polymer and the dye molecules. Physical absorption takes place between the binding sites on the polymer.[173]

1.3.6 Treatment of Water Contaminated With Various Pharmaceuticals

A common definition for the presence of pharmaceuticals in water is a high concentration of organic components, microbial activity, difficult to biodegrade, microbial toxicity, high amount of salt, *etc.* Removal of pharmaceutical contaminants requires purpose-built high-temperature incineration and cleaning of the passing flue gas.

Reverse osmosis and nanofiltration are also effective processes for the removal of pharmaceuticals from waste water. The pharmaceuticals enter the water supply by excretion.

1.4 Conclusions

In this chapter, chitosan supported with various nanomaterials, composites, metal oxides, and other substances has been discussed, along with its importance in the development of polymeric membranes and applications

Table 1.5 Removal efficiency of cationic and anionic dyes using various types of chitosan composites under different operating conditions. Reproduced from ref. 174, https://doi.org/10.3390/polym13173009, under the terms of the CC BY 4.0 license, https://creativecommons.org/licenses/by/4.0/.

Chitosan Composites	Cationic/ Anionic dye	pH dosage	Bio sorbent concentration	Initial dye (mg L^{-1})	Contact time (min)	Temperature (°C)	Efficiency (%)	Reference
Chitosan/Zeolite	Cationic	9	2.0	100	138.65	30	84.85	175
Chitosan/ZnO	Cationic	8	2.4	50	180	30	98.50	176
Chitosan/Activated carbon	Cationic	11	1.0	400	60	30	91.02	177
Chitosan/Bentonite	Cationic	10	0.2	100	360	25	88.00	178
Chitosan/Fe_3O_4/GO[a]	Cationic	11	1.0	100	70	27	87.6	179
Chitosan/MgO	Anionic	6.8	9.3	100	120	30	79.50	180
Chitosan/PVA[a]	Anionic	6	2.0	50	40	30	86.70	181
Chitosan/Cellulose	Anionic	6.6	2.5	500	625	30	95.00	182
Chitosan/Kaolin clay	Anionic	4	0.6	140	30	30	99.50	183
Chitosan/GO[a]	Anionic	2	1.0	250	1440	25	86.00	184

[a]GO, graphene oxide; PVA, poly vinyl alcohol.

in the purification of water. This approach aims to incorporate the effectiveness of the biological, renewable source chitosan and increase its efficiency in the removal of most metals, including heavy metals, and dyes from waste water by incorporation of various sized (0D, 1D, 2D and 3D) nanomaterials and nanocomposites. The role, mechanism, and stability of the polymeric membrane has been discussed. Chitosan-supported and composite membranes have been found to be efficient for the removal of contaminants from waste water, especially dyes, pharmaceuticals, and heavy metals.

Authors Contribution

Dr Vinayak Adimule and Dr Shanshanka Rajendrachari conceived the ideas, wrote the paper, and carried out grammatical corrections in the manuscript. Miss Nidhi Manhas was involved in the manuscript arrangement and further corrections, such as typographical and grammatical errors.

Data Availability

Data from the present study are presented in the chapter and further data can be obtained from the corresponding author on request by e-mail.

Funding

All authors declare that they have not received any funding from any source or organization.

Conflict of Interest

All authors declare that they do not have any conflict of interest.

Acknowledgements

All the authors thank Bartin University, Bartin, Turkey and School of Sciences, IGNOU, New Delhi for constant encouragement and support.

References

1. G. Crini and P. M. Badot, *Prog. Polym. Sci.*, 2008, **33**, 399.
2. Y. Pan, Y. Li, H. Zhao, J. Zheng, H. Xu, G. Wei, J. Hao and F. Cui, *Int. J. Pharm.*, 2002, **249**, 139.
3. R. Muzzarelli and C. Muzzarelli, *Adv. Polym. Sci.*, 2005, **186**, 151.
4. H. L. Zhang, S. H. Wu, Y. Tao, L. Q. Zang and Z. Q. Su, *J. Nanomater.*, 2010, **2010**, 1.
5. I. Younes and M. Rinaudo, *Mar. Drugs*, 2015, **13**, 1133.

6. L. Zhang, Y. Zeng and Z. Cheng, *J. Mol. Liq.*, 2016, **214**, 175.
7. K. A. Northcott, I. Snape, P. J. Scales and G. W. Stevens, *Chem. Eng. Sci.*, 2005, **60**, 6835.
8. Y. C. Chung, H. L. Wang, Y. M. Chen and S. L. Li, *Bioresour. Technol.*, 2003, **88**, 179.
9. G. Crini, *Prog. Polym. Sci.*, 2005, **30**, 38.
10. J. Berger, M. Reist, J. M. Mayer, O. Felt and R. Gurny, *Eur. J. Pharm. Biopharm.*, 2004, **57**, 35.
11. Y. Jiang, J. L. Gong, G. M. Zeng, X. M. Ou, Y. N. Chang, C. H. Deng, J. Zhang, H. Y. Liu and S. Y. Huang, *Int. J. Biol. Macromol.*, 2016, **82**, 702.
12. S. C. Motshekga, S. S. Ray, M. S. Onyango and M. N. B. Momba, *Appl. Clay Sci.*, 2015, **114**, 330.
13. T. Kamal, M. Ul-Islam, S. B. Khan and A. M. Asiri, *Int. J. Biol. Macromol.*, 2015, **81**, 584.
14. C. E. Onate and E. M. Frances, A Review of Chitosan-Based Materials for the Removal of Organic Pollution from Water and Bioaugmentation, *Chitin-Chitosan—Myriad Functionalities in Science and Technology*, InTech., London, UK, 2018.
15. V. K. Mourya and N. N. Inamdar, *React. Funct. Polym.*, 2008, **68**, 1013.
16. W. D. Wanrosli, R. Rohaizu and A. Ghazali, *Carbohydr. Polym.*, 2011, **84**, 262.
17. S. S. Kim, W. Y. Jeong, B. C. Shin, S. Y. Oh, H. W. Kim and J. M. Rhee, *J. Biomed. Mater. Res.*, 1998, **40**, 401.
18. M. Igura and M. Okazaki, *J. Hazard. Mater.*, 2010, **178**, 686.
19. D. M. Sulflet, G. C. Chitanu and V. I. Popa, *React. Funct. Polym.*, 2006, **66**, 1240.
20. A. M. Abu El-Soad, G. Lazzara, M. O. Abd El-Magied, G. Cavallaro, J. S. Al-Otaibi, M. I. Sayyed and E. G. Kovaleva, *Int. J. Mol. Sci.*, 2022, **23**, 2396.
21. T. Marino, M. Boerrigter, M. Faccini, C. Chaumette, L. Arockiasamy, J. Bundschuh and A. Figoli, Photocatalytic Activity and Synthesis Procedures of TiO_2 Nanoparticles for Potential Applications in Membranes, *Application of Nanotechnology in Membranes for Water Treatment*, CRC Press, London, 1st edn, 2017, pp. 127–146.
22. S. S. Madaeni, N. Ghaemi and H. Rajabi, Advances in Polymeric Membranes for Water Treatment, *Advances in Membrane Technologies for Water Treatment*, Woodhead Publishing Limited, Cambridge, UK, 2015, pp. 3–41.
23. R. Abedini, S. M. Mousavi and R. Aminzadeh, *Desalination*, 2011, **277**, 40.
24. K. Chaloupka, Y. Malam and A. M. Seifalian, *Trends Biotechnol.*, 2010, **28**, 580.
25. X. Cao, M. Tang, F. Liu, Y. Nie and C. Zhao, *Colloids Surf., B*, 2010, **81**, 555.
26. M. Byakodi, N. S. Shrikrishna, R. Sharma, S. Bhansali, Y. Mishra, A. Kaushik and S. Gandhi, *Biosens. Bioelectron.*, 2022, **12**, 100284.

27. M. S. Haider, G. N. Shao, S. M. Imran, S. S. Park, N. Abbas, M. S. Tahir, M. Hussain, W. Bae and H. T. Kim, *Mater. Sci. Eng., C*, 2016, **62**, 732.
28. J. Xu, X. Feng, P. Chen and C. Gao, *J. Membr. Sci.*, 2012, **413**, 62.
29. T. A. Saleh and V. K. Gupta, *Sep. Purif. Technol.*, 2012, **89**, 245.
30. Y. M. Mojtahedi, M. R. Mehrnia and M. Homayoonfal, *Desalin. Water Treat.*, 2013, **51**, 6736.
31. S. Zinadini, A. A. Zinatizadeh, M. Rahimi, V. Vatanpour and H. Zangeneh, *J. Membr. Sci.*, 2014, **453**, 292.
32. N. Niksefat, M. Jahanshahi and A. Rahimpour, *Desalination*, 2014, **343**, 140.
33. E. Fontananova, V. Grosso, S. A. Aljlil, M. A. Bahattab, D. Vuono, F. P. Nicoletta, E. Curcio, E. Drioli and G. Profio, *J. Membr. Sci.*, 2017, **541**, 198.
34. E. Celik, H. Park and H. Choi, *Water Res.*, 2011, **45**, 274.
35. J. N. Shen, C. C. Yu, H. M. Ruan, C. J. Gao and B. van der Bruggen, *J. Membr. Sci.*, 2013, **442**, 18.
36. S. S. Homaeigohar and M. Elbahri, *J. Colloid Interface Sci.*, 2012, **372**, 6.
37. S. F. Pan, X. X. Ke, T. Y. Wang, Q. Liu, L. B. Zhong and Y. M. Zheng, *Ind. Eng. Chem. Res.*, 2019, **58**, 984.
38. W. G. Kim and S. Nair, *Chem. Eng. Sci.*, 2013, **104**, 908.
39. A. Aghigh, V. Alizadeh, H. Y. Wong, M. S. Islam, N. Amin and M. Zaman, *Desalination*, 2015, **365**, 389.
40. C. Gerente, V. K. C. Lee, P. L. Cloirec and G. McKay, *Crit. Rev. Environ. Sci. Technol.*, 2007, **37**, 41.
41. J. H. Jhaveri and Z. V. P. Murthy, *Desalination*, 2016, **379**, 137.
42. A. M. Croitoru, A. Ficai, D. Ficai, R. Trusca, G. Dolete, E. Andronescu and S. C. Turculet, *Materials*, 2020, **13**, 1687.
43. M. Yang, C. Zhao, S. Zhang, P. Li and D. Hou, *Appl. Surf. Sci.*, 2017, **394**, 149.
44. Q. Xiaowei, L. Na, W. Qinzhuo and S. Ji, *Desalination*, 2018, **438**, 83.
45. P. Gao, M. H. Tai and D. D. Sun, *ChemPlusChem*, 2013, **78**, 1475.
46. P. S. Goh, K. C. Wong and A. F. Ismail, *Membranes*, 2020, **10**, 1.
47. L. X. Dong, X. C. Huang, Z. Wang, Z. Yang, X. M. Wang and C. Y. Tang, *Sep. Purif. Technol.*, 2016, **166**, 230.
48. H. Huang, X. Qu, X. Ji, X. Gao, L. Zhang, H. Chen and L. Hou, *J. Mater. Chem. A*, 2013, **1**, 11343.
49. A. L. Ahmad, M. A. Majid and B. S. Ooi, *Desalination*, 2011, **268**, 266.
50. S. Kumar, K. Prasad, J. M. Gil, A. Sobral and J. Koh, *Carbohydr. Polym.*, 2018, **198**, 401.
51. V. Adimule, S. Batakurki, B. C. Yallur, C. Bathula and L. Parashuram, *J. Mater. Sci.: Mater. Electron.*, 2022, **33**, 18699.
52. V. Adimule, S. Batakurki, V. S. Bhat, B. C. Yallur, G. Hegde and C. Bathula, *J. Energy Storage*, 2023, **57**, 106318.
53. V. Adimule, B. C. Yallur, R. Keri, C. Bathula and S. Batakurki, *Electron. Mater. Lett.*, 2022, 1.

54. V. Adimule, S. Rajendrachari, R. Mahale, S. Batakurki, B. C. Yallur, S. Nandi and G. Bagihalli, *Silicon*, 2022, 1.
55. V. Adimule, S. Batakurki, B. C. Yallur and R. Keri, *J. Mater. Res.*, 2022, **15**, 2396.
56. A. Spoial, C. I. Ilie, D. Ficai, A. Ficai and E. Andronescu, *Materials*, 2021, **14**, 2091.
57. S. Liang, K. Xiao, Y. Mo and X. Huang, *J. Membr. Sci.*, 2012, **394**, 184.
58. X. Zhang, Y. Wanga, Y. Liu, J. Xu, Y. Han and X. Xu, *Appl. Surf. Sci.*, 2014, **316**, 333.
59. J. Hong and Y. He, *Desalination*, 2012, **302**, 71.
60. Y. T. Chung, M. M. Ba-Abbad, A. W. Mohammad and A. Benamor, *Desalin. Water Treat.*, 2015, **57**, 7801.
61. J. E. Ghoul, I. Ghiloufi, A. S. A. L-Hobaib and L. E. Mir, *J. Ovonic Res.*, 2017, **13**, 83.
62. Y. J. Jo, E. Y. Choi, N. W. Choi and C. K. Kim, *Ind. Eng. Chem. Res.*, 2016, **55**, 7801.
63. S. Pintilie, L. G. Tiron, I. Birsan and S. Balta, *Mater. Plast.*, 2017, **54**, 257.
64. A. Ronen, R. Semiat and C. Dosoretz, *Water Res.*, 2013, **47**, 6628.
65. H. Bai, Z. Liu and D. D. Sun, *Appl. Catal., B*, 2012, **111**, 571.
66. S. B. Khan, K. A. Alamry, E. N. Bifari, A. M. Asiri, M. Yasir, L. Gzara and R. Z. Ahmad, *J. Ind. Eng. Chem.*, 2015, **24**, 266.
67. I. Akin and M. Ersoz, *Desalin. Water Treat.*, 2014, **57**, 3037.
68. X. Zhao, J. Li and C. Liu, *Desalination*, 2017, **408**, 102.
69. H. Isawi, M. H. El-Sayed, X. Feng, H. Shawky and M. S. A. Mottaleb, *Appl. Surf. Sci.*, 2016, **385**, 268.
70. Y. Zhang, Y. Wan, Y. Shi, G. Pan, H. Yan, J. Xu, M. Guo, L. Qin and Y. Liu, *J. Polym. Res.*, 2016, **23**, 105.
71. M. Ben-Sasson, X. Lu, E. Bar-Zeev, K. R. Zodrow, S. Nejati, G. Qi, E. P. Giannelis and M. Elimelech, *Water Res.*, 2014, **62**, 260.
72. Z. Yang, Y. Wu, J. Wang, B. Cao and C. Y. Tang, *Environ. Sci. Technol.*, 2016, **50**, 9543.
73. A. Ahmad, F. Jamshed, T. Riaz, S. E. Gul, S. Waheed, A. Sabir, A. A. AlAnezi, M. Adrees and T. Jamil, *Carbohydr. Polym.*, 2016, **149**, 207.
74. Y. Liao, R. Wang and A. G. Fane, *J. Membr. Sci.*, 2013, **440**, 77.
75. S. Zhang, G. Qiu, Y. P. Ting and T. S. Chung, *Colloids Surf., A*, 2013, **436**, 207.
76. A. Alpatova, E. S. Kim, X. Sun, G. Hwang, Y. Liu and M. G. El-Din, *J. Membr. Sci.*, 2013, **444**, 449.
77. Z. A. Rehan, L. Gzara, S. B. Khan, K. A. Alamry, M. S. El-Shahawi, M. H. Albeirutty, A. Figoli, E. Drioli and A. M. Asiri, *Recent Pat. Nanotechnol.*, 2016, **10**, 231.
78. M. Sile-Yuksel, B. Tas, D. Y. Koseoglu-Imer and I. Koyuncu, *Desalination*, 2014, **347**, 120.
79. D. Y. Koseoglu-Imer, B. Kose, M. Altinbas and I. Koyuncu, *J. Membr. Sci.*, 2013, **428**, 620.

80. E. M. V. Hoek, A. K. Ghosh, X. Huang, M. Liong and J. I. Zink, *Desalination*, 2011, **283**, 89.
81. I. C. Escobar and B. V. D. Bruggen, *J. Appl. Polym. Sci.*, 2015, **132**, 42002.
82. P. F. Andrade, A. F. de Faria, F. J. Quites, S. R. Oliveira, O. L. Alves, M. A. Z. Arruda and M. D. C. Gonçalves, *Cellulose*, 2015, **22**, 3895.
83. M. Zhang, R. W. Field and K. Zhang, *J. Membr. Sci.*, 2014, **471**, 274.
84. S. Liu, F. Fang, J. Wu and K. Zhang, *Desalination*, 2015, **375**, 121.
85. S. Liu, M. Zhang, F. Fang, L. Cui, J. Wu, R. Field and K. Zhang, *Desalin. Water Treat.*, 2015, **57**, 10560.
86. E. M. V. Hoek, A. K. Ghosh, X. Huang, M. Liong and J. I. Zink, *Desalination*, 2011, **283**, 89.
87. N. Akar, B. Asar, N. Dizge and I. Koyuncu, *J. Membr. Sci.*, 2013, **437**, 216.
88. M. Ben-Sasson, K. R. Zodrow, Q. Genggeng, Y. Kang, E. P. Giannelis and M. Elimelech, *Environ. Sci. Technol.*, 2014, **48**, 384.
89. S. S. Madaeni, S. Zinadini and V. Vatanpour, *J. Membr. Sci.*, 2011, **380**, 155.
90. F. Shi, Y. Ma, J. Ma, P. Wang and W. Sun, *J. Membr. Sci.*, 2012, **389**, 522.
91. B. Rajaeian, A. Heitz, M. O. Tade and S. Liu, *J. Membr. Sci.*, 2015, **485**, 48.
92. J. P. Mericq, J. Mendret, S. Brosillon and C. Faur, *Chem. Eng. Sci.*, 2015, **123**, 283.
93. A. Mollahosseini and A. Rahimpour, *J. Ind. Eng. Chem.*, 2014, **20**, 1261.
94. R. Abedini, S. M. Mousavi and R. A. Aminzadeh, *Desalination*, 2011, **277**, 40.
95. T. H. A. Ngo, D. T. Nguyen, K. D. Do, T. T. M. Nguyen, S. Mori and D. T. Tran, *J. Sci.: Adv. Mater. Devices*, 2016, **1**, 468.
96. M. Amini, A. Rahimpour and M. Jahanshahi, *Desalin. Water Treat.*, 2015, **57**, 14013.
97. F. Moghadam, M. R. Omidkhah, E. V. Farahani, M. Z. Pedram and F. Dorosti, *Sep. Purif. Technol.*, 2011, **77**, 128.
98. S. J. Kim, P. S. Lee, S. Bano, Y. I. Park, S. E. Nam and K. H. Lee, *J. Appl. Polym. Sci.*, 2016, **133**, 43383.
99. E. S. Kim, G. Hwang, M. G. El-Din and Y. Liu, *J. Membr. Sci.*, 2012, **394**, 37.
100. F. Ahmad, C. M. Santos, J. Mangadlao, R. Advincula and D. F. Rodrigues, *Water Res.*, 2013, **47**, 3966.
101. E. S. Kim, Y. Liu and M. G. El-Din, *J. Membr. Sci.*, 2013, **429**, 418.
102. P. Shah and C. N. Murthy, *J. Membr. Sci.*, 2013, **437**, 90.
103. J. N. Shen, C. C. Yu, H. M. Ruan, C. J. Gao and B. van der Bruggen, *J. Membr. Sci.*, 2013, **442**, 18.
104. E. Celik, H. Park and H. Choi, *Water Res.*, 2011, **45**, 274.
105. M. Sianipar, S. H. Kim, C. Min, L. D. Tijing and H. K. Shon, *J. Ind. Eng. Chem.*, 2016, **34**, 364.
106. A. Khalid, A. Abdel-Karim, M. A. Atieh, S. Javed and G. McKay, *Sep. Purif. Technol.*, 2018, **190**, 165.
107. J. Mulopo, *J. Environ. Chem. Eng.*, 2017, **5**, 4381.

108. C. Ursino, R. Castro-Munoz, E. Drioli, L. Gzara, M. H. Albeirutty and A. Figoli, *Membranes*, 2018, **8**, 18.
109. C. Zhao, X. Xu, J. Chen and F. Yang, *Desalination*, 2014, **334**, 17.
110. S. Xia, L. Yao, Y. Zhao, N. Li and Y. Zheng, *Chem. Eng. J.*, 2015, **280**, 720.
111. J. Lee, H. R. Chae, Y. J. Won, K. Lee, C. H. Lee, H. H. Lee, I. C. Kim and J. M. Lee, *J. Membr. Sci.*, 2013, **448**, 223.
112. L. M. Pastrana-Martinez, S. M. Torres, J. L. Figueiredo, J. L. Faria and A. M. T. Silva, *Water Res.*, 2015, **77**, 179.
113. S. A. Kiran, Y. L. Thuyavan, G. Arthanareeswaran, T. Matsuura and A. F. Ismail, *Chem. Eng. J.*, 2016, **286**, 528.
114. K. Goh, L. Setiawan, L. Wei, R. Si, A. G. Fane, R. Wang and Y. Chen, *J. Membr. Sci.*, 2015, **474**, 244.
115. J. Wang, C. Zhao, T. Wang, Z. Wu, X. Li and J. Li, *RSC Adv.*, 2016, **6**, 82174.
116. L. He, L. F. Dumee, C. Feng, L. Velleman, R. Reis, F. She, W. Gao and L. Kong, *Desalination*, 2015, **365**, 126.
117. L. Shen, S. Xiong and Y. Wang, *Chem. Eng. Sci.*, 2016, **143**, 194.
118. C. A. Crock, A. R. Rogensues, W. Shan and V. V. Tarabara, *Water Res.*, 2013, **47**, 3984.
119. Y. Han, Z. Xu and C. Gao, *Adv. Funct. Mater.*, 2013, **23**, 3693.
120. T. Gotoh, K. Matsushima and K. Kikuchi, *Chemosphere*, 2004, **55**, 135.
121. W. Wang, Y. Zhao, H. Bai, T. Zhang, V. Ibarra-Galvan and S. Song, *Carbohydr. Polym.*, 2018, **198**, 518.
122. K. C. Petkar, S. Chavhan, N. Kunda, I. Saleem, S. Somavarapu, K. M. G. Taylor and K. K. Sawant, *AAPS PharmSciTech*, 2018, **19**, 1758.
123. A. Cooper, R. Oldinski, H. Ma, J. D. Bryers and M. Zhang, *Carbohydr. Polym.*, 2013, **92**, 254.
124. Y. Andres, L. Giraud, C. Gerente and P. le Cloirec, *Environ. Technol.*, 2007, **28**, 1357.
125. S. Jagtap, M. K. Yenkie, S. Das and S. Rayalu, *Desalination*, 2011, **273**, 267.
126. M. M. Beppu, R. S. Vieira, C. G. Aimoli and C. C. Santana, *J. Membr. Sci.*, 2007, **301**, 126.
127. C. Qi, L. Zhao, Y. Lin and D. Wu, *J. Colloid Interface Sci.*, 2018, **517**, 18.
128. M. Amaike, Y. Senoo and H. Yamamoto, *Macromol. Rapid Commun.*, 1998, **19**, 287.
129. L. L. Min, L. M. Yang, R. X. Wu, L. B. Zhong, Z. H. Yuan and Y. M. Zheng, *J. Colloid Interface Sci.*, 2019, **535**, 255.
130. K. Gregor, P. Mateja, K. Željko and L. Maja, *Molecules*, 2019, **24**, 1960.
131. R. Khaydarov, S. Evgrafova, S. Wagner and S. Cho, Environmental and Human Health Issues of Silver Nanoparticles Applications, *Environ. Sec. Ecoterr.*, Springer, Dordrecht, The Netherlands, 1970, p. 117.
132. Q. Xu, X. Ke, N. Ge, L. Shen, Y. Zhang, F. Fu and X. Liu, *Fibers Polym.*, 2018, **19**, 1004.
133. N. Verma and N. Kumar, *ACS Biomater. Sci. Eng.*, 2019, **5**, 1170.
134. J. Chapman, L. Nor, R. Brown, E. Kitteringham, S. Russell, T. Sullivan and F. Regan, *J. Mater. Chem. B*, 2013, **1**, 6194.

135. T. Dankovich and J. Smith, *Water Res.*, 2014, **63**, 245.
136. A. T. Le, S. Y. Pung, S. Sreekantan, A. Matsuda and D. P. Huynh, *Helios*, 2019, **5**, e01440.
137. K. Nalwa, A. Thakur and N. Sharma, *Adv. Mater.*, 2017, **2**, 697.
138. M. Acheampong and D. Antwi, *J. Environ. Sci. Eng. A*, 2016, **9**, 13.
139. R. X. Zhang, L. Braeken, T. Y. Liu, P. Luis, X. L. Wang and B. van der Bruggen, *Appl. Sci.*, 2017, **7**, 81.
140. T. L. Khoo, A. S. Halim, A. M. Saad and A. A. Dorai, *Burns*, 2010, **36**, 897.
141. S. L. Chen, R. H. Fu, S. F. Liao, S. P. Liu, S. Z. Lin and Y. C. Wang, *Cell Transplant.*, 2018, **27**, 275.
142. W. S. W. Ngah, C. S. Endud and R. Mayanar, *React. Funct. Polym.*, 2002, **50**, 181.
143. W. C. Tsen, *Polym. Eng. Sci.*, 2020, **60**, 1832.
144. F. Hu, T. Li, F. Zhong, S. Wen, G. Zheng, C. Gong, C. Qin and H. Liu, *J. Appl. Polym. Sci.*, 2020, **137**, 49079.
145. G. Sharifzadeh, M. Soheilmoghaddam, H. Adelnia, M. U. Wahit, M. R. D. Arzhandi and A. Moslehyani, *Polym. Eng. Sci.*, 2020, **60**, 1169.
146. A. M. Croitoru, A. Ficai, D. Ficai, R. Trusca, G. Dolete, E. Andronescu and S. C. Turculet, *Materials*, 2020, **13**, 1687.
147. A. Sundramoorthy and S. Gunasekaran, *TrAC, Trends Anal. Chem.*, 2014, **60**, 36.
148. P. Ortega, J. Trigueiro, G. Silva and R. Lavall, *Electrochim. Acta*, 2015, **188**, 809.
149. L. Y. Yu, Z. L. Xu, H. M. Shen and H. Yang, *J. Membr. Sci.*, 2009, **337**, 257.
150. R. Abdeen and N. Salahuddin, *J. Chem.*, 2013, **2013**, 1.
151. Y. Kasirga, A. Oral and C. Caner, *Polym. Compos.*, 2012, **33**, 1874.
152. Z. G. Wu, X. J. Huang, Y. C. Li, H. Z. Xiao and X. Y. Wang, *Carbohydr. Polym.*, 2018, **199**, 210.
153. S. Bibi, T. Yasin, S. Hassan, M. Riaz and M. Nawaz, *Mater. Sci. Eng., C*, 2015, **46**, 359.
154. L. Fan, C. Luo, X. Li, F. Lu, H. Qiu and M. Sun, *J. Hazard. Mater.*, 2012, **215**, 272.
155. S. B. Khan, F. Ali and K. Akhtar, *Carbohydr. Polym.*, 2019, **207**, 650.
156. S. Herman, E. Saputra, M. R. Siregar, D. N. I. Panjaitan and J. A. Pinem, *J. Appl. Mater. Technol.*, 2019, **1**, 31.
157. N. Rosdi, M. N. M. Sokri, N. M. Rashid, M. S. C. Chik and M. S. Musa, *J. Appl. Membr. Sci. Technol.*, 2018, **23**, 63.
158. T. Truong, K. Takaomi and H. Bui, *J. Serb. Chem. Soc.*, 2019, **84**, 83.
159. H. Sashiwa and S. I. Aiba, *Prog. Polym. Sci.*, 2004, **29**, 887.
160. D. C. da Silva Alves, B. Healy, L. A. Pinto, T. R. Cadaval and C. B. Breslin, *Molecules*, 2021, **26**, 594.
161. A. M. Saad, M. R. Abukhadra, S. A. K. Ahmed, A. M. Elzanaty, A. H. Mady, M. A. Betiha, J. J. Shim and A. M. Rabie, *J. Environ. Manage.*, 2020, **258**, 110043.
162. V. K. B. Hoang, P. Duckshin and Y. C. Lee, *Polymers*, 2017, **9**, 21.
163. T. Velempini, E. Prabakaran and K. Pillay, *Mater. Today Chem.*, 2021, **19**, 100380.

164. S. Iftekhar, A. Deb, G. Heidari, M. Sillanpaa, V. P. Lehto, B. Doshi, M. Hosseinzadeh and E. N. Zare, *Environ. Sci. Pollut. Res.*, 2023, 1.
165. V. Adimule, S. S. Nandi, B. C. Yallur, D. Bhowmik and A. H. Jagadeesha, *Mater. Today Chem.*, 2021, **20**, 100438.
166. V. Adimule, S. Batakurki, B. C. Yallur and R. Keri, *J. Mater. Res.*, 2022, **37**, 2396.
167. V. Adimule, V. S. Bhat, B. C. Yallur, A. H. J. Gowda, P. De Padova, G. Hegde and A. Toghan, *Nanomater. Nanotechnol.*, 2022, **12**, 18479.
168. V. Adimule, B. C. Yallur and A. Gowda, *Anal. Bioanal. Electrochem.*, 2022, **14**, 1.
169. V. Adimule, G. B. Bagihalli and S. N. Unki, *Adv. Mater. Res.*, 2022, **1173**, 3.
170. N. F. Alharby, R. S. Almutairi and N. A. Mohamed, *Polymers*, 2021, **13**, 3659.
171. E. Salehi, P. Daraei and A. A. Shamsabadi, *Carbohydr. Polym.*, 2016, **152**, 419.
172. L. You, C. Huang, F. Lu, A. Wang, X. Liu and Q. Zhang, *Int. J. Biol. Macromol.*, 2018, **107**, 1620.
173. P. Arabkhani, A. Asfaram and M. Ateia, *J. Water Process Eng.*, 2020, **38**, 101651.
174. Y. L. Pang, J. H. Tan, S. Lim and W. C. Chong, *Polymers*, 2021, **13**(17), 3009.
175. A. U. Metin, H. Çiftçi and E. Alver, *Ind. Eng. Chem. Res.*, 2013, **52**(31), 10569.
176. V. M. Muinde, J. M. Onyari, B. Wamalwa and J. N. Wabomba, *Environ. Chem. Ecotoxicol.*, 2020, **2**, 115.
177. J. K. Fatombi, E. A. Idohou, S. A. Osseni, I. Agani, D. Neumeyer, M. Verelst, R. Mauricot and T. Aminou, *Fibers Polym.*, 2019, **20**, 1820.
178. G. L. Dotto, F. K. Rodrigues, E. H. Tanabe, R. Fröhlich, D. A. Bertuol, T. R. Martins and E. L. Foletto, *J. Environ. Chem. Eng.*, 2016, **4**(3), 3230.
179. H. V. Tran, L. T. Bui, T. T. Dinh, D. H. Le, C. D. Huynh and A. X. Trinh, *Mater. Res. Express*, 2017, **4**(3), 035701.
180. N. K. Nga, N. T. T. Chau and P. H. Viet, *J. Sci.: Adv. Mater. Devices*, 2020, **5**(1), 65.
181. T. Anitha, *J. Water Process Eng.*, 2016, **13**, 127.
182. Y. Wang, H. Wang, H. Peng, Z. Wang, J. Wu and Z. Liu, *Fibers Polym.*, 2018, **19**, 340.
183. J. Xie, C. Li, L. Chi and D. Wu, *Fuel*, 2013, **103**, 480.
184. M. A. Kamal, S. Bibi, S. W. Bokhari, A. H. Siddique and T. Yasin, *React. Funct. Polym.*, 2017, **110**, 21.

CHAPTER 2

Sustainable Synthesis of Green Novel Materials for Water Purification

AMAL I. HASSAN[a] AND HOSAM M. SALEH[a*]

[a]Radioisotope Department, Nuclear Research Center, Egyptian Atomic Energy Authority, Cairo, Egypt
*E-mail: hosam.saleh@eaea.org.eg, hosamsaleh70@yahoo.com

2.1 Introduction

Water is one of the most important factors of sustainable development, so there is an urgent need to rationalize the available use of water resources, especially in countries that suffer from scarcity of available water resources.[1] As a result, many countries around the world have adopted comprehensive planning and management approaches to facilitate the reuse of treated wastewater, thereby discontinuing the outdated practice of discharging wastewater into natural water bodies. Water covers over 70% of the Earth's surface. Water from the seas and oceans, which accounts for around 97% of the Earth's water bodies, is saline (rich in salt) and cannot be utilized for drinking, industrial, or agricultural purposes. On the other hand, fresh water accounts for only around 3% of the world's water.[2] Around 30% of this potential fresh water for human and animal use is locked in locations such as hilly regions, Greenland, ice caps, and glaciers, with approximately 69% being buried underground, which is usually quite deep and difficult to reach for ease of use. Just less than 1% of fresh water that is available as

Chemistry in the Environment Series No. 12
Novel Materials and Water Purification: Towards a Sustainable Future
Edited by Grigorios L. Kyriakopoulos and Miltiadis G. Zamparas

Published by the Royal Society of Chemistry, www.rsc.org

a water supply suitable for human and animal use is located in water bodies such as lakes, rivers, ponds, streams, and reservoirs; therefore, maintaining these water bodies as clean sources should be our top priority for survival on Earth.[3] Water quality has been deteriorating, while demand for pure water that is free of impurities and beneficial to health has been growing. Water purification is the removal of unwanted chemicals, biological pollutants, suspended particles, and gases from polluted water.[4] It is critical to cleanse water before using it for residential purposes. To discover the best quality water purifier, to enhance wastewater water quality, to eliminate all toxins from water, to make water useful for other purposes, and to avoid hazards that cause illness, the quality of water has been achieved or improved by utilizing only natural materials and no chemical materials. Treatment techniques and methods have evolved markedly, especially in recent years, and the success of industrial wastewater treatment and reuse depends on a set of environmental criteria that are related to the nature of the water and the purpose of treatment and reuse, which must ensure protection of the environment and of individuals, and take into account the economic situation.[5]

Water remediation is the method of removing contamination from water. The purpose of water remediation is to convert polluted water into clean water, or to sequester polluted water so that people are not exposed to risk and contaminants are not spread.[6] Biological, chemical, and physical treatment approaches are used in water remediation. The significance of water remediation has lately given rise to much research concentrating on various aspects of this discipline. Innovative and environmentally friendly materials are required to lower treatment system running and maintenance costs, and to enable the transfer of active treatment methods to the design of passive treatment approaches.[7] It is necessary to develop new materials with larger or more selective affinity for heavy metals and organic compounds. Similarly, innovative materials that may be synthesized to increase the efficacy of existing treatment procedures, such as microlenses to improve water pollutant photodegradation, should be studied.[8] Such materials should also provide operational stability under various process conditions and be inexpensive.

Water contamination occurs when waste from agricultural, residential, and industrial sources (*e.g.*, textile manufacture) are released directly or indirectly into bodies of water (*e.g.*, rivers, lakes, aquifers, and groundwater) without first filtering out dangerous compounds.[9] These toxic compounds (organic pollutants) can linger in water for an extended length of time and are absorbed by plants and animals in the lower part of the food chain, causing damage to living organisms higher up the food chain. As a result, water pollution is becoming a serious environmental concern that necessitates ongoing review and assessment of water resource policies globally.[10] According to recent studies, water contamination is the leading cause of sickness and death worldwide. Microbiological pollutants are the most dangerous to human health, since they can cause infectious illnesses that affect all demographic groups, many of which can create epidemics and are lethal. With the exception of a few compounds, such as cyanide and nitrate, chemical pollution poses a longer term health danger.[11,12] Charcoal is a good adsorbent for removing heavy metal contaminants from water. However, substances in water that change the purity, colour, or taste of water may make it unappealing to customers, thus reducing the capacity to recoup expenditure. The goal of one study was to find a low-cost, environmentally friendly water purifier that used non-biodegradable materials for treatment.[13] By eliminating unwanted chemicals, materials, and biological pollutants, the water is made suitable for home use as drinking water.[14] Conventional adsorbents, such as activated carbon and polymers, have long been used to eliminate heavy metals, organic material, and biological pollutants from water and waste water.[15] However, they are not always efficient in removing contaminants, and their manufacture is both expensive and energy intensive. Producing activated carbon from municipal wastewater treatment sludge has the potential to be a cost-effective waste management and adsorbent manufacturing method. Similarly, because of their huge surface area, hydrophilicity, extensive surface chemical modification capability, and lack of toxicity, cellulose nanofibers are intriguing alternatives to conventional water treatment adsorbents.[16]

This chapter reviews the scientific objective of assessing the potential of green nanoparticles for water treatment, which is a novel and effective substitute for traditional methods.

2.2 Innovative Approaches Using Nanotechnology in the Remediation of Water

2.2.1 Photocatalysis: Transforming Water Treatment with Nanomaterials

Recent research in nanotechnology has allowed the creation of novel photocatalytic nanomaterials. In recent years, solar-generated or natural sunlight-driven photocatalysis has emerged as a game-changing technology for decomposing organic pollutants.[17] Traditional photocatalysts, on the other hand, have significant limitations, such as the need for a solar source and side effects of the catalyst nanoparticles.[18] Ultraviolet (UV) light interacting with semiconductors frequently initiates the photocatalytic process, which results in electron transport from the occupied valence band to the unoccupied conductor band.[19] Negative electron (e^-) and positive hole (h^+) pairs will form on the photocatalyst surface as a result. The e^-–h^+ couples either participate in redox reactions that mineralize the organic pollutants, or are recombined, resulting in diminished photocatalytic activity (Figure 2.1). Mineralization is caused by interactions between the target pollutant and radicals, such as peroxide and superoxide, which are formed when a positive hole and electrons combine with water and oxygen.[20]

More research is needed to create effective, ecologically friendly, real-time treatment approaches for improved environmental adaptation. Photocatalytic advanced oxidation (PAO) technologies for water treatment have a long

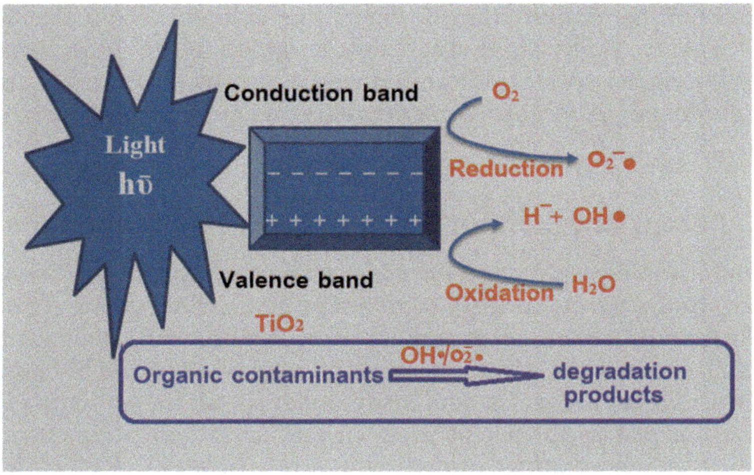

Figure 2.1 The photocatalytic process for the TiO_2 photocatalyst.

history of technical advancement, having been the focus of intensive study for the past 35 years.[21] Conventional PAO uses precursor chemical oxidants, such as O_3 and H_2O_2, in conjunction with an energy input (*e.g.*, UV irradiation) to generate reactive oxygen species for the oxidative annihilation of pollutants in water. Photolysis and photocatalysis are potential strategies for the elimination of persistent and hazardous contaminants from the environment, since sunlight is a free, renewable, and globally available source of energy. Both photolysis and photocatalysis are based on interactions between electromagnetic radiation and the existing molecules; however, in wastewater treatment, heterogeneous photocatalytic degradation is mostly studied.[22] This method relies on the use of several semiconductor materials, the most prevalent being TiO_2 and ZnO, as photocatalysts. Brookite, anatase, and rutile are the three crystalline phases of TiO_2. Because the brookite phase of titanium dioxide is unstable, it is not used in photocatalytic applications.[23] The anatase and rutile phases of titanium dioxide are thermodynamically stable, and because of its ease of production and superior catalytic efficacy the rutile phase is commonly employed in photocatalytic applications.[23] Mixed phases of titanium dioxide have been shown to be effective photocatalytic catalysts. The migration of photo-induced electrons from the rutile to the anatase phase of titanium dioxide is thought to provide mixed-phase titanium dioxide with better catalytic activity.[24] This electron mobility inhibits the recombination of positive holes and electrons, thus improving photocatalytic efficiency. Nevertheless, electrons have been demonstrated to migrate from the anatase phase to the rutile phase in mixed-phase photocatalysts.[25] The bandgap energy of the anatase and rutile phases of titanium dioxide is 3.2 eV and 3.0 eV, respectively. When titanium dioxide is exposed to photons with energies equal to or greater than 3.2 eV, one electron is excited from the valence band to the conduction band.[26] An electron–hole pair is formed because of this excitation. These photo-induced charges migrate to the surface of the titanium dioxide where they initiate a chain of redox processes.[26] Positive holes cause vacancies to develop in titanium dioxide and ignite the reduced species. Photo-induced electrons generate $O_2^{\bullet}$ free radicals. Subsequently, because these free radicals are extremely reactive and unstable, they react.[27]

2.2.2 Recent Advances in Water Purification Nanocomposites

As a new technology supported by recent research, Jagodić *et al.*[28] stated that Fe_2O_3/TiO_2/polyvinyl chloride (PVC) might be an excellent choice for environmentally friendly wastewater treatment. Specifically, heterogeneous photocatalysis collects sunlight, which is a free and renewable source of energy, and the high activity of the stated composite under simulated sunlight decreases operational expenses further because an artificial irradiation source is not required. Different iron(III) oxide/titanium(IV) oxide/PVC (Fe_2O_3/TiO_2/PVC) nanocomposites in tablet form were explored in the photodegradation of methylene blue (MB) under simulated sunshine, as well as their potential antibacterial properties. Scanning electron microscopy (SEM), X-ray diffraction (XRD), UV–vis diffuse reflectance spectroscopy, and Raman spectroscopy were used to evaluate the newly produced nanocomposites. In the case of the $Fe_2O_3{}^2$ and Fe_2O_3 samples, the findings revealed a hematite crystal form, whereas the $Fe_2O_3{}^1$ sample revealed a mix of hematite and the synthetic mineral akaganeite. When 70.6% of MB was eliminated in the presence of Fe_2O_3/TiO_2/PVC, maximum photocatalytic efficiency was attained. In addition, the possibility of photocleaning and reusing the indicated photocatalyst was investigated.[28]

When these materials are exposed to radiation, they can absorb photons only if their energy (hv) is greater or equal to that of the semiconductor bandgap (E_g). This mechanism results in the production of electron–hole pairs.[29] In subsequent phases of photocatalytic degradation, photogenerated holes can oxidise H_2O to highly reactive hydroxyl radicals, while electrons convert dissolved oxygen to superoxide radical anions and then to hydroxyl radicals. These radicals target existing contaminants, causing mineralization, which is desired (*i.e.*, formation of simple, harmless substances). Unfortunately, heterogeneous photocatalysis has significant drawbacks, such as the possibility of photogenerated electron–hole pair recombination and the lack of a sufficient electron acceptor in the system.[30] In the absence of scavengers, this recombination results in heat loss and heat dissipation. An additional disadvantage of heterogeneous photocatalysis is the large bandgap energy of the most widely employed semiconductors, which results in only a tiny proportion of overlap between the absorption spectra of semiconductors and of solar radiation that reaches the Earth's surface.[31] As a result, the use of artificial UV sources, which are more expensive, is required, and UV radiation requires protective aids for its use. This problem can be overcome by employing photocatalysts with bandgaps of 3 eV, known as visible-light-active photocatalysts.[31]

For many years, water and waste water were treated using ozone and chlorine chemicals. However, the chlorine residues are harmful to marine life. As a result, a new material (TiO_2), using Ti as the parent element, was created due to its good stability, economics, photoactivity, and non-toxicity.[32,33] TiO_2 has the ability to destroy microorganisms while also helping in the degradation of organic compounds. TiO_2 is frequently used as a catalyst when suspended in water.[33]

TiO_2 has demonstrated capabilities in a variety of areas of environmental concern apart from air and water pollution. When TiO_2 is combined with a membrane made from polyvinylidene fluoride (PVDF)[33] its hydrophilicity increases, while it also prevents biofouling. Because Ti is harmless to people, it is recommended for use as a photocatalyst in the presence of UV radiation in areas such as desalination, pharmaceutical degradation, and oily fluid treatment.[34]

The effective coating of TiO_2 particles on transparent plastic used in photocatalyst processes was reported by Utami *et al.*[35] using a spray technique. Under solar exposure, this approach was used to decompose MB-contaminated organic waste water. In the photocatalyst test, 300 mL of MB was employed.[36] The absorbance spectra and degradation rate of MB demonstrated that it effectively decomposed by around 98% with 16 hours of irradiation.[36] The findings point to a quicker rate of disintegration than in previous research. This technology was chosen because it is simple, ecologically benign, inexpensive, and has the potential to be further enhanced.

A novel low anode bias voltage photo-assisted electrochemical system was built by Liu *et al.* in 2019.[37] Then, a 20-cycle test for bisphenol degradation was carried out in the open air. According to the results, the developed system performed well and remained stable under low bias voltage, without any obvious oxygen evolution.[38] A Pt cluster-loaded TiO_2 (110) photocatalyst was prepared and catalyzed by Walenta *et al.*[39] using Auger electron spectroscopy, temperature-programmed desorption/reaction, isotope labelling, and isothermal photoreaction. They discovered that under photocatalytic circumstances the species on the catalyst surface might be catalytically processed.[39] This approach might also be used with different noble metal promoters and other TiO_2 alterations.[39] A preparation technique for a photocatalytic coating based on nano-TiO_2 and acrylate photopolymerization resin was created by Islam *et al.*[40] They carried out research on photocatalytic degradation under UV-B and sunlight. The results showed that it took 120 min for around 80% of a 6 ppm MB solution to deteriorate under UV-B light. In contrast, it took 60 min for the identical MB solution to break down by 90% in sunshine; the mechanism was also examined.[40] TiO_2 nanoparticles and an oxygen- or nitrogen-linked heptazine-based polymer (ONLH) were successfully mixed by Zhang *et al.*[41] to create a TiO_2/ONLH nanocomposite that reacted with sunlight. The management and purification of contaminants in water bodies is crucial for the sustainable growth of society and the economy because water resources are becoming increasingly scarce. In accordance with the green and sustainable development concept, Liu *et al.*[42] prepared novel TiO_2 nanoparticles using the sol–gel method, graphene oxide (GO) using the modified Hummers' method, and, finally, a new N-TiO_2/GO photocatalyst composite material by hydrothermal synthesis. They analyzed the adsorption and photocatalysis characteristics of the prepared GO and its composite materials using rhodamine B (RhB) as simulated pollutant.[42] Their findings should serve as experimental guidelines for future water resource treatment and purification.[42] Due to the superior performance of GO in suppressing the recombination of photogenerated electrons and holes, Liu's research has significantly expanded the photoresponse range of TiO_2. The vast specific surface area of GO can raise the ratio of photocatalysis materials to pollutants in contact with it. TiO_2, N-TiO_2, and TiO_2/GO were prepared as control groups and SEM, XRD, Fourier-transformed infrared (FTIR) spectroscopy, Raman spectroscopy, and X-ray photoelectron spectroscopy were used to investigate the materials produced. In addition, the photocatalysis degradation efficiency of the RhB solution was investigated under ultraviolet and visible sunlight.

2.2.3 ZnO: An Alternative Semiconductor for Water Treatment

ZnO is a biocompatible semiconductor with a broad bandgap energy that is suitable for everydayuse. ZnO has a similar bandgap energy to TiO_2, but its ability to absorb a broader range of electromagnetic radiation makes it more appealing for practical applications.[39] Furthermore, ZnO powder has a substantially higher quantum efficiency than TiO_2.[43] However, because of its broad bandgap, its use in natural sunlight (visible part of the spectrum) is limited, and results in photogenerated electron–hole pair recombination.[39] When the physical and chemical properties of ZnO are examined, it is clear that this semiconductor possesses distinct traits, such as chemical stability, electrochemical coupling coefficient, paramagnetic nature, and high photostability.[44] Longer irradiation, on the other hand, may cause photocorrosion, while in basic media self-oxidation and dissolution of ZnO occur, reducing the efficiency of this semiconductor.[39] The synthesis of diverse nanocomposites has recently been the most active area in the world of nanomaterials. Much progress has been accomplished as a result of several investigations. A unique nanoscale adsorbent, for example, was created by chemical deposition of nanozerovalent iron (nZVI) on carbon nanotubes (CNTs). According to the findings, the adsorbent has a high potential for rapid and efficient nitrate removal in water. Furthermore, the adsorbent may be easily removed from the solution by a magnet due to its peculiar magnetic features (Table 2.1). The presence of a significant number of chemical components, primarily phosphorus, in water has contributed to the emergence of environmental issues, such as eutrophication of water bodies.[45] As a result, among the advances in the field of water and effluent treatment processes with higher efficiency for aquatic contaminants, adsorption with nanomaterials stands out, due to the catalytic, optical, electronic, magnetic, and bactericidal properties of nanomaterials, as well as the high surface area in the adsorption sites.[46]

In the 21st century, the achievement of producing nanoparticles has been of significant interest due to their availability, good characteristics, biocompatibility, and size. Nanoparticles have a significant impact in many fields, including electronics, industry, and medicine, with promising applications in the allocation and managed delivery of medicines, imaging diagnostics, and remedies for diseases, such as lung, liver, cervical, prostate, and

Table 2.1 Nanomaterials in water remediation.

Nanomaterial	Application	Reference
Nanocatalyst polychlorinated biphenyls (PCB)	Removal of PCBs.	48
Nanozerovalent iron (nZVI)	Has given a fresh method for remediating polluted soil and groundwater. May also efficiently remove a significant quantity of polychlorinated organic contaminants, inorganic anions, and dissolved heavy metals from solutions.	49
Silver nanoparticles (Ag NPs)	Removal of bacteria and microorganisms. Colloidal Ag NPs improve filter effectiveness, and the filters can remove *Escherichia coli* in the range 97.8% to 100%.	50
Magnetic nanoparticles (MNPs)	Magnetic nanoparticles with high stability are extremely useful in catalysis. MNPs are recognised as a useful instrument for the removal of numerous sorts of pollutants from water.	51
Carbon nanotubes (CNTs)	CNTs have a strong metal ion adsorption capability. CNTs are a comparatively recent substance for removing trace contaminants from water.	52
Molybdenum sulfide nanosheets (MoS_2)	Molybdenum disulfide nanosheets are a promising alternative two-dimensional (2D) material with numerous great physicochemical, biological, and mechanical characteristics, which differ considerably from those of carbon nanomaterials, potentially resulting in new environmental phenomena and unique applications. Adsorption is regarded as a flexible water treatment process with a wide range of applications and a low cost. MoS_2 might be used as an effective adsorbent for the removal of heavy metal ions.	53–55
Nanoscale metal oxides	Have an enormous potential for the elimination of organic pollutants. Removal of heavy metals.	56
Bioactive nanoparticles	Eradication of bacteria and fungi.	
Nanoscale calcium peroxide	CaO_2 is an ecologically benign oxidising agent that not only degrades toxins without harming the environment, but also provides microorganisms with oxygen, which is advantageous for their development, increasing the efficacy of bioremediation. CP/Fe(II) technology in treatment of glyphosate waste water to remove total phosphorus.	57 and 58
Gold nanoparticles (Au NPs)	Can operate as sensors for the detection of a range of pollutants, such as heavy metals, pesticides, and pathogenic bacteria, through phenomena such as aggregation or size-dependent colour change.	59 and 60

colorectal cancer, while also exhibiting fungicidal, antibacterial, and antiparasitic action.[47] Adsorbent functionalization with magnetic nanoparticles enables recovery and reuse of the adsorbent, reducing environmental and economic costs owing to toxicity. The utilization of such materials, in addition to offering low-cost adsorbent production, allows reuse of waste. These benefits have attracted much attention from the scientific and industrial sectors, primarily owing to their environmental elements.

2.2.4 Innovative Membrane Separation Techniques

Membrane separation is a revolutionary technique for pollutant removal that requires no phase shift, uses less energy, makes it simple to separate particles even from very diluted solutions, and is more effective.[61] Reverse osmosis and nanofiltration membranes are typically used in membrane filtration processes for colour separation; however, nanofiltration membranes appear to perform the best due to their nanoporous structure, which prevents most organic debris from passing through.

2.2.5 Polymer Membranes for Enhanced Filtration

Polymeric membranes are currently the most used membrane type for water purification due to their benefits, including high degree of flexibility, low cost, smaller footprint, and easy-to-understand pore-forming mechanism.[62] In liquid–solid, gas–gas, and liquid–liquid separations, nanocomposite membranes made of dispersed nanoparticles in a polymer matrix are employed. Because of their biodegradability, high specific strength, stiffness, aspect ratio, renewability, low cost, and outstanding thermal stability, cellulose nanocrystals have a lot of potential for this purpose. The process of sequential vacuum filtering is widely used to create nanocellulose membranes. It is a common method for separating a solid–liquid mixture in order to keep the solid.[63] Vacuum filtration is analogous to gravity filtration in that the mixture is separated using a filter, but is distinguished by the assistance of a vacuum pump placed beneath the filter paper. Because the solvent, solution, and air are driven through the filter paper by the application of reduced pressure, vacuum filtration is quicker than gravity filtration. Moreover, it is more effective at eliminating leftover liquid, thus producing a more pure solid. Unfortunately, because of the force of the suction, tiny crystals may be drawn through the pores of the filter paper, causing some material to be lost in the filtrate, as well as some that cannot be retrieved from the filter paper. Particularly for nanocellulose, vacuum filtration of aqueous nanocellulose dispersions over a filter paper substrate is typically used to make nanocellulose membranes (together with a composite material). The effectiveness of the filtering of the membrane is also influenced by the drying techniques used. Research findings have shown that nanocellulose exhibits significant adsorption efficiency for dyes, nitrates, humic acid, metal ions, and other pollutants from industrial effluents, whether it has additional surface functionalization, such as enzymatic phosphorylation,

[(2,2,6,6-tetramethylpiperidin-1-yl)oxyl] (TEMPO) oxidation, cationization, and so forth, or even in its native form.[64] Manufacturing a membrane with controlled pore size and structure, as well as layering multiple membranes on top of one another, can be challenging while maintaining the eco-friendliness, flux, and rejection capacity of the nanocellulose.[65]

The process of producing membranes with a large surface area and an open, interconnected porous structure using electrospinning is gaining popularity. At the micrometre scale, electrospun membranes have been utilized to absorb contaminants from water. It is possible to create chitin nanocrystal, biobased, and independent electrospun membranes made of cellulose acetate with sufficient mechanical characteristics, water flux, and antifouling performance. It is also possible to create high flux, sustainable membranes for water filtering. On a cellulose acetate electrospun mesh template, cellulose nanocrystals were inserted. The membranes worked well at rejecting contaminating particles with a size range of 0.5 to 2.0 μm, and they had remarkable antifouling qualities. For the creation of biobased membranes that combine adsorption, size exclusion, and hydrophilicity, the functionalization of cellulose acetate membranes employing cellulose nanocrystals produced a highly effective surface treatment method.[66] The water flux was found to be high (22 000 $L\ m^2\ h^1$), but was reduced as the concentration of nanocrystals increased and the network grew more continuous.[66]

2.3 Green Materials for Bioremediation of Water

Innovative and environmentally friendly materials are required to lower running and maintenance costs for treatment systems and to enable the transfer of active treatment methods to designed passive treatment approaches.[67] It is necessary to develop new materials with larger or more selective affinity for heavy metals and organic compounds. Similarly, innovative materials that may be synthesized to increase the efficacy of existing treatment procedures, such as microlenses to improve water pollutant photodegradation, should be studied. Such materials should provide operational stability under various process conditions and be inexpensive.[68] These innovative materials could be mixed constituents or single constituent of biological waste.[108–111]

2.3.1 Bacteria

Bioremediation is an environmentally friendly remediation technology that takes advantage of the innate capacity of microorganisms to degrade/detoxify organic and inorganic contaminants in industrial waste waters.[69] Waste is transformed into inorganic chemicals, such as carbon dioxide, water, and methane, during the bioremediation process, resulting in mineralization and detoxification. Pollutant non-bioavailability tends to slow bioremediation. The addition of a biosurfactant enhances the solubility of a sorbed and/or hydrophobic contaminant, increasing mass transfer to microorganisms. By generating structural changes in the bacterial cell surface, it also improves microbial accessibility to pollution.[70] These measures hasten and improve the efficiency of the bioremediation process. Furthermore, bioremediation is a relatively simple and cost-effective process. Land farming and monitored natural attenuation are two noteworthy examples of its application. Both bioremediation procedures need less equipment and labour to operate and depend mostly on microbial aerobic and anaerobic processes to break down pollution. Importantly, bioremediation approaches are adaptable and may work in combination.[71] This simultaneous application of more than one bioremediation technology tends to boost bioremediation process efficiency by overcoming the limitations of a single technique.[72] Furthermore, bioremediation can be combined with other chemical or physical remediation approaches.[73] Finally, the process and efficacy of pollutant removal during bioremediation may be evaluated. Bacteria may be utilized to clean up our environmental problems, such as contaminated groundwater, oil spills or abandoned oil and gas sites, and waste water. A variety of studies conducted in recent years have emphasized microorganisms, which may biodegrade and biosorb colouring chemicals found in waste water. With growing concern about pollution, biological decolorization is gaining prominence in effluent treatment.[74] With economic restrictions on the pollution control process, cost-effective and efficient approaches have become essential. Because spent wash is high-strength waste water, it is usually subjected to anaerobic digestion in massive methane reactors to convert organics into methane, which is then used as fuel.

The anaerobically biodigested distillery effluent is subsequently aerobically processed in activated sludge treatment plants to decrease biochemical oxygen demand/chemical oxygen demand further and then utilized for composting press mud, a byproduct of sugar mills. Even after biomethanation, the hue of the wasted wash remains.[75] Numerous microbial species have been studied for their ability to break down and decolorize harmful chemical contaminants found in different industrial waste waters, including distilleries. For bioremediation of distillery effluent, free or immobilised cells have been extensively researched. When compared to free cells, immobilising microorganisms on inert support materials, such as alginate, polyacrylamide, agar, polystyrene, and polyurethane, is more favorable. The compact physical structure of carrier pellets, excellent biomass retention, reusability of culture, and ease of separation are just a few of the benefits.[76]

Su *et al.*[77] developed five segregates of photosynthetic bacteria for the removal of Cr from aqueous solutions, and it was shown that the segregates SC01 and SC05 displayed reduced oxidative damage to the cell membrane, leading to optimal removal capacities of 85 and 81% for Cr, respectively. The removal of Cr(VI) from an aqueous solution was studied using the newly discovered photosynthetic bacteria *Rhodobacter sphaeroides* SC01 in combination with various phosphor salts. It was shown that the culture conditions had a substantial effect on Cr(VI) remediation. The study also found that the bacterial cell membrane lost 99% of its Cr(VI), which was improved by 15% with the addition of phosphoric salts, leading the researchers to draw the conclusion that the bacterial cell in combination with phosphoric salts might be a strong contender for wastewater management.

2.3.2 Protein Nanofibrils

When it comes to adsorbing materials for water purification, proteins hold a position of utmost significance. They are excellent green materials for water purification because they are water-borne natural materials with characteristics of "universal molecular adhesives", which allow them to bind a wide range of pollutants in water. Also, by turning them into thin films or nanofibrils with extreme aspect ratios, a crucial component in adsorption processes, their specific area can be easily controlled and improved. Protein nanofibrils, a brand new class of nanomaterials, have an exceptionally high surface-to-volume ratio and special surface functionality for use as freshwater purifiers. They are made up of repeating core sequences, such as hydrophilic and hydrophobic amino acid segments, folded into fibrillar, frequently twisted structures, with lengths of up to several micrometres and widths typically ranging from 5 to 10 nm. As supramolecular polymers rich in β-sheet secondary structures, protein nanofibrils are packed in a cross-like configuration with the β-strands parallel to the fibrils in the case of silk nanofibrils and perpendicular to the fibril axis in the case of amyloids.[78] The structure is stabilized by a robust network of hydrogen bonds and van der Waals forces. The protein nanofibrils display exceptional nanomechanical qualities and stability against enzymatic degradation when compared to other protein materials. They stand out because of their chirality, polarity, and charge, in addition to their inherent rigidity.[78] Protein nanofibrils (PNFs) are dynamic scaffolds that can be developed as building blocks for creating smart functional materials for tissue engineering, biomedicine, materials science, nanotechnology, renewable energy, and environmental applications based on their unique properties.[79] Several techniques can be used to create PNFs, including amyloid and non-amyloid varieties. Monomers, such as β-lactoglobulin, whey and lysozyme, can be rapidly transformed to amyloid nanofibrils by unfolding, hydrolysis, and aggregation under conditions of low pH, high temperature, and several hours of incubation.[79,80] PNFs have primarily been applied as hybrid membranes and aerogels to remove phenolic chemicals, radioactive waste, heavy metals, metalloids, pesticides, dyes, and medications from water. PNF water solution, either pure or coupled with other materials, is poured on top of a filter support to create membranes.[80] The thin layer of protein nanofibrils that is still present on top of the support after the water phase has dried up serves as the membrane active layer. For instance, a homogeneous aqueous mixture of lactoglobulin fibril (2 wt%) and activated carbon (10 wt%) solution can be used to quickly produce a hybrid amyloid fibril-activated carbon membrane on the surface of a supporting cellulose filter. In order to create diverse organic/inorganic hybrid membranes for water purification, amyloid fibrils can be replaced by silk nanofibrils, and activated carbon by metallic molybdenum disulfide. This is possible because the fabrication method is general and applies to other types of protein nanofibril hybrids.

Conventional adsorbents, such as activated carbon and polymers, have been used to remove heavy metals, organics, and biological pollutants from water and waste water. However, they are not always efficient in removing contaminants, and their manufacture is both expensive and energy intensive.[80]

2.3.3 Seaweed

On a wider scale, red, green, and brown seaweeds are typically used to manufacture seaweed-based products that are rich in organic polymers composed of proteins and carbohydrates. We can acquire insight into material design and be inspired to create high-performance materials by examining their structure and the diversity of their chemical makeup.[81] The chemical makeup of the cell walls of seaweeds affects their capacity for biosorption. Large levels of hydroxyl, acetamide, amine, amide, sulfate, and phosphate functional groups are typically present in seaweed, which greatly promotes metal–ligand interaction and aids in the removal of heavy metal ions from aqueous solutions.[82] Alginate, carrageenan, and agarose are the specific components responsible for biosorption in brown algae, red seaweed, and green seaweed, respectively.[83] *Sargassum*, with notable biosorption capacity, has recently been investigated for the adsorption of lead, copper, and cadmium ions, and has several benefits, including high efficiency and affinity for toxic metals, ease of operation, reduction of the amount of chemical and biological sludge that needs to be disposed of, and low operating costs.[83] Researchers have also shown that *Ecklonia radiata* and *Durvillaea potatorum* have better biosorption capacities for nickel, copper, and lead ions than *Sargassum*. In reality, biosorption happens in two stages: (1) quick exchange of ions with surfaces that have functions, and (2) slow absorption into the multi-structured material.[84] It is important to remember that only cationic contaminants can be absorbed by seaweed-based products. Seaweeds with an anionic surface charge have poor biosorption and resist anionic contaminants. Investigators have also shown chemical activation, such as an acid/alkali wash to activate surface function and improve biosorption, followed by further soaking in calcium, magnesium, and potassium salts to prevent alginate leaching. Biomass waste that cannot currently be dealt with was also investigated for biosorption, and copper ion separation was accomplished with remarkable effectiveness. This could be as a result of the voids left behind by alginic extraction. Curiously, alginic beads immobilised in seaweed demonstrated superior biosorption for radioactive cesium.

Excellent agar/agarose-based hydrogels and aerogels for water purification applications have been prepared using a variety of techniques. Rani *et al.* looked into the capacity of an agar-based hydrogel for biosorption.[85]

2.3.4 Hydrogels and Aerogels

The hydrogel was created using the free radical copolymerization of acrylamide and *N*,*N*′-methylene bis-acrylamide, which demonstrated an impressive swelling behaviour of up to 550% and superhydrophilicity, which was attributed to the hydrophilic nature of the agar molecules. For solar-assisted water evaporation, an effective double-layer silver-poly(sodium-*p*-styrenesulfonate)-agarose gel/agarose gel has recently been developed.[86] While the bottom hydrophilic agarose gel serves as an effective water-transport medium, the top layer of active agarose that has been functionalized with silver induces a solar light–thermal transition that results in outstanding water vaporisation of up to 2.1 kg m^2 h^1 under one sun irradiation.[86] For the separation of organic pollutants, as well as for catalytic reduction, agarose-based hybrid materials with carrageenan, polyurethane, polyvinyl alcohol, and maltodextrin have also been produced. Materials based on agar or agarose have also been investigated for use in degrading applications. The hybrid aerogel made of agar and carbon nitride displayed efficient photodegradation of phenol, antibiotics, and dyes.

2.3.5 Carbonaceous Materials

Because they are inexpensive, environmentally benign, and can be manufactured on a large scale, carbonaceous materials derived from biomass have been intensively investigated for a variety of applications.[87] Also, it is advantageous to use functional materials made from biomass rather than from petroleum, and these can be readily marketed in the near future. Additionally, biomass-derived carbon compounds are biodegradable, and research suggests that they may be able to replace expensive carbon nanotubes, graphene and graphene oxide derivatives, and several metal-based nanocomposites. Additionally, by carefully selecting a suitable biomass with additional functionalities of interest, one can design and produce highly porous, heteroatom-doped, and metal-doped graphitic carbonaceous materials. These are typically processed using high-temperature pyrolysis, hydrothermal carbonisation, or both. Most biochars documented in the literature come from terrestrial lingocellulosic feedstocks, which typically produce a carbonaceous output with a C content of more than 70% and little to no nutritional value.[88]

However, a study found that compared to biochar made from red seaweed, carbonised products from brown seaweed-based species have higher C and H contents and lower S levels. The physicochemical characteristics of carbonaceous materials, including porosity, metal ion and heteroatom doping, and their effectiveness in adsorbing dyes, have been tuned using a variety of techniques.[89] As carbonisation of raw biomass only requires minimal pretreatments, such as washing and drying, it has major economic implications. *Ulva fasciata* and *Sargassum tenerrimum* were used to create porous carbon with a significant specific surface area that can be used to adsorb dyes, heavy metal ions, and fluoride from an aqueous solution. *Salicornia brachiata* has recently been used to produce a one-step production of exfoliated carbon nanosheets with a noteworthy surface area of 700 m^2 g^1.[90] Since recent data reveal massive output of these crops, managing the lignocellulosic leftovers of sugarcane bagasse, maize husk, and rice husk is a significant environmental concern.

2.3.6 Green Nanoparticles for Environmental Remediation

Because of the numerous chemicals employed in their manufacture, nanomaterial-based procedures may cause environmental difficulties, thereby negating the end goal. Green nanoparticles can be highly valuable in this area, both with regard to sustainability and reliability.[91] Table 2.2 displays the benefits and drawbacks of conventional and innovative materials in the treatment of water.

The capability and adsorption capacity of rice husk are frequently increased through pretreatment techniques, which also enable recovery of the adsorbent. Despite the fact that agro-industrial wastes like biosorbents are already utilized to remove pollutants, some physicochemical treatments, such as alkali treatment, also recognised as mercerization, can increase their adsorption ability.[92] This process frequently causes some lignin to be removed, changes the crystallinity and particular areas of the biopolymer, and makes the active groups more accessible for adsorption.

In research by Ponce *et al.*[92] more than 90% of methylene blue was successfully adsorbed after just two hours of contact. According to the Box–Behnken design for the three biosorbents under consideration, the adsorption process can be adequately described by quadratic mathematical models that are applicable across a wide range of pH values, dose, and dye concentration. The alkali treatment method thus offers an appealing technology to convert affordable waste materials into effective biosorbents for wastewater treatment.[92,93]

2.4 Future Prospects of Sustainable Materials for Water Purification

The growing quantity of research devoted to water decontamination using hybrid materials is the best evidence of their appropriateness and high performance, as shown by the positive outcomes compared with traditional materials.[104] Despite this, hybrid materials testing has thus far been done on a limited laboratory scale, necessitating further thought and study to assess their prospective use on a wider scale on the one hand, and to define appropriate techniques in addition to risk management measures on the other.[104] The remaining significant issues for transitioning hybrid composites from the laboratory to the industrial wastewater field are varied, *i.e.*, mass scale synthesis should preferably be continuous and scalable for high yield sustainable production, enhancing the specificity of these substances towards contaminant, and assessing their effectiveness over prolonged time using representative samples of industrial effluents. Hybrid materials would fulfil several functions, the most important of which would be to treat polluted water in order to remove contaminants.[105] Based on studies so far, hybrid materials offer numerous amazing qualities that make them effective tools for water decontamination, such as a large surface area, extraordinary durability, high porosity, adjustable properties, and a better affinity for a wide range of water contaminants.[105] As a result, hybrid materials provide an

Table 2.2 The benefits and drawbacks of conventional and innovative materials in the treatment of water.

Material	Benefit	Drawbacks	Reference
Activated carbon	High effectiveness at removing organics simple to prepare, simple to maintain, and inexpensive.	Inorganics have a moderate removal efficiency. Production using the fossil fuel coal. Thermal activation requiring a lot of energy. High CO_2 equivalent emissions during setup and activation. Ineffective against viruses and dangerous bacteria, and can contain bacteria, encouraging bacterial growth. Non-specific adsorption mechanisms that cause poor selectivity. Short life span.	94–96
Nanocellulose (NC)	NC with modified surfaces has a high removal effectiveness for both organic and inorganic materials.	Pure NC has a moderate removal efficiency.	97 and 98
	Derived from plentiful biomaterials.	High production and modification costs. Prone to biological deterioration.	
	High processability and mechanical stability.	High propensity to form hydrogen bonds with one another, which reduces the number of accessible adsorption sites.	
	Following regeneration, acceptable performance efficiency.	There is little information on removing coexisting contaminants of different sorts.	
Protein nanofibrils (PNF)	Excellent organic and inorganic removal effectiveness.	Very little selectivity.	78 and 99
	Produced using inexpensive biomaterials.	Possible infection with microorganisms.	
	Easy preparation.	There is not much information about adsorbent regeneration.	
	Cheap (by utilizing whey or other raw vegetable proteins, for example).	Perhaps expensive (in the case of spider silk and other silk nanofibrils).	
	Extended service life.		
Seaweeds	Due to their abundance, seaweeds are advantageous for commercial production in terms of labour and cost.	Poor physical and chemical stability results in chemical moieties leaching into the aqueous medium, which restricts their usefulness. The effective recycling of seaweeds is still a problem.	100 and 101
Lignocellulosic biosorbents (LB)	Given their abundance and abundance of organic components with active adsorption sites, agroindustrial wastes are excellent candidates for use as biosorbents.	The biological pretreatment based on LB is considered a slow process. Partial hemicellulose hydrolysis and potential health risks.	102 and 103

extraordinarily effective means of satisfying the expanding need for sustainable water purification. Small organic molecules or organic macromolecules, such as polymers/biopolymers, might be utilized as organic components in organic–inorganic hybrid materials.105,106 These latter, together with inorganic particles such as metals, are the most commonly employed components in the synthesis of organic–inorganic hybrid composites. Each substance has a distinct emphasis and function. Polymer/biopolymer matrices, for example, are frequently responsible for the structural qualities, component cohesion, and form flexibility, of the hybrid material, whereas inorganic reinforcements are essential for rigidity, thermal and mechanical durability, and electric and magnetic properties.

In addition, nanotechnology offers enormous promise for enhancing the effectiveness of water pollution prevention and treatment procedures.[107] Conventional decontamination techniques, such as chlorination and ozonation, use an excessive quantity of chemical agents and can generate harmful by-products, known as disinfection by-products. The development of nanotechnology for sustainable water treatment is relevant in these circumstances. Nanomaterials have ecologically fascinating chemical and physical features that make them appropriate for a wide range of applications. They are useful for wastewater and polluted soil treatment due to their higher resilience *versus* mechanical stress or weathering, large specific surface area, and strong chemical reactivity. Pesticides, herbicides, azo dyes, and phenols are among the persistent organic chemicals found in wastewater discharges from businesses, factories, laboratories, and other sources. Because these pollutants pose a major hazard to living creatures, they must be treated before being released into the environment. Traditional treatment methods, including oxidation processes, bioremediation, combustion, flocculation, adsorption on activated carbon, air stripping, and precipitation, have limitations, such as high cost and ineffectiveness in completely degrading organic chemicals, which can lead to secondary contaminant formation. The fundamental restriction of human health effects owing to nanomaterial toxicity can be removed by selecting less hazardous carbon-based nanomaterials. Nanomaterials based on fullerene, carbon nanotubes, and graphenes are now widely applied in many areas of sustainable desalination. It has been experimentally demonstrated that nanoporous graphene membranes (NGMs) can have water permeability with effective salt retention up to two to three times that of conventional membranes. As a result, there is enormous potential for functionalized NGM in the creation of high-capacity membranes for use in water desalination.

2.5 Conclusion

Environmental contamination has several causes, but one prominent issue is water pollution, which is a result of the rapid rise of the chemical industry. To address this issue, numerous water purifying methods and materials have been created. The sluggish adsorption rate and high heat energy needed for recycling are constraints for carbon-based porous materials employed in current adsorption processes. It has been challenging to produce materials that concurrently fulfil outstanding recyclability, high efficiency, inexpensive raw material, and potential for industrialization. Several materials have been developed to improve the efficiency of pollutant removal. Also, the potential of additional biomaterials with marine origins and their mixtures can be assessed for use in water filtration systems. Carbonaceous materials with high cost, such as graphene derivatives and carbon nanotubes, could potentially be replaced by carbonaceous materials generated from marine-derived biomaterials because of their graphitic character, high porosity, and surface functionality. Although significant progress has been made in the creation of functional biomaterials derived from marine sources for water filtration, much more work is needed to determine how they might be applied to satisfy present demand. Photocatalysis based on nanotechnology has the potential to be a green and sustainable technique for the full mineralization of dangerous organic compounds to water, carbon dioxide, and simple mineral acids. Another potential nano-based material for metal removal *via* electrostatic attraction and chemical precipitation is carbon nanotubes.

References

1. S. O. Sulaiman, A. H. Kamel, K. N. Sayl and M. Y. Alfadhel, Water resources management and sustainability over the Western desert of Iraq, *Environ. Earth Sci*, 2019, **78**, 1–15.
2. M. Ejeian and R. Z. Wang, Adsorption-based atmospheric water harvesting,*Joule*, 2021, **5**(7), 1678–1703.
3. C. J. Rhodes, Plastic pollution and potential solutions, *Sci. Prog.*, 2018, **101**(3), 207–260.
4. V. R. Bhagwat, Safety of water used in food production, in *Food Safety and Human Health*, Elsevier, 2019, pp. 219–47.
5. H. Asgharnejad, E. Khorshidi Nazloo, M. Madani Larijani, N. Hajinajaf and H. Rashidi, Comprehensive review of water management and wastewater treatment in food processing industries in the framework of water-food-environment nexus, *Compr. Rev. Food Sci. Food Saf.*, 2021, **20**(5), 4779–4815.
6. I. C. Vasilachi, D. M. Asiminicesei, D. I. Fertu and M. Gavrilescu, Occurrence and fate of emerging pollutants in water environment and options for their removal, *Water*, 2021, **13**(2), 181.
7. A. Nagar and T. Pradeep, Clean Water through Nanotechnology: Needs, Gaps, and Fulfillment, *ACS Nano*, 2020, **14**(6), 6420–6435.
8. M. Yew, Y. Ren, K. S. Koh, C. Sun and C. Snape, A review of state-of-the-art microfluidic technologies for environmental applications: Detection and remediation, *Global Chall.*, 2019, **3**(1), 1800060.
9. N. Akhtar, M. I. Syakir Ishak, S. A. Bhawani and K. Umar, Various natural and anthropogenic factors responsible for water quality degradation: A review, *Water*, 2021, **13**(19), 2660.
10. E. S. Okeke, C. O. Okoye, E. O. Atakpa, R. E. Ita, R. Nyaruaba and C. L. Mgbechidinma, *et al.*, Microplastics in agroecosystems-impacts on ecosystem functions and food chain, *Resour., Conserv. Recycl.*, 2022, **177**, 105961.

11. M. Haseena, M. F. Malik, A. Javed, S. Arshad, N. Asif and S. Zulfiqar, *et al.*, Water pollution and human health, *Environ. Risk Assess. Rem.*, 2017, **1**(3), 16–19.
12. R. E. Baker, A. S. Mahmud, I. F. Miller, M. Rajeev, F. Rasambainarivo and B. L. Rice, *et al.*, Infectious disease in an era of global change, *Nat. Rev. Microbiol.*, 2022, **20**(4), 193–205.
13. F. S. A. Khan, N. M. Mubarak, Y. H. Tan, M. Khalid, R. R. Karri and R. Walvekar, *et al.*, A comprehensive review on magnetic carbon nanotubes and carbon nanotube-based buckypaper for removal of heavy metals and dyes, *J. Hazard. Mater.*, 2021, **413**, 125375.
14. S. Sharma and A. Bhattacharya, Drinking water contamination and treatment techniques, *Appl. Water Sci.*, 2017, **7**(3), 1043–1067.
15. A. Power, S. Chandra and J. Chapman, Graphene, electrospun membranes and granular activated carbon for eliminating heavy metals, pesticides and bacteria in water and wastewater treatment processes, *Analyst*, 2018, **143**(23), 5629–5645.
16. W. S. Chai, J. Y. Cheun, P. S. Kumar, M. Mubashir, Z. Majeed and F. Banat, *et al.*, A review on conventional and novel materials towards heavy metal adsorption in wastewater treatment application, *J. Cleaner Prod.*, 2021, **296**, 126589.
17. M. Bartoli, P. Jagdale and A. Tagliaferro, A short review on biomedical applications of nanostructured bismuth oxide and related nanomaterials, *Materials*, 2020, **13**(22), 5234.
18. M. B. Tahir, M. Sohaib, M. Sagir and M. Rafique, Role of nanotechnology in photocatalysis, *Reference Module in Materials Science and Materials Engineering*, 2020.
19. M. Ai, J. Zhang, Y. Wu, L. Pan, C. Shi and J. Zou, Role of vacancies in photocatalysis: a review of recent progress, *Chem. – Asian J.*, 2020, **15**(22), 3599–3619.
20. E. S. M. Mouele, J. O. Tijani, K. O. Badmus, O. Pereao, O. Babajide and O. O. Fatoba, *et al.*, A critical review on ozone and co-species, generation and reaction mechanisms in plasma induced by dielectric barrier discharge technologies for wastewater remediation, *J. Environ. Chem. Eng.*, 2021, **9**(5), 105758.
21. Y. Shang, X. Xu, B. Gao, S. Wang and X. Duan, Single-atom catalysis in advanced oxidation processes for environmental remediation, *Chem. Soc. Rev.*, 2021, **50**(8), 5281–5322.
22. W. S. Koe, J. W. Lee, W. C. Chong, Y. L. Pang and L. C. Sim, An overview of photocatalytic degradation: photocatalysts, mechanisms, and development of photocatalytic membrane, *Environ. Sci. Pollut. Res.*, 2020, **27**, 2522–2565.
23. G. Reenamole, Metal Oxide Nanomaterials for Visible Light Photocatalysis, in *Emerging Nanomaterials for Catalysis and Sensor Applications*, CRC Press, 2023, pp. 103–30.
24. W. Chairungsri, A. Subkomkaew, P. Kijjanapanich and Y. Chimupala, Direct dye wastewater photocatalysis using immobilized titanium dioxide on fixed substrate, *Chemosphere*, 2022, **286**, 131762.

25. J. Li, Z. Jin, Y. Zhang, D. Liu, A. Ma and Y. Sun, *et al.*, Ag-induced anatase-rutile TiO2− x heterojunction facilitating the photogenerated carrier separation in visible-light irradiation, *J. Alloys Compd.*, 2022, **909**, 164815.
26. H. H. Alwan, Photo catalysis desulfurization at copper oxides/titanium oxide nanotubes under UV and visible light irradiation, *Iran. J. Catal.*, 2021, **11**(4), 363–375.
27. T. L. De Jager, A. E. Cockrell and S. S. Du Plessis, Ultraviolet light induced generation of reactive oxygen species, in *Ultraviolet Light in Human Health, Diseases and Environment*, ed. S. I. Ahmad, Advances in Experimental Medicine and Biology, Springer, Cham, 2017, vol. 996, pp. 15–23.
28. I. Jagodić, I. Guth, S. Lukić-Petrović, D. Tamindžija, D. Šojić Merkulov and N. Finčur, *et al.*, Reusable Fe_2O_3/TiO_2/PVC Photocatalysts for the Removal of Methylene Blue in the Presence of Simulated Solar Radiation, *Nanomaterials*, 2023, **13**(3), 460.
29. X. Wei, X. He, P. Wu, F. Gong, D. Wang and S. Wang, *et al.*, Recent advances in the design of semiconductor hollow microspheres for enhanced photocatalytic water splitting, *Int. J. Hydrogen Energy*, 2021, **46**(55), 27974–27996.
30. P. Singh, P. Shandilya, P. Raizada, A. Sudhaik, A. Rahmani-Sani and A. Hosseini-Bandegharaei, Review on various strategies for enhancing photocatalytic activity of graphene based nanocomposites for water purification, *Arabian J. Chem.*, 2020, **13**(1), 3498–3520.
31. A. A. Ansari and M. Sillanpää, Advancement in upconversion nanoparticles based NIR-driven photocatalysts, *Renewable Sustainable Energy Rev.*, 2021, **151**, 111631.
32. K. P. Gopinath, N. V. Madhav, A. Krishnan, R. Malolan and G. Rangarajan, Present applications of titanium dioxide for the photocatalytic removal of pollutants from water: A review, *J. Environ. Manage.*, 2020, **270**, 110906.
33. D.-E. Lee, V. Devthade, S. Moru, W.-K. Jo and S. Tonda, Magnetically sensitive TiO_2 hollow sphere/Fe_3O_4 core-shell hybrid catalyst for high-performance sunlight-assisted photocatalytic degradation of aqueous antibiotic pollutants, *J. Alloys Compd.*, 2022, **902**, 163612.
34. Y. Mokhbi, M. Korichi and Z. Akchiche, Combined photocatalytic and Fenton oxidation for oily wastewater treatment, *Appl. Water Sci.*, 2019, **9**(2), 35.
35. F. D. Utami, D. Y. Rahman, Sutisna, Kamirul, D. O. Margareta and M. Abdullah, Photocatalyst based on TiO_2 and its application in organic wastewater treatment using simple spray method, *J. Phys.: Conf. Ser.*, 2019, **1204**(1), 012086.
36. P. O. Oladoye, T. O. Ajiboye, E. O. Omotola and O. J. Oyewola, Methylene blue dye: Toxicity and potential elimination technology from wastewater, *Results Eng.*, 2022, **16**, 100678.
37. D. Liu, J. C. Liu, W. Cai, J. Ma, H. B. Yang, H. Xiao, J. Li, Y. Xiong, Y. Huang and B. Liu, Selective photoelectrochemical oxidation of glycerol to high value-added dihydroxyacetone, *Nat. Commun.*, 2019, **10**(1), 1779.

38. Y. B. Xie and X. Z. Li, Degradation of bisphenol A in aqueous solution by H_2O_2-assisted photoelectrocatalytic oxidation, *J. Hazard. Mater.*, 2006, **138**(3), 526–533.
39. C. A. Walenta, C. Courtois, S. L. Kollmannsberger, M. Eder, M. Tschurl and U. Heiz, Surface Species in Photocatalytic Methanol Reforming on Pt/TiO_2(110): Learning from Surface Science Experiments for Catalytically Relevant Conditions, *ACS Catal.*, 2020, **10**(7), 4080–4091.
40. M. T. Islam, A. Dominguez, R. S. Turley, H. Kim, K. A. Sultana and M. A. I. Shuvo, *et al.*, Development of photocatalytic paint based on TiO_2 and photopolymer resin for the degradation of organic pollutants in water, *Sci. Total Environ.*, 2020, **704**, 135406.
41. Q. Zhang, R. Du, C. Tan, P. Chen, G. Yu and S. Deng, Efficient degradation of typical pharmaceuticals in water using a novel TiO2/ONLH nano-photocatalyst under natural sunlight, *J. Hazard. Mater.*, 2021, **403**, 123582.
42. S. Liu, Z. L. Su, Y. Liu, L. Y. Yi, Z. L. Chen and Z. Z. Liu, Mechanism and purification effect of photocatalytic wastewater treatment using graphene oxide-doped titanium dioxide composite nanomaterials, *Water*, 2021, **13**(14), 1915.
43. F. Kamranvand, C. J. Davey, H. Sakar, O. Autin, E. Mercer and M. Collins, *et al.*, Impact of fouling, cleaning and faecal contamination on the separation of water from urine using thermally driven membrane separation, *Sep. Sci. Technol.*, 2018, **53**(9), 1372–1382.
44. H. Rezaei, M. Haghshenasfard and A. Moheb, Optimization of dye adsorption using Fe_3O_4 nanoparticles encapsulated with alginate beads by Taguchi method, *Adsorpt. Sci. Technol.*, 2017, **35**(1–2), 55–71.
45. Z. Ni, Y. Li and S. Wang, Cognizing and characterizing the organic phosphorus in lake sediments: Advances and challenges, *Water Res.*, 2022, **220**, 118663.
46. M. Anjum, R. Miandad, M. Waqas, F. Gehany and M. A. Barakat, Remediation of wastewater using various nano-materials, *Arabian J. Chem.*, 2019, **12**(8), 4897–4919.
47. A. Kaplan, The nanocomposites designs of phytomolecules from medicinal and aromatic plants: promising anticancer-antiviral applications, *Beni-Suef Univ. J. Basic Appl. Sci.*, 2022, **11**(1), 17.
48. R. Jing, S. Fusi and B. V. Kjellerup, Remediation of Polychlorinated Biphenyls (PCBs) in contaminated soils and sediment: State of knowledge and perspectives, *Front. Environ. Sci.*, 2018, **6**(JUL), 1–17.
49. X. Li and L. Liu, Recent advances in nanoscale zero-valent iron/oxidant system as a treatment for contaminated water and soil, *J. Environ. Chem. Eng.*, 2021, **9**(5), 106276.
50. R. Dhanker, S. Rawat, V. Chandna, Deepa, R. Kumar and S. Das, *et al.*, Recovery of silver nanoparticles and management of food wastes: Obstacles and opportunities, *Environ. Adv.*, 2022, **9**, 100303.
51. S. Shukla, R. Khan and A. Daverey, Synthesis and characterization of magnetic nanoparticles, and their applications in wastewater treatment: A review, *Environ. Technol. Innovation*, 2021, **24**, 101924.

52. B. Arora and P. Attri, Carbon Nanotubes (CNTs): A Potential Nanomaterial for Water Purification, **4**, *J. Compos. Sci.*, 2020, 135.
53. M. M. Halmann, *Photodegradation of Water Pollutants*, 1995, p. 320. Available from: https://books.google.com/books?id=xkqDymaQ7AcC&pgis=1
54. Y. Liu, C. Fang, S. zhang, W. Zhong, Q. Wei and Y. Wang, *et al.*, Effective adsorption of uranyl ions with different MoS_2-exposed surfaces in aqueous solution, *Surf. Interfaces*, 2020, **18**, 100409.
55. A. Ishag and Y. Sun, Recent Advances in Two-Dimensional MoS_2 Nanosheets for Environmental Application, *Ind. Eng. Chem. Res.*, 2021, **60**(22), 8007–8026.
56. T. Naseem and T. Durrani, The role of some important metal oxide nanoparticles for wastewater and antibacterial applications: A review, *Environ. Chem. Ecotoxicol.*, 2021, **3**, 59–75.
57. A. Tomer, R. Singh, S. K. Singh, S. A. Dwivedi, C. U. Reddy and M. R. A. Keloth, *et al.*, Role of fungi in bioremediation and environmental sustainability, *Mycoremediation and Environmental Sustainability*, 2021, vol. 3, pp. 187–200.
58. J. Zhai and C. H. Jiang, Synthesis of calcium peroxide microparticles in aqueous at room temperature and its application in heavy metal ions removal from waste liquid of COD determining, *Adv. Mater. Res.*, 2014, 648–653.
59. H. Chen, K. Zhou and G. Zhao, Gold nanoparticles: From synthesis, properties to their potential application as colorimetric sensors in food safety screening, *Trends Food Sci. Technol.*, 2018, **78**, 83–94.
60. Z. Sadiq, S. H. Safiabadi Tali, H. Hajimiri, M. Al-Kassawneh and S. Jahanshahi-Anbuhi, Gold Nanoparticles-Based Colorimetric Assays for Environmental Monitoring and Food Safety Evaluation, *Crit. Rev. Anal. Chem.*, 2022, 1–36.
61. J. Yang, M. Monnot, L. Ercolei and P. Moulin, Membrane-Based Processes Used in Municipal Wastewater Treatment for Water Reuse: State-Of-The-Art and Performance Analysis, **10**, *Membranes*, 2020, 131.
62. K. C. Khulbe and T. Matsuura, *Nanotechnology in Membrane Processes*, 2021, p. 364.
63. B. Joseph, V. K. Sagarika, C. Sabu, N. Kalarikkal and S. Thomas, Cellulose nanocomposites: Fabrication and biomedical applications, *J. Bioresour. Bioprod.*, 2020, **5**(4), 223–237.
64. P. Liu, C. Zhu and A. P. Mathew, Mechanically robust high flux graphene oxide - nanocellulose membranes for dye removal from water, *J. Hazard. Mater.*, 2019, **371**(March), 484–493.
65. Y. S. Khoo, W. J. Lau, Y. Y. Liang, M. Karaman, M. Gürsoy and A. F. Ismail, Eco-friendly surface modification approach to develop thin film nanocomposite membrane with improved desalination and antifouling properties, *J. Adv. Res.*, 2022, **36**, 39–49.
66. L. A. Goetz, N. Naseri, S. S. Nair, Z. Karim and A. P. Mathew, All cellulose electrospun water purification membranes nanotextured using cellulose nanocrystals, *Cellulose*, 2018, **25**(5), 3011–3023.
67. L. R. Glicksman, Promoting sustainable buildings, *HVAC&R Res.*, 2003, **9**(2), 107–109.

68. N. H. Mohd Yusoff, C. H. Chong, Y. K. Wan, K. H. Cheah and V.-L. Wong, Optimization strategies and emerging application of functionalized 3D-printed materials in water treatment: A review, *J. Water Process Eng.*, 2023, **51**, 103410.
69. N. Tarfeen, K. U. Nisa, B. Hamid, Z. Bashir, A. M. Yatoo and M. A. Dar, *et al.*, Microbial Remediation: A Promising Tool for Reclamation of Contaminated Sites with Special Emphasis on Heavy Metal and Pesticide Pollution: A Review, **10**, *Processes*, 2022, 1358.
70. E. Kaczorek, A. Pacholak, A. Zdarta and W. Smułek, The Impact of Biosurfactants on Microbial Cell Properties Leading to Hydrocarbon Bioavailability Increase, **2**, *Colloids Interfaces*, 2018, 35.
71. M. McClain, M. Ashby, K. Sajwan and M. Samuel, *A Scientific and Technical Review of The Use of Bioremediation as a Clean Up Technique At the Savannah River Site*, September 2005.
72. C. C. Azubuike, C. B. Chikere and G. C. Okpokwasili, Bioremediation techniques–classification based on site of application: principles, advantages, limitations and prospects, *World J. Microbiol. Biotechnol.*, 2016, **32**(11), 1–18.
73. E. T. Alori, A. I. Gabasawa, C. E. Elenwo and O. O. Agbeyegbe, Bioremediation techniques as affected by limiting factors in soil environment, *Front. Soil Sci.*, 2022, **2**(August), 1–13.
74. K. Meerbergen, *Decolorization of Textile Wastewater, with an Emphasis on Microbial Treatment Processes*, March 2018.
75. Y. Satyawali and M. Balakrishnan, Wastewater treatment in molasses-based alcohol distilleries for COD and color removal: A review, *J. Environ. Manage.*, 2008, **86**(3), 481–497.
76. T. T. Bezuneh, The Role of Microorganisms in Distillery Wastewater Treatment: A Review, *J. Biorem. Biodegrad.*, 2016, **7**(6), 375.
77. Y.-Q. Su, Y.-J. Zhao, N. Wu, Y.-E. Chen, W.-J. Zhang and D.-R. Qiao, *et al.*, Chromium removal from solution by five photosynthetic bacteria isolates, *Appl. Microbiol. Biotechnol.*, 2018, **102**(4), 1983–1995.
78. M. Peydayesh and R. Mezzenga, Protein nanofibrils for next generation sustainable water purification, *Nat. Commun.*, 2021, **12**(1), 3248.
79. C. Lendel and N. Solin, Protein nanofibrils and their use as building blocks of sustainable materials, *RSC Adv.*, 2021, **11**(62), 39188–39215.
80. L. J. G. Hoppenreijs, L. Fitzner, T. Ruhmlieb, T. R. Heyn, K. Schild and A.-J. van der Goot, *et al.*, Engineering amyloid and amyloid-like morphologies of β-lactoglobulin, *Food Hydrocolloids*, 2022, **124**, 107301.
81. H. S. Kalasariya, V. K. Yadav, K. K. Yadav, V. Tirth, A. Algahtani and S. Islam, *et al.*, Seaweed-Based Molecules and Their Potential Biological Activities: An Eco-Sustainable Cosmetics, **26**, *Molecules*, 2021, 5313.
82. H. Znad, M. R. Awual and S. Martini, The Utilization of Algae and Seaweed Biomass for Bioremediation of Heavy Metal-Contaminated Wastewater, *Molecules*, 2022, **27**(4), 1275.
83. K. Alba and V. Kontogiorgos, Seaweed polysaccharides (agar, alginate carrageenan), *Encyclopedia of Food Chemistry*, Elsevier, 2018, pp. 240–250.

84. D. A. Devault, R. Pierre, H. Marfaing, F. Dolique and P. J. Lopez, Sargassum contamination and consequences for downstream uses: a review, *J. Appl. Phycol.*, 2021, **33**, 567–602.
85. G. U. Rani, A. K. Konreddy, S. Mishra and G. Sen, Synthesis and applications of polyacrylamide grafted agar as a matrix for controlled drug release of 5-ASA, *Int. J. Biol. Macromol.*, 2014, **65**, 375–382.
86. S. L. Loo, L. Vásquez, A. Athanassiou and D. Fragouli, Polymeric Hydrogels—A Promising Platform in Enhancing Water Security for a Sustainable Future, *Adv. Mater. Interfaces*, 2021, **8**(24), 2100580.
87. M. S. Soffian, F. Z. Abdul Halim, F. Aziz, M. A. Rahman, M. A. Mohamed Amin and D. N. Awang Chee, Carbon-based material derived from biomass waste for wastewater treatment, *Environ. Adv.*, 2022, **9**, 100259.
88. A. A. Ţurcanu, E. Matei, M. Râpă, A. M. Predescu, G. Coman and C. Predescu, Biowaste Valorization Using Hydrothermal Carbonization for Potential Wastewater Treatment Applications, **14**, *Water*, 2022, 2344.
89. J. Quintana-Najera, A. J. Blacker, L. A. Fletcher and A. B. Ross, The effect of augmentation of biochar and hydrochar in anaerobic digestion of a model substrate, *Bioresour. Technol.*, 2021, **321**, 124494.
90. P. Lawtae and C. Tangsathitkulchai, The use of high surface area mesoporous-activated carbon from longan seed biomass for increasing capacity and kinetics of methylene blue adsorption from aqueous solution, *Molecules*, 2021, **26**(21), 6521.
91. B. Bhardwaj, P. Singh, A. Kumar, S. Kumar and V. Budhwar, Eco-Friendly GreenerSynthesisofNanoparticles,*Adv.Pharm.Bull.*,2020,**10**(4),566–576.
92. J. Ponce, J. G. da S. Andrade, L. N. dos Santos, M. K. Bulla, B. C. B. Barros and S. L. Favaro, *et al.*, Alkali pretreated sugarcane bagasse, rice husk and corn husk wastes as lignocellulosic biosorbents for dyes, *Carbohydr. Polym. Technol. Appl.*, 2021, **2**(November), 100061.
93. S. İnan, T. Mumcu and S. Seyhan Bozkurt, Box-Behnken design for removal of uranium(VI) from aqueous solution using poly(ethylene glycol) based dicationic ionic liquid impregnated chitosan, *Turk. J. Chem.*, 2020, **44**(3), 756–774.
94. S. K. Shukla, N. R. S. Al Mushaiqri, H. M. Al Subhi, K. Yoo and H. Al Sadeq, Low-cost activated carbon production from organic waste and its utilization for wastewater treatment, *Appl. Water Sci.*, 2020, **10**(2), 1–9.
95. A. G. Olabi, T. Wilberforce, E. T. Sayed, N. Shehata, A. H. Alami and H. M. Maghrabie, *et al.*, Prospect of Post-Combustion Carbon Capture Technology and Its Impact on the Circular Economy, *Energies*, 2022, **15**(22), 1–38.
96. S. Fahimirad, Z. Fahimirad and M. Sillanpää, Efficient removal of water bacteria and viruses using electrospun nanofibers, *Sci. Total Environ.*, 2021, **751**, 141673.
97. A. Qiao, M. Cui, R. Huang, G. Ding, W. Qi and Z. He, *et al.*, Advances in nanocellulose-based materials as adsorbents of heavy metals and dyes, *Carbohydr. Polym.*, 2021, **272**(May), 118471.
98. D. Trache, A. F. Tarchoun, M. Derradji, T. S. Hamidon, N. Masruchin and N. Brosse, *et al.*, Nanocellulose: From Fundamentals to Advanced Applications, *Front. Chem.*, 2020, **8**, 392.

99. Q. Wang, P. McArdle, S. L. Wang, R. L. Wilmington, Z. Xing and A. Greenwood, *et al.*, Protein secondary structure in spider silk nanofibrils, *Nat. Commun.*, 2022, **13**(1), 4329.
100. C. Rebours, E. Marinho-Soriano, J. A. Zertuche-González, L. Hayashi, J. A. Vásquez and P. Kradolfer, *et al.*, Seaweeds: An opportunity for wealth and sustainable livelihood for coastal communities, *J. Appl. Phycol.*, 2014, **26**(5), 1939–1951.
101. S. M. Hassan, M. Ashour, N. Sakai, L. Zhang, H. A. Hassanien and A. Gaber, *et al.*, Impact of Seaweed Liquid Extract Biostimulant on Growth, Yield, and Chemical Composition of Cucumber (*Cucumis sativus*), **11**, *Agriculture*, 2021, 320.
102. L. Cruz-Lopes, M. Macena, B. Esteves and I. Santos-Vieira, Lignocellulosic Materials Used as Biosorbents for the Capture of Nickel (II) in Aqueous Solution, **12**, *Appl. Sci.*, 2022, 933.
103. C. S. Osorio-González, K. Hegde, S. K. Brar, A. Kermanshahipour and A. Avalos-Ramírez, Challenges in lipid production from lignocellulosic biomass using Rhodosporidium sp.; A look at the role of lignocellulosic inhibitors, *Biofuels, Bioprod. Biorefin.*, 2019, **13**(3), 740–759.
104. M. Rigoletto, P. Calza, E. Gaggero and E. Laurenti, Hybrid materials for the removal of emerging pollutants in water: classification, synthesis, and properties, *Chem. Eng. J. Adv.*, 2022, **10**, 100252.
105. T. s. Algarni and A. M. Al-Mohaimeed, Water purification by adsorption of pigments or pollutants *via* metaloxide, *J. King Saud Univ. Sci.*, 2022, **34**(8), 102339.
106. M. Zahid, N. Nadeem, N. Tahir, Farid-Un-Nisa, M. I. Majeed and Asim-Mansha, *et al.*, Hybrid nanomaterials for water purification. *Multifunctional Hybrid Nanomaterials for Sustainable Agri-Food and Ecosystems*. 2020. pp. 155–188.
107. M. P. Ajith, M. Aswathi, E. Priyadarshini and P. Rajamani, Recent innovations of nanotechnology in water treatment: A comprehensive review, *Bioresour. Technol.*, 2021, **342**, 126000.
108. A. A. Abdelhamid, M. H. Badr, R. A. Mohamed and H. M. Saleh, Using Agricultural Mixed Waste as a Sustainable Technique for Removing Stable Isotopes and Radioisotopes from the Aquatic Environment, *Sustainability*, 2023, **15**, 1600.
109. H. M. Saleh and M. Koller, Recent Remediation Technologies for Contaminated Water, in *Water Engineering Modeling and Mathematic Tools*, ed. P. Samui, H. Bonakdari, and R. Deo, Elsevier, 2021, ch. 6, pp. 115–131.
110. M. M. A. Dawoud, M. M. Hegazy, W. K. Helew and H. M. Saleh, Overview of Environmental Pollution and Clean Management of Heavy Metals and Radionuclides by Using Microcrystalline Cellulose, *J. Nucl. Energy Sci. Power Gener. Technol.*, 2021, **10**(3), 2.
111. M. M. A. Dawoud, M. M. Hegazi, H. M. Saleh and W. K. El Helew, Removal of Stable and Radio Isotopes from Wastewater by Using Modified Microcrystalline Cellulose Based on Taguchi L16, *Int. J. Environ. Sci. Technol.*, 2023, **20**, 1289–1300.

CHAPTER 3

MXene-based Separation Membranes for Water Purification and Desalination

AKBAR ALI[a,b], AYAZ ALI MEMON[c], JUN YANG[a,b*], JAVEED AKHTAR[d] AND KHALID HUSSAIN THEBO[b,d,e*]

[a]State Key Laboratory of Multi-phase Complex Systems, Institute of Process Engineering (IPE), Chinese Academy of Sciences, Beijing 100190, China; [b]University of the Chinese Academy of Sciences, 19A Yuquan Road, Beijing 100049, China; [c]National Centre for Excellence in Analytical Chemistry, University of Sindh, Jamshoro, Pakistan; [d]Functional Nanomaterials Laboratory (FNL) Department of Chemistry, Mirpur University of Science and Technology (MUST), Mirpur, Azad Jammu and Kashmir, Pakistan; [e]Institute of Metal Research (IMR), Chinese Academy of Sciences, Shenyang, China
*E-mail: jyang@ipe.ac.cn, khalidthebo@yahoo.com

3.1 Introduction

One of the most pressing challenges of our time is the increasing need for fresh water.[1–5] Several technologies have been investigated for purification and desalination of sea water, industrial waste water, and brackish water. Among them, membrane technology has long been vital to numerous separation processes compared to traditional separation techniques, such as absorption, adsorption, distillation, filtration, *etc.*[6–8] Membrane technology has a number of benefits, including affordable cost, energy savings, good selectivity, compactness, ease of fabrication, and environmental friendliness.[9,10] Depending on its characteristics and the nature of the species, a

Chemistry in the Environment Series No. 12
Novel Materials and Water Purification: Towards a Sustainable Future
Edited by Grigorios L. Kyriakopoulos and Miltiadis G. Zamparas

Published by the Royal Society of Chemistry, www.rsc.org

membrane serves as a selective barrier to segregate one or more species, such as ions, molecules, nanoparticles, and tiny particles, while allowing other species to flow through. The membrane's ability to separate substances depends on the size of its pores, its physicochemical characteristics, and careful control over its size and shape.[11] So far, the main issue for membrane technology for water purification is antifouling, which seriously affects commercial membranes, and which raises operating expenses and shortens the membrane lifespan.[12,13] Therefore, it is currently difficult to synthesize and fabricate membranes that are affordable and have advanced separation qualities, including high selectivity, good permeance, and strong stability under adverse conditions. These limitations of membrane technology have all been improved by the use of various new materials that have been synthesized to date. Consequently, choosing one of them as the ideal material for membrane technology is difficult because of their various microstructural characteristics.[14] Polymeric-based membranes are the best option in the history of membrane technology for water desalination due to their low cost, high efficiency, simplicity of production, and wide range of applications.[15] Although polymeric membranes are a more affordable option, they have significant disadvantages, such as poor chemical and thermal stability, which need to be addressed. The best solution for these problems has been to use inorganic membranes, but their future applications are limited by high production costs and thermal expansion. Due to their customizable microstructural characteristics, carbon nanotubes[16,17] have received extensive study for nanofiltration (NF) applications with high separation efficiency.[18] However, due to their poor permeability and stability, such membranes have only received a small amount of interest for large-scale applications.

Two-dimensional (2D materials), *i.e.*, graphene oxide (GO), graphene, covalent organic frameworks (COFs), hexagonal boron nitride (h-BN), metal–organic frameworks (MOFs), and transition metal dichalcogenides (TMDCs),[19–35] have recently attracted a lot of interest for applications in desalination and water filtration due to their exceptional qualities. Among them, MXene is young member of the 2D family with the common formula $M_{n+1}X_nT_x$ (n = 1, 2, 3), where M represents a transition metals, T_x denotes surface termination, and X for nitrogen or carbon atoms, and is shown in Figure 3.1.[36] Most of the time, these materials are made by removing stacked "A" atomic layers from layered ternary carbide precursors (MAX phases) using top-down wet chemical or electrochemical methods, as shown in Figure 3.1. MXene with surface terminations such as O, F, and OH have negative surface charges. These charges are important for keeping MXene stable in water and organic solvents, so they can be used in different separation methods. So, based on these facts, MXene has been considered as a possible candidate for separation applications because it is only one atom thick, has a high strength, is stable against flux, and can filter feed molecules.[37–39]

In this chapter, we discuss how to make membranes out of MXene. Then we focus on the different ways to change their physical and chemical qualities, and highlight the different ways to clean and desalinate water. Last, we

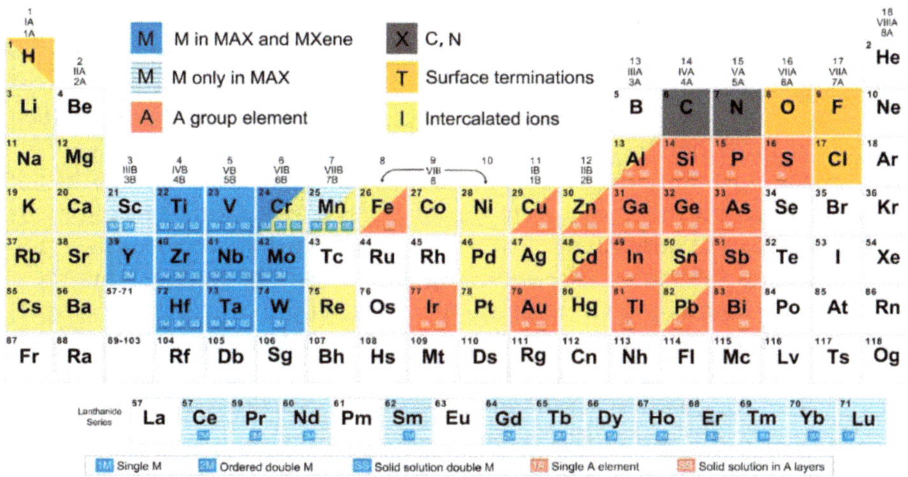

Figure 3.1 MXene family: MAX and MXene constituent elements and their intercalation ions. The elements in yellow-coloured backgrounds represent the cationic part of MXene. The elements with blue-coloured backgrounds show the MAX phase (MXene precursor) and were not prepared in MXenes, while the elements in red backgrounds are the "A" element of MAX (may be selectively etched to prepare MXenes). Reproduced from ref. 40 with permission from American Chemical Society, Copyright 2019.

will give possible solutions for the remaining problems and future research directions needed to improve the performance of MXene laminates, based on current research progress and the needs of real-world applications.

3.2 Methods for Preparation of MXene Membranes

MXene can be easily fabricated into membranes due to its good dispersion in water and other solvents. Generally, two types of membranes, *i.e.*, pure nanosheets and lamellar nanostructures, are fabricated from 2D MXenes.[41] MXene-based lamellar membranes are mostly made using wet chemical techniques, such vacuum filtration, spary coating, spin coating, layer-by-layer (LBL), drop casting, and Langmuir–Blodgett.[42–44] Among these methods, vacuum filtration is the most common and easiest way to prepare 2D MXene laminates.[45] It is important to note that this method does not alter the physicochemical characteristics of the materials because there is no covalent bonding between the 2D sheets, despite the presence of interactions such as stacking, electrostatic repulsion hydrogen bonding, and van der Waals attractive forces. Mostly, the microporous polymeric or anodic aluminium oxide substrate is used to fabricate uniform MXene membranes. Initially, Gogotsi and co-workers made 2D $Ti_3C_2T_x$-based membranes supported by polyvinylidine fluoride (PVDF).[46] The authors also made free-standing membranes for comparative studies. Due to the presence of sufficient functional groups such MXene-based membranes show more hydrophilic properties. Additionally, they also exhibit good flexibility and mechanical strength. Later, Ding and co-workers[44] synthesized $Ti_3C_2T_x$ laminates by vacuum filtration on a porous Ti_3AlC_2 substrate. These lamellar membranes feature a range of functional groups, including O, OH, F, and others. Kang *et al.*[47] mixed GO and MXene ($Ti_3C_2T_x$) to make an ultrathin $Ti_3C_2T_x$@GO hybrid membrane with a thickness of 90 nm on a porous substrate. In a distinct study, Wang and colleagues[48] also demonstrated a MXene-based composite membrane fabricated by combining $Ti_3C_2T_x$ with a polymer matrix to improve the membrane's microstructural and physicochemical characteristics. Vacuum filtration was also used by Feng and co-workers to prepare kelvar@$Ti_3C_2T_x$ nanofiber composite membranes for nanofluidic osmotic power generators, as shown in Figure 3.2a–d.[49] The authors prepared MXene membranes by dissolving $Ti_3C_2T_x$ MXene in dimethyl sulfoxide solution followed by addition of a suspension of aramid nanofibers (ANF) in a specific quantity (Figure 3.2a). The dispersed ion was sonicated for 15 minutes and stirred for 2 hours. The MXene@ANF mixture was vacuum filtered using anodized aluminium filter paper and repeatedly rinsed with deionized water (Figure 3.2b–d). The MXene@ANF hybrid membrane was simple to peel off the substrate after 24 hours of air drying (Figure 3.2b). Xu *et al.*[50] prepared

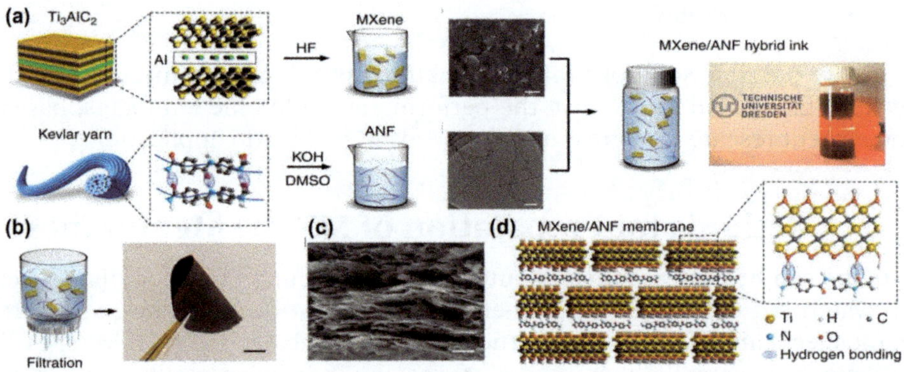

Figure 3.2 (a) Method for making Ti_3C_2Tx/ANF composite membranes: MXene is made from Ti_3AlC_2 precursors using HF etching method (and scanning electron microscopy (SEM) picture (scale bar, 1 μm)). Here, ANF used for preparation of membranes is synthesized from industrial Kevlar yarns through the KOH@DMSO exfoliation method (and transition electron microscopy (TEM) study of Kevlar (scale bar, 200 nm)). (b) Using vacuum-assisted filtration, a flexible MXene@ANF hybrid membrane with 11% ANF weight content is synthesized (scale bar, 5 mm). (c) A cross-sectional SEM image of the composite membrane reveals the lamellar microstructure (scale bar, 400 nm). (d) Schematic showing how the inside of the MXene@ANF hybrid membrane is put together. Reproduced from ref. 49, https://doi.org/10.1038/s41467-019-10885-8, under the terms of the CC BY 4.0 license https://creativecommons.org/licenses/by/4.0/.

$Ti_3C_2T_x$ composite-based membranes using a vacuum filtration method. The microstructure of the $Ti_3C_2T_x$-based membrane was improved by putting reduced GO (rGO) between the MXene layers. The surface of the membrane was gradually hydroxylated to make $Ti_3C_2T_x$ more accessible, improve the membrane's ability to absorb and reduce heavy metal ions, and make the membrane easier to clean. Recently, Wang and co-workers developed $Ti_3C_2T_x$ lamellar membranes by employing multivalent ions as hydrogel pillars in the interlayer gap.[51] The authors utilized a sodium alginate (SA) and $Ti_3C_2T_x$ MXene solution to obtain a consistent composite response. The SA-$Ti_3C_2T_x$ composite was also used to create the lamellar SA-$Ti_3C_2T_x$ membranes. Ii is interesting to note that the SA molecule forms a hydrogen connection with the MXene sheets and exerts a van der Waals force. To construct pillared SA-$Ti_3C_2T_x$ laminates, multivalent cations, such Ba^{2+}, Ca^{2+}, Mn^{2+}, and Al^{3+}, were dissolved in a solution of SA-$Ti_3C_2T_x$ membranes. Metal cations acted as hydrogel pillars and cross-linked the nanosheets in the interlayer space. The consistent nacre-like texture of the pillar membranes considerably decreased swelling. These studies show that vacuum filtration is the main technique used for producing MXene-based membranes. This approach offers a number of advantages, but it also requires a lot of liquid, takes a while, and is challenging to scale up. Therefore, it is essential to build a defect-free and controllable pore size MXene-based membrane for large-scale applications with outstanding separation performance. The development of an intelligent method for preparing MXene lamellar membranes is seriously needed since without it these materials would not be suitable for widespread application.

Besides vacuum filtration, fabrication of MXene membranes has been accomplished in a number of different ways, such as LB,[52] LBL,[44,53] spin coating,[54] spray coating, and dip coating.[55] Optimizing the thickness of MXene thin-film membranes requires the use of a variety of different approaches. It is usually necessary to enhance mechanical, chemical, physical, and electrical properties of the MXene thin films. Although vacuum filtration, spin coating, and spray coating are the most frequent fabrication processes for producing MXene membranes,[56] the dip-coating method is another way to make MXene membranes of different thicknesses. This method has attracted a lot of attention because of its simplicity and low cost.[57] In this method, the substrate is immersed into the MXene solution by hand, and thin films of different thickness are obtained. In the dip-coating method, the process of making a membrane is affected by the nature of the substrate, how the surface of the substrate is changed, how long it is dipped, and how much MXene solution is used. For layer deposition to work well, the base must always be dipped in an MXene solution for at least five minutes. To obtain uniform and defect-free films, the substrate needs to be dipped and washed repeatedly in cycles. Dip-coated MXene membranes have been employed in a variety of applications, including energy storage devices, electromagnetic interference shielding, flame retardancy, and others.[57] During use of this method, a lot of material is wasted and the membrane is not consistent.[58] Controlling the layers is one of the most difficult jobs in thin-film production. Therefore, this method is unsuitable for water purification and desalination applications. Furthermore, the liquid–liquid interface approach[59] has been used to fabricate MXene membranes. To create the thin-film membrane at the interface of two liquids, two immiscible liquids are used in this procedure. Most of the time, organic solvents (hexane, toluene) and inorganic solvents (water) have been utilized, and the gradual evaporation of the organic solvent induces the assembly of various nanomaterials at the interface. This approach is not generally appropriate for the manufacture of reverse osmosis (RO) and NF membranes. Therefore, obtaining monolayer MXene thin-film membranes using spray-, spin-, vacuum-, and dip-coating techniques is a more difficult challenge. Shear-induced alignment method,[60] wet spinning,[61–64] printed method,[65] and continuous centrifugal casting[66] are a few more of the methods commonly used to fabricate membranes based on 2D materials. This necessitates the exploration of such techniques for the production of MXene membranes.

3.3 Applications for Water Purification and Desalination

Water pollution and desalination are presently the most significant issues affecting the global standard of living.[67–71] Commercial RO and NF membranes used for desalination and water purification suffer from antifouling issues that shorten membrane life and raise operating costs. Therefore, in order to reduce operating costs and energy consumption, researchers are always working to introduce new materials that will improve the antifouling capabilities of the membrane. The use of MXene-based membranes in desalination, water purification, and effluent treatment has received considerable attention in recent years.[72] Initially, free-standing and PVDF-supported MXene membranes (Figure 3.3a and b) with an interlayer spacing of 6 Å were reported by Gogotsi and co-workers,[46] which could separate specific ions, including Mg^{2+}, Ca^{2+}, Li^{+}, Al^{3+}, Ni^{2+}, Na^{+}, and K^{+}. The as-fabricated membrane displayed good permeability, with a value of 37.4 L m^{-2} h^{-1} bar^{-1}. In comparison to single-charged cations, sluggish permeability was suggested by metal ions with larger charges and smaller hydration radii than the distance between MXene interlayers (6 Å), as shown in Figure 3.3c. These membranes have also been examined for the antibacterial and biofouling effects of colloidal solutions containing single and multilayer $Ti_3C_2T_x$ MXene flakes. Han *et al.*[73] showed Congo red (CR) dye separation using polyethersulfone (PES)/MXene ultrafiltration membranes. These membrane exhibited 92.3% rejection for CR with pure water permeance of ~115 L m^{-2} h^{-1} using 0.1 MPa. Ding and colleagues used the vacuum filtration technique to develop a novel 2D $Ti_3C_2T_x$-based laminate on an anodic aluminium oxide support.[44] The as-prepared membranes have excellent water permeability for deionized (DI) water up to 1000 L m^{-2} h^{-1} bar^{-1}. The authors also estimated the separation efficiency (>90%) of laminates containing molecules of varying diameters.

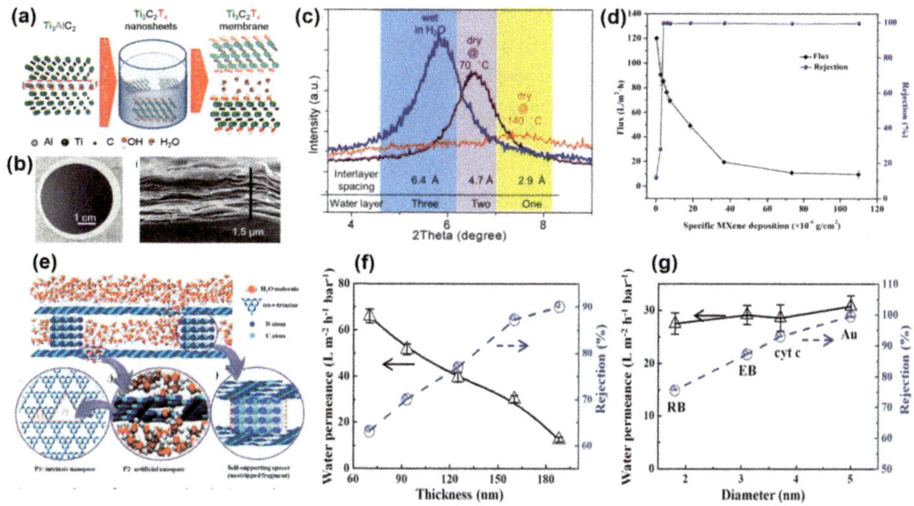

Figure 3.3 (a–d) Ti_3C_2Tx-based membrane. (a) Preparation of Ti_3C_2Tx membranes using Ti_3AlC_2 precursor. (b) Digital and cross-sectional images of MXene membrane. (c) X-ray diffraction (XRD) pattern of membranes at various moisture levels. Reproduced from ref. 46, https://doi.org/10.1021/acs.jpclett.5b01895, with permission from American Chemical Society, Copyright 2015. (d) The effect of how much MXene is deposited on the pervaporation desalination performance of an MXene membrane for a 3.5 wt% NaCl solution at 65 °C. Reproduced from ref. 74 with permission from Elsevier, Copyright 2018. (e) A diagram showing how synthetic nanopores in g-C_3N_4 nanosheets move water through them. Unstripped pieces that are adhered generate a space between the nanosheets. The self-supporting spacer between g-C_3N_4 nanosheets (*i.e.*, the unstripped pieces produced by partial exfoliation) is made up of intrinsic nanopores (P1) and an artificial nanopore (P2), which transport molecules of water. (f and g) The separation performance of the g-C_3N_4 membrane as a function of its thickness, and as a function of the largest cross-section of each rejected molecule, respectively. Reproduced from ref. 75 with permission from John Wiley & Sons, Copyright © 2017 Wiley-VCH Verlag GmbH & Co. KGaA, Weinheim.

Liu *et al.*[74] synthesized a polyacrylonitrile-based $Ti_3C_2T_x$ membrane for NaCl salt separation. As shown in Figure 3.3d, a membrane with a thickness of 60 nm has a separation efficiency of 99.5% and a maximal water permeance of 85 L m^{-2} h^{-1} bar^{-1} at 65 °C. When it comes to molecular separation, water purification using 2D g-C_3N_4 nanosheet membranes has been demonstrated by Wang and co-workers, as shown in Figure 3.3e. Several chemicals were removed using these nanosheet membranes, including cytochrome c (Cyt c), rhodamine B (RhB), gold nanoparticles (Au NPs), and Evans blue (EB) dye. Water permeance across the 160 nm thick membrane was up to ~29 L m^{-2} h^{-1} bar^{-1}, and the rejection percentages for EB, RhB, Au NPs, and Cyt c were 87.2%, 75.5%, 99.5%, and 93.1%, respectively (Figure 3.3f and g).[75]

To improve antifouling, water permeance, and rejection, Mahmoud *et al.*[76] produced an Ag@MXene membrane (470 nm thick) with a 21% loading of Ag@MXene. Under the same conditions, the as-prepared composite membrane exhibited three to five times more water permeance of 420 L m^{-2} h^{-1} bar^{-1}, compared to pristine MXene laminates (118 L m^{-2} h^{-1} bar^{-1}). The 21% Ag@MXene membrane also displayed outstanding flux recovery and strong rejection efficiency for organic compounds. In addition, the 21% Ag@MXene composite membrane inhibited *Escherichia coli* growth by more than 99% compared to the control hydrophilic PVDF-based membrane, but the MXene membrane only inhibited bacteria growth by 60%. Furthermore, in comparison to previously reported membranes, the 21% Ag@MXene membrane demonstrated outstanding rejection of organic foulants, such as bovine serum albumin (BSA) and methyl green.

In addition to membrane rejection, water permeation is essential for commercial applications. MXene membranes have been observed to have a lower permeance than commercial polymeric membranes, and thus cannot replace them. In this regard, Ding and colleagues increased the water permeance of $Ti_3C_2T_x$ membranes up to ~1000 L m^{-2} h^{-1} bar^{-1}, while maintaining excellent separation for numerous probe molecules.[44] However, researchers still have significant difficulty in optimizing separation efficiency in tandem with permeance. To satisfy this requirement, it is crucial to combine the right cross-linking agent and MXene. Similar to other 2D materials, MXene exhibits water swelling characteristics and is quickly delaminated due to hydrogen bonding and electrostatics, which impacts the effectiveness of membrane permeance as well as separation capability.[77,78] Wang *et al.*[51] used pillared lamellar membranes to improve separation properties, as demonstrated in Figure 3.4a. The authors used Mn^{2+} and Ca^{2+} cations to cross-link MXene nanosheet membranes, which enhanced the *d*-spacing from 13.8 Å to 15.2 Å. An ultrathin Mn-SAT pillared lamellar membrane showed complete rejection of Na_2SO_4 salt thanks to the increased *d*-spacing. Kang and colleagues prepared a $Ti_3C_2T_x$@GO composite membrane with a 90 nm thickness.[47] Due to expansion in water, the interlayer spacing of the membrane is approximately 5 Å. These membranes demonstrated 99% rejection for feed molecules with radii greater than 5 Å. Furthermore, these membranes demonstrated excellent rejection for brilliant blue (BB, hydrated radius 7.98 Å), rose Bengal (RB, 5.88 Å), methylene blue (MLB, 5.04 Å), and methyl blue (MB, 4.87 Å), with respective percentages of 100%, 93.5%, 99.5%, and 68%. Lu *et al.*[79] separated different monovalent ions using MXene membranes with high efficiency. When pristine MXene membrane was immersed into water solution, the interlayer spacing of the MXene nanosheets increased up to 16.6 Å due to the swelling effect of terminal hydroxyl groups. Therefore, the authors controlled the swelling of MXene membranes by cross-linking the terminal functional groups (–OH + –OH = –O– + H_2O) of the MXene nanosheets with each other and improved interlayer spacing up to 15.4 Å, as shown in Figure 3.4b. In addition, these membranes showed relatively low rates of permeability for monovalent ions, with penetration rates for Li^+, Na^+, and K^+ of 0.283, 0.222,

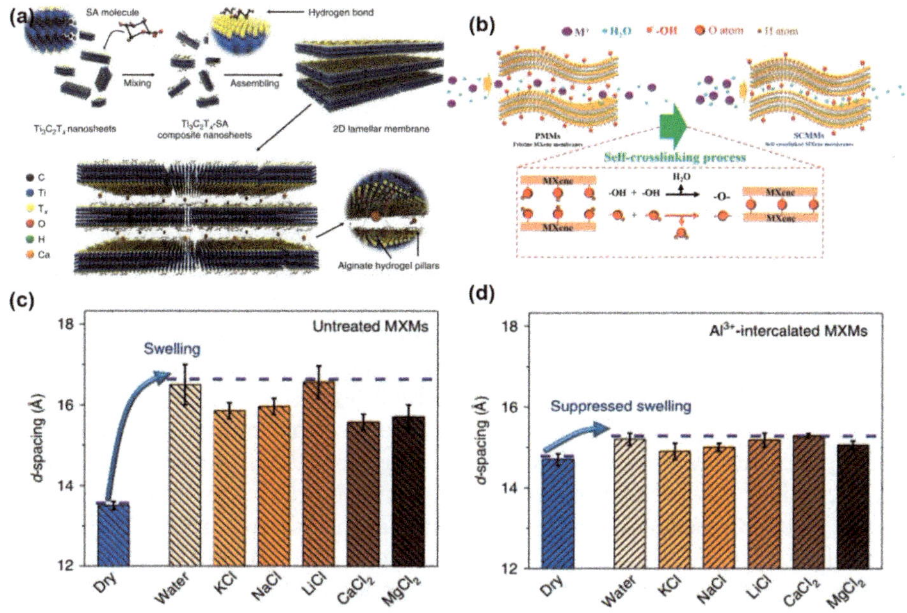

Figure 3.4 (a) The preparation of 2D lamellar Ti_3C_2Tx@alginate cross-linked membrane with hydrogel pillars in the interlayer space. Several multivalent Mn^+ cross-linking solutions (Al^{3+}, Mn^{2+}, Ca^{2+}, and Ba^{2+}) were poured over the SA-Ti_3C_2Tx membrane. Data from ref. 51. (b) The MXene membranes prepared by cross-linking terminal functional groups (hydroxyl) of MXene nanosheets with each other. Reproduced from ref. 79 with permission from American Chemical Society, Copyright 2019. (c and d) The *d*-spacing of MXMs with and without Al^{3+} intercalation. The *d*-spacing data of MXMs with (c) and without (d) Al^{3+} intercalated calculated by XRD in dry and wet state using different salt solutions with 0.2 M concentration, respectively. Reproduced from ref. 80 with permission from Springer Nature, Copyright 2020.

and 0.232 mol h^{-1} m^{-2}, respectively, for ions with hydrated diameters of 7.65, 7.16, and 6.62 Å, respectively. Furthermore, such self-cross-linking between MXene nanosheets showed excellent long-term stability of up to 70 hours during ion separation, with promising anti-swelling properties. In addition, in order to control the swelling effect, Ding *et al.*[80] introduced Al^{3+} between MXene 2D sheets to control the swelling effect. These ions show strong electrostatic interaction with the MXene functional groups. Such membranes show high stability up to 400 hours, and low swelling in water. Figure 3.4 shows that these laminates had superior water permeance of 1.1–8.5 L m^{-2} h^{-1} bar^{-1} and rejection against NaCl of 89.5–99.6%.

Furthermore, Shen and co-workers[81] demonstrated MXene/polysulfone ($Ti_3C_2T_x$) membranes with 99% rejection for the BSA molecule and up to ~218 L m^{-2} h^{-1} bar^{-1} water permeability. Ma *et al.*[82] achieved 99% rejection of gentian violet dye using P84-copolyimide@MXene composite membrane, along with a superb 268 L m^{-2} h^{-1} bar^{-1} water permeability. Liu *et al.*[83] achieved >99% rejection for BB, MLB, neutral red (NR), and crystal violet (CV) dyes using MXene/GO membrane. Such a membrane also demonstrated good rejection of humic acid (HA) and BSA molecules. Recently, Xing and colleagues[84] synthesized MXene membrane by modifying MXene nanosheets with polyetheneimine (PEI) in varying proportions to control swelling of membranes, and deposited on polyacrylonitrile (PAN) substrate (Figure 3.5a and b). Figure 3.5d shows the X-ray diffraction (XRD) patterns of PEI@MXene@PAN membrane, which exhibits the interlayer spacing of membranes. The interlayer spacing of PEI@MXene@PAN membrane decreased compared to pristine MXene@PAN membranes. The membrane composed of PEI@MXene@/PAN with 0.5:1 ratio exhibited the lowest interlayer spacing (14.15 Å). However, interlayer spacing increased to 14.52 Å when the PEI ratio increased. Interestingly, the PEI@MXene@PAN membranes produced when the interlayer spacing of the top MXene layers is set between *ca.* 14.15 Å and *ca.* 14.52 Å demonstrate an exceptional capacity for size sieving (Figure 3.5a). Moreover, the membrane PEI@MXene@PAN with a 1:1 ratio demonstrates an excellent rejection of 99.82%, 99.3%, and 99.9% for CR, acid blue 90, and direct red (DR), respectively, as well as a high water permeance of 882.6, 563.9, and 633.6 L m^{-2} h^{-1}, as shown in Figure 3.5c–f. The membrane PEI@MXene@PAN with

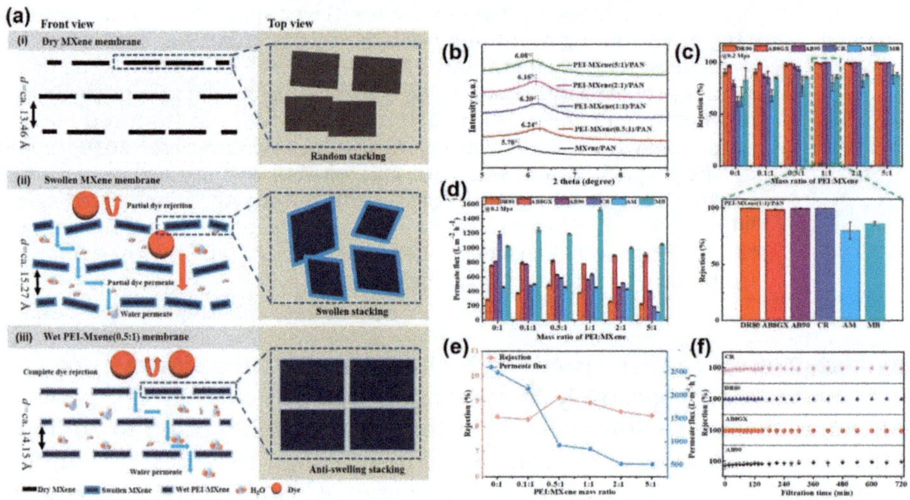

Figure 3.5 (a) Schematic representation (front and top view) of (i) dry, (ii) swollen, and (iii) wet PEI-MXene membrane. (b) The XRD patterns of different PEI-MXene membranes. (c) Rejection efficiency and (d) permeance of PEI-MXene membranes to various dyes of 100 ppm at 0.2 MPa. (e) The rejection performance of vitamin B12 through PEI-MXene membranes with different ratios at 100 ppm at 0.2 MPa. (f) Long-term separation efficiency of four dyes, DR80, AB8GX, AB90, and CR, at 100 ppm concentration at 0.2 MPa. Reproduced from ref. 84, https://doi.org/10.1002/eom2.12300, under the terms of the CC BY 4.0 license https://creativecommons.org/licenses/by/4.0/.

a 0.5 : 1 ratio showed a remarkable rejection rate of 99.9% to the AB8GX dye, with permeance of ~1528.1 L m^{-2} h^{-1} (Figure 3.5d). Figure 3.5e indicates that during the 720 min of operation, the rejection rates of all dyes remained high (>99%), confirming the remarkable stability of the PEI@MXene@PAN membranes.

Additionally, MXene membranes have also been utilized for recovery of precious metals. Liu *et al.*[85] recovered precious metal gold using the $Ti_3C_2T_x$/CNT hybrid membrane. The membranes show excellent capability for recovery of gold(III) up to 99.8%, and exhibited water permeance of up to 437.6 L m^{-2} h^{-1} bar^{-1}, which is 200 times higher than previously reported for pristine MXene membranes.

3.4 Concluding Remarks and Prospective Views

The MXene laminates for water filtration and desalination have made tremendous progress. The ideal membrane would be thin enough to have adequate separation efficiency, yet thick enough to sustain the pressure that would be applied to it. Ongoing efforts to address various challenges will have a significant impact on the effectiveness of these membranes. To meet commercial demand, separation efficiency for small size ions or molecules, as well as permeability, must be improved. It is also necessary to enhance the antifouling properties of these membrane, in order to strike a balance between fouling rejection and flux reduction. To do this, it is possible to improve the mechanical and antibacterial properties of the free-standing laminates.

The first challenge is the materials cost. The synthesis cost of MXene still needs to be lowered, else it cannot capture the market for commercial applications. Defect-free MXene with precisely sized pores and shaped interlayer distances is the goal of several ongoing efforts to perfect the art of MXene preparation. The cost of materials has been significantly lowered in recent years as a direct result of significant contributions made by newly created technologies. Therefore, there is still a huge need for improvement in manufacturing processes. According to reported studies, most MXene membranes are prepared by vacuum filtration, which calls for huge quantities of materials, takes longer, and has alignment and scalability problems. Therefore, time- and money-saving techniques are required for industrial use and large-scale production. In the future, mixing MXene with a suitable substrate will also be necessary to prevent dispersion in the liquid phase and to guarantee water flux. To date, most studies have only focused on a single MXene, *e.g.*, $Ti_3C_2T_x$. Other members of the same family must be investigated at the same time, to harness fully their potential in separation applications. Additionally, only certain ions and dye molecules are employed with MXene-based membranes. MXene still faces significant difficulties in separating small molecules (such as ions or dyes) with diameters smaller than 5 Å. MXene is significantly behind graphene in terms of uses for selectivity, permeance, and separation. Therefore, MXene will be researched for innovative kinds of separation and to increase its permeance and selectivity.

Another critical concern is the stability of MXene in water-based applications; it is rapidly destroyed in humid environments. MXene has a variety of functional groups, which contribute to its good hydrophilicity, similar to GO membranes. The hydrophilic characteristics of MXenes make it easy to create hydrogen bonds with water molecules, which causes them to break down and delaminate in water. This can be performed by combining functional groups from membranes with the appropriate cross-linking agents by the addition of a variety of materials to MXenes. These materials include 2D nanosheets, nanoparticles polymers, or three-dimensional (3D) materials. Making an effort to strengthen membrane stability is something that needs to be done. To further enhance the separation performance of MXenes, consideration should be paid to regulated interlayer spacing, shape, size, and atomic defects. Overall, there are several obstacles to the scientific community's ability to harness the power of this wonder substance properly. MXene is still in its infancy compared to graphene, and much more study is required, especially in desalinating sea water, brackish water, and industrial waste water.

List of Abbreviations

2D	Two-dimensional
ANF	Aramid nanofibers
AB8GX dye	Alcian blue 8GX
BSA	Bovine serum albumin
BB	Brilliant blue
CR	Congo red
COFs	Covalent organic frameworks
C (Cyt c)	Cytochrome c
CV	Crystal violet
DI	Deionized
DR	Direct red
EB	Evans blue
HA	Humic acid
LB	Langmuir–Blodgett
LBL	Layer-by-layer
MXene	Metal carbides and nitrides
NR	Neutral red
GO	Graphene oxide
Au NPs	Gold nanoparticles
MOFs	Metal–organic frameworks
MB	Methyl blue
MLB	Methylene blue
NF	Nanofiltration
h-BN	Boron nitrides
PAN	Polyacrylonitrile
PEI	Polyetheneimine

PVDF	Polyvinylidine fluoride
PES	Polyethersulfone
RB	Rose Bengal
RhB	Rhodamine B
RO	Reverse osmosis
SA	Sodium alginate
SEM	Scanning electron microscopy
XRD	X-ray diffraction
TEM	Transmission electron microscopy

Acknowledgements

A.A. is thankful for a financial grant and research support from CAS-TWAS president fellowship, Institute of Process Engineering, Chinese Academy of Sciences Beijing. J.Y. is thankful for financial support from State Key Laboratory of Multiple Complex Systems, Institute of Process Engineering, Chinese Academy of Sciences (MPCS-2021-A-05). K.H.T. is thankful for a postdoctoral fellowship from MUST, Pakistan.

References

1. K. H. Thebo, X. Qian, Q. Zhang, L. Chen, H. M. Cheng and W. Ren, *Nat. Commun.*, 2018, **9**, 1486.
2. C. He, Z. Liu, J. Wu, X. Pan, Z. Fang, J. Li and B. A. Bryan, *Nat. Commun.*, 2021, **12**, 4667.
3. C. Y. Tang, Z. Yang, H. Guo, J. J. Wen, L. D. Nghiem and E. Cornelissen, *Environ. Sci. Technol.*, 2018, **52**, 10215–10223.
4. M. R. M. Ramli, A. L. Ahmad and C. P. Leo, *ACS Omega*, 2021, **6**, 4609–4618.
5. F. Rehman, K. H. Thebo, M. Aamir and J. Akhtar, *Nanotechnology in the Beverage Industry*, ed. A. Amrane, S. Rajendran, T. A. Nguyen, A. A. Assadi, and A. M. Sharoba, Elsevier, 2020, pp. 207–240.
6. A. Ali, R. Pothu, S. H. Siyal, S. Phulpoto, M. Sajjad and K. H. Thebo, *Mater. Sci. Energy Technol.*, 2019, **2**, 83–88.
7. A. M. Nasir, M. R. Adam, S. N. E. A. M. Kamal, J. Jaafar, M. H. D. Othman, A. F. Ismail, F. Aziz, N. Yusof, M. R. Bilad, R. Mohamud, M. A. Rahman and W. N. W. Salleh, *Sep. Purif. Technol.*, 2022, **286**, 120454.
8. P. S. Goh, N. A. Ahmad, T. W. Wong, L. T. Yogarathinam and A. F. Ismail, *Chemosphere*, 2022, **307**, 136018.
9. A. Ali, M. Aamir, K. H. Thebo and J. Akhtar, *Chem. Rec.*, 2020, **20**, 344–354.
10. A. H. Jatoi, K. H. Kim, M. A. Khan, F. H. Memon, M. Iqbal, D. Janwery, S. N. Phulpoto, A. Samantasinghar, K. H. Choi and K. H. Thebo, *RSC Adv.*, 2023, **13**, 12695–12702.

11. F. Rehman, F. H. Memon, S. Ullah, M. A. J. Mazumder, A. Al-Ahmed, F. Khan and K. H. Thebo, *Chem. Rec.*, 2022, **22**, e202200107.
12. F. H. Memon, F. Rehman, J. Lee, F. Soomro, M. Iqbal, S. M. Khan, A. Ali, K. H. Thebo and K. H. Choi, *Sep. Purif. Rev.*, 2023, **52**, 43–57.
13. X. Zhao, N. Jia, L. Cheng, R. Wang and C. Gao, *ACS Omega*, 2019, **4**, 2320–2330.
14. M. K. Shahzad, F. H. Memon, F. Soomro, M. Iqbal, A. Ibrar, A. A. Memon, J. H. Lim, K. H. Choi and K. H. Thebo, *J. Environ. Chem. Eng.*, 2023, **11**, 109329.
15. R. K. Joshi, S. Alwarappan, M. Yoshimura, V. Sahajwalla and Y. Nishina, *Appl. Mater. Today*, 2015, **1**, 1–12.
16. Z. Ali, M. Mehmood, J. Ahmed, A. Majeed and K. H. Thebo, *Mater. Lett.*, 2020, **259**, 126831.
17. Z. Ali, M. Mehmood, J. Ahmed, A. Majeed and K. H. Thebo, *Mater. Res. Express*, 2019, **6**, 105627.
18. S. Sharif, K. S. Ahmad, F. Rehman, Z. Bhatti and K. H. Thebo, *J. Environ. Chem. Eng.*, 2021, **9**, 105605.
19. D. Janwery, F. H. Memon, A. A. Memon, M. Iqbal, F. N. Memon, W. Ali, K.-H. Choi and K. H. Thebo, *ACS Omega*, 2023, **8**, 7648–7656.
20. A. Ali, F. Rehman, M. Ali Khan, F. H. Memon, F. Soomro, M. Iqbal, J. Yang and K. H. Thebo, *ACS Omega*, 2022, **7**, 32410–32417.
21. S. Sharif, K. S. Ahmad, F. H. Memon, F. Rehman, F. Soomro and K. H. Thebo, *Mater. Res. Innovations*, 2021, 1–9.
22. I. Chandio, Y. Ai, L. Wu and Q. Liang, *Nano Res.*, 2023, DOI: 10.1007/s12274-023-5770-3.
23. S. Yuan, X. Li, J. Zhu, G. Zhang, P. Van Puyvelde and B. Van der Bruggen, *Chem. Soc. Rev.*, 2019, **48**, 2665–2681.
24. X. Li, Y. Liu, J. Wang, J. Gascon, J. Li and B. Van der Bruggen, *Chem. Soc. Rev.*, 2017, **46**, 7124–7144.
25. Z. Li, Y. Zheng, T. Gu, X. Meng, H. Wang, K. Xu, L. Cheng, R. Kasher, R. Zhang and Z. Jiang, *J. Membr. Sci.*, 2023, **675**, 121551.
26. H. Fan, M. Peng, I. Strauss, A. Mundstock, H. Meng and J. Caro, *Nat. Commun.*, 2021, **12**, 38.
27. C. Chen, J. Wang, D. Liu, C. Yang, Y. Liu, R. S. Ruoff and W. Lei, *Nat. Commun.*, 2018, **9**, 1902.
28. I. Maqbool, F. Rehman, F. Soomro, Z. Bhatti, U. Ali, A. H. Jatoi, B. Lal, M. Iqbal, S. Phulpoto, A. Ali and K. H. Thebo, *ChemBioEng Rev.*, 2021, **8**, 67–77.
29. F. A. Janjhi, D. Janwery, I. Chandio, S. Ullah, F. Rehman, A. A. Memon, J. Hakami, F. Khan, G. Boczkaj and K. H. Thebo, *ChemBioEng Rev.*, 2022, **9**, 574–590.
30. F. A. Janjhi, I. Chandio, A. A. Memon, Z. Ahmed, K. H. Thebo, A. A. A. Pirzado, A. A. Hakro and M. Iqbal, *Sep. Purif. Technol.*, 2021, **274**, 117969.
31. F. A. Janjhi, I. Ihsanullah, M. Bilal, R. Castro-Muñoz, G. Boczkaj and F. Gallucci, *Water Resour. Ind.*, 2023, **29**, 100202.
32. I. Chandio, F. A. Janjhi, A. A. Memon, S. Memon, Z. Ali, K. H. Thebo, A. A. A. Pirzado, A. A. Hakro and W. S. Khan, *Desalination*, 2021, **500**, 114848.

33. X. Qian, L. Chen, L. Yin, Z. Liu, S. Pei, F. Li, G. Hou, S. Chen, L. Song, K. H. Thebo, H.-M. Cheng and W. Ren, *Science*, 2020, **370**, 596–600.
34. M.-U.-N. Khilji, N. A. Nahyoon, M. Mehdi, K. H. Thebo, N. Mahar, A. A. Memon, N. Memon and N. Hussain, *Opt. Mater.*, 2023, **135**, 113260.
35. N. A. Nahyoon, L. Liu, K. Rabe, K. H. Thebo, L. Yuan, J. Sun and F. Yang, *Electrochim. Acta*, 2018, **271**, 41–48.
36. L. M. Dong, C. Ye, L. L. Zheng, Z. F. Gao and F. Xia, *Nanophotonics*, 2020, **9**, 2125–2145.
37. Z. Chen, H. Yang, Y. Li, L. Yao, Y. Fu, S. Kuang, G. Wang, Q. Deng and L. Xiong, *J. Mater. Res.*, 2023, **38**, 2655–2665.
38. I. Mahar, F. H. Memon, J.-W. Lee, K. H. Kim, R. Ahmed, F. Soomro, F. Rehman, A. A. Memon, K. H. Thebo and K. H. Choi, *Membranes*, 2021, **11**, 869.
39. I. Mahar, F. K. Mahar, N. Mahar, A. A. Memon, A. A. A. Pirzado, Z. Khatri, K. H. Thebo and A. Ali, *Chem. Eng. Res. Des.*, 2023, **191**, 462–471.
40. Y. Gogotsi and B. Anasori, *ACS Nano*, 2019, **13**, 8491–8494.
41. T. Yousaf, A. Areeb, M. Murtaza, A. Munir, Y. Khan and A. Waseem, *ACS Omega*, 2022, 7, 19502–19512.
42. Z. Ahmed, F. Rehman, U. Ali, A. Ali, M. Iqbal and K. H. Thebo, *ChemBioEng Rev.*, 2021, **8**, 110–120.
43. L. Huang, L. Ding and H. Wang, *Small Sci.*, 2021, **1**, 2100013.
44. L. Ding, Y. Wei, Y. Wang, H. Chen, J. Caro and H. Wang, *Angew. Chem., Int. Ed.*, 2017, **56**, 1825–1829.
45. J. Wang, J. He, D. Kan, K. Chen, M. Song and W. Huo, *Crystals*, 2022, **12**, 1034.
46. C. E. Ren, K. B. Hatzell, M. Alhabeb, Z. Ling, K. A. Mahmoud and Y. Gogotsi, *J. Phys. Chem. Lett.*, 2015, **6**, 4026–4031.
47. K. M. Kang, D. W. Kim, C. E. Ren, K. M. Cho, S. J. Kim, J. H. Choi, Y. T. Nam, Y. Gogotsi and H.-T. Jung, *ACS Appl. Mater. Interfaces*, 2017, **9**, 44687–44694.
48. Y. Liu, J. Zhang, X. Zhang, Y. Li and J. Wang, *ACS Appl. Mater. Interfaces*, 2016, **8**, 20352–20363.
49. Z. Zhang, S. Yang, P. Zhang, J. Zhang, G. Chen and X. Feng, *Nat. Commun.*, 2019, **10**, 2920.
50. X. Xie, C. Chen, N. Zhang, Z.-R. Tang, J. Jiang and Y.-J. Xu, *Nat. Sustainability*, 2019, **2**, 856–862.
51. J. Wang, Z. Zhang, J. Zhu, M. Tian, S. Zheng, F. Wang, X. Wang and L. Wang, *Nat. Commun.*, 2020, **11**, 3540.
52. J. Wang, G. Li, G. Zhang, J. Lv, W. Zhang, F. Guo, Y. Liu, G. Cheng and X. Li, *Infrared Phys. Technol.*, 2021, **119**, 103945.
53. J. Lipton, G.-M. Weng, J. A. Röhr, H. Wang and A. D. Taylor, *Matter*, 2020, **2**, 1148–1165.
54. F. Rehman, F. H. Memon, A. Ali, S. M. Khan, F. Soomro, M. Iqbal and K. H. Thebo, *Rev. Inorg. Chem.*, 2023, **43**, 13–31.
55. K. P. A. Kumar, K. Ghosh, O. Alduhaish and M. Pumera, *Electrochem. Commun.*, 2021, **122**, 106890.

56. I. Ali, M. F. Ud Din and Z.-G. Gu, *Molecules*, 2022, **27**, 4925.
57. K. Hantanasirisakul, M.-Q. Zhao, P. Urbankowski, J. Halim, B. Anasori, S. Kota, C. E. Ren, M. W. Barsoum and Y. Gogotsi, *Adv. Electron. Mater.*, 2016, **2**, 1600050.
58. F. Rehman, F. H. Memon, Z. Bhatti, M. Iqbal, F. Soomro, A. Ali and K. H. Thebo, *Rev. Inorg. Chem.*, 2022, **42**, 327–336.
59. J. D. Cain, A. Azizi, K. Maleski, B. Anasori, E. C. Glazer, P. Y. Kim, Y. Gogotsi, B. A. Helms, T. P. Russell and A. Zettl, *ACS Nano*, 2019, **13**, 12385–12392.
60. A. Akbari, P. Sheath, S. T. Martin, D. B. Shinde, M. Shaibani, P. C. Banerjee, R. Tkacz, D. Bhattacharyya and M. Majumder, *Nat. Commun.*, 2016, **7**, 10891.
61. Z. Liu, Z. Li, Z. Xu, Z. Xia, X. Hu, L. Kou, L. Peng, Y. Wei and C. Gao, *Chem. Mater.*, 2014, **26**, 6786–6795.
62. T. Huang, B. Zheng, Z. Liu, L. Kou and C. Gao, *J. Mater. Chem. A*, 2015, **3**, 1890–1895.
63. L. Kou, Z. Liu, T. Huang, B. Zheng, Z. Tian, Z. Deng and C. Gao, *Nanoscale*, 2015, **7**, 4080–4087.
64. X. Hu, S. Rajendran, Y. Yao, Z. Liu, K. Gopalsamy, L. Peng and C. Gao, *Nano Res.*, 2016, **9**, 735–744.
65. M. Fathizadeh, H. N. Tien, K. Khivantsev, J.-T. Chen and M. Yu, *J. Mater. Chem. A*, 2017, **5**, 20860–20866.
66. J. Zhong, W. Sun, Q. Wei, X. Qian, H.-M. Cheng and W. Ren, *Nat. Commun.*, 2018, **9**, 3484.
67. Q. Zhang, X. Qian, K. H. Thebo, H.-M. Cheng and W. Ren, *Sci. Bull.*, 2018, **63**, 788–794.
68. M. Mehdi, W. Jiang, Q. Zeng, K. H. Thebo, I.-S. Kim, Z. Khatri, H. Wang, J. Hu and K.-Q. Zhang, *Langmuir*, 2022, **38**, 6376–6386.
69. K. H. Thebo, X. Qian, Q. Wei, Q. Zhang, H.-M. Cheng and W. Ren, *J. Mater. Sci. Technol.*, 2018, **34**, 1481–1486.
70. S. A. Bhatti, F. H. Memon, F. Rehman, Z. Bhatti, T. Naqvi and K. H. Thebo, *Rev. Inorg. Chem.*, 2022, **42**, 283–295.
71. F. Soomro, F. H. Memon, M. A. Khan, M. Iqbal, A. Ibrar, A. A. Memon, J. H. Lim, K. H. Choi and K. H. Thebo, *Membranes*, 2023, **13**(1), 64.
72. M. Rezakazemi, A. A. Shamsabadi, H. Lin, P. Luis, S. Ramakrishna and T. M. Aminabhavi, *Renewable Sustainable Energy Rev.*, 2021, **143**, 110878.
73. R. Han, X. Ma, Y. Xie, D. Teng and S. Zhang, *RSC Adv.*, 2017, **7**, 56204–56210.
74. G. Liu, J. Shen, Q. Liu, G. Liu, J. Xiong, J. Yang and W. Jin, *J. Membr. Sci.*, 2018, **548**, 548–558.
75. Y. Wang, L. Li, Y. Wei, J. Xue, H. Chen, L. Ding, J. Caro and H. Wang, *Angew. Chem., Int. Ed.*, 2017, **56**, 8974–8980.
76. R. P. Pandey, K. Rasool, V. E. Madhavan, B. Aissa, Y. Gogotsi and K. A. Mahmoud, *J. Mater. Chem. A*, 2018, **6**, 3522–3533.
77. V. Natu, R. Pai, O. Wilson, E. Gadasu, H. Badr, A. Karmakar, A. J. D. Magenau, V. Kalra and M. W. Barsoum, *Chem. Mater.*, 2022, **34**, 678–693.
78. Z. Yin, Z. Lu, Y. Xu, Y. Zhang, L. He, P. Li, L. Xiong, L. Ding, Y. Wei and H. Wang, *Membranes*, 2021, **11**, 621.

79. Z. Lu, Y. Wei, J. Deng, L. Ding, Z.-K. Li and H. Wang, *ACS Nano*, 2019, **13**, 10535–10544.
80. L. Ding, L. Li, Y. Liu, Y. Wu, Z. Lu, J. Deng, Y. Wei, J. Caro and H. Wang, *Nat. Sustainability*, 2020, **3**, 296–302.
81. Z. Shen, W. Chen, H. Xu, W. Yang, Q. Kong, A. Wang, M. Ding and J. Shang, *Int. J. Environ. Res. Public Health*, 2019, **16**(23), 4659.
82. R. Han, Y. Xie and X. Ma, *Chin. J. Chem. Eng.*, 2019, **27**, 877–883.
83. T. Liu, X. Liu, N. Graham, W. Yu and K. Sun, *J. Membr. Sci.*, 2020, **593**, 117431.
84. C. Xing, Y. Zhang, R. Huang, S. Li, J. He, R. Fu, C. Lai, C. Liu, L. Shen and S. Zhang, *EcoMat*, 2023, **5**, e12300.
85. C. Wang, R. Cheng, P.-X. Hou, Y. Ma, A. Majeed, X. Wang and C. Liu, *ACS Appl. Mater. Interfaces*, 2020, **12**, 43032–43041.

CHAPTER 4

Advanced Nanomaterials for Removal of Emerging Organic Pollutants From Water

ADEDAPO O. ADEOLA[a] AND PATRICIA B.C. FORBES[a*]

[a]Department of Chemistry, Faculty of Natural and Agricultural Sciences, University of Pretoria, Lynnwood Road, Hatfield, Pretoria 0002, South Africa
*E-mail: patricia.forbes@up.ac.za

4.1 Introduction

Great concern has been expressed worldwide over the occurrence of emerging organic pollutants in water systems. Emerging pollutants are a variety of substances that are newly controlled or unregulated, but have the potential to have negative effects on human health and the ecosystem.[1] Organic trace contaminants, such as cosmetic products, disinfection by-products, pharmaceutical agents, industrial additives, pesticides, steroids and endocrine disruptors, *etc.*, are several examples of these trace, hazardous, emerging contaminants.[2,3] According to a survey, around 1.8 billion kg of waste is dumped into aquatic bodies every day and approximately 350 000 chemicals are estimated to be in use globally.[4] The availability of potable water remains a significant concern, as highlighted by the United Nations Sustainable Development Goals (UN SDGs), particularly in developing countries.[5] To develop appropriate, effective, and environmentally acceptable remediation techniques for emerging pollutants, significant efforts have been made by researchers globally.

Chemistry in the Environment Series No. 12
Novel Materials and Water Purification: Towards a Sustainable Future
Edited by Grigorios L. Kyriakopoulos and Miltiadis G. Zamparas

Published by the Royal Society of Chemistry, www.rsc.org

Due to their potential for practical use, and benefits that include low cost, safety, and tailorability, advanced nanomaterials have drawn a lot of attention in comparison to other approaches.[6–9] Organic molecules bind to the surface of an adsorbent by physisorption or chemisorption in a process known as adsorption.[10] Using a green synthesis strategy for preparing adsorbents could end the loop of substituting dangerous compounds with substances that are just as harmful. Hence, the use of carbon nanomaterials for water purification mainly hinges on the concept of sustainable chemical processes and green chemistry. Advanced carbon nanomaterials, including graphene, carbon nanotubes, carbon dots, and fullerenes, have been explored for the adsorption of emerging chemical pollutants.[11,12]

Several drawbacks still exist with the application of carbon nanomaterials, including potential health risks, production cost, specificity/selectivity, recyclability, and sustainability, yet these advanced nanomaterials have their distinct merits, which can be harnessed for water purification applications (see Figure 4.1). This chapter highlights the utilization of carbon nanomaterials,

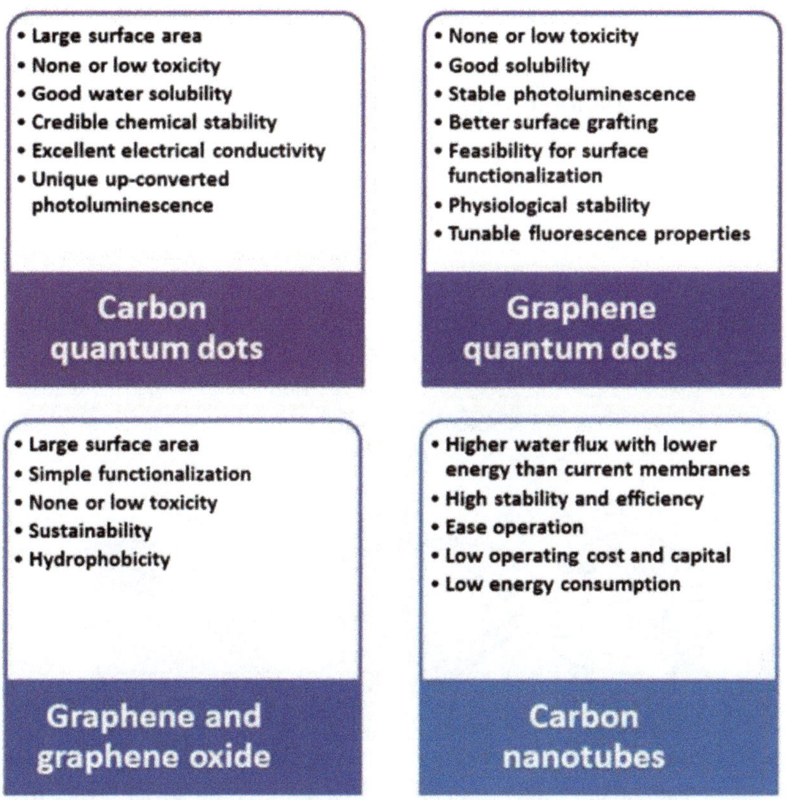

Figure 4.1 Merits of several advanced carbon-based nanomaterials. Adapted from ref. 12 with permission from Elsevier, Copyright 2022.

such as carbon nanotubes, graphene, carbon-based composites, and other derivatives, for the removal of various organic contaminants from water.

4.2 Graphene and Graphene-based Composites

Graphene is a sheet of carbon atoms that is only one atom thick and is structured in a two-dimensional (2D) honeycomb arrangement.[13] Figure 4.2 reveals the structure of graphene as a single sheet or stack of sheets. Because of their considerable relevant physicochemical properties, including large specific surface area, hydrophobicity, high porosity, surface functional groups and ease of functionalization, graphene nanostructures have been used to remediate waste water. Graphene nanocomposites are suitable for the removal of harmful and dangerous mixtures of various contaminants.[14,15] There have been several reports in recent years that cover the synthesis and characteristics of graphene-based materials.[16,17] Two routes can be used to prepare graphene, as well as other nanomaterials: namely, top-down or bottom-up approaches.[18,19] The bottom-up method entails the assembly/growth of graphene from low molecular weight hydrocarbon or smaller units, whereas the top-down method requires the delamination of graphite or graphitic compounds. Bottom-up strategies include pyrolysis, chemical vapour deposition, and epitaxial growth, whilst top-down techniques frequently use reduction and exfoliation procedures.[16,20–22]

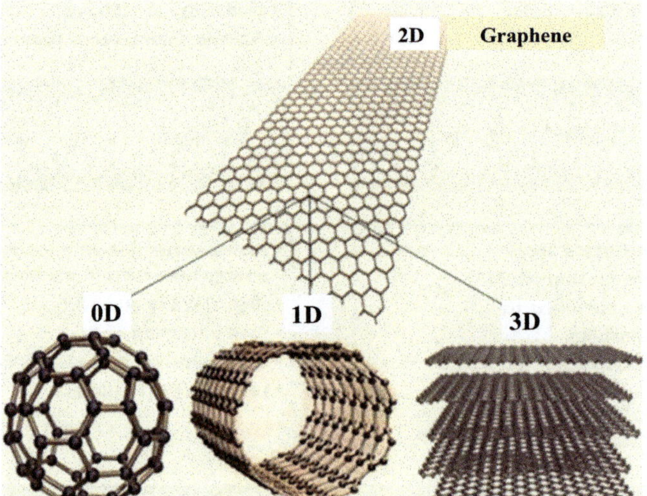

Figure 4.2 A single layer of carbon atoms in a honeycomb lattice structure (fullerene), nanotube structure, and stack of layers of graphite. Reproduced from ref. 22, https://doi.org/10.3390/polym13172869, under the terms of the CC BY 4.0 license https://creativecommons.org/licenses/by/4.0/.

Porous graphene that is both superhydrophobic and superoleophilic has been synthesised as a nanoabsorbent for a variety of compounds, with absorption capabilities of over 90%, good selectivity, and reusability.[23] With two basal planes available for contaminant adsorption, single-layered graphene or graphene nanosheets provide a number of benefits for the aqueous elimination of hazardous trace organics.[24] Graphene-based materials interact with a vast number of chemical contaminants in ionic and/or molecular states *via* mechanisms such as electrostatic, hydrogen bonding, π-interactions, van der Waals and hydrophobic interactions, *etc.*[25] Thus, graphene has received a lot of attention as a powerful nanosorbent for pollution remediation.

Graphenic materials have been utilized as effective adsorbents for treating waste water containing emerging organic pollutants.[23] By altering the experimental setup, one can change the adsorption capability of these materials. The adsorption capacity of graphenic materials can be influenced by variables such as the initial dosage, the properties of the graphene-based materials and the target contaminant, temperature, pH, contact time, salinity, and presence of competing ions.[12,26,27] Table 4.1 shows the various organic chemical pollutants that have been removed from water using pristine graphene, functionalized graphene, or doped graphene-based nanomaterials, with most of these showing a multilayer adsorption mechanism when applied for the decontamination of aqueous media. In a number of cases organic dyes, such as malachite green or methylene blue, have been used as proxies for the determination of adsorption efficiencies.

4.3 Carbon Nanotubes and Composites

Carbon atoms with a diameter of 3 to 15 nm are organized in hexagonal and pentagonal shapes to form carbon nanotubes (CNT), a carbon allotrope.[44] A hollow cylindrical structure called a CNT is composed of one or more layers of graphene sheets. CNTs may be subcategorized as shown in Figure 4.3 into single-walled carbon nanotubes (SWCNTs), double-walled carbon nanotubes (DWCNTs), and multi-walled carbon nanotubes (MWCNTs), based on the number of exterior layers or walls.[45,46]

A list of CNT-based nanomaterials that have been applied to remove organic compounds from contaminated aqueous media is shown in Table 4.2. The fact that different carbon nanotube-based materials have been shown to have adsorption capacities up to 679.6 mg g^{-1} (rhodamine B) for dyes, 227.3 mg g^{-1} (chipton) for pesticides, 221.2 mg g^{-1} (meloxicam) for pharmaceuticals, and 46.2 mg g^{-1} (bisphenol A (BPA)) suggests the strong affinity of CNTs and its composites for various classes of organic compounds. In addition to the pollutants in Table 4.2, it has also been claimed that CNT-based materials may remove polyfluoroalkyl substances and other endocrine-disrupting pollutants.[47] Additional perfluorinated compounds (PFCs) with the same functional group were able to bind to MWCNTs as the carbon–fluorine chain length increased. When compared to pristine CNTs, the removal effectiveness of CNTs containing hydroxyl and carboxyl groups was considerably lower.[48]

Table 4.1 Graphene-based adsorbents for the removal of selected organic chemical pollutants in aqueous solution.

Material	Compound	Adsorption capacity (mg g^{-1})	Isotherm and kinetic models	Reference
Modified magnetic graphene oxide	Acid red	524.2	Freundlich model, pseudo-second order	28
Graphene-modified cyclodextrin (GD-TAC)	Roxarsone	153.59	Sips, pseudo-second order	29
	Methyl orange	445.60		
	Bisphenol A	237.90		
Reduced graphene oxide	Bisphenol A	181.82	Langmuir, pseudo-second order	30
GO-silica-coated magnetic nanoparticles	Chlorpyrifos; parathion; malathion	25.6; 135; 61.9	Sips, pseudo-second order	31
Graphene oxide	Chlorpyrifos; malathion	98; 1667	Langmuir, pseudo-second order	32
Graphene wool	Phenanthrene; pyrene; efavirenz; nevirapine	5.0; 20.0; 4.4; 48.3	Sips, pseudo-second order	33 and 10
Graphene-coated materials	Phenanthrene	1.74	Freundlich	34
Graphene oxide–FeO–Fe_2O_3	Naphthalene	2.63	Freundlich	35
		1.05		
Graphene; graphene oxide	Biphenyl	102.6; 59.0	Freundlich	36
Graphene oxide/calcium alginate	Tetracycline	131.6	Freundlich model, pseudo-second order	37
Reduced graphene oxide	Malachite green	476.2	Langmuir model, pseudo-second order	38
Graphene oxide/PD	Methylene blue	2000.1	Langmuir model, pseudo-second order	39
Graphene oxide/ionic liquid-modified	Methylene blue	448.0	Langmuir model, pseudo-second order	40
Graphene oxide/hollow mesoporous silica	Methylene blue; methylene orange	476.19; 279.27	Langmuir model, pseudo-second order	41
Graphene oxide/4 Å molecular sieve	Rhodamine B	62.81	Langmuir model, pseudo-second order	42
Graphene oxide/amine-amide-functionalized	Basic blue 41; methyl orange	199.5; 64.0	Freundlich (BB41), Langmuir model (MO), pseudo-second order	43

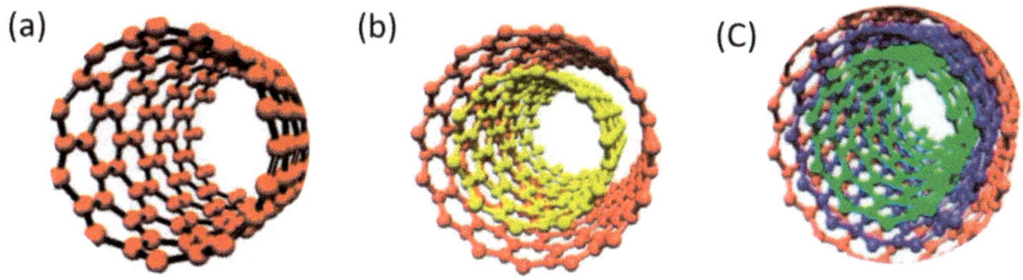

Figure 4.3 Structure of a single-walled carbon nanotube (a), double-walled carbon nanotube (b), and a multi-walled carbon nanotube (c). Reproduced from ref. 46, https://doi.org/10.3390/app11209550, under the terms of the CC BY 4.0 license https://creativecommons.org/licenses/by/4.0/.

The primary adsorption process was dominated by hydrophobic interactions. Magnetic MWCNTs coated with chitosan were used to treat water containing BPA, another endocrine disruptor, with an adsorption capacity of 46.2 mg g^{-1}.[49] The adsorption capacities for various classes of pollutants reported for graphene-based materials were typically higher than many of those obtained for carbon nanotubes (see Tables 4.1 and 4.2), indicating that carbon structure and atomic configurations may also be important in the binding mechanism and capacity of different carbon-based nanomaterials/nanocomposites.[72]

Many methods, such as spectroscopic characterization, zeta potential measurement, imaging techniques, computational modelling, and theoretical calculations, have been employed in the mechanistic study of interactions between pollutants and carbon-based nanomaterials. Potential interaction processes include electrostatics, complexation, π-overlapping interactions, hydrogen bonding, hydrophobic interactions, covalent chemisorption, van der Waals interactions, *etc.* (Figure 4.4).

4.4 Concluding Remarks

Hazardous organic pollutants in water bodies are a cause for concern because of the possibility of bioaccumulation, non-target exposures, and negative health impacts. Owing to the rising demand for potable and clean water, wastewater treatment and reusability must be considered. As a result, wastewater treatment facilities should be upgraded to accommodate the rapidly rising industrial and environmental demands for clean water. The majority of developing nations lack stringent guidelines for monitoring, handling, recycling, or disposal of several waste chemicals. The lack of, or inadequacies in, detailed environmental monitoring programmes for chemical pollutants makes it impossible to establish the threshold levels for a number of toxicants/contaminants in water systems. In order to establish allowed thresholds for concentrations in effluents discharged into water bodies, it

Table 4.2 Carbon nanotube-based materials developed for the removal of selected organic chemical pollutants in aqueous solution.

Material	Compound	Adsorption capacity (mg g^{-1})	Isotherm and kinetic models	Reference
Polyaniline-multiwalled carbon nanotubes (MWCNTs)	Malachite orange	149	Pseudo-second order/Langmuir and Freundlich models	50
4-Aminosalicylic acid-functionalized MWCNTs	Crystal violet dye	440	Pseudo-second order	51
Single-walled carbon nanotube (SWCNTs)	Reactive yellow dye 15 and reactive yellow dye 42	179.9 and 156.1	Pseudo-second order/Langmuir and Freundlich models	52
SWCNTs	Acid blue 92	86.91	Pseudo-second order/Langmuir	53
Sulfonic acid-functionalized carbon nanotubes (CNT)	Malachite blue	236.5	Pseudo-second order/Freundlich (multilayer adsorption)	54
Hydrogel nanocomposite	Malachite blue	647	Pseudo-second order/linear Redlich–Peterson	55
MWCNTs	Violet 2R	76.92	Pseudo-second order/multiple models	56
CNT/MgO/$CuFe_2O_4$	Methyl violet dye (MVD) and Nile blue dye (NBD)	>35	Pseudo-second order/Freundlich (MVD), Langmuir (NBD)	57
Poly(acrylonitrile-*co*-styrene)/MWCNTs	Methyl orange	121.95	Pseudo-second order/multiple models	58
WS_2/Fe_3O_4/CNTs-nanocomposite	Amaranth and brilliant blue	174.8 and 166.7	Pseudo-second order/Langmuir models	59
Nano-cobalt wrapped by nitrogen-doped CNT	Rhodamine B	679.56	Pseudo-second order/Langmuir models	60
Asparagine-functionalized MWCNTs	Malachite green and blue	637 and 500	Pseudo-second order/Langmuir models	61
Alginate/f-CNT-CD MFA hydrogel composite beads	Malachite blue	10.26	Pseudo-second order/Langmuir models	62
MWCNTs and Ox-MWCNTs	Diuron	39.59 and 48.60	Pseudo-second order/Freundlich, Langmuir and Polanyi–Manes	63
MWCNTs from plastic waste	Diuron	40.375	Pseudo-second order/Temkin, Hill, and Koble–Corrigan	64
UiO_{66}-NH_2@MPCA	Chipton	227.3	Pseudo-second order/Langmuir models	65
Polyanilline/MWCNTs	Meloxicam	221.2	Pseudo-second order/Elovich and Langmuir–Freundlich	66
Granular CNTs/alumina	Diclofenac and carbamazepine	106.5 and 157.4 μmol g^{-1}	Langmuir	67

Acid modified-MWCNTs	Diclofenac	24	Pseudo-second order/Freundlich	68
MWCNTs	Amoxicillin	22.9	Langmuir	69
SWCNTs	Amoxicillin	122.8	Pseudo-second order/Langmuir	
Acid-treated MWCNTs	Trihalomethanes	0.92–2.41	Freundlich–Langmuir	70
MWCNTs	Trihalomethanes	10.98, 6.85, 6.57, 5.95	Pseudo-second order/Sips	71
Magnetic MWCNTs/SiO_2/CS	BPA	46.2	Pseudo-second order/Langmuir	49

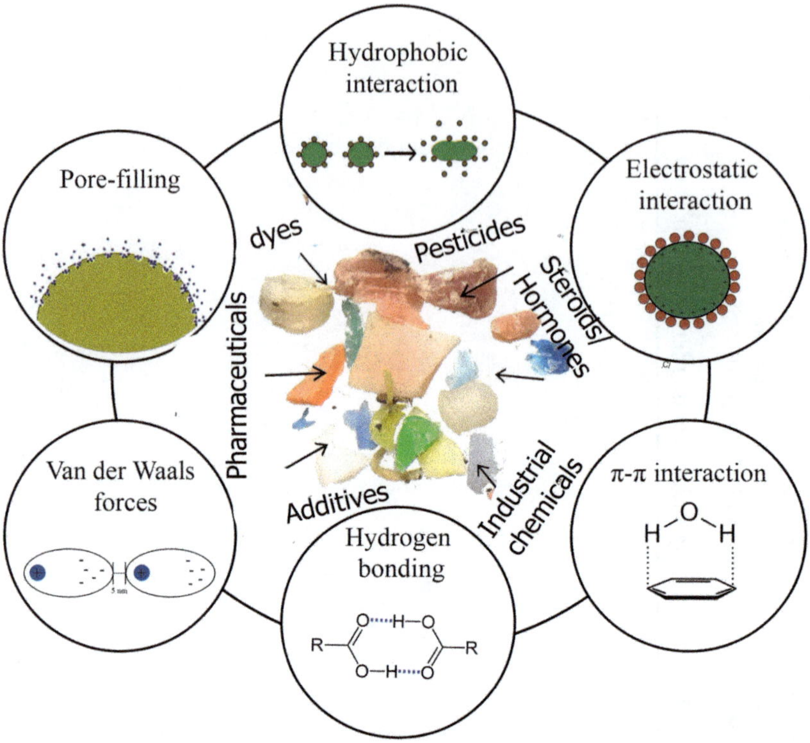

Figure 4.4 Various adsorption mechanisms of interaction between sorbents (in this case microplastics are given as an example) and organic chemical pollutants in water. Reproduced from ref. 72 with permission from Elsevier, Copyright 2021.

is necessary to conduct thorough risk assessments of a number of contaminants in water bodies.

Over time, there have been significant constraints due to the cost effectiveness and ease of commercial application of carbon nanomaterials for the purification of polluted water. Emerging technologies must therefore work to overcome these obstacles and deliver solutions that are better suited for widespread practical use in the world's developing nations. There are drawbacks to the use of carbon-based nanomaterials, including high operational costs, clogging, difficult regeneration, prolonged treatment times, production of a lot of sludge or secondary pollutants, *etc.* Models are also needed to forecast the longevity, regenerability, and reusability of carbon-based nanomaterials and membranes.

Research and development activities on the presence of organic contaminants in water resources, and the water treatment removal efficiencies thereof, have been limited by the challenges in analyzing these trace organic water contaminants. There is a need to conduct techno-economic feasibility analysis for field-based and industrial-scale uses of carbon nanomaterials, as so far the majority of studies have been proof-of-principle ones at laboratory scale. In order to improve current water treatment systems and mitigate negative impacts on our environment, there is a need for collaboration between different scientific fields, improved research funding, public sensitization, and collaborations between industry and government agencies globally.

References

1. N. T. Tavengwa, in *Emerging Freshwater Pollutants: Analysis, Fate and Regulations*, ed. T. Dalu, and N. T. Tavengwa, Academic Press, London, 2021.
2. E. Sanganyado, *Emerging Freshwater Pollutants*, Elsevier, 2021.
3. A. Tyhali and P. B. C. Forbes, *TrAC, Trends Anal. Chem.*, 2023, **158**, 116873.
4. Z. Wang, G. W. Walker, D. C. G. Muir and K. Nagatani-Yoshida, *Environ. Sci. Technol.*, 2020, **54**, 2575.
5. UNSDG, Sustainable Development Goal 6, *Synthesis Report on Water and Sanitation*, 2018, https://sustainabledevelopment.un.org/content/documents/19901SDG6_SR2018_web_3.pdf
6. G. Ucankus, M. Ercan, D. Uzunoglu and M. Culha, in *New Polymer Nanocomposites for Environmental Remediation*, ed. C. M. Hussain, and A. K. Mishra, Elsevier, 2018.
7. A. A. Oladipo, in *New Polymer Nanocomposites for Environmental Remediation*, ed. C. M. Hussain, and A. K. Mishra, Elsevier, 2018.
8. J. O. Ighalo, P.-S. Yap, K. O. Iwuozor, C. O. Aniagor, T. Liu, K. Dulta, F. U. Iwuchukwu and S. Rangabhashiyam, *Environ. Res.*, 2022, **212**, 113123.
9. K. O. Iwuozor, K. G. Akpomie, J. Conradie, K. A. Adegoke, K. O. Oyedotun, J. O. Ighalo, J. F. Amaku, C. Olisah and A. O. Adeola, *J. Water Process Eng.*, 2022, **49**, 103059.
10. A. O. Adeola, J. de Lange and P. B. C. Forbes, *Appl. Surf. Sci. Adv.*, 2021, **6**, 100157.
11. V. Dutta, R. Verma, C. Gopalkrishnan, M.-H. Yuan, K. M. Batoo, R. Jayavel, A. Chauhan, K.-Y. A. Lin, R. Balasubramani and S. Ghotekar, *Inorganics*, 2022, **10**, 169.
12. M. Nasrollahzadeh, M. Sajjadi, S. Iravani and R. S. Varma, *Chemosphere*, 2021, **263**, 128005.
13. M. Xu, T. Liang, M. Shi and H. Chen, *Chem. Rev.*, 2013, **113**, 3766.
14. G. Ersan, O. G. Apul, F. Perreault and T. Karanfil, *Water Res.*, 2017, **126**, 385.
15. A. Azizi, A. Torabian, E. Moniri, A. H. Hassani and H. A. Panahi, *Desalin. Water Treat.*, 2016, **57**, 28806.
16. G. Schoonraad, M. J. Madito, N. Manyala and P. B. C. Forbes, *J. Mater. Sci.*, 2020, **55**, 545.
17. A. A. Mirghni, K. O. Oyedotun, B. A. Mahmoud, A. Bello, S. C. Ray and N. Manyala, *Composites, Part B*, 2019, **174**, 106953.
18. V. B. Mbayachi, E. Ndayiragije, T. Sammani, S. Taj, E. R. Mbuta and A. U. Khan, *Results Chem.*, 2021, **3**, 100163.
19. N. Baig, I. Kammakakam and W. Falath, *Mater. Adv.*, 2021, **2**, 1821.

20. C. L. Yap, S. Gan and H. K. Ng, *Chem. Eng. J.*, 2012, **200–202**, 247.
21. M. Coroş, F. Pogăcean, L. Măgeruşan, C. Socaci and S. Pruneanu, *Front. Mater. Sci.*, 2019, **13**, 23.
22. A. Ibrahim, A. Klopocinska, K. Horvat and Z. A. Hamid, *Polymers*, 2021, **13**, 2869.
23. T. A. Tabish, F. A. Memon, D. E. Gomez, D. W. Horsell and S. Zhang, *Sci. Rep.*, 2018, **8**, 1817.
24. J. Zhao, Z. Wang, J. C. White and B. Xing, *Environ. Sci. Technol.*, 2014, **48**, 9995.
25. F. Perreault, A. F. de Faria and M. Elimelech, *Chem. Soc. Rev.*, 2015, **44**, 5861.
26. A. O. Adeola and P. B. C. Forbes, *Water Environ. Res.*, 2021, **93**, 343.
27. G. Kubheka, A. O. Adeola and P. B. C. Forbes, *RSC Adv.*, 2022, **12**, 23922.
28. X. Tang, P. Tang and L. Liu, *J. Braz. Chem. Soc.*, 2018, **29**(22), 334.
29. Q. Liu, J. Wang, C. Duan, T. Wang and Y. Zhou, *J. Hazard. Mater.*, 2022, **426**, 128074.
30. J. Xu, L. Wang and Y. Zhu, *Langmuir*, 2012, **28**, 8418.
31. V. W. O. Wanjeri, C. J. Sheppard, A. R. E. Prinsloo, J. C. Ngila and P. G. Ndungu, *J. Environ. Chem. Eng.*, 2018, **6**, 1333.
32. S. Yadav, N. Goel, V. Kumar and S. Singhal, *Anal. Chem. Lett.*, 2019, **9**, 291.
33. A. O. Adeola and P. B. C. Forbes, *Water Sci. Technol.*, 2019, **80**, 1931.
34. K. Yang, B. Chen and L. Zhu, *Sci. Rep.*, 2015, **5**, 11641.
35. X. Yang, J. Li, X. Ren, Y. Huang and X. Wang, *Colloids Surf., A*, 2013, **422**, 118.
36. O. G. Apul, Q. Wang, Y. Zhou and T. Karanfil, *Water Res.*, 2013, **47**, 1648.
37. H. Zhu, T. Chen, J. Liu and D. Li, *RSC Adv.*, 2018, **8**, 2616.
38. Z. Dong, D. Wang, X. Liu, X. Pei, L. Chen and J. Jin, *J. Mater. Chem. A*, 2014, **2**, 5034.
39. J. Wang, J. Zhang, L. Han, J. Wang, L. Zhu and H. Zeng, *Adv. Colloid Interface Sci.*, 2021, **289**, 102360.
40. K. Gupta, S. R. Yasa, A. Khan, O. P. Sharma and O. P. Khatri, *J. Colloid Interface Sci.*, 2022, **607**, 1973.
41. M. S. Yilmaz, *Microporous Mesoporous Mater.*, 2022, **330**, 111570.
42. X. Liu, Y. Guo, C. Zhang, X. Huang, K. Ma and Y. Zhang, *Sep. Purif. Technol.*, 2022, **286**, 120400.
43. S. Verma, K.-H. Kim, N. Kumar, S. S. Bhattacharya, M. Naushad and R. K. Dutta, *J. Hazard. Mater.*, 2022, **429**, 128308.
44. M. P. Spaolonzi, E. D. V. Duarte, M. G. Oliveira, H. P. S. Costa, M. C. B. Ribeiro, T. L. Silva, M. G. C. Silva and M. G. A. Vieira, *J. Cleaner Prod.*, 2022, **373**, 133961.
45. S. Rathinavel, K. Priyadharshini and D. Panda, *Mater. Sci. Eng., B*, 2021, **268**, 115095.
46. T. V. Patil, D. K. Patel, S. D. Dutta, K. Ganguly, A. Randhawa and K.-T. Lim, *Appl. Sci.*, 2021, **11**, 9550.
47. C. T. Vu and T. Wu, *Crit. Rev. Environ. Sci. Technol.*, 2022, **52**, 90.
48. S. Deng, Q. Zhang, Y. Nie, H. Wei, B. Wang, J. Huang, G. Yu and B. Xing, *Environ. Pollut.*, 2012, **168**, 138.

49. A. A. Mohammadi, M. H. Dehghani, A. Mesdaghinia, K. Yaghmaian and Z. Es'haghi, *Int. J. Biol. Macromol.*, 2020, **155**, 1019.
50. S. Pete, R. A. Kattil and L. Thomas, *Diamond Relat. Mater.*, 2021, **117**, 108455.
51. M. Saxena, N. Sharma and R. Saxena, *ChemistrySelect*, 2021, **6**, 10601.
52. A. Naghizadeh, A. Karimi, E. Derakhshani and A. Esform, *Environ. Qual. Manage.*, 2022, **31**, 133.
53. D. Balarak, M. Zafariyan, C. A. Igwegbe, K. K. Onyechi and J. O. Ighalo, *Environ. Processes*, 2021, **8**, 869.
54. Y. Lei, Q. Huang, J. Dou, H. Huang, G. Yang, F. Deng, M. Liu, X. Li, X. Zhang and Y. Wei, *Colloid Interface Sci. Commun.*, 2021, **40**, 100328.
55. S. Mallakpour and F. Tabesh, *Int. J. Biol. Macromol.*, 2021, **166**, 722.
56. K. M. Abualnaja, A. E. Alprol, M. Ashour and A. T. Mansour, *Appl. Sci.*, 2021, **11**, 4786.
57. R. Foroutan, S. J. Peighambardoust, Z. Esvandi, H. Khatooni and B. Ramavandi, *J. Environ. Chem. Eng.*, 2021, **9**, 104752.
58. K. M. Abualnaja, A. E. Alprol, M. A. Abu-Saied, A. T. Mansour and M. Ashour, *Nanomaterials*, 2021, **11**, 1144.
59. P. Arabkhani, H. Javadian, A. Asfaram, F. Sadeghfar and F. Sadegh, *J. Hazard. Mater.*, 2021, **420**, 126644.
60. G. Yang, Y. Li, S. Yang, J. Liao, X. Cai, Q. Gao, Y. Fang, F. Peng and S. Zhang, *Sep. Purif. Technol.*, 2021, **259**, 118098.
61. M. Saxena, A. Lochab and R. Saxena, *Surf. Interfaces*, 2021, **25**, 101187.
62. S. Mallakpour, V. Behranvand and F. Mallakpour, *J. Environ. Chem. Eng.*, 2021, **9**, 105170.
63. J. Deng, Y. Shao, N. Gao, Y. Deng, C. Tan, S. Zhou and X. Hu, *Chem. Eng. J.*, 2012, **193**, 339.
64. S. K. Deokar, G. S. Bajad, P. Bhonde, R. P. Vijayakumar and S. A. Mandavgane, *J. Polym. Environ.*, 2017, **25**, 165.
65. W. Liang, B. Wang, J. Cheng, D. Xiao, Z. Xie and J. Zhao, *J. Hazard. Mater.*, 2021, **401**, 123718.
66. F. V. A. Dutra, B. C. Pires, T. A. Nascimento and K. B. Borges, *J. Environ. Manage.*, 2018, **221**, 28.
67. H. Wei, S. Deng, Q. Huang, Y. Nie, B. Wang, J. Huang and G. Yu, *Water Res.*, 2013, **47**, 4139.
68. X. Hu and Z. Cheng, *Chin. J. Chem. Eng.*, 2015, **23**, 1551.
69. A. Mohammadi, M. Kazemipour, H. Ranjbar, R. B. Walker and M. Ansari, *Fullerenes, Nanotubes Carbon Nanostruct.*, 2015, **23**, 165.
70. C. Lu, Y.-L. Chung and K.-F. Chang, *Water Res.*, 2005, **39**, 1183.
71. M. H. Dehghani, M. Mohammadi, M. A. Mohammadi, A. H. Mahvi and K. Yetilmezsoy, *Water, Air, Soil Pollut.*, 2016, **227**, 332.
72. F. G. Torres, D. C. Dioses-Salinas, C. I. Pizarro-Ortega and G. E. De-la-Torre, *Sci. Total Environ.*, 2021, **757**, 143875.

CHAPTER 5

Magnetic Adsorbents/Photocatalysts for Water Purification: Progress and Challenges

M. G. GALLONI[a*], H. M. ADEEL SHARIF[b], A. GRAINCA[a], M. RIZWAN HAIDER[c] AND R. DJELLABI[d]

[a]Dipartimento di Chimica, Università degli Studi di Milano, Via Golgi 19, 20133 Milano, Italy; [b]Research Center for Eco-Environmental Engineering, Dongguan University of Technology, Dongguan 523808, China; [c]College of Marine Sciences, South China Agricultural University, Guangzhou 510642, China; [d]Department of Chemical Engineering, Universitat Rovira i Virgili, 43007, Tarragona, Spain
*E-mail: melissa.galloni@unimi.it

5.1 Introduction

Purification of wastewater is one of the challenges of modern society, as a result of large-scale human and industrial activities that have led to contamination of water bodies with different biological and chemical pollutants.[1,2] Indeed, international environmental organizations are taking this issue into full consideration and many strict rules have been drawn up to guide both industrial and scientific communities to overcome problems of water pollution. It is worth mentioning that the huge extent of

Chemistry in the Environment Series No. 12
Novel Materials and Water Purification: Towards a Sustainable Future
Edited by Grigorios L. Kyriakopoulos and Miltiadis G. Zamparas

Published by the Royal Society of Chemistry, www.rsc.org

environmental pollution and ecosystem disequilibrium is the consequence of the linear economy that has been used for many decades.[3,4] In fact, this concept of a linear economy aims to maximise industrial production while ignoring management of the waste produced, and therefore this route leads to discharge of a wide range of pollutants into our environment. In recent decades, a new concept of a circular (bio)economy has been realised to tackle the issues facing the environment and society.[5,6] It aims mostly to boost the application of green production routes and use of natural resources towards achieving sustainable development.[7] For water pollution, European and international environmental organizations recommend the use of sustainable materials and processes to achieve purification of polluted waters. On a large scale, the selection of technologies for use for water treatment relies on several factors and contemporary conditions. Usually, wastewater is passed through three treatment stages (primary, secondary, and tertiary) using adequate technologies, such as physical, chemical, biological, electrochemical, and membrane filtration. To further boost the treatment in terms of efficiency, cost, and sustainability, the scientific community is investigating alternative emerging technologies to replace existing technologies. Photocatalytic strategies are one of the most studied technologies as they provide several advantages and fit well with aspects of the circular economy, as solar light can be exploited for photocatalytic water remediation.[8,9] This approach focuses on the excitation of photocatalytic semiconductors to produce e^-/h^+ charges, followed by generation of reactive oxygen species (ROS) and photoproduced free electrons, capable of oxidizing/reducing pollutant species in water. ROS are non-selective towards most organic pollutants, whereas generated electrons can reduce heavy metals to lower oxidative states, thus reducing toxicity.[10,11] Photocatalytic technology has been studied from different viewpoints, such as mechanisms, photoreactors, and materials design.[12–14] Recently, photocatalytic technology has received a lot of criticism from both scientific and industrial communities because it has failed to be transferred to the large scale. In fact, several factors limit its transfer as an alternative to existing technologies.[15] Djellabi *et al.* have critically discussed the pros and cons of photocatalytic materials, and also the possible opportunities to use this technology at large scale (Figure 5.1a).[16] In short, photocatalysis shows several advantages as well as unique applications, such as solar light exploitation, low production of secondary waste, and wastewater valorisation into high-value products and fuels. On the other hand, the technology suffers from some drawbacks that must be overcome (*e.g.*, slow kinetics, production of toxic by-products, and nanoparticle (NP) recovery). To narrow the gap between laboratory research and real application, several recommendations can be listed, including the design of materials having upper fast photocatalytic kinetics, solving the problem of NP recovery, and combination of photocatalysis with other technologies for the purpose of synergism and to solve technological issues. Loed *et al.* discussed the reasons for the gap between laboratory research and real-world application for photocatalysis,

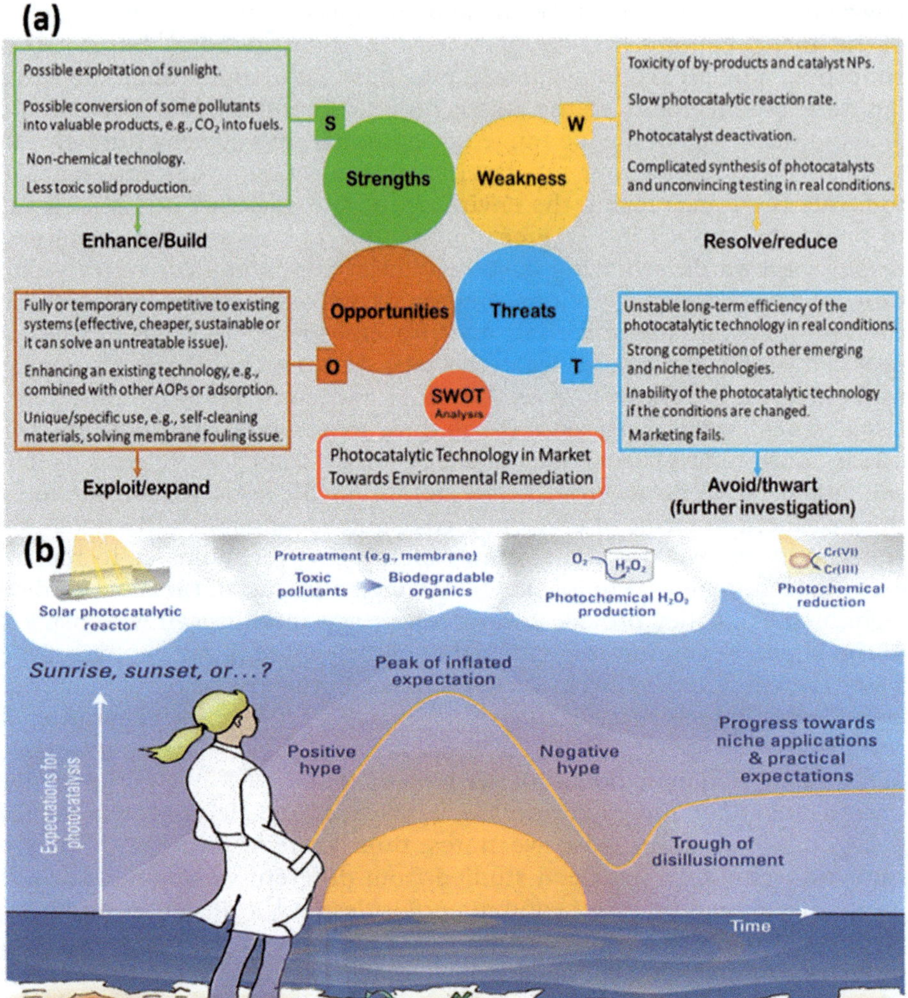

Figure 5.1 (a) SWOT (strengths, weaknesses, opportunities, threats) analysis of photocatalytic materials. Reproduced from ref. 16 with permission from Elsevier, Copyright 2022. (b) Scheme of the scientific community future vision to transfer photocatalytic water treatment to real application. Reproduced from ref. 17 with permission from American Chemical Society, Copyright 2019.

concluding that it requires further investigation under quasi-real conditions using pilot systems and design of niche technologies to understand the process from the technological viewpoint, and also involvement of this technology in large-scale systems (Figure 5.1b).[17]

With regard to the application of photocatalysis, photocatalytic materials in suspension exhibit the highest photocatalytic activity due to excellent mass transfer, thus allowing better pollutant adsorption and radical oxidation. However, suspension recovery is very difficult following treatment, and also NPs cannot be totally recovered, thus raising toxicity in the treated water.[18] Therefore, the use of photocatalysts supported on objects, glass plates, cellulose papers, and metals has been suggested to avoid the issue of NP recovery and toxicity.[13,19–21] Although supported photocatalysts overcome the issue of NP filtration and toxicity, this approach shows very slow kinetics and the production of more by-products due to the weak mass transfer. Therefore, to resolve the balance between excellent mass transfer and easy separation of photocatalytic materials, the scientific community has suggested the use of magnetic materials that can be easily recovered after treatment through application of an external magnetic field.[21] This route is interesting from the technological point of view.

This chapter discusses the most recent advances and challenges regarding the application of magnetic materials in photocatalytic technology for water treatment. To maintain excellent activity and stability with material recycling, the synthesis of magnetic materials is very challenging. Indeed, the synthetic route has a direct effect on the performance of the magnetic materials. This chapter discusses the following aspects:

- the synthesis routes of different magnetic materials: nanozerovalent iron (nZVI), maghemite (γ-Fe_2O_3) and magnetite (Fe_3O_4);
- the catalytic activity of magnetic materials to catalyse activate Fenton agents and sodium borohydride ($NaBH_4$);
- the photocatalytic activity of raw magnetic materials;
- the photocatalytic performance and mechanistic pathways of core–shell magnetic heterostructure systems;
- the role of the use of insulator to boost the photocatalytic activity and long-term stability.

5.2 Synthesis of Magnetic Materials

The following sections discuss the synthetic routes of the magnetic materials that can be exploited as supports for photocatalytic materials.

5.2.1 Synthesis of Nanozerovalent Iron

Physical or chemical methods (grinding, phase separation, abrasion, lithography, nucleation from homogeneous solutions or gas, or annealing at high temperature) have been developed for preparing nZVI.[22,23]

Apart from these, chemical reduction is a widely used and simple strategy, resulting in a homogeneous product of high reactivity.[24,25] Alternatively, nZVI can also be obtained by: (1) goethite (α-FeOOH) or hematite (α-Fe_2O_3)

reduction by H_2 at high temperature,[26] or (2) iron pentacarbonyl ($Fe(CO)_5$) decomposition in organic solvents or argon.[27]

Generally speaking, the synthetic methods for nZVI preparation (Table 5.1) are classified in "top down" (*i.e.*, physical, chemical methods, resulting in a nanomaterial obtained from building blocks, *e.g.*, atom, molecule), or "bottom up" approaches aimed at decomposing and reconstructing a material at the nano-scale.[23]

Among these possibilities, precision milling relies on iron milling, being carried out in a high-speed rotating chamber and produces nZVIs of well-defined morphology,[29] whereas chemical reduction with $NaBH_4$ leads to irregular-shaped nZVI particles described by a pronounced tendency to aggregate.[25,29]

Carbothermal reduction of iron compounds is the reduction of iron oxide nanoparticles,[33] or hydrous Fe^{2+} in salt form at high temperatures,

Table 5.1 Methods classification for nZVI preparation. Adapted from ref. 28 with permission from Elsevier, Copyright 2016.[a]

Method	Description	Pros and cons	Diameter (nm)	Surface area ($m^2\ g^{-1}$)	Reference
Top down					
Lithography, grinding	Decomposition of bulk iron materials	Cheap, but limited morphology control	n.r.	n.r.	22
Precision milling	Steel beads in a rotating chamber	No toxic reagents, low energy consumption	10–50	39	29
Bottom up					
Chemical reduction	Iron salts are reduced	Simple, but toxic reactants used	1–100	34	25
Carbothermal reduction	Fe^{2+} reduced to nZVI at high temperatures in the presence of gaseous reducing agents	Spherical particles obtained	20–150	130	30
Ultrasound method	Use of ultrasound reducing agent	Nanoparticles are formed, but toxic reactants used	10	34–42	24
Electrochemical method	Iron salt reduction with electrodes	Cheap, but clusters formed	1–20	25	31
Green synthesis	Nanoparticle biosynthesis by plant extracts	Sustainable, but irregular shape	20–120	6	32

[a]n.r.: not reported.

in the presence of H_2, CO_2, or CO, coming from thermal decomposition of carbon-based materials.[30] As a result, Fe_0 is obtained through the endothermic reaction at temperatures higher than 500 °C,[30] according to the following cheap routes:

$$Fe(C_2H_3O_2)_2 + C \rightarrow Fe^0 + 2CH_2CO + CO + H_2O \tag{5.1}$$

$$Fe_3O_4 + 2C \rightarrow 3Fe^0 + 2CO_2 \tag{5.2}$$

Alternatively, nZVI particles can also be obtained from the petroleum industry.[34]

Ultrasound can also be exploited for obtaining nZVI particles, but with reduced particle size (*ca.* 10 nm), as first reported by Tao *et al.*[35] Here, $NaBH_4$ is used[24] and, at the same time, the oxidation reaction of nZVI particles takes place,[24] decreasing the process efficiency, as reported:

$$4Fe^{2+}_{aq} + BH^{-}_{4aq} + 3H_2O \rightarrow 3Fe^0 + H_3BO_{3aq} + 7H^{+}_{aq} \tag{5.3}$$

$$Fe^0 + 2H^+ \rightarrow 2Fe^{2+} + H_2 \uparrow \tag{5.4}$$

To avoid this, addition of ammonium hydroxide to the $NaBH_4$ solution can be considered. This method is quite versatile because it strongly depends on the ultrasound frequency: indeed, nZVI particles with different morphology (spherical, plate, needle shapes) can be obtained.[36]

Electrochemical methods exploit the principles of electrolysis,[31] so are simple and cheap.[37] In this case, the produced iron atoms that are deposited at the cathode tend to aggregate as clusters, so stabilizing agents (*e.g.*, cationic surfactants)[29] and ultrasonic waves of *ca.* 20 kHz are used.

Green synthesis is an easily scalable biosynthesis where plant extracts responsible for reduction of metal compounds are used under mild conditions.[38–40] In this case, the control of nZVI morphology depends strongly on the type of plant extract used; iron reduction to nZVI is not complete and so iron oxides and hydroxides can also be obtained during the process.

Hydrothermal synthesis with plant extracts, synthesis with plant biomass, or biotemplate-assisted synthesis are additional routes for the preparation of nZVI, but they are affected by easy destruction of the plants. To overcome this issue, a possible solution is the use of agrowastes[32] or extracts from various residues of fruits (*e.g.*, lemons, oranges, or vine pomace).[39,41–43]

5.2.2 Synthesis of Magnetite (Fe_3O_4)

Different methods are used for magnetite preparation.[44] Physical strategies are considered top-down technologies, consisting of the mechanical fractionation of metal into smaller pieces, whereas chemical and biological methods belong to the bottom-up-category and they exploit organic solvents and plants or microorganisms, respectively (Figure 5.2). The detailed description of these methods is reported below.

TOP DOWN APPROACH

Physical methods
- Ball milling
- Laser ablation
- Electron beam deposition
- Sputtering
- Aerosol spray pyrolysis

BOTTOM UP APPROACH

Chemical methods
- Co-precipitation
- Thermal decomposition
- Sol - gel
- Microemulsion
- Sonochemical, hydrothermal
- Microwave assisted
- Chemical reduction
- Electrochemical, solvothermal, polyol

Biological methods
- Bacteria - based
- Plant - based

Figure 5.2 Main routes for magnetite synthesis.

Physical methods are the most suitable and simple choice for use in large-scale applications, even if they are expensive, time consuming, and result in fairly dispersed particle size distributions.[44,45] Among these methods, the most common industrial route is ball milling,[46] which is recognized to be affected by powder contamination from steel balls, and mechanical and chemical effects, resulting in amorphization of the magnetite crystals, high energy requirements, and long milling times.[47–49] Alternatively, electron beam deposition, consisting of electron beam emission towards a pure iron material, can be used.[47,50] Laser ablation is when the solid material is put under a thin layer and irradiated by laser,[49,51] allowing selective control of phase composition and morphology.[52] Sputtering is another approach, consisting of atom ejection from the bulk material surface through particle bombardment,[51] which leads to the formation of nanoparticles described by a composition equal to that of the targeted material.[49] However, the morphology and optical properties of the resulting nanoparticles are strongly influenced by the sputtering gas used.[53,54] Finally, another physical method is aerosol spray pyrolysis, which is advantageous for cost and easy scalability, and which gives spherical nanoparticles.[49,55]

Chemical methods consist of nanoparticle formation from iron condensates under specific experimental conditions. There are different strategies that belong to this category.[47,56] First, co-precipitation is the simultaneous precipitation of Fe(II) and Fe(III) salts in aqueous medium at basic pH.[48,57] It is a simple method, ensuring high yields, simple scale-up, and reduced time,[58] and in which a strong base is necessary for the synthesis and optimizing the ratio between the two iron salts is the key to properly tune the properties of the final products.[47,59–61]

Thermal decomposition is a strategy aimed at synthesizing crystalline monodisperse magnetite nanoparticles,[47] which consists of the decomposition of iron precursors (*e.g.*, iron acetylacetonate, oleate, *etc.*) dissolved in organic solvents of high melting point at high temperature when a surfactant is present.[48,62] The resulting magnetite particles are highly crystalline and uniform, and nano-magnetite as plates, octahedra, tetrahedra, multi-branched crystals can be obtained by tuning the operating synthetic conditions (*e.g.*, temperature, reactants ratio, *etc.*).[63,64] Its main drawbacks are the energy and time required for production, and the use of high cost and dangerous substances.[57,58] Another common method is the sol–gel route, generally involving the formation of a stable colloidal solution as a result of hydrolysis.[48] It represents a valid cheap alternative to chemical processes.[65] Magnetite nanoparticles with reduced size can be easily obtained through the microemulsion method,[56] which necessitates use of a macroscopically homogeneous solution (polar, non-polar phases, and surfactant), in which the reactant exchange occurs during the droplet collision in the microemulsion.[51,66] In this way, magnetite nanoparticles with controlled size and superparamagnetic properties can be obtained.[56] The sonochemical process is another possibility to successfully synthesize magnetite nanoparticles,[48,67] in which ultrasound is used to produce cavitation bubbles, thus giving rise to nanoparticles.[47] However, this strategy produces amorphous materials, so a further sintering step is necessary to obtain the crystalline features.[62] The hydrothermal process, the oldest method to obtain uniform magnetite nanoparticles,[62] involves dissolution of iron precursors in an aqueous environment at high temperature and pressure and is carried out in autoclaves. It ensures highly crystalline nanorods or nanospheres starting from simple operations and cheap raw materials,[48,62,68] but with low reproducibility and expensive equipment.[49,69] Microwave-assisted synthesis is an efficient, simple, and highly selective way to prepare magnetite nanoparticles in short reaction times, by exploiting an alternative thermal energy source, but requires expensive equipment.[57,61,65,70]

Chemical reduction is the easiest way to prepare magnetite nanoparticles: the metal precursor is dissolved in the solvent in the presence of a reducing agent and a surfactant in a stirred batch reactor.[45,49] Besides its simplicity, the main drawback to this method is the use of toxic and expensive reducing agents.[49] Alternative methods that can be used are electrochemical and solvothermal ones. The former is based on the reduction and/or oxidation of an iron-based electrode immersed in an electrolyte,[62,71] whereas the latter is carried out in a pressurized vessel, allowing the solvent to be heated above its boiling temperature,[49] thus forming high-quality crystallized nanoparticles.[49,62,72,73] Finally, as an even more sustainable chemistry, biological methods are interesting alternatives to physical and chemical routes.[74] Here, metal precursors are used to form nanoparticles in green substrates, overcoming high costs, the use of toxic chemicals, and the formation of hazardous by-products.[47,75] Depending on the biological substrate used, two approaches can be identified: bacteria- and plant-based syntheses.[44] The former consists of the production of single domains of magnetite nanoparticles under anaerobic conditions using iron-reducing bacteria.[47] Thanks to this approach the rapid multiplication and easy manipulation and cultivation of bacteria are exploited.[49] In addition, bacteria fermentation can be easily scaled up, is reproducible, and results in high-value products with lower defects than those obtained by conventional technologies.[76] However, there are some drawbacks (*e.g.*, small magnetosome yield not sufficient for commercialization) that should be noted.[77]

5.2.3 Synthesis of Maghemite (γ-Fe_2O_3)

Several methods have been reported over the years for the preparation of maghemite nanoparticles, including co-precipitation, sol–gel, microemulsion, flow injection synthesis, flame spray pyrolysis, organic precursor decomposition at high temperature, and oxidation of magnetite nanoparticles.[78] Maghemite is a stable ferric oxide phase at room temperature, and can be transformed into the stable α-phase at temperatures over 450 °C. It can be prepared at temperatures below 450 °C *via* magnetite oxidation or ferric hydroxide dehydration. According to this route, crystalline particles, in which Fe^{3+} cations and vacancies are heterogeneously distributed, are obtained.[79,80]

Generally, the synthetic strategy strongly influences the purity and morphology of the resulting maghemite. Indeed, although the co-precipitation route is the simplest and cheapest way to produce maghemite, several researchers have worked towards finding alternatives. Woo *et al.* developed a novel method combining the sol–gel reaction in reverse micelles with reflux crystallization for developing maghemite as nanorods.[81] Belin *et al.* prepared ultrafine particles of maghemite by iron chloride precursors,[82] employing different ratios of iron/oleic acid to properly tune the particle size.[83] Dispersed maghemite nanoparticles of *ca.* 10 nm size can be easily prepared in only 30 min reaction at a temperature of 130 °C *via* a fast and inexpensive chemical procedure using iron pentacarbonyl thermal decomposition in the presence of dimethyl sulfoxide and oleic acid as oxidant and surfactant, respectively.[84] Maghemite thin films are largely used in magnetic storage because of their interesting properties: in this context, numerous investigations have reported solutions to obtain maghemite thin films.[85–88]

Finally, from an industrial viewpoint, maghemite particles can be prepared by co-precipitation of ferric and ferrous ions in an alkali solution, followed by magnetite oxidation at 200–400 °C.[88]

5.3 Catalytic Activation of $NaBH_4$ Towards Organic Pollutant Reduction

The catalytic reduction of organic pollutants contributes to reduced pollution by converting toxic organic pollutants into non-toxic substances.[89–92] The advantage of catalytic reduction is the conversion of pollutants into valuable products. 4-Nitrophenol is an organic contaminant that can be converted into 4-aminophenol with the aid of a catalyst and reducing substance, which is then used in medicine production. On account of this, Luo *et al.* designed graphene/polydopamine (PDA)-Au NPs for conversion of 4-nitrophenol using $NaBH_4$.[93] This material showed good catalytic performance and incorporated the benefits of graphene and Au NPs. Similarly, Lu *et al.*[94] showed a synergistic impact between graphene oxide (GO) support and $NaBH_4$ to increase the catalytic activity when 4-nitrophenol (4-NP) was reduced by immobilized

Au NPs on the TWEEN-functionalized GO surface. According to this report, the large surface area of Au NPs led to greater selectivity and activity in the catalytic reduction. Because Au NPs have a high surface energy, they prefer to aggregate and this greatly reduces their catalytic effectiveness.[89,95] As a result, researchers are still looking into the need to substitute for nanoparticles with greater potential and lower costs (*e.g.*, metal, metal oxide, carbon nanotube, and polymer-based nanomaterials).[96] Recently, Zhu *et al.*[97] prepared Fe_3O_4@ polyaniline (PANI)@Au NPs by electrostatic self-assembly and seed-growing techniques. $NaBH_4$ was then used to test the catalytic effectiveness of the materials for the oxidation of dyes and 4-nitrophenol. The rate constant K_{app} for the catalytic conversion of 4-nitrophenol into 4-aminophenol was determined as 8.63×10^{-3} s^{-1} at room temperature with a high concentration of $NaBH_4$ as the model system, demonstrating exceptional catalytic stability and efficiency. The Fe_3O_4@PANI@Au exhibited substantially enhanced catalytic activity, recycling, and stability compared to the Au NPs.

The adsorption model applied during the reduction reaction has been reported as the main pathway for nitrophenol reduction. According to the literature, the catalytic reduction of nitrophenol occurs by the Langmuir–Hinshelwood (L–H) adsorption model in almost all investigations reported. At the beginning of the reaction the dye (4-NP or 2-NP) and the reductant ($NaBH_4$) are introduced into the reaction solution containing a nanocatalyst.[91] Both dye and reductant are deposited on the NP surface according to the L–H mechanism. The nitrophenol molecule will be deposited on the NP surface in its current form, but the $NaBH_4$ molecule will be changed into the ions Na^+ and BH_4^-. The BH_4^- discharges an electron and functions as an electron-donating species. This electron will move from the nanocatalyst surface to the pigment molecules *via* an electron-transfer surface, where it will be accepted and transformed into an aminophenol (Figure 5.3). The reduction using $NaBH_4$ can simultaneously produce H_2 gas, which can boost stirring in the medium, and thus improving the catalyst reduction ability. In fact, H_2 can remove oxides and impurities that could be adsorbed onto the catalyst surface, thus allowing *in situ* regeneration. H_2 production can be obtained by the following:

$$NaBH_4 + 2H_2O \rightarrow NaBO_2 + 4H_2 \quad (5.5)$$

Zhao *et al.*[98] used Fe_3O_4@PDA as support to produce Au/Fe_3O_4@PDA when *p*-nitrophenol was reduced with $NaBH_4$. They discovered that the catalyst had outstanding performance and excellent stability and reusability. Similarly, Zeng *et al.*[99] also examined the catalytic activity and recycling of Fe_3O_4@ PDA-Au for *o*-nitroaniline reduction to benzenediamine in the presence of $NaBH_4$. Adeel *et al.*[100] studied dye reduction by easily recoverable Fe_3O_4@ EDTA-Ag in the presence of $NaBH_4$, which can reduce dye within a few minutes and in addition the materials can be recycled up to ten times. Figure 5.3 summarizes the Fe_3O_4@EDTA-Ag synthetic process and the catalytic process for dye reduction.

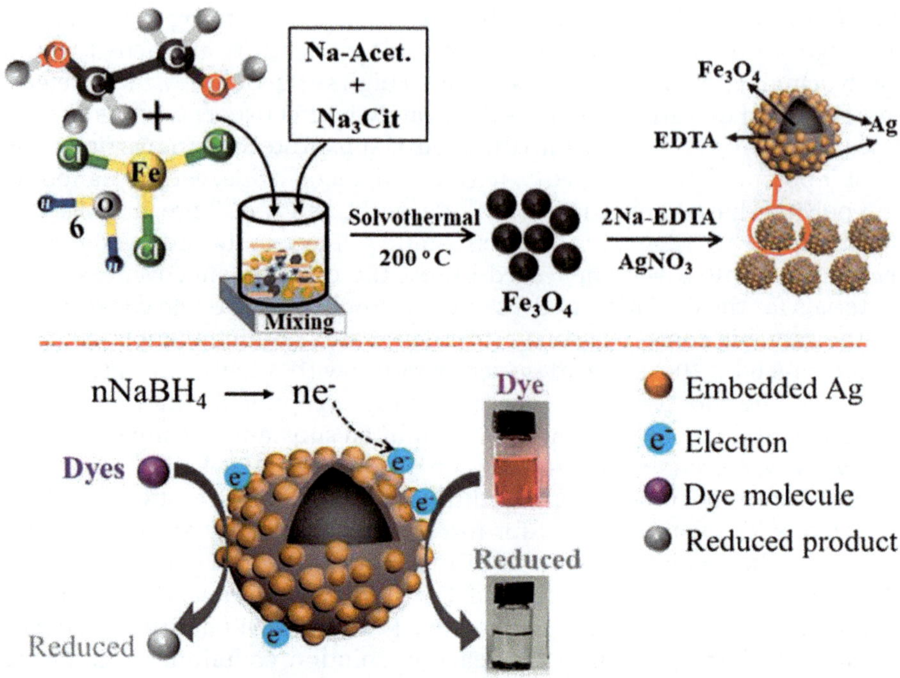

Figure 5.3 Synthesis of Fe_3O_4@EDTA-Ag and its catalytic mechanism in the presence of $NaBH_4$. Adapted from ref. 100 with permission from American Chemical Society, Copyright 2019.

5.4 Fenton and Photo-Fenton of Magnetic Materials

The traditional Fenton process relies on the activation of H_2O_2 by ferrous iron (Fe^{2+}) to produce hydroxyl radicals ($^{\bullet}OH$) as a non-selective, powerful oxidant (eqn (5.6)), which can attack most of organics until complete mineralization. However, the regeneration of Fe^{2+} from Fe^{3+} (eqns (5.7) and (5.8)) is considerably slower than the consumption of Fe^{2+} in eqn (5.6). Consequently, the process stops once either of the reactants in eqn (5.6) is consumed, which results in a significant amount of metal sludge being generated as a secondary pollutant.[101] Furthermore, the carboxylic acids produced from the degradation of organic pollutants can form stable metal–organic complexes, which inhibit the metal catalyst even in the heterogeneous Fenton-like process,[102] where a solid metal catalyst replaces the dissolved Fe^{2+}. Among numerous Fe-based Fenton-like catalysts, such as zerovalent iron (ZVI), iron oxides and oxyhydroxides (Fe_3O_4, γ-Fe_2O_3, α-Fe_2O_3 and α-FeOOH), iron disulfide, and iron oxychloride, magnetite (Fe_3O_4) has gained more popularity owing to its low cost and good optical and magnetic properties, which allows easy recovery of the catalyst simply using an external magnetic field.[103]

$$H_2O_2 + Fe^{2+} \rightarrow \cdot OH + HO^- + Fe^{3+} \qquad (5.6)$$

$$H_2O_2 + Fe^{3+} \rightarrow HO_2^{\bullet} + H^+ + Fe^{2+} \tag{5.7}$$

$$HO_2^{\bullet} + Fe^{3+} \rightarrow O_2 + H^+ + Fe^{2+} \tag{5.8}$$

Light irradiation of metal–organic complexes induces ligand-to-metal charge transfer reactions that can break the metal–organic bond, resulting in regeneration of the metal catalyst. Therefore, a combination of the Fenton process and photocatalysis (photo-Fenton) offers economic and performance advantages over either of the separate processes, owing to a higher concentration of powerful oxidants and the least amount of secondary metal pollution.[104] As magnetite alone suffers from low surface area and poor electron–hole separation, it is usually used as a composite material with other semiconductors for enhanced separation and excellent photo response. For instance, metal–organic frameworks (MOFs) as highly porous and photo-responsive materials are combined with magnetite for enhanced photo-Fenton performance and easy recovery of the catalyst at the process end.[105,106]

First, the Fe_3O_4 in the photocatalyst itself activates the added H_2O_2 to give $^{\bullet}OH$. In the presence of light irradiation, electron–hole pairs are generated *via* ligand–cluster charge transfer or direct excitation of Fe–O clusters. The photogenerated electrons are transferred to the surface of Fe_3O_4, where they can facilitate the regeneration of Fe^{3+} from Fe^{2+}, which enhances the H_2O_2 activation for $^{\bullet}OH$ formation for the degradation of organic contaminants (Figure 5.4). Additionally, the photogenerated electrons are consumed by direct activation of H_2O_2 to $^{\bullet}OH$. This high consumption of electrons in a combined photo-Fenton process can suppress the rapid recombination of electron–hole pairs, which is a serious drawback in a single photocatalytic process. Meanwhile, this generates high amounts of positively charged holes as additional oxidants, which can contribute to the degradation of target compounds. These catalysts show efficient recovery and reuse over several

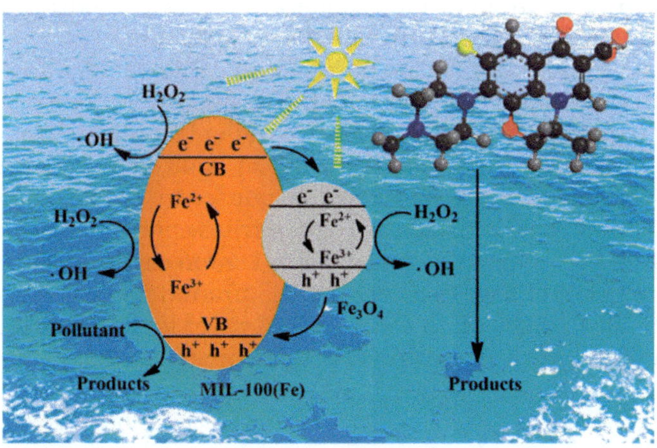

Figure 5.4 Mechanism of the photo-Fenton process. Reproduced from ref. 103 with permission from Elsevier, Copyright 2021.

cycles with enhanced degradation of pollutants in wastewater. Thus, integrating photocatalysis and the Fenton process offers a promising opportunity for high-performance advanced oxidation and potential applications in refractory wastewater treatment.

5.5 Photocatalytic Oxidation of Pollutants

5.5.1 Photoactivity of Bare Magnetic Materials

In this section, we discuss the synthesis and application of single-phase magnetic materials in photocatalysis, with a focus on iron oxides, such as magnetite (Fe_3O_4), maghemite (γ-Fe_2O_3 and β-Fe_2O_3), and hematite (α-Fe_2O_3). These have garnered attention due to their unique combination of photoactivity and magnetic properties, offering the potential for efficient catalyst recovery and reuse. However, single-phase photocatalysts often exhibit low visible-light response, resulting in poor catalytic efficiency. Many studies have reported that binary and ternary ferrites can be used to extend the visible-light absorption.

A study investigating the photocatalytic activity of nanometric Fe_3O_4 in the degradation of polyethylene terephthalate (PET) materials under UV-A irradiation found that Fe_3O_4 nanoparticles exhibited photoactivity and promoted PET degradation through photolysis and oxidation by radicals in the treatment medium.[107] The nanoparticle magnetic properties facilitated easy separation and recovery from the treatment medium, making them suitable for wastewater treatment. However, when Fe_3O_4 nanoparticles were coated with a silica layer (Fe_3O_4@SiO_2), the catalytic performance was significantly reduced, probably due to insufficient porosity in the silica coating.[107] Another study reported the synthesis of monodisperse hollow Fe_2O_3 nano-ovals (NMHFNs) and found that they demonstrated superior photocatalytic performance for salicylic acid oxidation compared to solid Fe_2O_3 nano-ovals.[108] The composite showed good efficiency due to the high porosity and hollow structure.

Binary and ternary ferrites are designed to enhance light absorption and photoactivity. Ferrites, in the form of MFe_2O_4 (M: Mn, Ba, or Ni), exhibit high saturation magnetization, making them excellent candidates for use in magnetic fields. On top of this, the inclusion of additional metal narrows the bandgap, enhancing the visible-light response. In photocatalytic systems featuring a potent oxidant, such as H_2O_2, ferrites contribute to a unique type of photocatalytic system known as a photo-Fenton catalyst. A good magnetic photocatalyst should have excellent high saturation magnetization (Ms) and low remanence to allow rapid dispersion in water for catalytic reactions and fast magnetic recovery. These properties can be achieved by designing superparamagnetic materials composed of small, single-domain particles without magnetic memory (remanence).[109] Many studies have been devoted to developing new routes to such magnetic materials, emphasizing the advantages of separation and dispersion properties exhibited by superparamagnetic materials.

The control of MFe_2O_4 size and use of stabilizing agents (*e.g.*, citric acid and ethylene glycol) can lead to narrow bandgaps for effective photocatalytic performance.[110] Bismuth ferrites (*e.g.*, $BiFeO_3$,[110–115] Bi_25FeO_40,[111] and Bi_25FeO_40–$BiFeO_3$[111]) have garnered significant attention as magnetic photocatalysts due to their defects. Furthermore, $CoFe_2O_4$ has been extensively studied as a visible-light-driven photocatalyst due to its narrow bandgap (1.6–2.2 eV) and robust magnetic properties. Several methods have been employed to synthesize $CoFe_2O_4$ nanoparticles, such as hydrothermal, sol-gel, and co-precipitation techniques. They allow control of particle size, morphology, and magnetic properties.[111] Ferrites, particularly nanoferrites, have shown great potential for photocatalytic applications under visible-light irradiation. Their unique magnetic properties allow for efficient separation and dispersion in the liquid phase, whereas the inclusion of additional metals narrows the bandgap, thus enhancing their photocatalytic activity. Several synthesis methods (*e.g.*, sol–gel, hydrothermal, and co-precipitation techniques) have been employed to fabricate these ferrites, with control over particle size, morphology, and magnetic properties.

In conclusion, ferrites and their nanostructured counterparts have shown great promise in various applications, including magnetic data storage, microwave devices, and photocatalysis. The unique magnetic properties of ferrites enable efficient separation and dispersion in the liquid phase, which, when combined with other metals, results in narrowed bandgaps and enhanced photocatalytic activity. A range of synthetic methods has been employed to fabricate ferrites with controlled particle size, morphology, and magnetic properties.

5.5.2 Heterojunction-based Magnetic Photocatalysts

The development of photocatalytic heterojunctions is regarded as one of the hottest research topics in photocatalytic technology, in which different semiconductors combine between two sections with varying band structures, resulting in an interfacial band arrangement.[116] Because of a physical division of pairs of photogenerated electrons and holes, correctly designed heterojunction photocatalysts can have more significant photocatalytic activity.[117] As a result, $CoFe_2O_4$ and other metal oxide-based heterostructure photocatalysts require a mixture of $CoFe_2O_4$ and other metal oxides that have higher photocatalytic efficiency than bare $CoFe_2O_4$.[118] It has been found that forming a suitable heterojunction between $CoFe_2O_4$ and the metal oxide is a superior strategy for reducing electron–hole pair recombination, by improving the charge separation rate and achieving enhanced photocatalytic efficiency. Liu *et al.*[119] designed TiO_2/polythiophene/γ-Fe_2O_3 (TiO_2/PTh/γ-Fe_2O_3) *via* polymerization to remove As(III) from water. Indeed, the removal rate under visible light was more than twice that in the dark, indicating the photocatalytic oxidation of As(III) to As(V). In the TiO_2/PTh/γ-Fe_2O_3 system, it was reported that photogenerated holes (h^+) and superoxide free radical ($^{\bullet}O_2^-$) species were responsible for As(III) oxidation. Veronica *et al.*[120] reported the

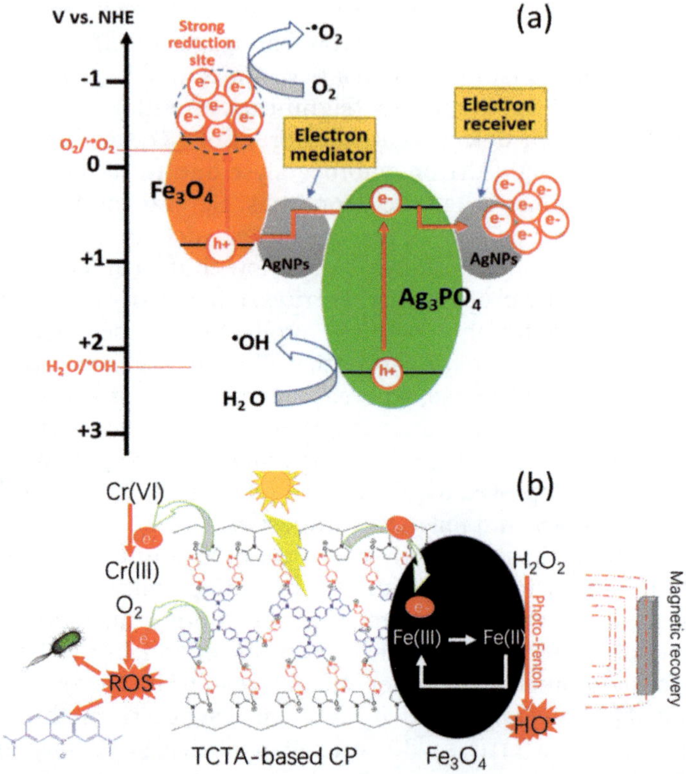

Figure 5.5 (a) Synthesis of Ag_3PO_4/Fe_3O_4 for the photocatalytic oxidation of pollutants and Cr(VI) reduction. Reproduced from ref. 120 with permission from Elsevier, Copyright 2022. (b) Magnetic photoactive polymer for dye oxidation, bacteria inactivation, and Cr(VI) reduction. Reproduced from ref. 121 with permission from Elsevier, Copyright 2020.

design of magnetic Ag_3PO_4/Fe_3O_4 for the oxidation of diclofenac, acid green 25, and Cr(VI) reduction. They reported that Ag NPs can be produced during the heterojunction composite synthesis and detected by X-ray diffraction. It was claimed that Ag NPs play the role of electron mediator between Ag_3PO_4 and Fe_3O_4, boosting the photoredox abilities (Figure 5.5a). Djellabi *et al.*[121] reported that in photoactive polymer-coated Fe_3O_4, the composite can also catalyse the Fenton reaction on top of the photocatalytic process, providing a synergistic system for water treatment and disinfection (Figure 5.5b).[121]

5.5.3 Role of Insulator

The magnetic photocatalysts consist of two parts, *i.e.*, a magnetic core with exterior magnetic force separation function, and a semiconductor shell with a photocatalytic role under light irradiation. Some of the biggest issues in magnetic materials are the low stability and the charge recombination.

The continuous transfer of photogenerated electrons from the semiconductor shell to the magnetic core leads to destruction of iron oxides. On top of this, in many magnetic heterojunction systems, lower activity is due to the role of iron oxides serving as recombination sites for e^-/h^+ couples.[122,123] Using an insulating intermediate coating to prevent contact between these materials is widely documented in the literature as being able overcome the disadvantages of poor activity and photo-dissolution.[124–126] SiO_2 is regarded as one of the best insulators because of its chemical inertia, limiting direct electrical interaction between both phases.[127,128] Figure 5.6 simplifies the role of SiO_2 in $Fe_3O_4/SiO_2/mTiO_2$.[129] It is important to point out that a core–shell–shell structure using silica as insulator leads to a decrease in the magnetic response of the composite.[130] The layer of silica as an insulator is a very important factor, needing

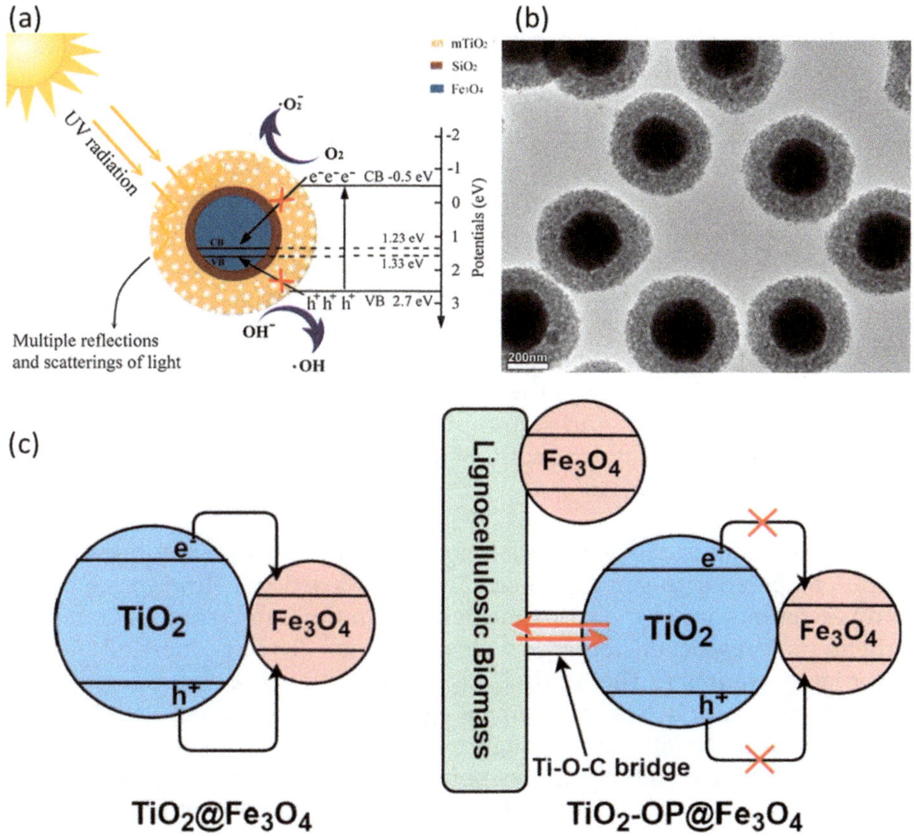

Figure 5.6 (a) Scheme showing the role of SiO_2 and (b) transmission electron microscopy image of $Fe_3O_4/SiO_2/mTiO_2$. Reproduced from ref. 129 with permission from John Wiley & Sons, Copyright © 2017 WILEY-VCH Verlag GmbH & Co. KGaA, Weinheim. (c) Role of biomass as insulator in TiO_2/Fe_3O_4. Reproduced from ref. 132 with permission from Elsevier, Copyright 2019.

optimization to provide a balance between the magnetic response and the photocatalytic activity.[131] Other compounds can also be used as insulators, *e.g.*, PANI, poly(methyl methacrylate) and poly(sodium 4-styrene sulfonate). The best insulator is expected to prevent the photo-dissolution of magnetic cores while improving the photocatalytic activity and nanodevice stability. Djellabi *et al.*[132] reported that biomass used as a support of Fe_3O_4/SiO_2 might also play a role as insulator to reduce charge recombination (Figure 5.6c).

5.6 Conclusions

The use of magnetic materials for water treatment has proved to be an excellent approach as it combines high mass transfer and easy recovery. The following conclusions can be drawn.

- The route of synthesis of raw magnetic materials is the most important factor to obtain excellent magnetic features, *i.e.*, high surface area, stability, size and shape, catalytic and photocatalytic abilities.
- Magnetic materials are multifunctional because they combine good adsorptive ability, catalytic performance (activation of Fenton agents and $NaBH_4$), visible-light absorption, and photocatalytic activity.
- The main issue for core–shell heterojunction magnetic-based photocatalysts is the dissolution of magnetic materials. In fact, direct contact between magnetic materials and semiconductor leads to destruction of the magnetic shell by photoproduced redox species. In addition, many reports have shown that the effect on efficiency is significant because the magnetic core plays the recombination center role. Therefore, use of an insulator is highly recommended to obtain highly efficient, stable magnetic photocatalytic composites. SiO_2 is the most used insulator to design such materials.

Magnetic photocatalytic materials could be a unique option for water treatment compared with single photocatalytic nanoparticles. However, some recommendations are needed for consideration in the future studies. Stability is still an issue, especially for large-scale application. The thickness of the insulator layer requires further optimization to find a trade-off between stability and efficiency. Studies on designing single magnetic materials with high stability and efficiency at wide pH range are still needed. The analysis of magnetic materials from a technological viewpoint is recommended to evaluate the system in a proper way. The application of magnetic materials at pilot and large scale to treat real wastewaters should be taken into consideration.

References

1. M. Yadav, R. Gupta and R. K. Sharma, *Advances in Water Purification Techniques*, 2019, p. 355.

2. G. Crini and E. Lichtfouse, *Environ. Chem. Lett.*, 2019, **17**, 145.
3. A. Turner, *Environ. Int.*, 2018, **117**, 308.
4. R. A. Sheldon and M. Norton, *Green Chem.*, 2020, **22**, 6310.
5. H. Corvellec, A. F. Stowell and N. Johansson, *J. Ind. Ecol.*, 2022, **26**, 421.
6. M. Geissdoerfer, P. Savaget, N. M. Bocken and E. J. Hultink, *J. Cleaner Prod.*, 2017, **143**, 757–768.
7. R. Djellabi, D. Aboagye, M. G. Galloni, V. V. Andhalkar, S. Nouacer, W. Nabgan, S. Rtimi, M. Constantí, F. M. Cabello and S. Contreras, *Bioresour. Technol.*, 2022, **368**, 128333.
8. R. Djellabi, M. Fouzi Ghorab, A. Smara, C. L. Bianchi, G. Cerrato, X. Zhao and B. Yang, *Green Materials for Wastewater Treatment*, 2020, vol. 38, p. 291.
9. R. Djellabi, C. L. Bianchi, M. R. Haider, J. Ali, E. Falletta, M. F. Ordonez, A. Bruni, M. Sartirana and R. Geioushy, *Nanomaterials for Water Treatment and Remediation*, CRC Press, 1st edn, 2021, p. 217.
10. L. Xiong and J. Tang, *Adv. Energy Mater.*, 2021, **11**, 2003216.
11. R. Djellabi, P. Su, E. A. Elimian, V. Poliukhova, S. Nouacer, I. A. Abdelhafeez, N. Abderrahim, D. Aboagye, V. V. Andhalkar and W. Nabgan, *J. Water Process Eng.*, 2022, **50**, 103301.
12. B. A. Marinho, R. O. Cristóvão, R. Djellabi, J. M. Loureiro, R. A. Boaventura and V. J. Vilar, *Appl. Catal., B*, 2017, **203**, 18.
13. B. A. Marinho, R. Djellabi, R. O. Cristóvão, J. M. Loureiro, R. A. Boaventura, M. M. Dias, J. C. B. Lopes and V. J. Vilar, *Chem. Eng. J.*, 2017, **318**, 76.
14. R. Djellabi, M. F. Ordonez, F. Conte, E. Falletta, C. L. Bianchi and I. Rossetti, *J. Hazard. Mater.*, 2022, **421**, 126792.
15. M. G. Alalm, R. Djellabi, D. Meroni, C. Pirola, C. L. Bianchi and D. C. Boffito, *Catalysts*, 2021, **11**, 562.
16. R. Djellabi, R. Giannantonio, E. Falletta and C. L. Bianchi, *Curr. Opin. Chem. Eng.*, 2021, **33**, 100696.
17. S. K. Loeb, P. J. Alvarez, J. A. Brame, E. L. Cates, W. Choi, J. Crittenden, D. D. Dionysiou, Q. Li, G. Li-Puma and X. Quan, *Environ. Sci. Technol.*, 2019, **53**, 2937.
18. E. Friehs, Y. AlSalka, R. Jonczyk, A. Lavrentieva, A. Jochums, J.-G. Walter, F. Stahl, T. Scheper and D. Bahnemann, *J. Photochem. Photobiol., C*, 2016, **29**, 1.
19. R. Djellabi and M. Ghorab, *Desalin. Water Treat.*, 2015, **55**, 1900.
20. R. Djellabi and M. Ghorab, *Desalin. Water Treat.*, 2015, **53**, 3649–3655.
21. M. G. Galloni, E. Ferrara, E. Falletta and C. L. Bianchi, *Catalysts*, 2022, **12**, 923.
22. G. Shan, S. Yan, R. Tyagi, R. Y. Surampalli and T. C. Zhang, Applications of nanomaterials in environmental science and engineering, *Pract. Period. Hazard., Toxic, Radioact. Waste Manage.*, 2009, **13**, 110.
23. F. Fu, D. D. Dionysiou and H. Liu, *J. Hazard. Mater.*, 2014, **267**, 194.
24. M. R. Jamei, M. R. Khosravi and B. Anvaripour, *Asia-Pac. J. Chem. Eng.*, 2013, **8**, 767.
25. C.-B. Wang and W.-x. Zhang, *Environ. Sci. Technol.*, 1997, **31**, 2154.

26. J. T. Nurmi, P. G. Tratnyek, V. Sarathy, D. R. Baer, J. E. Amonette, K. Pecher, C. Wang, J. C. Linehan, D. W. Matson and R. L. Penn, *Environ. Sci. Technol.*, 2005, **39**, 1221.
27. Y.-P. Sun, X.-Q. Li, W.-X. Zhang and H. P. Wang, *Colloids Surf., A*, 2007, **308**, 60.
28. M. Stefaniuk, P. Oleszczuk and Y. S. Ok, *Chem. Eng. J.*, 2016, **287**, 618.
29. S. Li, W. Yan and W.-x. Zhang, *Green Chem.*, 2009, **11**, 1618.
30. L. B. Hoch, E. J. Mack, B. W. Hydutsky, J. M. Hershman, J. M. Skluzacek and T. E. Mallouk, *Environ. Sci. Technol.*, 2008, **42**, 2600.
31. S.-S. Chen, H.-D. Hsu and C.-W. Li, *J. Nanopart. Res.*, 2004, **6**, 639.
32. T. Wang, X. Jin, Z. Chen, M. Megharaj and R. Naidu, *Sci. Total Environ.*, 2014, **466**, 210.
33. M. Bystrzejewski, *J. Solid State Chem.*, 2011, **184**, 1492.
34. R. A. Crane and T. B. Scott, *J. Hazard. Mater.*, 2012, **211**, 112.
35. N. Tao, M. Sui, J. Lu and K. Lua, *Nanostruct. Mater.*, 1999, **11**, 433.
36. M. R. Jamei, M. R. Khosravi and B. Anvaripour, *Ultrason. Sonochem.*, 2014, **21**, 226.
37. B.-Y. Yoo, S. Hernandez, B. Koo, Y. Rheem and N. Myung, *Water Sci. Technol.*, 2007, **55**, 149.
38. S. Machado, S. Pinto, J. Grosso, H. Nouws, J. T. Albergaria and C. Delerue-Matos, *Sci. Total Environ.*, 2013, **445**, 1.
39. S. Machado, J. Grosso, H. Nouws, J. T. Albergaria and C. Delerue-Matos, *Sci. Total Environ.*, 2014, **496**, 233.
40. G. E. Hoag, J. B. Collins, J. L. Holcomb, J. R. Hoag, M. N. Nadagouda and R. S. Varma, *J. Mater. Chem.*, 2009, **19**, 8671.
41. S. Machado, W. Stawiński, P. Slonina, A. Pinto, J. Grosso, H. Nouws, J. T. Albergaria and C. Delerue-Matos, *Sci. Total Environ.*, 2013, **461**, 323.
42. M. Herlekar, S. Barve and R. Kumar, *J. Nanopart.*, 2014, **2014**, 1.
43. L. Huang, X. Weng, Z. Chen, M. Megharaj and R. Naidu, *Spectrochim. Acta, Part A*, 2014, **130**, 295.
44. A. D. Mihai, C. Chircov, A. M. Grumezescu and A. M. Holban, *Int. J. Mol. Sci.*, 2020, **21**, 7355.
45. E. Sánchez-López, D. Gomes, G. Esteruelas, L. Bonilla, A. L. Lopez-Machado, R. Galindo, A. Cano, M. Espina, M. Ettcheto and A. Camins, *Nanomaterials*, 2020, **10**, 292.
46. P. S. R. Kumar and S. J. Alexis, Synthesized carbon nanotubes and their applications, *Carbon-Based Nanofillers and Their Rubber Nanocomposites*, Elsevier, 2019, p. 109.
47. G. A. Marcelo, C. Lodeiro, J. L. Capelo, J. Lorenzo and E. Oliveira, *Mater. Sci. Eng., Proc. Conf.*, 2020, **106**, 110104.
48. S. Liu, B. Yu, S. Wang, Y. Shen and H. Cong, *Adv. Colloid Interface Sci.*, 2020, **281**, 102165.
49. P. G. Jamkhande, N. W. Ghule, A. H. Bamer and M. G. Kalaskar, *J. Drug Delivery Sci. Technol.*, 2019, **53**, 101174.
50. X. Fu, J. Cai, X. Zhang, W.-D. Li, H. Ge and Y. Hu, *Adv. Drug Delivery Rev.*, 2018, **132**, 169.

51. A. V. Rane, K. Kanny, V. Abitha and S. Thomas, *Synthesis of Inorganic Nanomaterials*, Elsevier, 2018, vol. 121.
52. X. Luo, A. H. Al-Antaki, T. M. Alharbi, W. D. Hutchison, Y.-c. Zou, J. Zou, A. Sheehan, W. Zhang and C. L. Raston, *ACS Omega*, 2018, **3**, 11172.
53. R. Chandra, A. Chawla and P. Ayyub, *J. Nanosci. Nanotechnol.*, 2006, **6**, 1119.
54. Q. Chen, H. Wang, S. Perero, Q. Wang and Q. Chen, *J. Non-Cryst. Solids*, 2015, **408**, 43.
55. P. Nie, G. Xu, J. Jiang, H. Dou, Y. Wu, Y. Zhang, J. Wang, M. Shi, R. Fu and X. Zhang, *Small Methods*, 2018, **2**, 1700272.
56. T. Gu, Y. Zhang, S. A. Khan and T. A. Hatton, *Colloid Interface Sci. Commun.*, 2019, **28**, 1.
57. M. James, R. A. Revia, Z. Stephen and M. Zhang, *Nanomaterials*, 2020, **10**, 2113.
58. C. Pereira, A. M. Pereira, C. Fernandes, M. Rocha, R. Mendes, M. P. Fernández-García, A. Guedes, P. B. Tavares, J.-M. Grenèche and J.o.P. Araújo, *Chem. Mater.*, 2012, **24**, 1496.
59. A. Sirivat and N. Paradee, *Mater. Des.*, 2019, **181**, 107942.
60. M. Salvador Fernández, G. Gutiérrez Cervelló, S. Noriega, A. Moyano Artime, M.d.C. Blanco López and M. Matos González, *Int. J. Mol. Sci.*, 2021, **22**, 427.
61. H. Hu, H. Yang, P. Huang, D. Cui, Y. Peng, J. Zhang, F. Lu, J. Lian and D. Shi, *Chem. Commun.*, 2010, **46**, 3866.
62. J. Wallyn, N. Anton and T. F. Vandamme, *Pharmaceutics*, 2019, **11**, 601.
63. W. Xie, Z. Guo, F. Gao, Q. Gao, D. Wang, B.-s. Liaw, Q. Cai, X. Sun, X. Wang and L. Zhao, *Theranostics*, 2018, **8**, 3284.
64. N. J. Orsini, B. Babić-Stojić, V. Spasojević, M. P. Calatayud, N. Cvjetićanin and G. F. Goya, *J. Magn. Magn. Mater.*, 2018, **449**, 286.
65. N. Basavegowda and K.-H. Baek, *Molecules*, 2021, **26**, 912.
66. Z. Cai, Q. Yao, X. Chen and X. Wang, *Environmental and Energy Applications*, Elsevier, 2019, p. 435.
67. A. U. Badnore, M. A. Salvi, N. L. Jadhav, A. B. Pandit and D. V. Pinjari, *Adv. Sci. Lett.*, 2018, **24**, 5681.
68. O. A. Attallah, E. Girgis and M. M. Abdel-Mottaleb, *J. Magn. Magn. Mater.*, 2016, **399**, 58.
69. E. M. Modan and A. G. Plăiașu, *The Annals of "Dunarea de Jos" University of Galati. Fascicle IX, Metallurgy and Materials Science*, 2020, vol. 43, p. 53.
70. S. Yaragalla, R. K. Mishra, S. Thomas, N. Kalarikkal and H. J. Maria, *Carbon-based Nanofillers and Their Rubber Nanocomposites: Fundamentals and Applications*, Elsevier, 2019.
71. L. Cabrera, S. Gutierrez, N. Menendez, M. Morales and P. Herrasti, *Electrochim. Acta*, 2008, **53**, 3436.
72. P. Tipsawat, U. Wongpratat, S. Phumying, N. Chanlek, K. Chokprasombat and S. Maensiri, *Appl. Surf. Sci.*, 2018, **446**, 287.

73. A. Bunge, A. S. Porav, G. Borodi, T. Radu, A. Pîrnău, C. Berghian-Grosan and R. Turcu, *J. Mater. Sci.*, 2019, **54**, 2853.
74. Y. C. López and M. Antuch, *Curr. Opin. Green Sustainable Chem.*, 2020, **24**, 32.
75. P. Singh, Y.-J. Kim, D. Zhang and D.-C. Yang, *Trends Biotechnol.*, 2016, **34**, 588.
76. J.-W. Moon, C. J. Rawn, A. J. Rondinone, L. J. Love, Y. Roh, S. M. Everett, R. J. Lauf and T. J. Phelps, *J. Ind. Microbiol. Biotechnol.*, 2010, **37**, 1023.
77. N. Prabhu, *J. Bacteriol. Mycol.*, 2018, **6**, 260.
78. S. Yoon, *J. Magn.*, 2014, **19**, 323.
79. H. Kronmüller and S. S. Parkin, *Handbook of Magnetism and Advanced Magnetic Materials*, Wiley, New York, 2007.
80. I. Nurdin, M. Johan, I. Yaacob, B. Ang and A. Andriyana, *Mater. Res. Innovations*, 2014, **18**, S6–S200.
81. K. Woo and H. J. Lee, *J. Magn. Magn. Mater.*, 2004, **272**, E1155.
82. T. Belin, N. Guigue-Millot, T. Caillot, D. Aymes and J. Niepce, *J. Solid State Chem.*, 2002, **163**, 459.
83. B. Choi, S. Kim, Y. Jeon, J. Moon, G. Lee, Y. Chang and J. Park, *Int. J. Mod. Phys. B*, 2006, **20**, 4422.
84. I. Kartswnakis, N. Papadopoulos, P. Tserotas and P. Švec, *Key Eng. Mater.*, 2013, 468.
85. T. Terashima and Y. Bando, *J. Appl. Phys.*, 1984, **56**, 3445.
86. Y. Ishii, A. Terada, O. Ishii, S. Ohta, S. Hattori and K. Makino, *IEEE Trans. Magn.*, 1980, **16**, 1114.
87. M. Langlet, M. Labeau, B. Bochu and J.-C. Joubert, *IEEE Trans. Magn.*, 1986, **22**, 151.
88. M. Abe, Y. Tamaura, M. Oishi, T. Saitoh, T. Itoh and M. Gomi, *IEEE Trans. Magn.*, 1987, **23**, 3432.
89. B. K. Ghosh and N. N. Ghosh, *J. Nanosci. Nanotechnol.*, 2018, **18**, 3735.
90. R. Kottappara, S. C. Pillai and B. K. Vijayan, *Inorg. Chem. Commun.*, 2020, **121**, 108181.
91. M. I. Din, R. Khalid, Z. Hussain, T. Hussain, A. Mujahid, J. Najeeb and F. Izhar, *Crit. Rev. Anal. Chem.*, 2020, **50**, 322.
92. A. Ramu, S. Salla, S. Chandrasekaran, P. Silambarasan, S. Gopi, S.-y. Seo, K. Yun and D. Choi, *Environ. Pollut.*, 2021, **270**, 116063.
93. J. Luo, N. Zhang, R. Liu and X. Liu, *RSC Adv.*, 2014, **4**, 64816.
94. W. Lu, R. Ning, X. Qin, Y. Zhang, G. Chang, S. Liu, Y. Luo and X. Sun, *J. Hazard. Mater.*, 2011, **197**, 320.
95. A. Corma, P. Concepción, M. Boronat, M. J. Sabater, J. Navas, M. J. Yacaman, E. Larios, A. Posadas, M. A. López-Quintela and D. Buceta, *Nat. Chem.*, 2013, **5**, 775.
96. M. E. El-Naggar and K. Shoueir, *J. Environ. Chem. Eng.*, 2020, **8**, 104175.
97. Y. Zhu, X. Zhou, D. Chen, F. Li, T. Xue and A. S. Farag, *Sci. China: Technol. Sci.*, 2017, **60**, 749.
98. Y. Zhao, Y. Yeh, R. Liu, J. You and F. Qu, *Solid State Sci.*, 2015, **45**, 9.

99. T. Zeng, H.-y. Niu, Y.-r. Ma, W.-h. Li and Y.-q. Cai, *Appl. Catal., B*, 2013, **134**, 26.
100. H. M. A. Sharif, A. Mahmood, H.-Y. Cheng, R. Djellabi, J. Ali, W.-L. Jiang, S.-S. Wang, M. R. Haider, N. Mahmood and A.-J. Wang, *ACS Appl. Nano Mater.*, 2019, **2**, 5310.
101. M. R. Haider, W.-L. Jiang, J.-L. Han, A. Mahmood, R. Djellabi, H. Liu, M. B. Asif and A.-J. Wang, Boosting Hydroxyl Radical Yield *via* Synergistic Activation of Electrogenerated $HOCl/H_2O_2$ in Electro-Fenton-like Degradation of Contaminants under Chloride Conditions, *Environ. Sci. Technol.*, 2023, DOI: 10.1021/acs.est.2c07752.
102. E. Brillas and S. Garcia-Segura, *Sep. Purif. Technol.*, 2020, **237**, 116337.
103. W. He, Z. Li, S. Lv, M. Niu, W. Zhou, J. Li, R. Lu, H. Gao, C. Pan and S. Zhang, *Chem. Eng. J.*, 2021, **409**, 128274.
104. R. Zhong, W. Yang, H. Gao, T. Wang, D. Zhang, H. Wu, R. Zhou, Y. Wu, C. Kong and Z. Yang, *Colloids Surf., A*, 2022, **655**, 130200.
105. S. Li, J. Cui, X. Wu, X. Zhang, Q. Hu and X. Hou, *J. Hazard. Mater.*, 2019, **373**, 408.
106. Q. Gong, Y. Liu and Z. Dang, *J. Hazard. Mater.*, 2019, **371**, 677.
107. V. Blanco-Gutiérrez, P. Li, R. Berzal-Cabetas and A. Dos santos-García, *J. Solid State Chem.*, 2022, **316**, 123509.
108. J. Zhong and C. Cao, *Sens. Actuators, B*, 2010, **145**, 651–656.
109. Y. Shin, J. M. Blackwood, I.-T. Bae, B. W. Arey and G. J. Exarhos, *Mater. Lett.*, 2007, **61**, 4297.
110. R. Sharma and S. Singhal, *J. Chem. Technol. Biotechnol.*, 2015, **90**, 955.
111. G. Q. Tan, Y. Q. Zheng, H. Y. Miao, A. Xia and H. J. Ren, *J. Am. Ceram. Soc.*, 2012, **95**, 280.
112. X. Wang, Y. Lin, X. Ding and J. Jiang, *J. Alloys Compd.*, 2011, **509**, 6585.
113. W. Wang, N. Li, Y. Chi, Y. Li, W. Yan, X. Li and C. Shao, *Ceram. Int.*, 2013, **39**, 3511.
114. C.-j. Teng, D. Xie, M.-x. Sun, J.-l. Xu, C.-s. Zhao, P. Yang, Y.-l. Sun, C. Zhang and X. Li, *RSC Adv.*, 2016, **6**, 67550.
115. S. Bharathkumar, M. Sakar and S. Balakumar, *Phys. Chem. Chem. Phys.*, 2015, **17**, 17745.
116. H. Wang, L. Zhang, Z. Chen, J. Hu, S. Li, Z. Wang, J. Liu and X. Wang, *Chem. Soc. Rev.*, 2014, **43**, 5234.
117. J. Low, J. Yu, M. Jaroniec, S. Wageh and A. A. Al-Ghamdi, *Adv. Mater.*, 2017, **29**, 1601694.
118. V. Dutta, S. Sharma, P. Raizada, A. Hosseini-Bandegharaei, V. K. Gupta and P. Singh, *J. Saudi Chem. Soc.*, 2019, **23**, 1119.
119. X. Liu, Y. Wang, T. C. Zhang, G. Xiang, X. Wang and S. Yuan, *Ind. Eng. Chem. Res.*, 2020, **60**, 528.
120. V. Bortolotto, R. Djellabi, A. Giordana, G. Cerrato, A. Di Michele and C. L. Bianchi, *Inorg. Chem. Commun.*, 2022, **141**, 109516.
121. R. Djellabi, J. Ali, B. Yang, M. R. Haider, P. Su, C. L. Bianchi and X. Zhao, *Sep. Purif. Technol.*, 2020, **250**, 116954.

122. A. Costa, B. Ballarin, A. Spegni, F. Casoli and D. Gardini, *J. Colloid Interface Sci.*, 2012, **388**, 31.
123. M. Ye, Q. Zhang, Y. Hu, J. Ge, Z. Lu, L. He, Z. Chen and Y. Yin, *Chem. – Eur. J.*, 2010, **16**, 6243.
124. Y. Fan, C. Ma, W. Li and Y. Yin, *Mater. Sci. Semicond. Process.*, 2012, **15**, 582.
125. W. Su, T. Zhang, L. Li, J. Xing, M. He, Y. Zhong and Z. Li, *RSC Adv.*, 2014, **4**, 8901.
126. S. Tawkaew and S. Supothina, *Mater. Chem. Phys.*, 2008, **108**, 147.
127. Q. Yuan, N. Li, W. Geng, Y. Chi and X. Li, *Mater. Res. Bull.*, 2012, **47**, 2396.
128. J. Joo, Y. Ye, D. Kim, J. Lee and S. Jeon, *Mater. Lett.*, 2013, **93**, 141.
129. D. Wang, J. Yang, X. Li, J. Wang, H. Zhai, J. Lang and H. Song, *Phys. Status Solidi A*, 2017, **214**, 1600665.
130. A. Zielińska-Jurek, Z. Bielan, I. Wysocka, J. Strychalska, M. Janczarek and T. Klimczuk, *J. Environ. Manage.*, 2017, **195**, 157.
131. J. C. Terra, A. Desgranges, C. Monnereau, E. H. Sanchez, J. A. De Toro, Z. Amara and A. Moores, *ACS Appl. Mater. Interfaces*, 2020, **12**, 24895.
132. R. Djellabi, B. Yang, H. M. A. Sharif, J. Zhang, J. Ali and X. Zhao, *J. Cleaner Prod.*, 2019, **233**, 841.

CHAPTER 6

Nanocellulose-based Membranes for Water Purification: Multifunctional Nanocellulose Extraction, Characterization, Modification Strategies, and Current Release in Water Treatment and Environmental Remediation

SALY R. EL DAKKONY[a*], MAHMOUD F. MUBARAK[a,b*] AND ADEL A.-H. ABDEL-RAHMAN[c]

[a]Egyptian Petroleum Research Institute (EPRI), 1 Ahmed El-Zomer, Nasr City, Box. No. 11727, Cairo, Egypt; [b]Civil Department, Faculty of Engineering, Suez Canal University, Egypt; [c]Menoufia University, Shebin El-Koam, Egypt

*E-mail: sally_refat48@yahoo.com, Fathy8753@yahoo.com

Chemistry in the Environment Series No. 12

Novel Materials and Water Purification: Towards a Sustainable Future

Edited by Grigorios L. Kyriakopoulos and Miltiadis G. Zamparas

Published by the Royal Society of Chemistry, www.rsc.org

6.1 Introduction

Distribution of water, a necessary ingredient for all forms of life, is highly inequitable. Population expansion, industrialization, and climate change are all increasing the demand for fresh water, which is putting pressure on the world's finite renewable water supply.[1]

While water accounts for 70% of the Earth's surface, just 3% is drinkable, and 83% of it is locked up in glaciers or is otherwise inaccessible. Fresh water is essential for human consumption, hygiene, industry, and agriculture (Figure 6.1).[2]

Uses of water encompass the water cycle and sanitation systems, and also help to make progress on many other Sustainable Development Goals (SDGs), such as health, education, economic growth, and protection of the environment[3] (Figure 6.2).

In response, several technologies have been created to treat and purify water, from age-old processes, such as coagulation and sedimentation, to cutting-edge ones, such as reverse osmosis and ultraviolet (UV) sterilization.[4]

However, another important issue is contamination of the water supply. Agricultural runoff and industrial waste are just two of the many types of

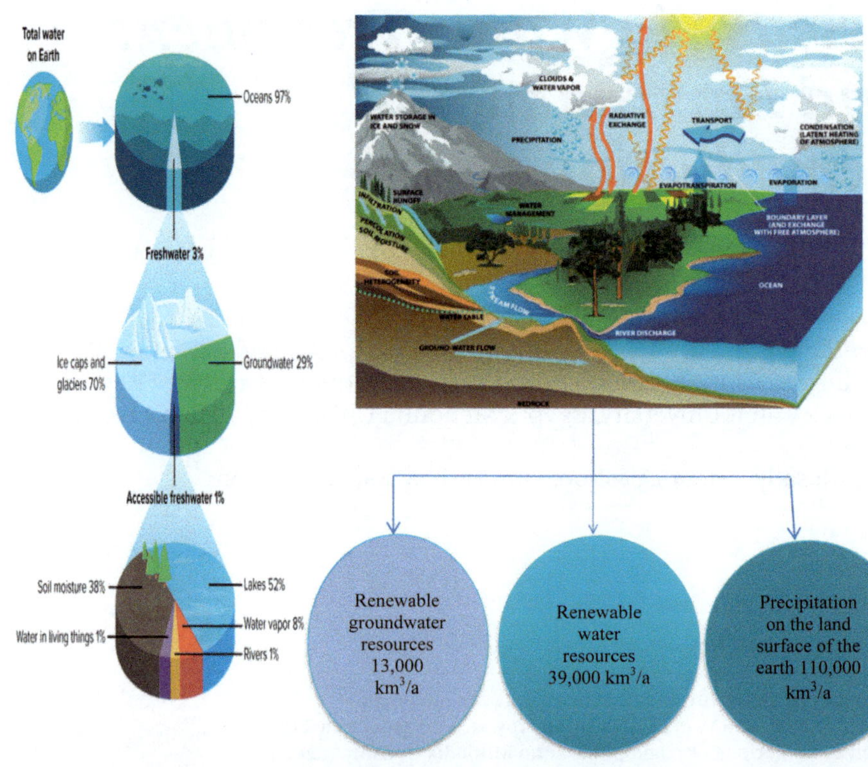

Figure 6.1 Global values for water resources.

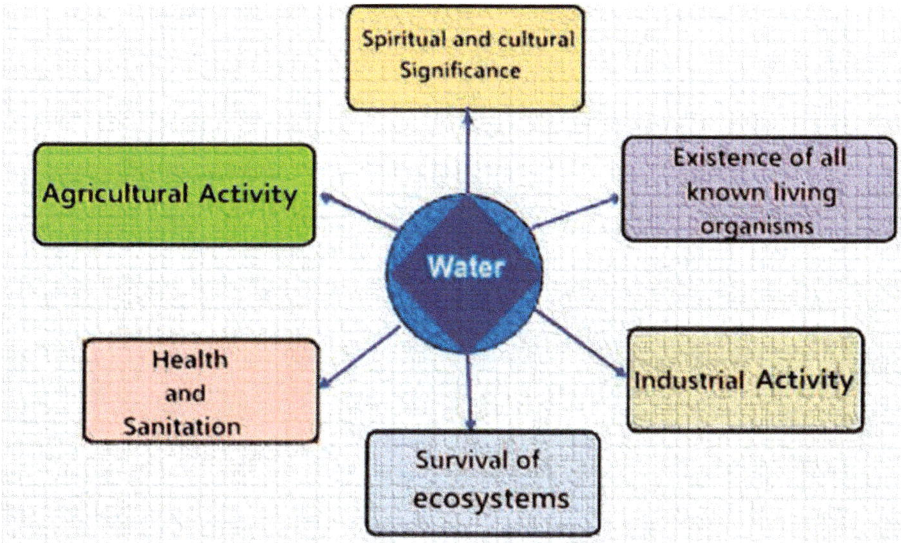

Figure 6.2 The importance of water and its applications.

water pollution that continue to pose a hazard to human and environmental health. Physical, chemical, biological, and radioactive contaminants are all examples of pollution that can make water unsafe for human consumption or other purposes, as shown in (Figure 6.3). Several methods, including adsorption, biofiltration, oxidation, and reduction, have been developed to purify water.[5]

Nanotechnology is now generally acknowledged as a major driving force behind a new industrial revolution in many different types of industries,[3] such as pharmaceutical, automotive, food, *etc*. Due to their recyclability, biodegradability, compatibility, and benign behavior, there has been a significant uptick in the study of eco-friendly, sustainable materials derived from natural sources.[6]

Nanomaterials differ from bulk materials in their chemical, physical, and biological properties because they have at least one dimension at the nanometer scale, of 100 nm or less. Nanocellulose is a naturally occurring nanomaterial with several desirable qualities, including a large surface area, nanoscale dimension, high modulus and strength, renewability, and excellent optical quality.[7]

Cellulose is extensively available in the cell wall of plant biomass and is the most abundant renewable polymer on Earth. It is also biodegradable and environmentally friendly. It is a white fibrous substance that has no discernible taste or smell and is insoluble in both organic solvents and water. Plants, mammals, algae, fungi, and even certain minerals contain cellulose. Cellulose makes up a larger portion of the composition of a plant than any other substance (*e.g.* cotton, wood).[8]

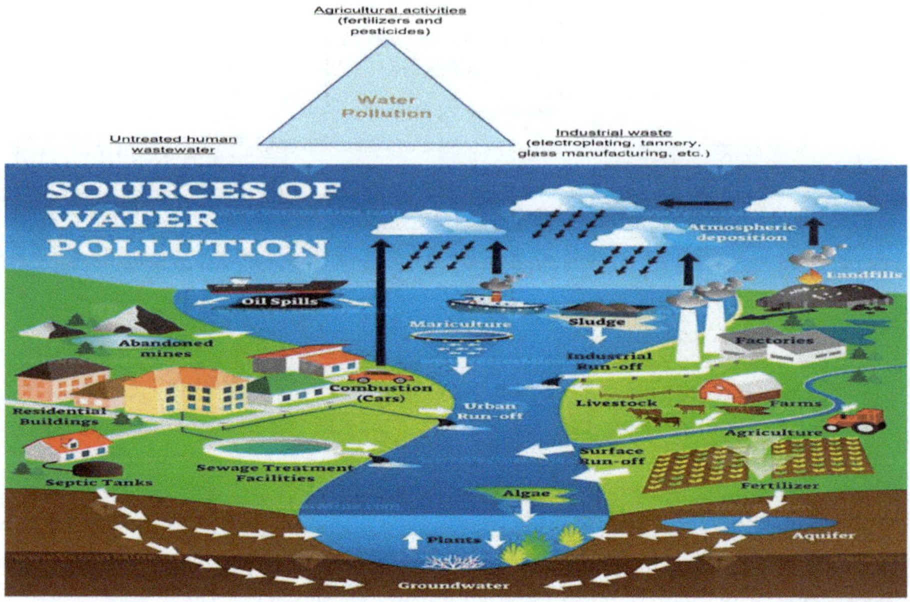

Figure 6.3 Selected sources of pollutants contributing to water contamination globally.

Cellulose production in nature is of the order of 10^{11}–10^{12} tonnes per year. Consistent with the current trend in the development of the global economy, energy, and novel materials, the development of cellulose-based functional materials and enhancement of their added value are essential.[9]

Cellulose makes up 90% of cotton fibers (not all components of the cotton plant), with the remaining 10% consisting of waxes, monosaccharides, and genetic material from biosynthesis. However, cellulose can occasionally be found in combination with hemicelluloses, lignin, and trace amounts of extractives. Not all animals contain cellulose, since their cells lack cell walls. However, it is present in several animal species, including the shells of tunicates, which are marine invertebrates.[10]

Cellulosic precursors of activated carbon (AC) include a wide range of by-products of aging and waste. These ingredient building blocks feature in coconut husks and wood, olive stones, sugarcane bagasse, pecan hulls, palm seeds, apple pulp, rubber seeds, palm seed coats, and molasses.[11]

One new type of carbon material is activated carbon thin film (ACTF), which is a monolayer of carbon similar to graphite. The structure and characteristics of a novel adsorbent for sodium ions from synthetic water were investigated by Fathy *et al.*, who synthesized ACTF from rice straw. By combining the carbon nanostructure with ground and mixed unused waste polystyrene bottles, Fathy *et al.* also produced waste polystyrene (WPS)/amorphous carbon thin film composite resin (ACTF) beads. Amorphous carbon thin film made from wood sawdust was used efficiently for oil removal with high adsorptive capacity, as reported by El-Sayed *et al.*[12]

Fabrics, pulp and paper, and wood structures have all benefited from, and continue to make use of, cellulose from plant sources. Higher order structures of cellulose, created by physical and/or chemical processing, also have the potential to be of industrial importance.

As a result of its nano-size effect, high elastic modulus, high mechanical strength, biocompatibility, regeneration, and degradability, nanocellulose has more uses than cellulose. An increasing number of high-quality composites based on nanocellulose have been developed in recent years.[13,57–60]

At present, there are primarily three types of nanocellulose-based superhydrophobic materials being studied: coatings, membranes, and aerogels. Polyacrylonitrile (PAN), polyimide (PI), poly(vinyl alcohol) (PVA), poly(vinylidene fluoride) (PVDF), cellulose acetate (CA), polylactic acid (PLA), polyurethane (PU), polycaprolactone (PCL), and polyethylene oxide (PEO) nanofibers have been produced using several different methods.[14,15] For example, nanofibers with desirable properties, such as high porosity, large specific surface area, and high size homogeneity, may find uses in the treatment of waste water.[16]

Water purification and desalination using membrane technology is on the rise. To separate and purify water, membrane processes employ a selectively permeable barrier, such as a polymer membrane. High efficiency, low energy consumption, and minimal maintenance costs are three of the key benefits of membrane technology. In conventional membrane technology, contaminants accumulate on the membrane surface, leading to a reduction in flux over time.[17]

In order to create continuous fibers in the submicrometer to nanometer scale range, the electrospinning process was developed in recent decades (from 2012). Various nanofibers, including those made from polymers, metals, ceramics, and composites, are fabricated using this technique. Scaffolds are created by combining nanoparticles with polymers and then electrospinning the resulting mixture. In order to lower the Gibbs free energy, nanoparticles can be assembled *via* electrospinning by aligning with fibers. An additional benefit of electrospinning is that it calls for no functionalization procedure beyond the use of a solvent that can both break down the polymer and distribute the nanoparticles.[18] For the most part, strong electrostatic forces are what make electrospinning possible. The success of an electrospinning project depends on a variety of variables including, but not limited to, polymer content, solution viscosity and flow rate, electric field intensity, the work distance, and air humidity.[19]

The electrospinning technique is a flexible and straightforward approach to producing nanoscale membranes, with promising applications in the amelioration of environmental and several energy issues. However, the inferior mechanical qualities of such films often restrict their application in industry. Due to its large specific surface area, good mechanical qualities, and renewable nature, nanocellulose makes for a great environmentally friendly reinforcing material.[20,21]

Researchers have recently started looking into nanocellulose-based membranes for use in water purification and desalination systems. These membranes are constructed from cellulose, a renewable material with the potential to be tailored to have an exceptionally high surface area relative to volume. That is why these nanocellulose membranes are superior to standard polymer membranes in various ways, including permeability, mechanical strength, and resistance to fouling. Nanocellulose-based membranes provide a number of benefits that make them competitive in the water treatment and desalination markets, and researchers are always looking for methods to enhance their efficiency and lower their cost.[7,22]

This chapter provides an overview of the current state of water distribution and water pollution, as well as the technologies used to address these issues. The focus is on the advantages and disadvantages of different technologies, with an emphasis on membrane technology and its applications in water treatment and desalination and the following topics will be discussed.

1. The promising future of nanocellulose due to its properties and applications in the field of water treatment and desalination.
2. The synthesis of nanocellulose using both bottom-up and top-down methods, resulting in different types of nanocellulose with unique properties.
3. Discussion of the properties of nanocellulose that make it a suitable material for antifouling and self-healing membranes, which can have a significant impact on water treatment and other industries.
4. An assembled ACTF composite membrane (CM) designed to succeed in the general performance of desalination, in addition to a nanocellulose filler that enhances the self-healing properties of the prepared membrane.
5. Thin-film composite (TFC)-based nanocellulose membranes composed of a polyamide active layer filled with ACTF with a natural polymer of cellulose. Subsequent nylon 6,12 thin-film layer casting onto polysulfone that has nanofillers on the surface, with a selective ultrathin layer, is extensively employed for desalination.
6. Some different characterization methods for the external morphology and chemical composition of nanocellulose and composite membranes of nanocellulose, such as scanning electron microscopy (SEM), mechanical testing, Fourier-transform infrared (FT-IR) spectroscopy, X-ray diffraction, thermogravimetric analysis (TGA), and atomic force microscopy (AFM).
7. The systematic investigation of the mechanical properties of the TFC@ACNCE membrane for effects on the ACTF and nanocellulose content.
8. The study of the prepared TFC@ACNCE-based nanocellulose membrane for water flux and salt rejection, to determine its antifouling efficiency.
9. Finally, the further research needed to fully understand the potential of nanocellulose and to develop cost-effective methods of production, and to optimize the properties of nanocellulose to make it more suitable for different applications.

6.2 Macro and Micro Sources of Cellulose

Nanocellulose compounds derive their cellulose from a wide range of different sources. Cellulose may be found in a wide variety of places, including wood pulp, agricultural waste (such straw or bagasse), and even cotton or hemp textiles. Microorganisms, such as bacteria and algae, can also potentially be a source. Softwoods, such as pine and spruce, and hardwoods, such as eucalyptus and birch, are the most prevalent cellulose sources for nanocellulose synthesis (Figure 6.4).[23]

The most common type of cellulose used to produce nanocellulose is called α-cellulose, which is a purified form of cellulose that has a high degree of crystallinity, of around 95%.[24] The other type of cellulose is β-cellulose, which is cellulose with a lower degree of crystallinity of around 60–70%, but which has a lower cost in terms of production and raw materials (Figure 6.5).[23,25]

6.3 Types of Nanocellulose

Nanocellulose is a new class of material made from cellulose, a natural polymer that is the main component of plant cell walls. There are several different types of nanocellulose (Figure 6.6), each with unique properties and potential applications.[26]

6.3.1 Cellulose Nanocrystals

Cellulose nanocrystals (CNCs) are small, needle-like particles that have a diameter of around 5–20 nanometers and a length of several micrometers. They are produced through acid hydrolysis of cellulose fibers, which breaks

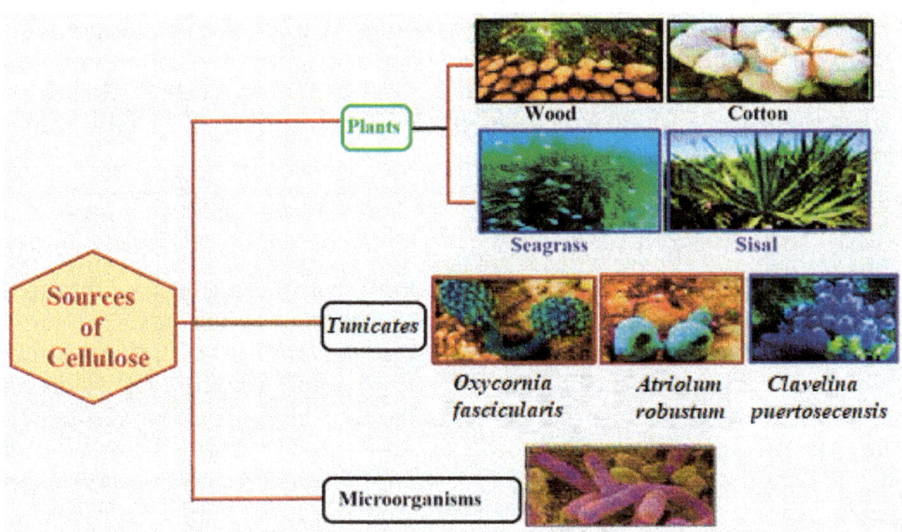

Figure 6.4 Macro- and microcellulose natural sources.

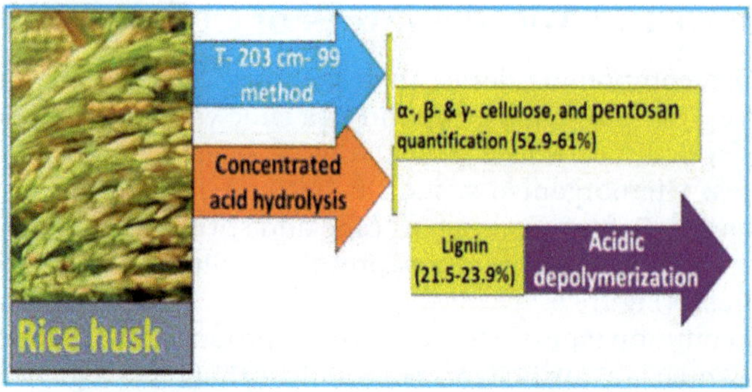

Figure 6.5 The common types of cellulose that produce different nanocellulose configurations.

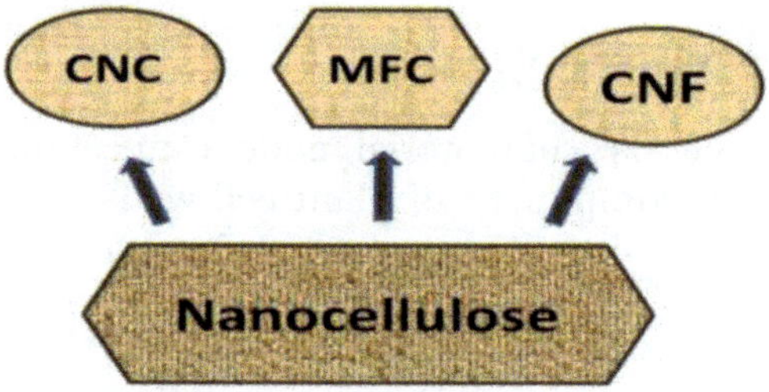

Figure 6.6 Different types of nanocellulose.

down the fibers into small crystalline particles. The most common acids used for this process are sulfuric and hydrochloric acid, but other acids, such as nitric and formic acids, can also be used. CNCs are highly crystalline, with a degree of crystallinity of around 90%. They have a high Young's modulus, a measure of how stiff a material is, and a high strength-to-weight ratio, which makes them well suited for use in composites, coatings, and other high-performance materials.[27]

6.3.2 Cellulose Nanofibrils

Cellulose nanofibrils (CNFs) are thin, elongated fibers that are obtained from cellulose fibers by mechanical and/or chemical methods. The most common method for producing CNFs is mechanical grinding, which involves breaking down cellulose fibers into smaller fibrils. Other methods that can be used to produce CNFs include high-pressure homogenization, sonication, and acid hydrolysis. CNFs have a diameter of around 20–200 nanometers and a length of several micrometers. They have a high specific surface area, high aspect ratio, and high mechanical properties, which make them interesting for use in filtration, reinforcement, and barrier applications.[28]

6.3.3 Microfibrillated Cellulose

Microfibrillated cellulose (MFC) is a cellulose material that is composed of small fibers that are less than one micrometer in diameter. It is obtained from cellulose fibers by using a combination of mechanical and chemical methods, such as grinding, high-pressure homogenization, and acid hydrolysis. MFC has high mechanical properties and high water-holding capacity, which make it suitable for use in various applications, including as a thickening agent, binder, and filler.[29]

6.4 Modern Technologies for Preparing Nanocellulose Membranes

There are two main approaches to synthesizing nanocellulose membranes: bottom-up and top-down methods.

6.4.1 Bottom-up Methods

Nanocellulose may be grown from the bottom up, with the help of microorganisms, such as bacteria or algae. This method, which is also known as biosynthesis, can be carried out by the culture of microorganisms or the use of artificial cells. Bottom-up approaches typically include the growth of bacteria to produce nanocellulose. Using a technique known as microbial cellulose fermentation, the bacteria are manipulated to generate nanocellulose. This approach provides a cost-effective and environmentally friendly way to make nanocellulose, with the added benefit of allowing for extremely high levels of purity and uniformity in the final product (Figure 6.7).[30]

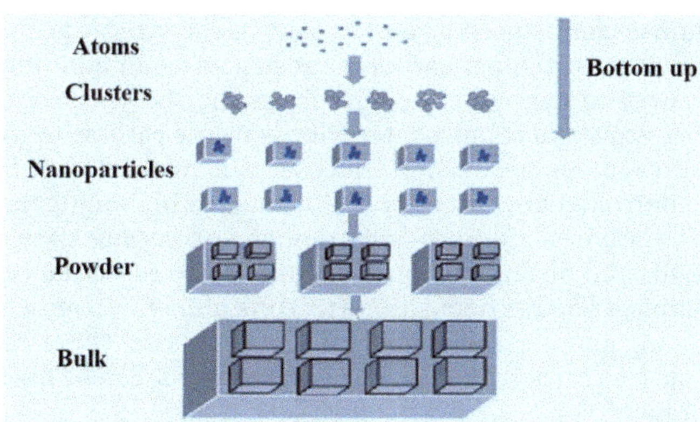

Figure 6.7 The bottom-up approach.

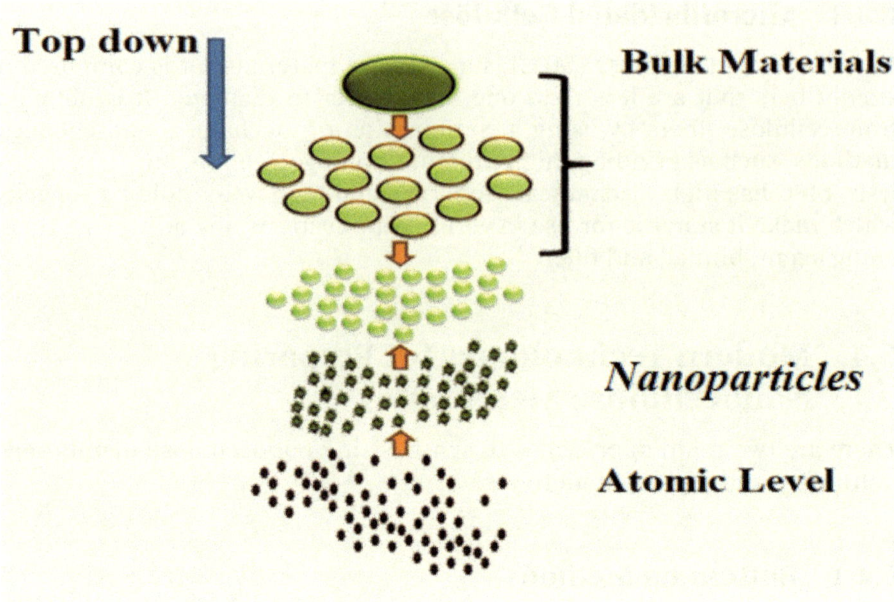

Figure 6.8 The top-down approach.

6.4.2 Top-down Methods

In contrast, nanocellulose fibers can be created from cellulose strands using top-down approaches. Mechanical processes, such as grinding, high-pressure homogenization, and sonication, and chemical methods, such as acid hydrolysis, are common examples of top-down approaches. Nanocrystals, nanofibrils, and microfibrillated cellulose are all possible outcomes of these processes, among others. Multiple nanocellulose varieties, each with their own unique features, may be produced by combining top-down and bottom-up approaches (Figure 6.8).[31]

Nanocellulose can be used as a precursor in the synthesis of a wide variety of dimensionally distinct carbon nanostructures in addition to its own production. Carbon nanotubes and graphene may be synthesized from cellulose nanocrystals and cellulose nanofibrils by the carbonization synthesis method. A carbon source, such as glucose, is combined with the nanocellulose in a controlled environment and heated to high temperature (about 600–800 °C). Carbon nanotubes and graphene of varying sizes and characteristics are formed during this process, which also turns the cellulose into carbon. Chemical vapour deposition (CVD) is a chemical process that can be used to bring about carbonization of nanocellulose. Other carbon nanostructures, such as carbon fibers and carbon aerogels, can be made using the same technique.[32]

It is important to remember that manufacturing nanocellulose and other carbon nanostructures is a thriving field where novel approaches are

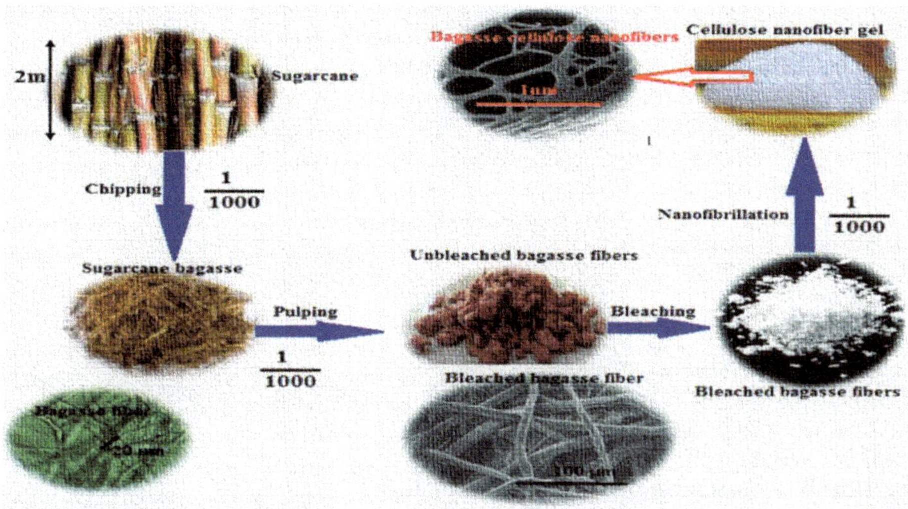

Figure 6.9 Top-down approach for preparation of cellulose nanofibers (CNFs) from bagasse.

constantly being explored[33,34] to improve non-manufacturing techniques, as shown in Figure 6.9.

6.5 Advantages of Using Nanocellulose in Membranes

Many of the benefits that nanocellulose offers make it a promising candidate for application in membrane technology. The high surface area to volume ratio is responsible for its high permeability, which is one of its main selling points. This improves the filtration efficiency of conventional polymer membranes by increasing the permeability of water and other liquids. Cellulose, the natural polymer from which nanocellulose-based membranes are derived, is renowned for its exceptional mechanical strength. As a result, they are extremely robust and hard to bend or break.[35]

Nanocellulose-based membranes also benefit from the excellent chemical stability of cellulose, a naturally occurring polymer that is very resistant to chemical degradation. They can survive a broad variety of pH and chemical conditions, making them useful in a wide variety of water treatment applications. Further, the extreme surface roughness of nanocellulose-based membranes makes it difficult for contaminants to cling to the membrane surface, reducing the likelihood of fouling.[36]

Nanocellulose-based membranes are environmentally friendly and sustainable since they are derived from cellulose, a harmless biodegradable substance and they can be manufactured utilizing efficient and low-cost processes. In addition, some types of nanocellulose may be changed chemically, physically, or through composite techniques to adjust their characteristics,

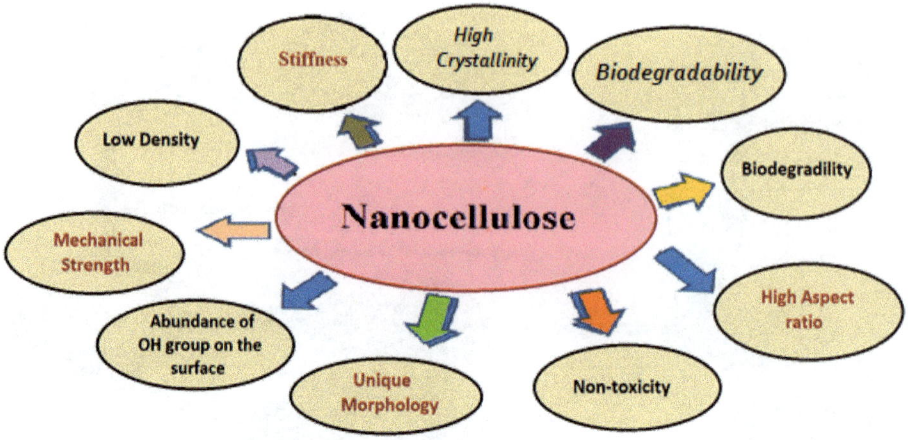

Figure 6.10 Unique properties of nanocellulose.

which enables a better-tailored membrane for the precise requirements of the application (Figure 6.10).[37]

Membranes based on nanocellulose are also thermally stable, meaning they can resist extremely high temperatures without losing their characteristics. Therefore, they may be utilized in processes that call for extremely high temperatures, such as the generation of biofuels. Furthermore, nanocellulose may be manufactured in flexible forms, such as films, which can lead to new opportunities for nanocellulose-based membranes in applications that demand flexibility, such as membrane contactors or wearable water purification systems.[38]

6.5.1 Antifouling Nanocellulose Membranes

Nanocellulose membranes with an antifouling property are those that prevent contaminants from sticking to their surface. Using methods such as chemical functionalization to alter the surface characteristics of nanocellulose membranes and lower impurity adsorption is one approach to this. A high roughness can make it difficult for contaminants to stick to the surface, which is one manner in which antifouling qualities can be achieved.[39]

Adding a nanocellulose layer on top of a larger coating of another polymer is another technique to obtain antifouling qualities; this can assist in preventing contaminants from adhering to the membrane. Impurities on the membrane's surface can be kept to a minimum by introducing antifouling chemicals, such as biocides or enzymes, within the nanocellulose membrane (Figure 6.11). The antifouling capabilities of a membrane may also be modified by utilizing different types of nanocellulose. These include cellulose nanocrystals, cellulose nanofibrils, and microfibrillated cellulose. In particular, nanocrystal-based nanocellulose has been shown to be more resistant to fouling than other nanocellulose variants.[40]

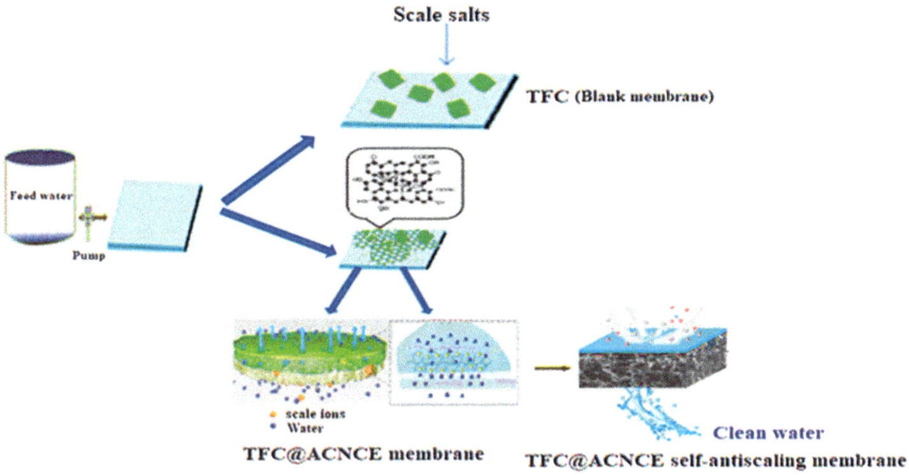

Figure 6.11 Mechanism of self-antiscaling membrane TFC@ACNCE.

Although promising, nanocellulose-based membranes require additional study to fully realize their potential, optimize their performance, and reduce their costs. In addition, although production costs are decreasing alongside progress in technology, nanocellulose is still not competitive with conventional polymer membranes in terms of price.[41]

In summary, nanocellulose is a biomaterial made from plant fibers, such as wood, straw, or grass; it is tiny in size; and has a high specific surface area. Mechanical or chemical procedures, including acid hydrolysis or high-pressure homogenization, are commonly used in its production.[42]

According to El-Dakkony *et al.*, a composite membrane was prepared that interconnected with nanocellulose, which binds to the top dense active polyamide layer chemically through coordinated and hydrogen bonding, an effect that has been characterized as a physicochemical surface adsorption process.[43,44]

The typical method used for characterizing nanocellulose membranes is infrared spectroscopy (IR), which provides insight into the chemical composition of the material. The crystallinity and substitution of nanocellulose may be calculated using IR data for the hydroxyl, carboxyl, and ether functional groups present in the material.[45]

The FT-IR spectrum region of the polysulfone (PSF) membrane shown in Figure 6.12 in the region 1580–1500 cm^{-1} is highly indicative of the benzene ring stretching mode. Spectral rocking and bending of the C–H bond is seen, which is characteristic of aromatic compounds.[44] FT-IR spectroscopy was used to study the nylon 6,12 composites and to describe the generation of PI *via* interfacial polymerization (Figure 6.12). The FT-IR spectra of polyethersulfone (PES) and TFC membranes show bands in the CM spectra between 1440 and 1620 cm^{-1} that are associated with the CN, C, and O; these peaks

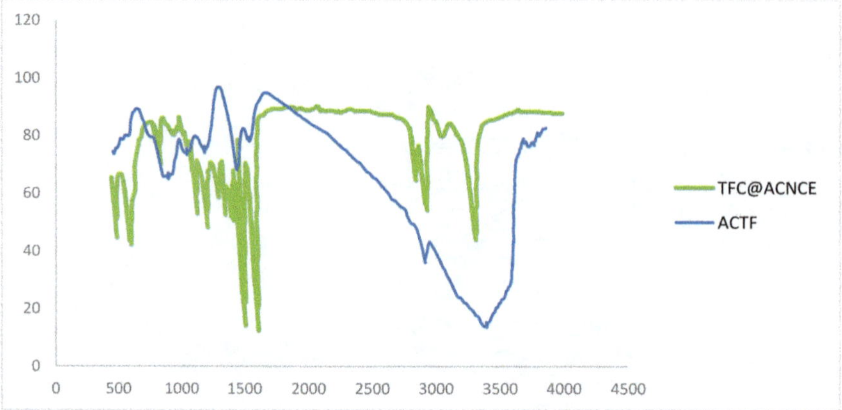

Figure 6.12 FT-IR spectra of ACTF and the TFC@ACNCE membranes. Reproduced from ref. 44 with permission from Springer Nature, Copyright 2022.

relate to the stretching and bending vibrations of the NH group. The absorption band at 1540 cm^{-1} is unique to the NH stretching vibrations, whereas the one at 1380 cm^{-1} is a combination of the CN stretching vibrations and the in-plane NH deformation of amide. The spreading vibration CH_2 in the polyamide backbone is responsible for the absorption band in the 2856 cm^{-1} region.[43,46]

Thermogravimetric analysis (TGA) measures the weight loss as a function of temperature to ascertain the thermal stability of nanocellulose, and the thermal breakdown behavior may be elucidated using TGA, providing details of the temperatures at which degradation begins and ends and how much weight is lost at different temperatures.[47]

TGA tests for ACTF and TFC@ACNCE membrane nanocomposites were carried out at a heating rate of 10 °C min^{-1} in an air environment, as shown in Figure 6.13. According to Figure 6.13, ACTF is shown to be thermally stable. However, at temperatures above 180 °C, the thermal stability of ACTF diminished due to the breakdown of the carboxylic acid groups and emission of CO_2 gas. ACTF lost about 20% of its weight at 162 °C due to pyrolysis of functional groups that contain oxygen, and by the oxidation of carbon at around 620 °C, which was connected to the oxidation of sp^2-hybridized carbon atoms, and so the total weight of ACTF was lowered.[44]

High temperatures, above 200 °C, caused ACTF weight loss to rise. This might be because the functional hydroxyl groups of ACTF have been dehydrated. Based on measurements of the effect of the nylon polyamide layer on the thermal characteristics of the TFC@ACNCE membrane, we know that the new PS breakdown process starts at roughly 305 °C and continues until the temperature reaches 390 °C. The process of dehydrating the hydroxyl functional groups present in ACTF leads to a decrease in its weight during the second stage, which typically takes place at temperatures above 200 °C. In other

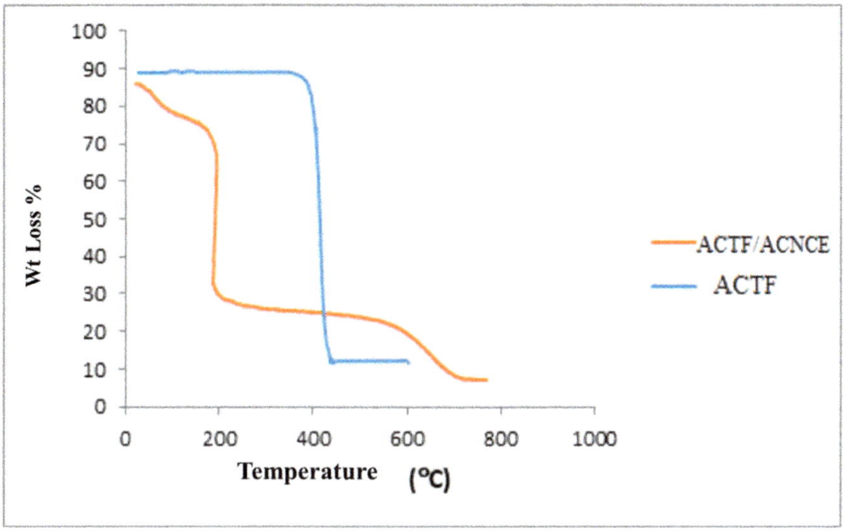

Figure 6.13 Thermogravimetric curves of ACTF, and the ACTF/ACNCE membranes. Reproduced from ref. 44 with permission from Springer Nature, Copyright 2022.

words, when ACTF is exposed to high temperatures, the hydroxyl functional groups lose water molecules, resulting in a reduction in the overall weight of ACTF. This weight loss is observed during the second stage of heating, which occurs at temperatures surpassing 200 °C.[44,48]

The thermal stability of the TFC@ACNCE membrane was improved by the addition of ACTF, due to the creation of an ACTF layer within the cellulose surface, which slows down thermal degradation. Since the ACTF/cellulose composite is present in this study, finger-like pores towards the top and the porous structure in the membrane are visible in the cross-section.[44] In spite of maintaining the ACTF membrane concentration at 1 wt%, the inclusion of polystyrene, a more hydrophobic polymer, hinders the interchange of non-solvents throughout the setup period. This demonstrated the efficacy of the TFC@ACNCE membrane produced by adding ACTF-containing structures onto cellulose.[44]

The surface morphology of nanocellulose can be observed using SEM, showing surface characteristics, such fibrils and fractures, along with the particle size, shape, and distribution.

There is already a way to synthesize polysulfone composite membranes and ACTF, and membranes of the cellulose nanoparticles and ACTF layers dispersed in the PSF matrix have been characterized using SEM. The overall structure comprises a thick outer skin layer and a more porous inner support layer. After incorporating ACTF and cellulose nanoparticles, a distinctive asymmetric structure manifests, with a deep porous layer sporting finger-like sponge holes at the base, and a dense skin layer on top. When ACTF and cellulose nanoparticles are loaded into the TFC@ACNCE

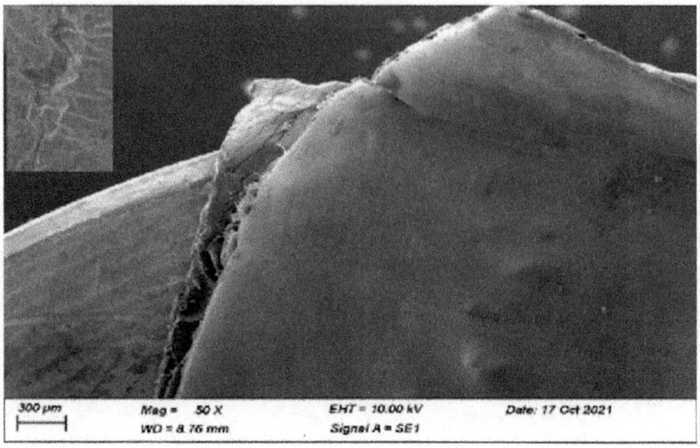

Figure 6.14 SEM image of the auto-self-healing process in the TFC@ACNCE membrane.

membrane, the finger-like channels transform into enormous, open-ended macrovoids, and the spongy fraction grows considerably. As can be seen in Figure 6.14, the wall thickness has reduced and the number of holes has grown in comparison to pure TFC membrane. Water easily penetrates these types of construction. The water permeability and self-healing abilities of the TFC@ACNCE membrane were enhanced by the appearance of horizontal channels.[43,49]

To put it another way, the faster the TFC@ACNCE membrane auto-self-heals during coagulation processes, the smaller its holes and the more spongy or non-voided its structure becomes. Through incorporation of ACTF and cellulose nanoparticles into the membrane matrix, substrates are thus effectively converted to stimulate the body's natural healing mechanisms.[43]

The surface roughness and topographical characteristics of nanocellulose can be studied using AFM. Mechanical characteristics, including size, shape, elasticity, and viscosity, can all be determined using AFM.

The self-healing characteristics of the TFC@ACNCE membrane are improved by the incorporation of hydrophilic ACTF and cellulose nanoparticles under the surface of the membrane, which increases surface contamination and adsorption of scale ions. In addition to other parameters, the hydrophilicity and surface roughness of the membrane might influence the performance of self-healing fillers. The fouling tendency of the TFC@ACNCE membrane is predicted to rise with an increase in surface roughness, in contrast to the TFC membrane. Figure 6.15A and B demonstrates the smooth nature of the membranes. Membrane fouling may have been caused by an insufficient supply of non-self-healing nanofillers present during the energy production phase of the membrane. It was determined that there were statistically significant variations in the roughness values of the membranes by analyzing three scanned regions for each membrane sample. It is possible that the cation content of the raw water (from elements such as calcium

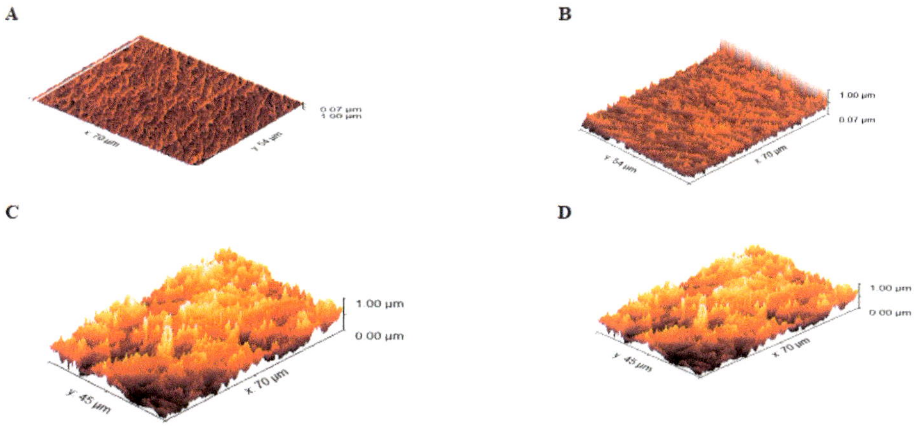

Figure 6.15 AFM analysis (A and B) of the TFC membrane, and (C and D) of the TFC@ACNCE composite self-healing membrane. Reproduced from ref. 44 with permission from Springer Nature, Copyright 2022.

and salt) aided the self-healing ability of the TFC@ACNCE membrane (Figure 6.15C and D). This occurs because the affinity of the membrane surface for other foulants is increased when other cations are present in the feed water.[44] The ability of the TFC@ACNCE membrane to host electrostatic and non-electrostatic interactions between cations and nanofillers was bolstered by the presence of different cations.[44,50]

The presence of nanocellulose and its efficacy in the self-healing system were also tested by the team, and the results showed that the system was very effective in preventing the prepared membranes from being cut directly or indirectly by the high pressures of water used in the desalination process.[44,51]

6.5.2 Auto-self-healing Property of the Membrane

Performance, self-healing, and the original membranes used in water purification were used to evaluate auto-self-healing properties. Figure 6.16 shows the membrane preparation method in general, while the details are explained in the paragraph. A decrease in the membrane's ability to purify water is noted, presumably as a result of minor injury. After that, the membrane's auto-self-healing process starts to work again.[52]

Figure 6.16 shows the method for preparing the membrane using the phase inversion system, and the stages of production based on nanocellulose.

Almost 40% of the water flowing through the initial membrane makes it to the 9 m day^{-1} levels.[61] In the meantime, the repaired membrane demonstrates impressive rejection ratios (91.0% for composite TFC@ACNCE membrane). These results show that the composite TFC@ACNCE membrane has remarkable self-healing properties. The self-healing mechanism should not

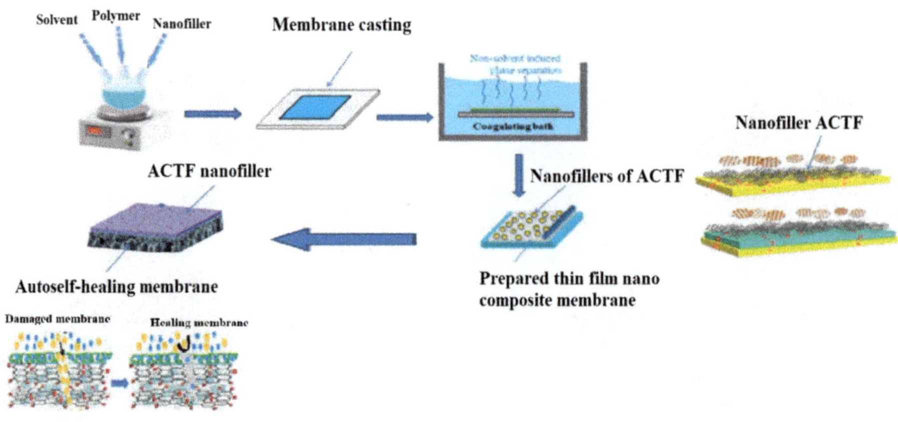

Figure 6.16 Preparation of auto-self-healing composite membranes.

restrict water passage through the repaired membrane. As a result, the TFC@ACNCE membrane is strengthened.[53,54,62,63]

6.5.2.1 *Self-healing Predicted Mechanism*

Figures 6.17 and 6.18 show that the main idea of auto-self-healing is related to the prepared membrane, which consists of the natural polymer with a support layer of the hybrid carbon nanostructure ACTF, and which is used in the desalination process. If a breach occurs, either in the lower or the prepared membrane surface, the natural polymer will rapidly absorb water, expand within the ACTF, and fill any voids that have formed.

Meanwhile, an ACTF layer with a carbon nanostructure is crucial to the separation and desalination process.[55] Constraining the membrane's native polymer chain is essential, as is improving membrane capacity for the investigated ion reserve method. Based on the results of this study, the presence of the hybrid carbon nanostructure ACTF during desalination with the prepared membrane acts as a protective layer in two distinct ways: (1) physically, by stabilizing the membrane structure at the time of injury; and (2) chemically, by providing a buffer against harmful ions.[49] The hybrid carbon nanostructure ACTF within the membrane as a whole, and the natural polymer used to make the membrane, are responsible for the van der Waals, electrostatic attractiveness deriving from the existence of functional groups on the surface, which facilitates the chemical process.[56] The presence of the hybrid carbon nanostructure ACTF generates a strong link between the top layer (active polyamide) of the membrane and the carboxylic groups on the surface, representing 10–20% of the covalent bond strength, which increases the likelihood of self-healing of the membrane, as shown in Figure 6.17. Desalination makes use of cuts in the membrane, either on the surface or inside, which are brought about by high pressure or rupture in the membrane's interior tissue.[44]

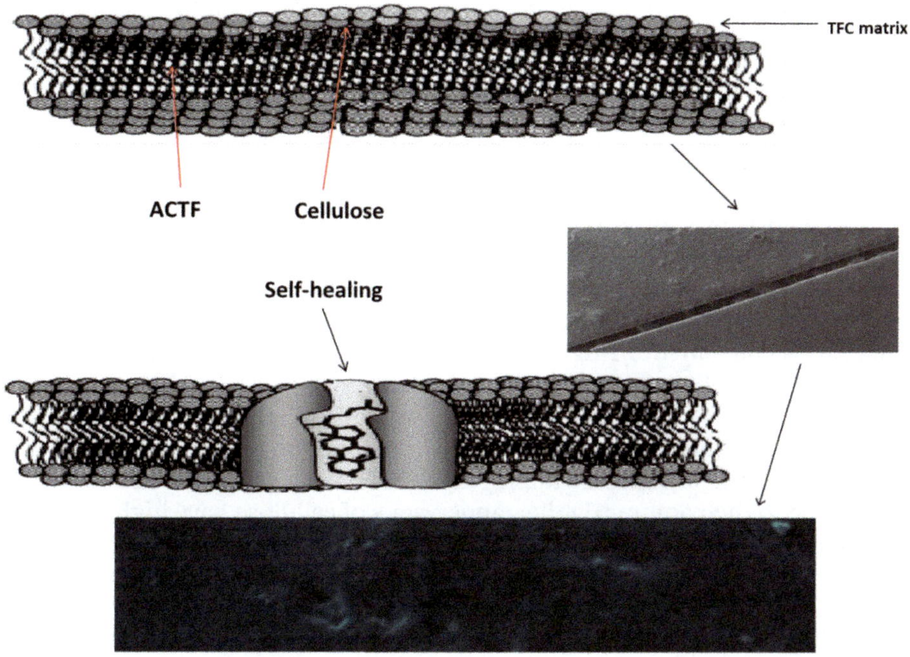

Figure 6.17 Mechanism of auto-self-healing in the prepared nanocomposite membrane.

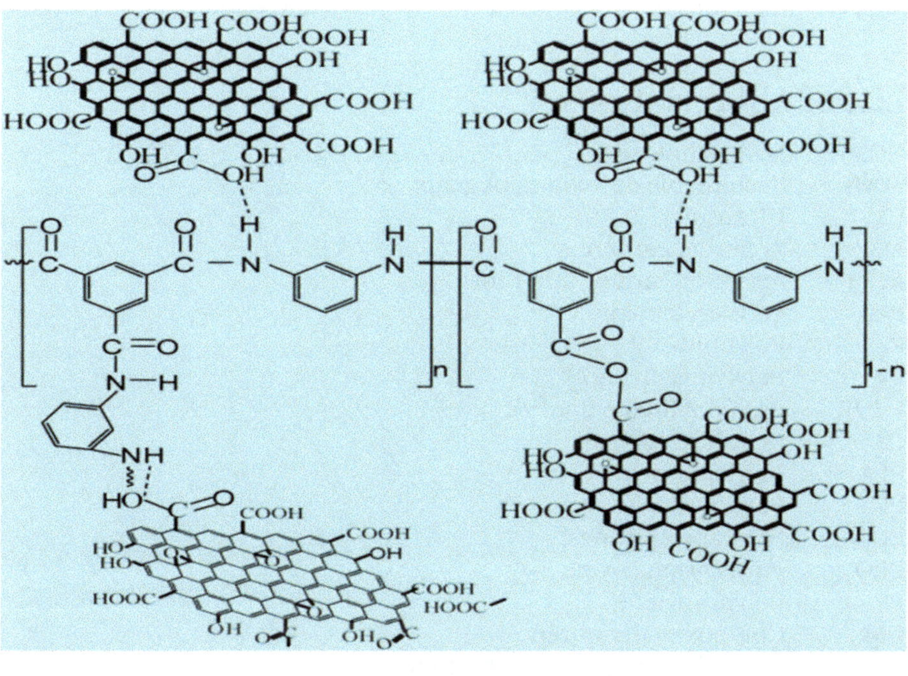

Figure 6.18 Mechanism of ACTF auto-self-healing with an active polyamide layer in the composite membrane. Reproduced from ref. 44 with permission from Springer Nature, Copyright 2022.

These methods of characterizing nanocellulose and its membranes add to our understanding of the material by shedding light on its unique behavior.[44]

6.6 Conclusion

Nanocellulose is a promising and flexible material with a wide range of potential uses, including desalination and water purification. Using either a bottom-up or a top-down approach is possible for nanocellulose production, and both have their merits and drawbacks. Nanocellulose can be grown from small molecules by utilizing microorganisms, or it can be fragmented from larger cellulose fibers into smaller nanocellulose fibers using a top-down approach. Nanocrystals, nanofibrils, and microfibrillated cellulose are all distinct forms of nanocellulose with their own sets of characteristics and possible uses. Nanofibrils of cellulose are highly flexible and have a large surface area in relation to their volume, while nanocrystals of cellulose have high mechanical strength and thermal stability. Antifouling and self-healing membranes are one of the most exciting potential uses of nanocellulose. Because buildup of contaminants on the surface of the membrane can lower its performance and necessitate frequent cleaning, antifouling membranes have been developed. Self-healing membranes, on the other hand, can mend minor tears and keep functioning normally. Surface modification methods, roughness modification, multilayer architectures, integration of antifouling chemicals, and selecting particular forms of nanocellulose can all be used to achieve these qualities.

List of Abbreviations

NC	Nanocellulose
SDGs	Sustainable development goals
UV	Ultraviolet
AC	Activated carbon
ACTF	Activated carbon thin film
PAN	Polyacrylonitrile
PI	Polyimide
PVA	Poly(vinyl alcohol)
PVDF	Poly(vinylidene fluoride)
CA	Cellulose acetate
PLA	Polylactic acid
PU	Polyurethane
PCL	Polycaprolactone
PEO	Polyethylene oxide
PP	Polypropylene
CM	Composite membrane
TFC	Thin-film composite membranes
CNCs	Cellulose nanocrystals

CNFs	Cellulose nanofibrils
MFC	Microfibrillated cellulose
CVD	Chemical vapour deposition
PSF	Polysulfone
PES	Polyethersulfone

Acknowledgements

We would like to thank everyone in the Water Desalination Group in the Petroleum Research Institute (EPRI) and Menoufia University for their efforts. Special thanks to Dr Adel Nassar and Dr Mahmoud Fathy Mubarak for their efforts in developing the new composites and interpreting new important results during work in their laboratory with the Water Treatment and Desalination Research Group.

References

1. L. Shi, *et al.*, Shared injustice, splintered solidarity: Water governance across urban-rural divides, *Global Environ. Change*, 2021, **70**, 102354.
2. M. Rodrigue and A. M. Romi, Environmental escalations to social inequities: Some reflections on the tumultuous state of Gaia, *Crit. Perspect. Account.*, 2022, **82**, 102321.
3. S. Lustenberger and R. Castro-Muñoz, Advanced biomaterials and alternatives tailored as membranes for water treatment and the latest innovative European water remediation projects: A review, *Case Stud. Chem. Environ. Eng.*, 2022, **5**, 100205.
4. S. Malato, *et al.*, Decontamination and disinfection of water by solar photocatalysis: Recent overview and trends, *Catal. Today*, 2009, **147**(1), 1–59.
5. I. LeviRam, *et al.*, Sustainable micropollutant bioremediation via stormwater biofiltration system, *Water Res.*, 2022, **214**, 118188.
6. C. N. Focacci and C. Perez, The importance of education and training policies in supporting technological revolutions: A comparative and historical analysis of UK, US, Germany, and Sweden (1830–1970), *Technol. Soc.*, 2022, **70**, 102000.
7. S. Moyo, *et al.*, A mini-review of nanocellulose-based nanofiber membranes incorporating carbon nanomaterials for dye wastewater treatment, *Environ. Nanotechnol., Monit. Manage.*, 2022, **18**, 100714.
8. Z. Jiang, *et al.*, Application of biodegradable cellulose-based biomass materials in wastewater treatment, *Environ. Pollut.*, 2021, **290**, 118087.
9. J. Wang, *et al.*, Pyrolysis-reforming of cellulose over Ni/mordenite for production of hydrogen and heavier organics: The distinct impacts of Ni and mordenite on nature of coke, *Mol. Catal.*, 2023, **538**, 112980.

10. S. Akhtar, *et al.*, Tissue specific expression of bacterial cellulose synthase (Bcs) genes improves cotton fiber length and strength, *Plant Sci.*, 2023, **328**, 111576.
11. D. Xie, *et al.*, Activated carbon derived from hydrochar of food waste for supercapacitor: Effect of components on electrochemical performance, *Fuel Process. Technol.*, 2023, **244**, 107691.
12. M. Fathy, *et al.*, Adsorption separation of condensate oil from produced water using ACTF prepared of oil palm leaves by batch and fixed bed techniques, *Egypt. J. Pet.*, 2018, **27**(3), 319–326.
13. P. Panchal, *et al.*, Photocatalytic and antibacterial activities of green synthesized Ag doped MgO nanocomposites towards environmental sustainability, *Chemosphere*, 2022, **297**, 134182.
14. A. Yaroshevich, A. Eirikh-Rose and H. Khalaily, Management of symbols: Weapons variability and the emergence of large tanged points in early pre-pottery neolithic context of Motza, Judean Hills, Israel, *J. Archaeol. Sci. Rep.*, 2023, **47**, 103790.
15. L.-X. Wang, *et al.*, Review of lignans from 2019 to 2021: Newly reported compounds, diverse activities, structure-activity relationships and clinical applications, *Phytochemistry*, 2022, **202**, 113326.
16. R. A. Bertels, L. M. Harteveld and N. A. Blom, Comments on "Risk Factors for Left Ventricle Enlargement in Children With Frequent Ventricular Premature Complexes" by Bo Chen et al, June 2020, *Am. J. Cardiol.*, 2021, **152**, 178–179.
17. M. G. Günay, *et al.*, Review of functionalized nano porous membranes for desalination and water purification: MD simulations perspective, *Environ. Res.*, 2023, **217**, 114785.
18. R. Revati, *et al.*, In vitro biodegradation, cytotoxicity, and biocompatibility of polylactic acid/napier cellulose nanofiber scaffold composites, *Int. J. Biol. Macromol.*, 2022, **223**, 479–489.
19. Y. Si, S. Shi and J. Hu, Applications of electrospinning in human health: From detection, protection, regulation to reconstruction, *Nano Today*, 2023, **48**, 101723.
20. C. Luo, *et al.*, Preparation and application of high performance PVDF/PS electrospinning film-based triboelectric nanogenerator, *Chem. Phys. Lett.*, 2023, **813**, 140276.
21. Y.-M. Zhao, *et al.*, Effects of ultrasonic-assisted pH shift treatment on physicochemical properties of electrospinning nanofibers made from rapeseed protein isolates, *Ultrason. Sonochem.*, 2023, **94**, 106336.
22. M. Nicolas, *et al.*, Grafting of polyamines onto periodate oxidized nanocellulose, and its application to the fabrication of ionic nanopapers, *Polymer*, 2023, **270**, 125760.
23. M.-H. Sun, *et al.*, Applications of hierarchically structured porous materials from energy storage and conversion, catalysis, photocatalysis, adsorption, separation, and sensing to biomedicine, *Chem. Soc. Rev.*, 2016, **45**(12), 3479–3563.

24. O. A. Hisseine, *et al.*, Nanocellulose for improved concrete performance: A macro-to-micro investigation for disclosing the effects of cellulose filaments on strength of cement systems, *Constr. Build. Mater.*, 2019, **206**, 84–96.
25. A. Firmanda, *et al.*, 3D printed cellulose based product applications, *Mater. Chem. Front.*, 2022, **6**(3), 254–279.
26. M. N. F. Norrrahim, *et al.*, Nanocellulose: the next super versatile material for the military, *Mater. Adv.*, 2021, **2**(5), 1485–1506.
27. J. L. Shamshina and N. Abidi, Cellulose nanocrystals from ionic liquids: a critical review, *Green Chem.*, 2021, **23**(17), 6205–6222.
28. J. Song, *et al.*, Fast and continuous preparation of high polymerization degree cellulose nanofibrils and their three-dimensional macroporous scaffold fabrication, *Nanoscale*, 2013, **5**(6), 2482–2490.
29. X. Qi, *et al.*, Microfibrillated cellulose-reinforced bio-based poly(propylene carbonate) with dual shape memory and self-healing properties, *J. Mater. Chem. A*, 2014, **2**(47), 20393–20401.
30. Á. Coogan and Y. K. Gun'ko, Solution-based "bottom-up" synthesis of group VI transition metal dichalcogenides and their applications, *Mater. Adv.*, 2021, **2**(1), 146–164.
31. H.-D. Yu, *et al.*, Chemical routes to top-down nanofabrication, *Chem. Soc. Rev.*, 2013, **42**(14), 6006–6018.
32. D. K. Patel, S. D. Dutta and K.-T. Lim, Nanocellulose-based polymer hybrids and their emerging applications in biomedical engineering and water purification, *RSC Adv.*, 2019, **9**(33), 19143–19162.
33. K. M. Tripathi, *et al.*, Temperature dependent, shape variant synthesis of photoluminescent and biocompatible carbon nanostructures from almond husk for applications in dye removal, *RSC Adv.*, 2016, **6**(35), 29545–29553.
34. S. R. Djafari, *et al.*, Recent Advances in Cellulose Nanofibers Preparation through Energy-Efficient Approaches: A Review, *Energies*, 2021, **14**(20), 6792.
35. Z. Shen, *et al.*, Nanocellulose-based composite phase change materials for thermal energy storage: status and challenges, *Energy Environ. Sci.*, 2023, **16**(3), 830–861.
36. Y. Ai, *et al.*, Toward cleaner production of nanocellulose: a review and evaluation, *Green Chem.*, 2022, **24**(17), 6406–6434.
37. C. Vilela, *et al.*, Nanocellulose-based materials as components of polymer electrolyte fuel cells, *J. Mater. Chem. A*, 2019, 7(35), 20045–20074.
38. L. Valencia, *et al.*, On the mineralization of nanocellulose to produce functional hybrid materials, *J. Mater. Chem. A*, 2022, **10**(17), 9248–9276.
39. L. Valencia, *et al.*, Fully bio-based zwitterionic membranes with superior antifouling and antibacterial properties prepared via surface-initiated free-radical polymerization of poly(cysteine methacrylate), *J. Mater. Chem. A*, 2018, **6**(34), 16361–16370.

40. R. Zhang, *et al.*, Antifouling membranes for sustainable water purification: strategies and mechanisms, *Chem. Soc. Rev.*, 2016, **45**(21), 5888–5924.
41. M. N. F. Norrrahim, *et al.*, Emerging development of nanocellulose as an antimicrobial material: an overview, *Mater. Adv.*, 2021, **2**(11), 3538–3551.
42. A. Barhoum, *et al.*, Plant celluloses, hemicelluloses, lignins, and volatile oils for the synthesis of nanoparticles and nanostructured materials, *Nanoscale*, 2020, **12**(45), 22845–22890.
43. S. R. El-Dakkony, *et al.*, Effective Antiscaling Performance of ACTF/Nylon 6, 12 Nanofiltration Composite Membrane: Adsorption, Membrane Performance, and Antifouling Property, *Arabian J. Sci. Eng.*, 2022, **47**(6), 6951–6962.
44. S. R. El-Dakkony, *et al.*, Correction to: Composite thin-film membrane of an assembled activated carbon thin film with autoself-healing and high-efficiency water desalination, *Environ. Dev. Sustainability*, 2022, **24**(2), 2542.
45. E. J. Foster, *et al.*, Current characterization methods for cellulose nanomaterials, *Chem. Soc. Rev.*, 2018, **47**(8), 2609–2679.
46. Z. Wang, *et al.*, Nanocellulose coupled flexible polypyrrole@graphene oxide composite paper electrodes with high volumetric capacitance, *Nanoscale*, 2015, **7**(8), 3418–3423.
47. K. Hua, *et al.*, Translational study between structure and biological response of nanocellulose from wood and green algae, *RSC Adv.*, 2014, **4**(6), 2892–2903.
48. J. Zhu, *et al.*, Graphene-coupled nitrogen-enriched porous carbon nanosheets for energy storage, *J. Mater. Chem. A*, 2017, **5**(32), 16732–16739.
49. A. N. El-hoshoudy, *et al.*, Enhanced oil recovery using polyacrylates/ACTF crosslinked composite: Preparation, characterization and coreflood investigation, *J. Pet. Sci. Eng.*, 2019, **181**, 106236.
50. N. K. Khadka, *et al.*, Membrane elasticity modulated by cholesterol in model of porcine eye lens-lipid membrane, *Exp. Eye Res.*, 2022, **220**, 109131.
51. Y. Cui and E. T. Tchetgen, On a necessary and sufficient identification condition of optimal treatment regimes with an instrumental variable, *Stat. Probab. Lett.*, 2021, **178**, 109180.
52. Z. Wang, *et al.*, Preparation, characterization and properties of intrinsic self-healing elastomers, *J. Mater. Chem. B*, 2019, **7**(32), 4876–4926.
53. S. Islam and G. Bhat, Progress and challenges in self-healing composite materials, *Mater. Adv.*, 2021, **2**(6), 1896–1926.
54. M. Keshawy, *et al.*, Synthesis, characterization, and evaluation of polymeric oil sorbent for remediation of hydrocarbons spillage, *J. Dispersion Sci. Technol.*, 2017, **38**(5), 729–736.
55. D. U. Pascoli, *et al.*, Lignocellulosic nanomaterials production from wheat straw via peracetic acid pretreatment and their application in plastic composites, *Carbohydr. Polym.*, 2022, **295**, 119857.

56. C. Dawes, Wang et al. Arch. Oral. Biol. (2015) 60: 385–392, *Arch. Oral Biol.*, 2016, **61**, 83.
57. J.-H. Kim, S. Shim, S.-Y. Kim, J. Park, J. Jung, Y.-W. Kim and S. Seo, Ultra-thin flexible graphene foam for the selective detection of toxic gases, *Sci. Rep.*, 2015, **5**, 10210.
58. M. Jonoobi, J. Harun, A. P. Mathew and K. Oksman, Mechanical properties of cellulose nanofiber (CNF) reinforced polylactic acid (PLA) prepared by twin screw extrusion, *Compos. Sci. Technol.*, 2015, **10**, 174–179.
59. Y. Liu, J. Chen, M. Gu, Z. Zhou, C. Liu, X. Niu, X. Bai, Y. Wang, S. Liu and Q. Li, Preparation of cellulose nanofibril and its application as adsorption material, *Cellulose*, 2018, **25**, 3551–3564.
60. H. Xie, S. Li, S. Zhang and Z. Zhao, Review on the production and application of carbon dots, *J. Mater. Sci. Technol.*, 2019, **35**(2), 245–255.
61. Y. Liu, J. Chen, M. Gu, Z. Zhou, C. Liu, X. Niu, X. Bai, Y. Wang, S. Liu and Q. Li, Preparation of cellulose nanofibril and its application as adsorption material, *Cellulose*, 2015, **22**(2), 1227–1241.
62. A. El-Moghny, Preparation and characterization of nanocomposite membranes for water treatment applications, MSc Thesis, Faculty of Science, Tanta University, 2016.
63. Y. Liu, J. Chen, M. Gu, Z. Zhou, C. Liu, X. Niu, X. Bai, Y. Wang, S. Liu and Q. Li, Preparation of cellulose nanofibril and its application as adsorption material, *Cellulose*, 2015, **22**(5), 2981–2992.

CHAPTER 7

Graphene-based Materials for Water Remediation: Recent Advances on Pollutant Sorption, Photodegradation and Filtration

BRUNO HENRIQUES[a,b*], NICOLE FERREIRA[a,b], MARTINA KOCIJAN[c], MILAN VUKŠIĆ[d], BURAK ÜNLÜ[e,f], SERBÜLENT TÜRK[e,f], MAHMUT ÖZACAR[f,g*] AND GIL GONÇALVES[h,i*]

[a]LAQV-REQUIMTE – Associated Laboratory for Green Chemistry, University of Aveiro, Aveiro, Portugal; [b]Department of Chemistry University of Aveiro, Aveiro, Portugal; [c]Department of Materials, Faculty of Mechanical Engineering and Naval Architecture University of Zagreb, Ivana Lučića 5, 10000 Zagreb, Croatia; [d]Department for Nanostructured Materials, Jožef Stefan Institute, Jamova cesta 39, 1000 Ljubljana, Slovenia; [e]Sakarya University, Biomedical, Magnetic and Semiconductor Materials Research Center (BIMAS-RC), 54187, Sakarya, Turkey; [f]Sakarya University, Biomaterials, Energy, Photocatalysis, Enzyme Technology, Nano and Advanced Materials, Additive Manufacturing, Environmental Applications and Sustainability Research and Development Group (BIOENAMS R&D Group), 54187, Sakarya, Turkey; [g]Sakarya University, Faculty of Science, Department of Chemistry, 54187, Sakarya, Turkey; [h]Department of Mechanical Engineering, Centre for Mechanical Technology and Automation, University of Aveiro, 3810-193 Aveiro, Portugal; [i]Intelligent Systems Associate Laboratory (LASI), 3810-193 Aveiro, Portugal
*E-mail: brunogalinho@ua.pt, mozacar@sakarya.edu.tr, ggoncalves@ua.pt

Chemistry in the Environment Series No. 12
Novel Materials and Water Purification: Towards a Sustainable Future
Edited by Grigorios L. Kyriakopoulos and Miltiadis G. Zamparas

Published by the Royal Society of Chemistry, www.rsc.org

7.1 Graphene and Its Derivatives

It is well known that new materials play an important role in the development of science and new technologies, since the implementation of new technologies often requires the support of new materials. Therefore, investigation of new materials with exceptional properties plays an important role in scientific research. Carbon-based materials are among the most widely represented groups of materials and have a wide range of applications due to their exceptional properties. Not only are there many allotropes of carbon, such as diamond, graphene, fullerene, and nanotubes, but also today there are many allotropes of graphene itself.[1] Graphite as a material has been known since the invention of the pencil in 1564, and got its name from the Greek words for 'to draw/write' in 1789 from the German chemist and mineralogist A. G. Werner.[26] Graphite is formed when carbon atoms are stacked in a layered honeycomb structure, *i.e.*, they have a hexagonal structure.[2] The graphite layers are arranged according to Bernal's model, with every third carbon layer in the same position. The high stability of each layer is attributed to strong covalent bonds (C–C) with a bond length of ~0.142 nm, while weaker van der Waals interactions between the layers occur at a distance of ~0.335 nm. Each carbon atom is bonded in-plane to three other carbon atoms by sp^2 hybridization at an angle of 120°.[3,4] The structure of graphite is shown in Figure 7.1. Less inert than diamond but more thermodynamically stable, graphite is used in electrodes, resins, adhesives, pencils, and lubricants. In addition, it features strong oxidation resistance, good thermal insulation, and corrosion resistance.[5,6] Despite its short history, thanks to its unique electrical, optical, and thermal properties, graphene is a "hot spot" material in science and has attracted attention worldwide since Andre Geim and Konstantin Novoselov separated a layer of graphene from graphite in 2004. Graphene is a single layer of graphite, as shown in Figure 7.1.[7]

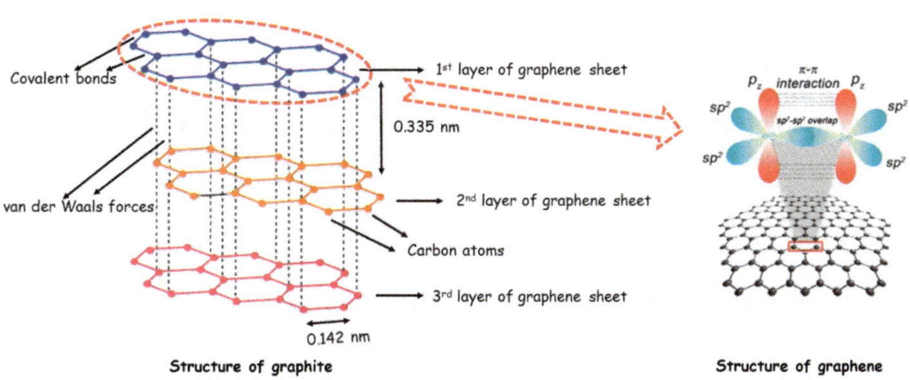

Figure 7.1 Structure of graphite and graphene. Adapted from ref. 8, https://doi.org/10.3390/s17102161, under the terms of the CC BY 4.0 license https://creativecommons.org/licenses/by/4.0/.

Graphene possesses high thermal conductivity (~5000 W m^{-1} K^{-1}), exceptional mobility of charge carriers at room temperature (200 000 cm^2 V^{-1} s^{-1}), high Young's modulus (~1 TPa), optical permeability (97.7%), and extremely high theoretical specific surface area (~2600 m^2 g^{-1}).[9,10] Bonding in the graphene plane is based on an sp^2-hybridized orbital composed of s, p_x and p_y, while p_z is free and perpendicular to the plane. The plane has a σ bond with a very short interatomic length of about 0.145 nm, making it a stronger bond than the sp^3-hybridized orbital in diamond. Such a strong C–C–C bond is the reason for the mechanical resistance of a single layer of graphene.[11] The half-filled π band causes the absence of a band gap between the valence and conduction bands, resulting in free movement of electrons (e^-) in one layer of graphene. Also, weak π bonds result in weak van der Waals interactions between the layers, which allows them to move with each other under very weak shear stress.[11,12] These properties make graphene an attractive material for various applications, such as membranes, sensors, solar, and electrochemical cells, and in biomedicine.[13–15] Polar derivatives of graphene are graphene oxide (GO) and reduced graphene oxide (rGO).

GO is an oxidized form of graphene containing aromatic rings (sp^2 carbon) and carbon atoms with single bonds (sp^3 carbon). It also contains oxygen functional groups in its basal planes and edges. Functional groups, such as carboxylic (COOH), hydroxyl (C–OH), carbonyl (C=O), and epoxy (C–O–C), are covalently attached to the structure. Oxygen functional groups are responsible for many advantages over graphene, including higher solubility and the possibility of surface functionalization, which opens up many possibilities for use in nanocomposite materials.[16] However, the oxidation of graphite in GO introduces defects in the carbon structure that result in significantly changed physicochemical properties.[17,18] GO is used in biological applications due to its interesting properties, such as hydrophilicity, high dispersion in aqueous medium, simplicity of synthesis, high biocompatibility, low cytotoxicity and surface functionalization due to the presence of oxygen functional groups.[19,20] Despite the high colloidal stability in water and the unique mechanical, colloidal, and optical properties, the methods used in GO synthesis are known to use strong oxidizing agents, such as potassium permanganate, resulting in a significant number of defects in its crystalline structure. The defects present affect the conductivity of GO, which becomes far lower than that of graphene. One option to restore graphene-like properties is the reduction of GO.[20,21] rGO is the form of GO that can be processed by various methods to reduce the oxygen content in the structure.

7.1.1 Preparation of Graphene and Its Derivatives

The desired structure and properties, which largely depend on the shape, size, and functional groups attached to the surface of the material, play an important role in choosing the most favourable synthesis method of graphene and its derivatives. Synthesis of GO is divided into two categories: bottom-up and top-down approaches (Figure 7.2).[22] In the top-down approach, larger structures

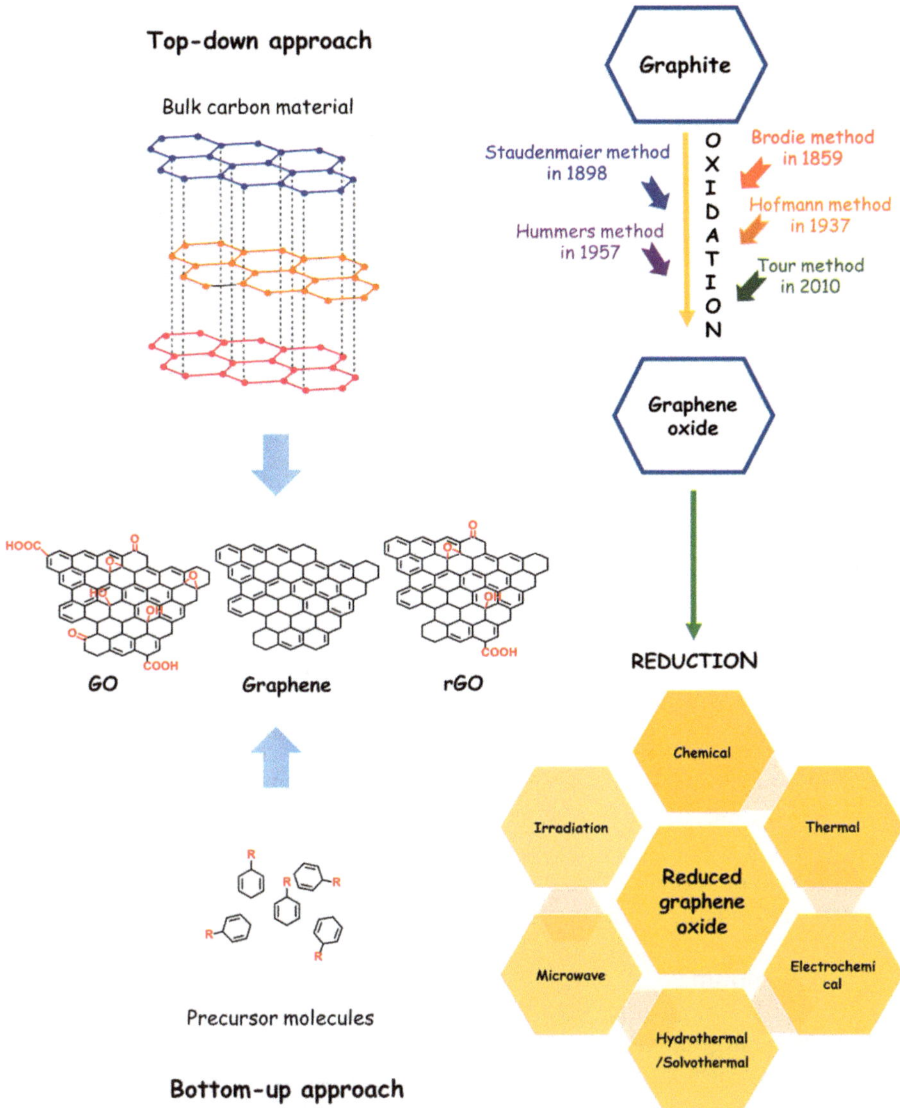

Figure 7.2 Scheme of bottom-up and top-down approaches for synthesis of GO and its reduction.

are reduced to nano sizes while retaining their original properties without control at the atomic level. Drexler *et al.* introduced a bottom-up approach in which materials are prepared from atomic or molecular components.[23] In the bottom-up approach, simple carbon molecules are used to prepare graphene. Because graphene can be assembled from smaller carbon precursors, this method can allow atomically precise control over the synthesis of graphene nanosheets.[24] Disadvantages of the bottom-up approach include higher costs, low reaction yield, long process time, and difficulties in method scalability compared to the top-down approach.[25] The advantages of the top-down approach are scalability, no substrate required, cost effectiveness, and high reproducibility. However, it is impossible to predict the exact structure of the graphene.[22,26] The most well-known approaches to fabricate graphene using the top-down method are chemical and mechanical exfoliation, while for bottom-up approaches they are chemical vapor deposition (CVD) and pyrolysis.[24]

Many top-down synthesis approaches are used to produce GO, and flakes of natural graphite with different particle sizes are used as precursors (Figure 7.2).[27] Although graphene received a lot of attention from scientists in 2004 after Geim and Novoselov isolated a graphene sheet in the form of a single layer, the synthesis of GO was explored much earlier. A scheme showing synthesis methods for GO and its reduction is shown in Figure 7.2. In 1859, Benjamin Brody synthesized GO from graphite flakes using $KClO_3$ and HNO_3.[28] Then, in 1898, Staudenmaier slightly modified Brody's method using concentrated H_2SO_4 in the presence of $KClO_3$ and HNO_3.[29] The methods mentioned are very simple for synthesis; however, a major drawback is the formation of toxic components, such as NO_2, N_2O_4, and ClO_2 during the manufacturing process.[30] In 1958, Hummers improved the existing synthesis process of GO by using strong oxidants, such as $KMnO_4$ and $NaNO_3$, resulting in greater utilization of the reaction with a higher yield and purity compared to the synthesis product from previous methods.[31]

Hummers made several improvements to the original two GO synthesis techniques to make GO safer to produce, including using $KMnO_4$ as the oxidant (instead of $KClO_3$, which evolves the toxic gas ClO_2) and adding sodium nitrate (to generate nitric acid *in situ* rather than using nitric acid as solvents).[32] In 2010, Marcano *et al.*[33] prepared a GO material by implementing several modifications of the Hummers method, called the improved Hummers method, using natural graphite flakes of ~150 μm in size. The improved Hummers method removes sodium nitrate by adding phosphoric acid and increased amounts of $KMnO_4$. No toxic gases are produced by the above modifications, allowing for easy temperature control, and the synthesis product is GO with a higher degree of oxidation. The size and shape of the graphite flakes determine the size and shape of the GO produced.[32,34,35] When modifying the Hummers method, it is necessary to consider the choice of chemicals used, and these depend on the application. The most frequently considered parameters are carbon source, oxidizing agent, choice of protonated solvent, and pretreatment methods, as they can significantly affect the C/O ratio in GO.[32,36,37]

7.1.2 Reduction of Graphene Oxide

GO consists of sp^2- and sp^3-hybridized carbon atoms, which are characterized by their optical and electronic properties. The controlled reduction of GO occurs by adjusting the ratio of sp^2 and sp^3 hybrid carbon atoms. To obtain rGO, *i.e.*, a material with properties similar to graphene, it is necessary to remove the oxygen functional groups (hydroxyl, carbonyl, carboxyl, and epoxy) from the structure of GO. Reduction of oxygen functional groups can be achieved by thermal annealing reduction, chemical, and electrochemical reduction (Figure 7.2). Depending on the reduction process used, the morphological, optical, and electronic properties change.[38,39] During the implementation of the GO reduction process, the ratio of carbon and oxygen, surface defects, and the choice of reducing agent play important roles, which consequently affect the physical and chemical properties of the rGO.[32,39] rGO can be easily prepared from cheap GO by using a variety of chemical, electrochemical, microwave, thermal, and photo-assisted methods. The reduction of GO by thermal annealing relies on the decomposition of oxygen functional groups into CO and CO_2 gases, which develop in the spaces between the graphene sheets at elevated temperatures.[53] In thermal reduction, a pressure of 40 MPa is created at 300 °C, while 130 MPa is created at 1000 °C.[40] With a rapid increase in temperature, the oxygen-containing functional groups attached to the carbon plane disintegrate into gases that create enormous pressure between the stacked layers. This results in small and wrinkled sheets of rGO because the decomposition of oxygen functional groups occurs, which consequently removes carbon atoms from the carbon plane; that is, the GO layers split into small sheets and this results in distortion of the carbon plane.[41,42] Thermal reduction can also be performed with non-conventional heating sources, such as microwaves.[43] Microwave reduction, along with the removal of oxygen functional groups, is particularly effective for removing the C=O group and removes defects on the graphene layers.[44,45] Reduction with chemical reducing agents is based on their chemical reactions with GO. Reduction can usually be carried out at room temperature or by moderate heating. The most well-known reducing agents are hydrazine, metal hydrides, and hydrohalic acids.[41,46] Reducing agents such as borohydride mainly reduce carbonyl groups, aluminium hydride removes carboxyl groups and esters, while hydrohalic acids remove hydroxyl and epoxy groups.[44] rGO can also be produced by photochemical reactions using photocatalysts, such as titanium dioxide (TiO_2). Williams *et al.* reported the reduction of GO in the colloidal state with the help of TiO_2 nanoparticles under ultraviolet irradiation.[47] This method leads to the elimination of hydroxyl and carbonyl functional groups.[44] Electrochemical reduction is another method of reducing GO to rGO and does not require the use of chemical agents. The reduction is driven solely by electron exchange between GO and electrodes in an electrochemical cell using an aqueous buffer solution at room temperature. For electrochemical reduction, GO can be in the form of sheets or films. The advantage of this method is that it avoids the use of dangerous reducing agents and eliminates by-products.[41,48] Solvothermal reduction is a process carried out in a closed vessel, so that the solvent can be brought to a temperature well above its boiling point by increasing the pressure, resulting in heating.[49,50] In the hydrothermal process, superheated supercritical water can act as a reducing agent and offers an environmentally friendly alternative to organic solvents. In addition, its physicochemical properties can change greatly with changes in pressure and temperature.[41] In addition to the hydrothermal reduction method, there are also numerous other environmentally acceptable reducing agents, such as ascorbic acid, sugar, amino acids, and microorganisms for the effective preparation of rGO.[38,51,52] Using microorganisms as a reducing agent removes epoxy and hydroxyl groups.[44] The properties of GO and rGO depend not only on the GO synthesis procedure selected, but also on the structure and size of the natural graphite used for the GO synthesis. Chee *et al.* studied the effect of particle size range of graphite, from 0.05 to 41 μm, on the synthesis of GO and its reduction. It was observed that the stated range of graphite particle sizes does not affect the defects and the amount of functional oxygen groups after thermally reducing GO.[53] As a result, the electrochemical properties of thermally reduced GO are independent of the graphite particle size. Smaller graphite particle sizes led to higher reduction of GO, resulting in high electrochemical properties of graphite-derived rGO with a particle size of less than 20 μm.[54,75] Dao *et al.* also studied graphite particle size and showed that smaller graphite particle sizes resulted in a higher degree of oxidation and better delamination. An optimal graphite particle size of about 50 μm was found to show the best exfoliation of GO and thermally reduced rGO with high specific surface area, low oxygen content, and better conductivity.[55] Graphite with a large crystallite diameter was found to be the most suitable material for the preparation of GO and rGO with a high degree of reduction and a small number of defects.[56,57] Bychko *et al.* studied different structures of graphite particles and their influence on the properties of the synthesis products, *i.e.*, GO and rGO. It was found that the degree of graphitization plays an important role in obtaining complete oxidation and exfoliation of graphite, which led to GO and rGO with a high content of sp^2-hybridized carbon atoms, where rGO showed improved catalytic activity.[58]

7.2 Graphene Composites for Photodegradation of Aqueous Organic Contaminants

Semiconductor materials have a high crystallinity and photosensitivity, low absorption capacity, wide band gap (E_g), and fast electron–hole (e^-–h^+) recombination. However, when semiconductors are combined with graphene-based materials, they display improved photocatalytic activity. Overcoming the limitations present in semiconductor-based photocatalysis, as well as the search for potentially efficient materials, is extremely important and the modification of semiconductors with carbon-based materials, such as graphene, is promising due to its unique properties. Graphene and its derivatives have stimulated interest in the preparation of highly efficient composite photocatalysts for increasing the photocatalytic efficiency of semiconductor materials used in the processes for degradation of pollutants.[59] Incorporation of graphene-based materials in the semiconductor structure may overcome these limitations due to its unique properties. Remarkable properties, such as structural, chemical, electrical, and optical, make graphene-based materials one of the main types of nanomaterials in a wide range of applications. The major factor that improves the photocatalytic process of graphene-based nanocomposites is the preparation method, which controls the morphology of the synthesized photocatalysts, and thus the recombination charge carriers are reduced and so is the band gap, E_g.[60]

GO with good charge mobility, large surface area, and good thermal conductivity has been widely used as an electron carrier to suppress e^-–h^+ recombination.[61] However, many defects occur during the oxidation of graphite to GO, which leads to a change in material properties, while the reduction of GO to rGO can significantly improve the electron mobility and charge capacity of GO.[17] Defects in rGO play an important role in the photocatalytic activity of nanocomposites because the carbon vacancies and low oxygen content in the rGO structure provide more bonding sites with semiconductor materials, potentially reducing the energy gap of the photocatalyst.[62,63] An attractive solution is the implementation of rGO as an electron carrier. The high conductivity of rGO facilitates electron transfer because it acts as an electronic bridge between semiconductor particles, which slows recombination and extends the lifetime of charge carriers.[64,65] Furthermore, rGO can promote photocatalytic activity towards degradation by adsorption of pollutants due to various possible mechanisms, including improved charge separation and reduction of e^-–h^+ recombination rates, but also through highly reactive oxidative species.[66,67]

7.2.1 Synthesis Methods of Graphene-based Composites

Depending on the desired properties of the nanoparticles, many known synthesis methods are used.[68–70] Hydrothermal and solvothermal synthesis using the sol–gel method is one of the most used methods for the preparation of graphene-based nanoparticles. With optimal control of parameters, including temperature (>100 °C) and pressure (>1 atm), it is possible to prepare nanomaterials of different morphologies and particle sizes.[71] The technique of solvothermal synthesis is analogous to hydrothermal synthesis. The main difference is that the hydrothermal process uses water as a solvent, while the solvothermal process uses organic substances as solvents.[72] However, the sol–gel method is costly because it involves expensive raw materials, and the drying process itself, to remove the organic material, leads to cracking and reduction in the volume of the prepared synthesis product.[73]

A sonochemical method, which uses ultrasound to mix, homogenize, and reduce particle sizes, with a short preparation time is often used.[74] Recently, the microwave-assisted method for nanoparticle synthesis has gained popularity because it involves homogeneity and rapid heating of the reaction mixture at the required temperature, which leads to time savings; however, microwave-assisted methods require high-power microwave heating with high energy consumption. Also, this method does not allow monitoring of the growth of nanoparticles over time. Moreover, microwave-assisted synthesis does not allow the preparation of a large number of nanoparticles.[75]

The mechanochemical method for the preparation of nanoparticles is obtained without the use of organic solvents or high temperatures, therefore it has the potential to reduce the negative impact on the environment. Nevertheless, high-energy ball milling may require high energy consumption.[76] Disadvantages of mechanochemical synthesis compared to other methods include that synthesis products are subject to contamination from by-product production, milling balls, or milling reactors. Synthesis products may be of low crystallinity due to defects caused by the input of mechanical energy. During the preparation of highly soluble metal oxides, they decompose after contact with washing liquids, and cannot be easily separated from the synthesis by-products.[76,77]

7.2.2 Graphene Binary Composites

Prepared binary composites show higher photocatalytic activity than individual photocatalysts due to the synergetic effect between graphene sheets and the semiconductor material. Therefore, much more effort is invested in the preparation of various graphene-based composites through developing different methods and fabricating diverse microstructures with improved properties compared with the bare photocatalyst. GO is used to reduce the probability of e^-–h^+ pair recombination by promoting electron transfer in semiconductor photocatalysts. The other benefits of using GO to achieve improved photocatalytic activity are increased surface area, as demonstrated by the enhanced absorption of light by the prepared composites. In this section, various graphene binary composites are presented.

Seema *et al.*[78] achieved uniform distribution of SnO_2 onto rGO sheets. The rGO–SnO_2 composite was successfully used in the photodegradation of methylene blue (MB) under sunlight and showed improved activity compared to SnO_2 nanoparticles, due to the electrical conductivity and enhanced charge separation of rGO in the composite. Chai *et al.*[79] synthesized rGO–WO_3 composites *via* a one-pot hydrothermal method. They reported an optimal amount of rGO in the composite of 15 wt%, tested with photodegradation of MB. Chen *et al.*[80] investigated photocatalytic activity of β-Bi_2O_3 nanocomposites. The composite with 7 wt% of graphene loading showed the highest activity, while further increase of graphene resulted in decreased photocatalytic activity, which was ascribed to high adsorption of MB on the surface of the catalyst during dark reaction. Li *et al.*[81] used a chemical deposition route to prepare a ZnO–GO nanocomposite. The flower-like ZnO nanoparticles obtained were evenly distributed onto GO sheets, which led to a favourable high surface area and hierarchical porosity, improving the adsorption and mass transfer during photodegradation of MB under visible-light irradiation. Yusoff *et al.*[82] decorated GO sheets with CuO nanoparticles *via* a hydrothermal process. They reported the possibility of tuning the size of deposited CuO nanoparticles by varying the amount of functionalized GO. The photocatalytic activity of GO–CuO nanocomposites was assessed by the degradation of MB in the presence of H_2O_2 using UV–vis spectroscopy, where the composite with a GO content up to 30 mg showed the highest photoactivity due to stronger interactions between CuO and GO. Wang *et al.*[83] prepared composite films of rGO–TiO_2 *via* post-treatment of as-prepared GO–TiO_2 films, which were used as photocatalyst for dye degradation in the photo-electrocatalytic process. The optimal rGO–TiO_2 mass ratio of 1 wt% was determined based on photo-electrocatalytic degradation of rhodamine B (RhB) and acid orange II (AO-II). In both cases the film used exhibited an enhanced degradation rate compared to bare TiO_2, while demonstrating satisfactory stability. Jahangir *et al.*[84] synthesized CdO–GO nanocomposites with varying weight ratios of GO *via in situ* deposition of CdO nanoparticles on GO sheets by applying a solution phase approach. The synthesized CdO–GO nanocomposite with a GO loading of 3.3 wt% showed the highest activity. The increase in the photocatalytic activity noticed was ascribed to the positive effect of GO, the presence of which resulted in high charge separation, suppressed recombination of photogenerated e^-–h^+ pairs, and boosted adsorption of pollutant molecules.

7.2.3 Graphene Tertiary Composites

Ternary graphene-based nanocomposites exhibit synergistic effects of their constituents at the nano level, which may lead to improved physical/chemical properties, in comparison with the individual components. One of those effects may be a higher electrocatalytic activity in the quantification of chemical species.[85]

Ojha *et al.*[86] prepared a photo-Fenton-type tertiary nanocomposite ZnO-Fe_3O_4/rGO, which demonstrated enhanced degradation of a modal pollutant, methylene blue dye. Furthermore, the proposed applied photo-Fenton reagent synthesis method was low cost and less toxic to fabricate various nanocomposites. Labhane *et al.*[87] prepared a Fe-CeO_2/rGO ternary photocatalyst by using a hydrothermal method. The catalyst showed remarkable photodegradation of organic dye (Congo red) under simulated UV light. The excellent photodegradation performance was explained by the synergistic effect between the reduced GO and the generated redox systems (Ce^{4+}/Ce^{3+}; Fe^{3+}/Fe^{2+}). Tahir *et al.*[88] evaluated the photocatalytic activity and antibacterial activity of the ternary nanocomposite Ag_2WO_4-MoS_2/GO synthesized by a hydrothermal method. The observed increase in photocatalytic activity and antibacterial activity under visible light was assigned to enhanced separation of charge carriers at the interfaces of Ag_2WO_6 and MoS_2 at the ternary heterojunction. Furthermore, the optimal photodegradation conditions (aqueous medium pH of 5, initial dye concentration of 10 mg L^{-1}, and catalyst dose of 50 mg/100 mL) were determined by applying the response surface methodology. Altin[89] reported the preparation of CuO-TiO_2/GO nanocomposite and its activity for the photodegradation of bisphenol A under visible light. The optimal addition of GO for the fastest degradation of the pollutant was 10 wt%. Zarrabi *et al.*[90]

used a solvothermal method to prepare ZnO–GO–Fe_3O_4 photocatalyst for dye removal from waste water. The composition with 25 wt% of Fe_3O_4 demonstrated better photoactivity, while increasing the Fe_3O_4 content resulted in a decrease in the photoactivity due to filling of the structural gaps, and consequently reduction in the available active sites for photodegradation of the dye. The photodegradation efficiency for MB was about 97%, whereas the same amount of photocatalyst for the photodegradation of eosin yellow achieved an efficiency of about 60% after 180 min. Zheng *et al.*[91] combined the advantages of photocatalysis and the Fenton reaction to achieve mineralization of organic pollutants by developing a novel Fe_2O_3/TiO_2/rGO nanocomposite synthesized using a solvothermal method. The mineralization of methyl orange (MO) was conducted under visible light over a wide pH range. The photodegradation was enhanced by visible-light absorbance and diminished e^-–h^+ recombination due to fast degradation of MO *via* Fenton reactions. Additionally, the strong interaction between the evenly distributed particles of TiO_2 and Fe_2O_3 on rGO prolonged the lifetime of the developed catalyst during photocatalytic/Fenton reactions over a wide pH range, from 3.4 to 9.2. Danish *et al.*[92] synthesized a ternary nanocomposite Fe_2O_3/g-C_3N_4/GO *via* the thermal impregnation technique, for removal of organic pollutants under visible light in an aqueous system. The sample with 10 mg of Fe_2O_3 to g-C_3N_4 with a fixed amount of GO (10 mg) showed the best photoactivity for the removal of RhB, MB, and 3-methyl-4-nitrophenol in aqueous suspension. The high photocatalytic performance was attributed to the decreased rate of recombination of e^-–h^+ pairs, as a consequence of heterojunctions formed between the nanocatalysts.

7.3 Graphene-based Composites for Sorption of Metal Ions

Metal contamination in aquatic environments due to the fast development of industries that generate a large amount of waste has been extensively reported.[93] Classical contaminants, such as Hg, Pb, and Cd, and emerging contaminants, such as Co, Sr, and rare earth elements, which are also critical raw materials, can be found in industrial waste waters, ranging from low to high contamination levels. Most of these contaminants are toxic at low concentration, which highlights the need to remove them from aquatic environments. The use of graphene-based materials as sorbents for water remediation has proven to be a promising and effective approach. Several studies have shown the capacity of GO, rGO, and their composites for the sorption of contaminants from water, such as metal ions,[94,95] organic pollutants,[96] microplastics,[97] and pharmaceuticals,[98] among others. GO can be easily dispersed in water, making it difficult to remove the material from solution post-sorption. In addition, due to its interplanar forces, GO can readily aggregate in solution, reducing its surface area and minimizing its adsorption capacity. To improve the adsorption capacity and minimize the drawbacks of GO, surface modifications of GO and GO composites have been developed.[99]

Among surface modification and functionalization of GO, some of the most important functional groups are the oxygen-containing functional groups, such as hydroxyl (–OH), carboxyl (–COOH), epoxy (–COC–), and carbonyl (–C=O), and the nitrogen-containing functional groups, such as amines ($-NH_2$), amide ($-RC=ONH_2$), and cyano groups (–CN). The functionalization of GO using nitrogen-containing functional groups was reported by Bessa *et al.*[100] using polyethylenimine (PEI) to modify GO for the efficient sorption of Hg from different natural waters. The study showed that Hg species were associated with nitrogen functional groups provided by the PEI modification.

Among the most studied composites it is possible to observe the tendency of using GO-polymers, such as GO-chitosan (GO-CS) and GO-alginate (GO-ALG), due to the addition of several organic functional groups, such as amino, carboxyl, and thiol groups, which improve metal removal. For example, GO-CS has been studied for the removal of Hg from different water matrices and demonstrated up to 97% efficiency in ultrapure water using very low doses (10 mg L^{-1}) of material. Although GO-CS composites have better water solubility and dispersity compared to single CS or GO, the formation of composites reduces active adsorption sites compared to GO, thus minimizing the adsorption capacity to remove metal ions in aquatic environments. On the other hand, several studies aimed at the removal of metals using GO-alginate composites have been reported,[101–103] especially regarding the gelling property of alginate to form hydrogels,[104] aerogels,[105] and beads,[106] which make the removal of GO-alginate from solution easier.

To overcome the drawback of separating GO adsorbents from solution, studies have also been done to incorporate magnetic properties into GO, to produce magnetic GO (MGO), using Fe(II and III) oxide (Fe_3O_4). Chandra *et al.* (2010)[107] studied the removal of As(III) with MGO nanoparticles, resulting in a high removal efficiency (~100%) at a low concentration (1 ppb). The removal efficiency of MGO for Cd(II) and Pb(II) was evaluated by Bao *et al.*,[99] resulting in a maximum adsorption capacity of 128.2 and 385.1 mg g^{-1} for Cd and Pb, respectively. Some examples of GO composites can be found in Table 7.1.

In sorption, it is important to evaluate the effect of the operating parameters (pH, initial metal concentration, GO dosage, competing ions, *etc.*) to achieve maximum efficiency, and to know the mechanisms behind metal sorption on GO composites for its effective use in real wastewater treatment. The mechanisms of metal sorption by GO composite include numerous types of interactions, such as ion exchange, surface complexation, electrostatic interaction, metal–π bond interaction, and physical adsorption.[108]

The potential of GO to remove positively charged metals owing to its negatively charged oxygen functional groups has been widely explored.[109] Nevertheless, some obstacles still need to be overcome before commercial application. GO composites for metal removal are still undergoing laboratory testing, so costs for production and scalability of the material are still not fully developed. Few studies have shown selectivity of GO and GO composites towards a specific metal. Additionally, more studies should be conducted to

Table 7.1 Comparison of performance parameters of some graphene-based composites for metal sorption[a]

Group	Element	Material	Sorption capacity (mg g^{-1})	Removal (%)	Sorbent dosage (g L^{-1})	pH	Concentration (mg L^{-1})	Time (h)	Surface area (m^2 g^{-1})	Matrix	Reference
Classical contaminants	Cd	GO-FG	1532		0.03	4	10–1100	3		Ultrapure water	110
		GO-AG	1793		0.03–0.06						
		CS/GO-SH	177		1	5	10–500	1.5		Ultrapure water	111
		MMSP-GO	167		0.1	7.1	10–100	24			112
	Mn	MGON	400	98		7	100–1000	4 min	224	Ultrapure water	113
	Zn	GO	8.5		0.1	3.5		0.08			114
		MGON	373	92		7	100–1000	6 min	224	Ultrapure water	113
	Cr	RGO/Ni	198		0.2	4	10–110	30			115
		MCGO-IL	145		1	2	60–200	1			116
	Ni	GO		91	0.2	8	20–120	24			117
		rGO		84	0.2	8	20–120	24			117
	Hg	rGO	109	73	0.22	6.4	10–100	1			112
		GO-EDTA-CS	324		1.25	5.1	100	24	1.33		118
		CS/GO	397	93	1	6	0–500	24			119
		Xanthate-Fe_3O_4-GO	119		0.15	7	5–80	3	30.13		120
	Pb	GOMO	190		0.4	6.5	20–160	0.5			121
Emerging contaminants (also critical raw materials)	Eu	GONS	175		0.2	6	50	24			122
	Sr	PVA/GO	20	51	0.5		22.16	8			123
			17	79		7	10	12			
	Co	GO	21.3	> 90%	1	5.5	2–25	1		Ultrapure water	124
		GO	66.7		0.03–0.3	6.8	10	24	2.8		125
		MGO-CS	15.24		2	5	20–200	5	27.15		126
		β-CGO	72.4		0.1	6	10	24			127

[a]Note: polyvinyl alcohol/GO (PVA/GO); magnetic GO-chitosan (MGO-CS); β-cyclodextrin-modified GO (β-CGO); GO-flaky graphite (GO-FG); GO-amorphous graphite (GO-AG); chitosan/sulfhydryl-functionalized GO composite (CS/GO-SH); polyethylenimine-modified magnetic mesoporous silica and GO (MMSP-GO); magnetic GO nanocomposite (MGON); rGO/NiO (RGO/NiO); magnetic chitosan and GO-ionic liquid (MCGO-IL); GO (GONS); nanohybrid of GO and MgO (GOMO).

improve the ease of separation of GO from water. Furthermore, the adsorption of different metal ions, column-scale, and pilot-scale experiments must be conducted to determine the future commercial application of GO and GO composites. The selection and functionalization of GO must be further investigated to make highly efficient and reusable adsorbents, minimizing the use of chemical reagents, chemical degradation, and the environmental exposure of these materials.

7.4 Graphene-based Membranes for Water Filtration

Graphene and its derivatives have been excellent candidates for water purification membranes because of their ultra-low thickness (one atom thickness for graphene), extremely high specific surface area, and high mechanical strength. Furthermore, graphene-based materials have low-cost production, which makes them advantageous for use in water purification. As for most nanomaterials, graphene-based membranes have been used for nanofiltration, ultrafiltration, forward osmosis, reverse osmosis, ion exchange, gas separation, and pervaporation.[128–131]

The properties of graphene mean that graphene-based membranes possess low transport resistance and high permeate flux. Having a high mechanical strength, with a 2D planar thin structure makes graphene-based membranes indispensable in processes requiring low pressure, such as liquid-phase and gas-phase separation processes in water treatment. However, the electron density of graphene causes graphene-based membranes to be impermeable even to gases as small as helium.[132] For using graphene as a separation membrane for water purification, selective nanopores should be introduced into graphene layers, whereupon it is called nanoporous graphene, thus converting graphene to a porous material that increases selective passage for water, ions, and gases.[133,134] GO is a graphene derivative hosting functional groups containing oxygen, such as carboxyl, hydroxyl, *etc.* Having these functional groups means that GO is defective when compared with an ideal graphene layer. However, these defects cause stacking of GO layers. Therefore, one-dimensional nanopores or two-dimensional nanochannels occur due to deformation and stacking, which makes GO advantageous when used as a membrane material for water treatment.[135] Another approach to use graphene-based materials for membrane fabrication is stacking of several graphene or graphene derivative layers to obtain laminates. Combining graphene or graphene derivatives with a polymer or inorganic material, the properties of the pristine membrane, such as permselectivity, can be increased.

Graphene-based membranes can be considered in three subgroups: combined graphene laminates, porous graphene layers, and graphene-based composites (Figure 7.3).

Although there are different synthesis routes for the production of nanoporous graphene, all of the graphene synthesis methods cannot be applied for the fabrication graphene-based membranes utilized in water

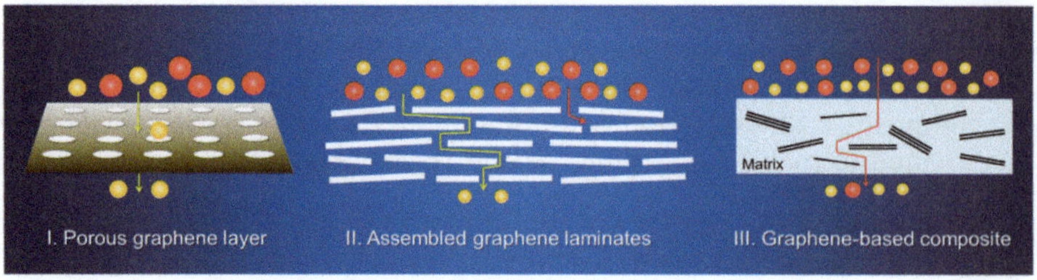

Figure 7.3 Structurally different graphene-based membrane types. Reproduced from ref. 132 with permission from the Royal Society of Chemistry.

treatment.[136] Due to the effect of pore size on the selectivity of the membrane, the pore size should be equal and homogeneously distributed on the graphene. Top-down production methods consist of removing one layer from bulk material. Therefore, top-down methods are complicated for transferring graphene layers and pore generation. Additionally, size, shape, and layer number cannot be controlled because of the randomness of top-down methods. Bottom-up methods, which use stacking of atoms to synthesize nanomaterial, are more suitable for pore generation onto graphene but are very demanding.[137] Furthermore, bottom-up methods cannot produce graphene on a large scale for water purification membranes.

Porous supports can be used for introducing nanopores on CVD-grown single-layer graphene using physical or chemical etching.[138] O'Hern *et al.* used accelerated gallium ions to generate pores on the graphene layer. Another nanoporous single-layer graphene was prepared and used as a water desalination membrane by Surwade *et al.*[139] After synthesizing single-layer graphene by ambient-pressure CVD, nanopores were introduced by oxygen plasma exposure. Raman spectra showed that even for 0.5 s of oxygen plasma exposure, the D peak increased to one-third of the G peak. By increasing the exposure time to 6.0 s, the 2D peak disappeared completely and the G and D peaks broadened due to defect formation. Transmission electron beam ablation lithography (TEBAL) is another method for preparing nanoporous graphene membranes. Merchant *et al.* used a nitride membrane that had ~1.5 μm diameter holes as support, and TEBAL was used for creating nanopores on graphene. The membrane obtained had nanopores of 5 to 10 nm diameter, which can be used for water treatment.[140] GO differs from graphene with respect to its synthesis. It can easily be synthesized *via* the modified Hummers method by oxidizing graphite with strong agents, such as $KMnO_4$ and $NaNO_3$, and the GO sheets obtained can be exfoliated *via* sonication. Furthermore, GO has some functional groups, such as –COOH and –OH, which impart hydrophilic properties to GO. These functional group also help in forming hydrogen bonds between the GO flakes, which can self-assemble and form membranes with excellent mechanical strength.[141] The simplest way to manufacture GO-based membranes is vacuum filtration of a GO suspension on a substrate, and the amount of GO filtered influences the thickness of the membranes. Tang *et al.* prepared GO-based membranes *via* a pressurized ultrafiltration method and used them as a pervaporation membrane.[142] In this study, it was observed that the interlaminar spacing was affected by the packing density of the GO sheets and the water content of the feed solution. Furthermore, it was found that the packing density is affected by the applied pressure.

Langmuir–Blodgett with layer-by-layer (LbL) self-assembly methods have been used for producing large-area graphene membranes by sequential deposition of layers. Covalent bonds, hydrogen bonds, dipole–dipole, or electrostatic interactions stabilize these layers. Kulkarni *et al.* used interaction between negatively charged GO, due to functional groups, and cationic polymers to prepare a GO-based nanomembrane using LbL *via* Langmuir–Blodgett method.[143] The GO nanomembranes obtained showed increased elastic modulus and toughness. Another GO nanosheet-based water separation membrane was fabricated by Hu *et al.*[144] The GO-based membrane was produced by LbL deposition of GO nanolayers with 1,3,5-benzenetricarbonyl trichloride as a cross-linker on a polydopamine-covered polysulfone support. By using a cross-linker, stability of the stacked GO nanosheets was achieved in water, as well as fine-tuning of the functionality and spacing of the GO nanosheet. The prepared GO-based water separation membrane showed 4–10-fold higher flux than several mercantile membranes. rGO is a derivative of GO that is obtained by reducing the functional groups on GO. While this reduction moderates the hydrophilicity of GO-based membranes, it also increases their stability by inhibiting dissociation of the membrane in the water environment. However, the reduction agent should be chosen carefully because the reductant can damage the membrane. Su *et al.* prepared free-standing membranes by reducing GO laminates using hydroiodic acid or ascorbic acid.[145] The rGO-based membranes were determined to be extremely impermeable to entire gases and aggressive chemicals, such as hydrofluoric acid. In particular, the hydroiodic acid-reduced rGO membrane displayed exceptional barrier properties towards gases, liquids, and salts.

Graphene and GO can be used with inorganic nanoparticles and polymers to obtain composite membranes. These types of membranes can show increased permeability and selectivity. In particular, functional groups on GO can act as active sites to bind nanomaterials. Yuan *et al.* prepared a polyvinylidene fluoride/graphene (PVDF/GE)-based composite membrane for oil/water emulsion separation.[146] The PVDF film was prepared by electrospinning as a nanofiber support material. The PVDF-doped solution consisting of graphene with 0.5 wt% PVDF was uniformly spread onto a nanofibrous film to obtain the membrane. As a result, a hydrophobic and antipollution membrane was obtained with high mechanical-tensile strength of 2.5 MPa and high flux for separation. Guan *et al.* investigated a UiO-66/rGO membrane for use in water purification.[147] The UiO-66/rGO membranes were fabricated *via* the pressure filtration method and cross-linking was conducted by adding ethylenediamine to the filtration solution. As can be seen in Figure 7.4,

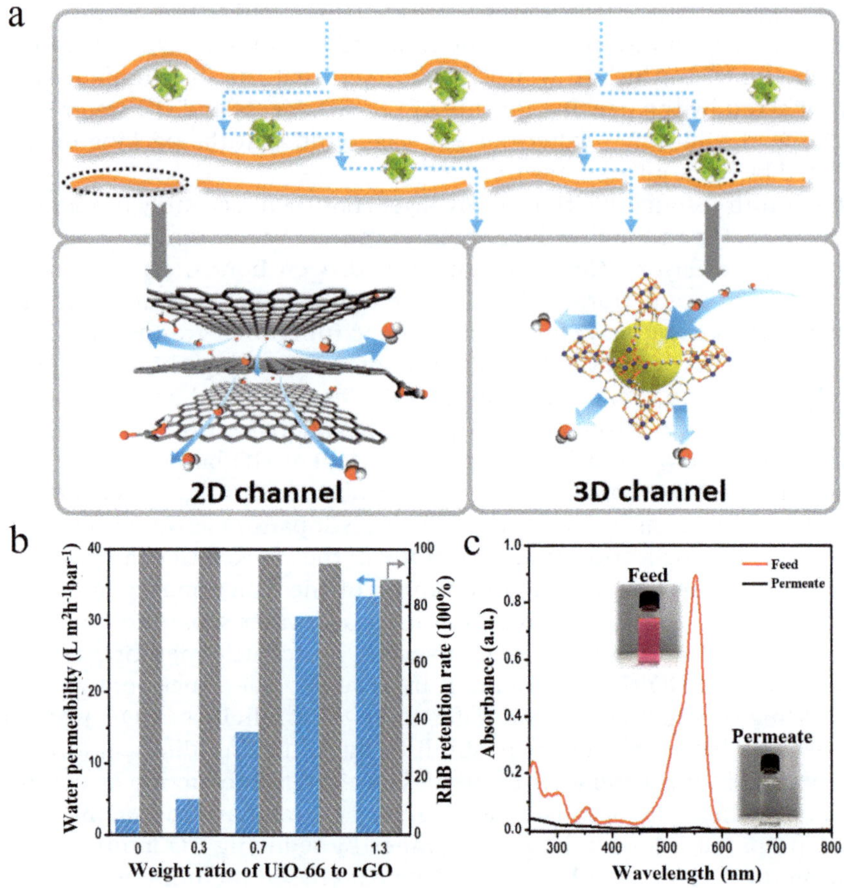

Figure 7.4 (a) Structure of fabricated UiO-66/rGO membrane, (b) filtration performance of obtained membrane, and (c) UV–vis. spectra of feed and permeate samples. Reproduced from ref. 147 with permission from Elsevier, Copyright 2017.

the obtained membrane consisted of three-dimensional (3D) nanoporous crystals incorporated into the 2D rGO channels. The prepared membrane showed 15-fold higher water permeability than the rGO membrane.

7.5 Separation Mechanism and Properties of Graphene-based Membranes

While large organic molecules are filtered by utilizing the exclusion effect, at the same time, by changing the distance between the GO layers, selectivity against different molecules can be improved. These can be used in water desalination when the layer spacing of GO membranes is less than 0.7 nm, in water treatment and wastewater reuse when it is 1–2 nm, and in biomedical implementations, such as artificial kidneys, when it exceeds 2 nm.[148] Due to the polarity on the GO sheet, the oxidized parts are hydrophilic, thus facilitating the flow of water molecules, while the unoxidized surface is hydrophobic. The membranes have a negative charge due to the oxygen-containing groups, such as carboxyl or hydroxyl, at the sides of the GO membranes, and thus small molecules and charged ions can be repudiated *via* the Donnan exclusion effect. Through the presence of an electrostatic attraction force towards the cations in solution, the membrane has selectivity against different salts thanks to the application of an electrostatic repulsion force towards anions (Figure 7.5). In addition to size and Donnan exclusions, selectivity for small hydrated ions can be acquired *via* adsorption, thanks to the interactions of oxygen-containing functional groups with transition metal cations. In addition, the selective ion permeability mechanism of the GO layers, through strong adsorption, is based on the sp^3 coordination interaction between the transition metal ions and alkali/alkaline earth cations and GO, as well as sp^2 cation–*p* interactions.

Due to the excellent impermeability of graphene, one of the methods for its use as a filtration material is to create a pore of appropriate size that allows water flow and prevents the permeability of particles and bacteria. The geometry of the nanopores on graphene can be created using the ion bombardment method, with variables of incidence angle and ion energy. Since it is known that the permeability of water and the permeability of ions and molecules increases with increase in pore size, the pore size should be defined so as to allow only the passage of water molecules. Thanks to functionalization of the pores along with the pore size, the charge and functional groups on the sides of the pores can also increase ion rejection. Because the charged groups in the membrane boost water flow, the hydrogenated pore has a lower water flow than a hydroxylated pore of the same size.[134] With stacked nanolayers of GO, which has good stability under pressure, the functional groups that can be formed on the surface form nanocapillaries that provide water flow and

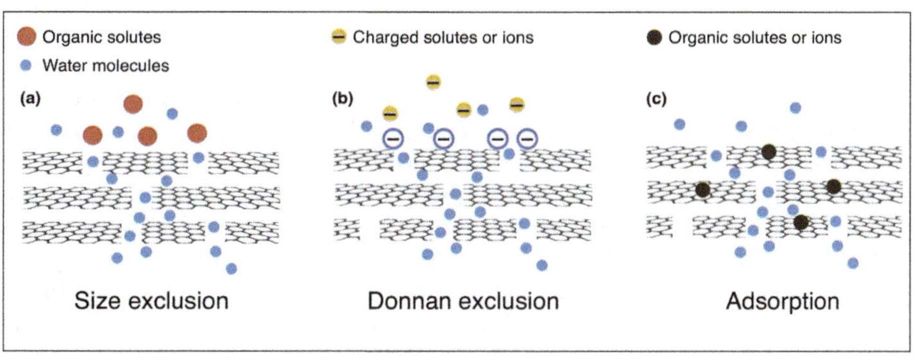

Figure 7.5 Different separation mechanisms for GO membranes: (a) size exclusion, (b) Donnan exclusion, and (c) adsorption. Reproduced from ref. 149 with permission from Elsevier, Copyright 2017.

block solutes in the water. Thanks to its amphiphilic character, GO adsorbs water molecules in its hydrophilic region and diffuses it from the hydrophobic walls of carbon. High electrostatic repulsion and high permeability can be achieved by reducing the gap between the nanolayers by covalent bonding of small molecules to the GO layers. The layer spacing of the GO layers, which is ~0.6 nm before immersion, increases after immersion due to diffusion of water molecules, and the water molecules reach high mobility. The water flow in the capillaries of multilayered GO structures is reduced because of hydrogen bonding between water molecules and functional groups, such as epoxy and hydroxyl. Due to deformation that can reverse the nanochannels, the water permeability and ion rejection of the membrane also depends on the applied pressure. Graphene is a very important material for the nanofiltration process due to these suitable properties of adjustable pore size and functionalization.

Membrane nanolayers, which are mechanically stable under dry conditions used for nanofiltration at high pressure and hydrophilic conditions, can be damaged by weakening of membranes due to their hydrophilic nature. The filter support only provides mechanical strength and does not play a role in the permeability of the membrane. Another method used to obtain membranes with high performance and resistance to cross-flow circumstances, which are frequently utilized in water treatment applications, is to incorporate GO layers into the polymer matrices instead of the surface coating. Due to the widespread use of dyes in different industries, dye waste waters that are harmful to health and the environment are produced. High concentrations of inorganic salts and organic compounds in waste water containing dyes make the treatment process challenging. This has contributed to the development of hybrid membranes and new methods in nanofiltration applications being applied for dye removal and heavy metal ion removal through modifying graphene-based membranes.

Another problem encountered in water treatment implementations of membranes is biological contamination, as it significantly affects the life of the membrane. Biological contamination can clog the pores of the membranes, reducing performance and leading to membrane rot. Membrane contamination can be hydraulically reversible or irreversible, but removal of contaminants causes increased cost, decreased performance, and shortened membrane life. Microorganisms cause biofilm formation by forming extracellular polymeric secretions in the dynamic process of biofouling, where colonies of microorganism are formed on the membrane. Microorganism colonies on the surface adhere to the membranes by generating strong hydrogen bonds and van der Waals forces. Modified membranes (such as those supplemented with copper and silver nanoparticles) are used to prevent this adhesion and biofouling. GO nanoplates have superior antifouling properties and GO-containing thin-film hybrid membranes are produced with selective and permeable membranes with superior antifouling features. For membranes with a support material, the selection of a support material with roughness and chemical compatibility is an important factor in the nanofiltration process. Incompatibility between the support and the active material can cause instability during filtration. A porous substrate is generally used so that the low mechanical strength of the GO membrane does not adversely affect the nanofiltration process. If the interaction between the support and the active substance (GO) is not sufficient, the active substance peels off the support surface over time and this affects the performance. To prevent this, ion beam applications can be used as a pretreatment, and ion-bombarded active ingredient (GO) is used for hybrid formation.[150] Similarly, compatibility is also achieved by adding functionalized polymers to the membrane structure.[151] Interfacial bonding between GO nanoplates and the substrate may also be evoked *via* surface alteration of the substrate materials.[152]

The stability of GO-based membranes in water is increased by covalent and non-covalent (van der Waals forces, hydrophobic interactions, hydrogen bonds, electrostatic interactions, and *p*–*p* interactions) cross-linking and protective coating methods. It is crucial to impart excellent stability to the GO nanofiltration membrane to keep a high permeability flux. Although the stability of the heavily cross-linked GO membranes is ensured, the nanochannels in the membrane prevent water flow due to the cross-linking density, and therefore water flow is reduced. It is very important for GO membranes to ensure both water permeability and excellent stability.[153,154] Thanks to the ionic interactions between the types of components that make up the membrane in the composite form, both the antipollution feature and the filtration efficiency can be increased.[155]

Because of the large number of oxygen-containing functional groups in the GO content, the hydrophilic layers of the GO membrane materials attract water molecules and thus the interlayer distance increases.[156] The van der Waals forces and hydrogen bonds that keep the GO layers together lose their effect, with the GO membrane swelling in water, which is supported by the electrostatic repulsion between the negatively charged functional groups on the membrane surface, and therefore the GO membrane can break down. In order to overcome this instability in water, it is important that GO membranes have swelling–limiting properties. Thanks to partial reduction, van der Waals forces (*p*–*p* stacking) dominate the GO nanolayer, which acts against the electrostatic repulsion and so the GO layer can be stabilized in the aqueous environment. Since the distance between the layers will decrease with excessive reduction and water flow will be prevented, this disadvantage can be eliminated by increasing the *d* distance by adding nanomaterials. The pH of the water environment also influences the stability of GO membranes, as it will increase the electrostatic repulsion. As an alternative to cross-linking, which is a strategy that prevents water flow in cross-linked channels, a protective coating on the outer surface is used for GO membranes. Thanks to this protective coating, both the stability and the water transport channels are improved.[157] Another performance-enhancing process is plasma treatment. By using laminated reduced GO in the hybrid membrane, the stability of the layered structure is preserved and, at the same time, the separation efficiency is increased.[158] When it comes to graphene membranes, the orientation in the membrane structure of graphene is a much more crucial factor. The effectiveness of nanofiltration depends on the layered and parallel-oriented structure. Overcoming this difficulty can be obtained by polymeric functionalization and polymerization of graphene, which is the membrane-active material, in a form that provides the filtration property of materials, that is, in parallel structures.[159]

The transportation mechanisms of graphene-based membranes have been investigated theoretically with analysis methods such as density functional theory, van der Waals distance theory, and breathing theory, and using variables such as defects size/shape, defect doping, and interlayer nanostructure.[160–174] In addition, water permeability measured by water flow can be obtained using the following equation:

$$Q = A \times \Delta P \times t \times \text{flux} \tag{7.1}$$

where A, ΔP, t, and Q are the effective filtration area of the membrane, the transmembrane pressure, the filtration time, and the volume of water, respectively. The water treatment parameters and some performance values of various graphene-based nanofiltration membranes are shown in Table 7.2.

Table 7.2 Comparison of performance parameters of some graphene-based composite membranes[a]

Membrane	Solute	Water flux	Rejection	Reference
GO/polyamide	Na_2SO_4	39.0 ± 1 L m^{-2} h^{-1}	$98.1 \pm 0.4\%$	175
GO/TiO_2/cellulose acetate	Methylene blue	16 L m^{-2} h^{-1} bar^{-1}	99%	176
PSF/GO/DMF	Cr^{2+}, Cu^{2+}, Cd^{2+}, Pb^{2+}	500 ppm, 4 bar	~90%	177
GO liquid crystals	Monovalent and divalent ions	71 ± 5 L m^{-2} h^{-1} bar^{-1}	30–40%	178
GO-starch/polyethersulfone	Safranin O	8.8 L m^{-2} h^{-1} bar^{-1}	92.4%	179
HPEI/GO/EDA	Pb^{2+}	1000 ppm, 2 bar	95.7%	180
	Ni^{2+}	1000 ppm, 2 bar	96.0%	
	Zn^{2+}	1000 ppm, 2 bar	97.4%	
	Cd^{2+}	1000 ppm, 2 bar	90.5%	
GO/polyamide	NaCl	22 L m^{-2} h^{-1}	> 88%	181
	$MgSO_4$		>97%	
GO/cyclodextrin	Eriochrome black	11 Lm^{-2} h^{-1} bar^{-1}	99%	182
PSF/PDA/IRMOF/GO	Cu^{2+}	200 ppm, 7 bar	~90%	183
GO/polysulfone	$MgCl_2$	—	62.1%	184
GO/polyethersulfone/polyamide	Na^+	30.3 L m^{-2} h^{-1} bar^{-1}	93.56%	185
GO/COF/nylon	Congo red	50 L m^{-2} h^{-1} bar^{-1}	99.82%	186

[a]PSF: sulfonated polysulfone; DMF: dimethylformamide; HPEI: hyperbranched polyethylenimine; EDA: ethylenediamine; PDA: polydopamine; IRMOF: isoreticular metal–organic framework; COF: covalent–organic framework.

7.6 Conclusion and Future Perspectives

With the increase in human activities and global warming, purification of water to remove pollutants is an increasingly important issue for future generations. The extraordinary properties of graphene-based materials and several simple synthesis routes lead to acceptable amounts of graphene materials for creating diverse nanocomposites. Graphene-based materials are shown to be highly attractive materials suitable for various engineering applications. From the examples presented here, the trend of preparing graphene-based nanocomposites to achieve synergetic effects for the newly developed composites are clearly visible. The advantages of the synergetic effects of composites usually manifest as enhanced properties for the newly obtained structure compared to the individual components. This approach has been widely accepted and utilized in photocatalytic, sorption, and membrane applications. To bring graphene-based composites to their full potential for application, novel or improved synthesis methods need to be designed, especially green synthesis methods. This could enable cost-effective and environmentally acceptable up-scaling of laboratory methods, which could then be tested in real scenarios for water purification, and potentially contribute to the sustainable development of society.

References

1. A. Razaq, F. Bibi, X. Zheng, R. Papadakis, S. H. M. Jafri and H. Li, *Materials*, 2022, **15**, 1012.
2. D. D. L. Chung, *J. Mater. Sci.*, 1987, **22**, 4190–4198.
3. D. G. Trikkaliotis, A. K. Christoforidis, A. C. Mitropoulos and G. Z. Kyzas, *ChemEngineering*, 2021, **5**, 64.
4. A. Khalid, R. A. Khushnood, S. Saleem, S. Z. Farooq and N. Shaheen, *IOP Conf. Ser.: Mater. Sci. Eng.*, 2018, **431**, 62005.
5. Y. Su, H. Ren, S. Zhao, N. Chen, X. Wang, Y. Zhao, Y. Bai and C. Li, *Ceram. Int.*, 2023, **49**, 7761–7767.
6. M. Inagaki, Y. Kaburagi and Y. Hishiyama, *Adv. Eng. Mater.*, 2014, **16**, 494–506.
7. K. S. Novoselov, A. K. Geim, S. V Morozov, D. Jiang, Y. Zhang, S. V Dubonos, I. V Grigorieva and A. A. Firsov, *Science*, (1979), 2004, **306**, 666–669.
8. P. Suvarnaphaet and S. Pechprasarn, *Sensors*, 2017, **17**(10), 2161.
9. Q. Xiang, J. Yu and M. Jaroniec, *Chem. Soc. Rev.*, 2012, **41**, 782–796.
10. Y. Zhu, S. Murali, W. Cai, X. Li, J. W. Suk, J. R. Potts and R. S. Ruoff, *Adv. Mater.*, 2010, **22**, 3906–3924.
11. M. Tahriri, M. Del Monico, A. Moghanian, M. Tavakkoli Yaraki, R. Torres, A. Yadegari and L. Tayebi, *Mater. Sci. Eng. C*, 2019, **102**, 171–185.
12. G. Yang, L. Li, W. B. Lee and M. C. Ng, *Sci. Technol. Adv. Mater.*, 2018, **19**, 613–648.

13. T. M. Magne, T. de Oliveira Vieira, L. M. R. Alencar, F. F. M. Junior, S. Gemini-Piperni, S. V Carneiro, L. M. U. D. Fechine, R. M. Freire, K. Golokhvast, P. Metrangolo, P. B. A. Fechine and R. Santos-Oliveira, *J. Nanostruct. Chem.*, 2022, **12**, 693–727.
14. Z. Li, D. Yang, H. Chu, L. Guo, T. Chen, Y. Mu, X. He, X. Zhong, B. Huang, S. Zhang, Y. Gao, Y. Wei, S. Wang and W. Zhou, *Nanomaterials*, 2022, **12**, 1474.
15. B. Aïssa, N. K. Memon, A. Ali and M. K. Khraisheh, *Front. Mater.*, 2015, **2**, 58.
16. Abid, P. Sehrawat, S. S. Islam, P. Mishra and S. Ahmad, *Sci. Rep.*, 2018, **8**, 3537.
17. R. Zouzelka, M. Remzova, J. Plsek, L. Brabec and J. Rathousky, *Catalysts*, 2019, **9**, 708.
18. A. T. Dideikin and A. Y. Vul', *Front. Phys.*, 2019, **6**, 148.
19. A. N. Ghulam, O. A. L. dos Santos, L. Hazeem, B. Pizzorno Backx, M. Bououdina and S. Bellucci, *J. Funct. Biomater.*, 2022, **13**, 77.
20. A. N. Ghulam, O. A. L. dos Santos, L. Hazeem, B. Pizzorno Backx, M. Bououdina and S. Bellucci, *J. Funct. Biomater.*, 2022, **13**, 77.
21. R. Tarcan, O. Todor-Boer, I. Petrovai, C. Leordean, S. Astilean and I. Botiz, *J. Mater. Chem. C*, 2020, **8**, 1198–1224.
22. J. M. Tour, *Chem. Mater.*, 2014, **26**, 163–171.
23. K. E. Drexler, C. Peterson and G. Pergamit, *Unbounding the Future: The Nanotechnology Revolution*, Quill, 1991.
24. C. K. Chua and M. Pumera, *Chem. Soc. Rev.*, 2014, **43**, 291–312.
25. X.-Y. Wang, A. Narita and K. Müllen, *Nat. Rev. Chem*, 2017, **2**, 100.
26. V. Eswaraiah, S. S. Jyothirmayee Aravind and S. Ramaprabhu, *J. Mater. Chem.*, 2011, **21**, 6800–6803.
27. D. Sahu, H. Sutar, P. Senapati, R. Murmu and D. Roy, *J. Comp. Sci.*, 2021, **5**, 181.
28. B. C. Brodie, *Philos. Trans. R. Soc. London*, 1859, **149**, 249–259.
29. L. Staudenmaier, *Ber. Dtsch. Chem. Ges.*, 1898, **31**, 1481–1487.
30. M. Al Kausor and D. Chakrabortty, *Inorg. Chem. Commun.*, 2021, **129**, 108630.
31. W. S. Hummers Jr and R. E. Offeman, *J. Am. Chem. Soc.*, 1958, **80**, 1339.
32. A. T. Smith, A. M. LaChance, S. Zeng, B. Liu and L. Sun, *Nano Mater. Sci.*, 2019, **1**, 31–47.
33. D. C. Marcano, D. V Kosynkin, J. M. Berlin, A. Sinitskii, Z. Sun, A. Slesarev, L. B. Alemany, W. Lu and J. M. Tour, *ACS Nano*, 2010, **4**, 4806–4814.
34. C. Botas, P. Álvarez, P. Blanco, M. Granda, C. Blanco, R. Santamaría, L. J. Romasanta, R. Verdejo, M. A. López-Manchado and R. Menéndez, *Carbon*, 2013, **65**, 156–164.
35. E. M. Deemer, P. K. Paul, F. S. Manciu, C. E. Botez, D. R. Hodges, Z. Landis, T. Akter, E. Castro and R. R. Chianelli, *Mater. Sci. Eng., B*, 2017, **224**, 150–157.
36. A. Alkhouzaam, H. Qiblawey, M. Khraisheh, M. Atieh and M. Al-Ghouti, *Ceram. Int.*, 2020, **46**, 23997–24007.

37. R. Muzyka, M. Kwoka, Ł. Smędowski, N. Díez and G. Gryglewicz, *New Carbon Mater.*, 2017, **32**, 15–20.
38. K. K. H. De Silva, H. H. Huang, R. K. Joshi and M. Yoshimura, *Carbon*, 2017, **119**, 190–199.
39. A. Alazmi, S. Rasul, S. P. Patole and P. M. F. J. Costa, *Polyhedron*, 2016, **116**, 153–161.
40. M. J. McAllister, J.-L. Li, D. H. Adamson, H. C. Schniepp, A. A. Abdala, J. Liu, M. Herrera-Alonso, D. L. Milius, R. Car, R. K. Prud'homme and I. A. Aksay, *Chem. Mater.*, 2007, **19**, 4396–4404.
41. S. Pei and H.-M. Cheng, *Carbon*, 2012, **50**, 3210–3228.
42. K. N. Kudin, B. Ozbas, H. C. Schniepp, R. K. Prud'homme, I. A. Aksay and R. Car, *Nano Lett.*, 2008, **8**, 36–41.
43. D. Yang and C. Bock, *J. Power Sources*, 2017, **337**, 73–81.
44. M. S. Ahmad and Y. Nishina, *Nanoscale*, 2020, **12**, 12210–12227.
45. Y. Zhu, S. Murali, M. D. Stoller, A. Velamakanni, R. D. Piner and R. S. Ruoff, *Carbon*, 2010, **48**, 2118–2122.
46. M. J. Fernández-Merino, L. Guardia, J. I. Paredes, S. Villar-Rodil, P. Solís-Fernández, A. Martínez-Alonso and J. M. D. Tascón, *J. Phys. Chem. C*, 2010, **114**, 6426–6432.
47. G. Williams, B. Seger and P. V Kamat, *ACS Nano*, 2008, **2**, 1487–1491.
48. M. Zhou, Y. Wang, Y. Zhai, J. Zhai, W. Ren, F. Wang and S. Dong, *Chem. – Eur. J.*, 2009, **15**, 6116–6120.
49. Y. Zhou, Q. Bao, L. A. L. Tang, Y. Zhong and K. P. Loh, *Chem. Mater.*, 2009, **21**, 2950–2956.
50. S. Dubin, S. Gilje, K. Wang, V. C. Tung, K. Cha, A. S. Hall, J. Farrar, R. Varshneya, Y. Yang and R. B. Kaner, *ACS Nano*, 2010, **4**, 3845–3852.
51. S. W. Chong, C. W. Lai and S. B. Abdul Hamid, *Ceram. Int.*, 2015, **41**, 9505–9513.
52. X. Jin, N. Li, X. Weng, C. Li and Z. Chen, *Chemosphere*, 2018, **208**, 417–424.
53. S. Y. Chee, H. L. Poh, C. K. Chua, F. Šaněk, Z. Sofer and M. Pumera, *Phys. Chem. Chem. Phys.*, 2012, **14**, 12794–12799.
54. M.-H. Tran, C.-S. Yang, S. Yang, I.-J. Kim and H. K. Jeong, *Curr. Appl. Phys.*, 2014, **14**, S74–S79.
55. T. D. Dao and H. M. Jeong, *Mater. Res. Bull.*, 2015, **70**, 651–657.
56. L. Shen, L. Zhang, K. Wang, L. Miao, Q. Lan, K. Jiang, H. Lu, M. Li, Y. Li, B. Shen and W. Zheng, *RSC Adv.*, 2018, **8**, 17209–17217.
57. R. Muzyka, S. Drewniak, T. Pustelny, M. Chrubasik and G. Gryglewicz, *Materials*, 2018, **11**, 1050.
58. I. Bychko, A. Abakumov, O. Didenko, M. Chen, J. Tang and P. Strizhak, *J. Phys. Chem. Solids*, 2022, **164**, 110614.
59. L. M. Pastrana-Martínez, S. Morales-Torres, J. L. Figueiredo, J. L. Faria and A. M. T. Silva, in *Multifunctional Photocatalytic Materials for Energy*, ed. Z. Lin, M. Ye, and M. Wang, Woodhead Publishing, 2018, pp. 79–101.
60. S. Singh, M. R. Hasan, P. Sharma and J. Narang, *Sens. Int.*, 2022, **3**, 100190.

61. R. Giovannetti, E. Rommozzi, M. Zannotti and C. A. D'Amato, *Catalysts*, 2017, **7**, 305.
62. Y. Duan, M.-L. Gou, Y. Guo, J. Cai, W. Song, Z. Liu and E. Zhou, *J. Mater. Res.*, 2021, **36**, 906–915.
63. A. Mondal, A. Prabhakaran, S. Gupta and V. R. Subramanian, *ACS Omega*, 2021, **6**, 8734–8743.
64. G. Žerjav, M. S. Arshad, P. Djinović, I. Junkar, J. Kovač, J. Zavašnik and A. Pintar, *Nanoscale*, 2017, **9**, 4578–4592.
65. B. R. Cruz-Ortiz, J. W. J. Hamilton, C. Pablos, L. Díaz-Jiménez, D. A. Cortés-Hernández, P. K. Sharma, M. Castro-Alférez, P. Fernández-Ibañez, P. S. M. Dunlop and J. A. Byrne, *Chem. Eng. J.*, 2017, **316**, 179–186.
66. P. Singh, P. Shandilya, P. Raizada, A. Sudhaik, A. Rahmani-Sani and A. Hosseini-Bandegharaei, *Arabian J. Chem.*, 2020, **13**, 3498–3520.
67. M. Minella, F. Sordello and C. Minero, *Catal. Today*, 2017, **281**, 29–37.
68. A. Haruna, F.-K. Chong, Y.-C. Ho and Z. M. A. Merican, *Environ. Sci. Pollut. Res. Int.*, 2022, **29**, 70706–70745.
69. N. C. Horti, M. D. Kamatagi, N. R. Patil, S. K. Nataraj, M. S. Sannaikar and S. R. Inamdar, *Optik*, 2019, **194**, 163070.
70. N. Baig, I. Kammakakam and W. Falath, *Mater. Adv.*, 2021, **2**, 1821–1871.
71. B. Gomathi Thanga Keerthana, T. Solaiyammal, S. Muniyappan and P. Murugakoothan, *Mater. Lett.*, 2018, **220**, 20–23.
72. D. Maruthamani, D. Divakar, M. Harshavardhan and M. Kumaravel, *J. Chem. Pharm. Res.*, 2016, **8**, 8.
73. R. Hamid-Reza and D. Mohsen, in *Applications of Nanobiotechnology*, ed. S. Margarita, and Z. Roumen, IntechOpen, Rijeka, 2019, ch. 1.
74. S. Arefi-Oskoui, A. Khataee, M. Safarpour, Y. Orooji and V. Vatanpour, *Ultrason. Sonochem.*, 2019, **58**, 104633.
75. M. A. Irshad, R. Nawaz, M. Z. ur Rehman, M. Adrees, M. Rizwan, S. Ali, S. Ahmad and S. Tasleem, *Ecotoxicol. Environ. Saf.*, 2021, **212**, 111978.
76. T. Tsuzuki, *Commun. Chem.*, 2021, **4**, 143.
77. O. Y. Posudievsky, A. S. Kondratyuk, O. A. Kozarenko, V. V Cherepanov, G. I. Dovbeshko, V. G. Koshechko and V. D. Pokhodenko, *Carbon*, 2019, **152**, 274–283.
78. H. Seema, K. Christian Kemp, V. Chandra and K. S. Kim, *Nanotechnology*, 2012, **23**, 355705.
79. B. Chai, J. Li, Q. Xu and K. Dai, *Mater. Lett.*, 2014, **120**, 177–181.
80. X. Chen, J. Dai, G. Shi, L. Li, G. Wang and H. Yang, *J. Alloys Compd.*, 2015, **649**, 872–877.
81. B. Li, T. Liu, Y. Wang and Z. Wang, *J. Colloid Interface Sci.*, 2012, **377**, 114–121.
82. N. Yusoff, N. M. Huang, M. R. Muhamad, S. V Kumar, H. N. Lim and I. Harrison, *Mater. Lett.*, 2013, **93**, 393–396.
83. D. Wang, X. Li, J. Chen and X. Tao, *Chem. Eng. J.*, 2012, **198–199**, 547–554.
84. J. Ahmad and K. Majid, *New J. Chem.*, 2018, **42**, 3246–3259.
85. C. H. B. Silva, M. Iliut, C. Muryn, C. Berger, Z. Coldrick, V. R. L. Constantino, M. L. A. Temperini and A. Vijayaraghavan, *Beilstein J. Nanotechnol.*, 2018, 2936–2946.

86. D. P. Ojha, M. K. Joshi and H. J. Kim, *Ceram. Int.*, 2017, **43**, 1290–1297.
87. P. K. Labhane and G. H. Sonawane, *Inorg. Chem. Commun.*, 2020, **113**, 107809.
88. N. Tahir, M. Zahid, A. Jillani, M. Yaseen, Q. Abbas, R. Abdul shakoor and I. shahid, *J. Photochem. Photobiol., A*, 2023, **436**, 114376.
89. I. Altin, *J. Mol. Struct.*, 2022, **1252**, 132199.
90. M. Zarrabi, M. Haghighi, R. Alizadeh and S. Mahboob, *Mater. Res. Bull.*, 2022, **153**, 111907.
91. R. Zheng, D. Yang, Y. Chen, Z. Bian and H. Li, *J. Environ. Sci.*, 2023, **134**, 11–20.
92. M. Danish, M. Saud Athar, I. Ahmad, M. Z. A. Warshagha, Z. Rasool and M. Muneer, *Appl. Surf. Sci.*, 2022, **604**, 154604.
93. Q. Zhang, Q. Hou, G. Huang and Q. Fan, *Environ. Sci. Pollut. Res.*, 2020, **27**, 190–209.
94. E. M. Domingues, G. Gonçalves, B. Henriques, E. Pereira and P. A. A. P. Marques, *Environ. Sci.: Nano*, 2022, **9**, 1156–1167.
95. A. Ahamed, N. Nusrathali, M. Mufeeth, K. Ranaweera and U. Majeed, *J. Bangladesh Agric. Univ.*, 2021, 1.
96. M. Kocijan, L. Ćurković, G. Gonçalves and M. Podlogar, *Sustainability*, 2022, **14**(19), 12703.
97. C. Sun, Z. Wang, H. Zheng, L. Chen and F. Li, *J. Hazard. Mater.*, 2021, **420**, 126599.
98. Y. Li, Z. Lai, Z. Huang, H. Wang, C. Zhao, G. Ruan and F. Du, *Appl. Surf. Sci.*, 2021, **550**, 149342.
99. S. Bao, W. Yang, Y. Wang, Y. Yu and Y. Sun, *J. Hazard. Mater.*, 2020, **381**, 120914.
100. A. Bessa, B. Henriques, G. Gonçalves, G. Irurueta, E. Pereira and P. A. A. P. Marques, *Chem. Eng. J.*, 2020, **398**, 125587.
101. Y. Yu, G. Zhang and L. Ye, *J. Appl. Polym. Sci.*, 2019, **136**, 47318.
102. W. M. Algothmi, N. M. Bandaru, Y. Yu, J. G. Shapter and A. V Ellis, *J. Colloid Interface Sci.*, 2013, **397**, 32–38.
103. L. Pan, Z. Wang, Q. Yang and R. Huang, *Nanomaterials*, 2018, **8**(11), 957.
104. L. Gan, H. Li, L. Chen, L. Xu, J. Liu, A. Geng, C. Mei and S. Shang, *Colloid Polym. Sci.*, 2018, **296**, 607–615.
105. J. M. Bloor, R. D. Handy, S. A. Awan and D. F. L. Jenkins, *Ecotoxicol. Environ. Saf.*, 2021, **208**, 111422.
106. Y. Fei, Y. Li, S. Han and J. Ma, *J. Colloid Interface Sci.*, 2016, **484**, 196–204.
107. V. Chandra, J. Park, Y. Chun, J. W. Lee, I.-C. Hwang and K. S. Kim, *ACS Nano*, 2010, **4**, 3979–3986.
108. M. Adel, M. A. Ahmed, M. A. Elabiad and A. A. Mohamed, *Environ. Nanotechnol., Monit. Manage.*, 2022, **18**, 100719.
109. R. Gusain, N. Kumar and S. S. Ray, *Coord. Chem. Rev.*, 2020, **405**, 213111.
110. Y. Zhang, W. Peng, L. Xia and S. Song, *J. Environ. Chem. Eng.*, 2017, **5**, 4157–4164.
111. X. Li, H. Zhou, W. Wu, S. Wei, Y. Xu and Y. Kuang, *J. Colloid Interface Sci.*, 2015, **448**, 389–397.

112. T. Tene, S. Bellucci, M. Guevara, E. Viteri, M. Arias Polanco, O. Salguero, E. Vera-Guzmán, S. Valladares, A. Scarcello, F. Alessandro, L. S. Caputi and C. Vacacela Gomez, *Nanomaterials*, 2022, **12**(3), 309.
113. C. Bulin, Z. Ma, T. Guo, B. Li, Y. Zhang, B. Zhang, R. Xing and X. Ge, *J. Phys. Chem. Solids*, 2021, **156**, 110130.
114. A. Yu. Romanchuk, A. S. Slesarev, S. N. Kalmykov, D. V Kosynkin and J. M. Tour, *Phys. Chem. Chem. Phys.*, 2013, **15**, 2321–2327.
115. K. Zhang, H. Li, X. Xu and H. Yu, *Microporous Mesoporous Mater.*, 2018, **255**, 7–14.
116. L. Li, C. Luo, X. Li, H. Duan and X. Wang, *Int. J. Biol. Macromol.*, 2014, **66**, 172–178.
117. G. Rajivgandhi, V. RTV, R. Nandhakumar, S. Murugan, N. S. Alharbi, S. Kadaikunnan, J. M. Khaled, K. F. Alanzi and W.-J. Li, *Environ. Res.*, 2021, **199**, 111322.
118. M. Verma, I. Lee, J. Oh, V. Kumar and H. Kim, *Chemosphere*, 2022, **287**, 132385.
119. G. Z. Kyzas, N. A. Travlou and E. A. Deliyanni, *Colloids Surf., B*, 2014, **113**, 467–476.
120. L. Cui, X. Guo, Q. Wei, Y. Wang, L. Gao, L. Yan, T. Yan and B. Du, *J. Colloid Interface Sci.*, 2015, **439**, 112–120.
121. S. Mohan, V. Kumar, D. K. Singh and S. H. Hasan, *J. Environ. Chem. Eng.*, 2017, **5**, 2259–2273.
122. Y. Sun, Q. Wang, C. Chen, X. Tan and X. Wang, *Environ. Sci. Technol.*, 2012, **46**, 6020–6027.
123. J. Huo, G. Yu and J. Wang, *Chemosphere*, 2021, **278**, 130492.
124. L. P. Lingamdinne, J. R. Koduru, H. Roh, Y.-L. Choi, Y.-Y. Chang and J.-K. Yang, *Hydrometallurgy*, 2016, **165**, 90–96.
125. H. Zhu, F. Nicot and F. Darve, *Int. J. Solids Struct.*, 2016, **96**, 25–37.
126. S. Zhuang and J. Wang, *Environ. Prog. Sustainable Energy*, 2019, **38**, S32–S41.
127. W. Song, J. Hu, Y. Zhao, D. Shao and J. Li, *RSC Adv.*, 2013, **3**, 9514–9521.
128. S. Manshad, M. R. Sazegar, M. G. M. Nawawi and H. b. Hassan, *RSC Adv.*, 2016, **6**, 103888–103894.
129. L. Shen, S. Xiong and Y. Wang, *Chem. Eng. Sci.*, 2016, **143**, 194–205.
130. H.-R. Chae, J. Lee, C.-H. Lee, I.-C. Kim and P.-K. Park, *J. Membr. Sci.*, 2015, **483**, 128–135.
131. S. Zinadini, A. A. Zinatizadeh, M. Rahimi, V. Vatanpour and H. Zangeneh, *J. Membr. Sci.*, 2014, **453**, 292–301.
132. G. Liu, W. Jin and N. Xu, *Chem. Soc. Rev.*, 2015, **44**, 5016–5030.
133. D. Jiang, V. R. Cooper and S. Dai, *Nano Lett.*, 2009, **9**, 4019–4024.
134. D. Cohen-Tanugi and J. C. Grossman, *Nano Lett.*, 2012, **12**, 3602–3608.
135. L. Ma, I. Jia, X. Guo and L. Xiang, *Chin. J. Catal.*, 2014, **35**, 108–119.
136. A. Boretti, S. Al-Zubaidy, M. Vaclavikova, M. Al-Abri, S. Castelletto and S. Mikhalovsky, *npj Clean Water*, 2018, **1**, 1–11.
137. C. Sun, M. Liu and B. Bai, *Carbon*, 2019, **153**, 481–494.
138. Y. T. Nam, J. H. Kang, J. D. Jang, J. H. Bae, H. T. Jung and D. W. Kim, *Membranes*, 2021, **11**(10), 793.

139. S. P. Surwade, S. N. Smirnov, I. V Vlassiouk, R. R. Unocic, G. M. Veith, S. Dai and S. M. Mahurin, *Nat. Nanotechnol.*, 2015, **10**, 459–464.
140. C. A. Merchant, K. Healy, M. Wanunu, V. Ray, N. Peterman, J. Bartel, M. D. Fischbein, K. Venta, Z. Luo, A. T. C. Johnson and M. Drndić, *Nano Lett.*, 2010, **10**, 2915–2921.
141. J. Chong, B. Wang and K. Li, *Curr. Opin. Chem. Eng.*, 2016, **12**, 98–105.
142. Y. P. Tang, D. R. Paul and T. S. Chung, *J. Membr. Sci.*, 2014, **458**, 199–208.
143. D. D. Kulkarni, I. Choi, S. S. Singamaneni and V. V Tsukruk, *ACS Nano*, 2010, **4**, 4667–4676.
144. M. Hu and B. Mi, *Environ. Sci. Technol.*, 2013, **47**, 3715–3723.
145. Y. Su, V. G. Kravets, S. L. Wong, J. Waters, A. K. Geim and R. R. Nair, *Nat. Commun.*, 2014, **5**, 4843.
146. X. Yuan, W. Li, H. Liu, N. Han and X. Zhang, *Compos. Commun.*, 2016, **2**, 5–8.
147. K. Guan, D. Zhao, M. Zhang, J. Shen, G. Zhou, G. Liu and W. Jin, *J. Membr. Sci.*, 2017, **542**, 41–51.
148. B. Mi, *Science*, 2014, **343**, 740–742.
149. Y. Zhang and T.-S. Chung, *Curr. Opin. Chem. Eng.*, 2017, **16**, 9–15.
150. Y. T. Nam, S. J. Kim, K. M. Kang, W.-B. Jung, D. W. Kim and H.-T. Jung, *Carbon*, 2019, **148**, 370–377.
151. Y. Qin, H. Liu, Y. Liu, M. Chen, K. Chen, Y. Huang and C. Xiao, *J. Membr. Sci.*, 2020, **604**, 118064.
152. Y. Lou, G. Liu, S. Liu, J. Shen and W. Jin, *Appl. Surf. Sci.*, 2014, **307**, 631–637.
153. K. H. Thebo, X. Qian, Q. Zhang, L. Chen, H.-M. Cheng and W. Ren, *Nat. Commun.*, 2018, **9**, 1486.
154. S. Wan, Y. Chen, Y. Wang, G. Li, G. Wang, L. Liu, J. Zhang, Y. Liu, Z. Xu, A. P. Tomsia, L. Jiang and Q. Cheng, *Matter*, 2019, **1**, 389–401.
155. Y. Song, Y. Sun, M. Chen, P. Huang, T. Li, X. Zhang and K. Jiang, *J. Water Process. Eng.*, 2020, **34**, 101086.
156. S. Zheng, Q. Tu, J. J. Urban, S. Li and B. Mi, *ACS Nano*, 2017, **11**, 6440–6450.
157. K. Huang, G. Liu, J. Shen, Z. Chu, H. Zhou, X. Gu, W. Jin and N. Xu, *Adv. Funct. Mater.*, 2015, 5809–5815.
158. S. Mohammed, H. M. Hegab, R. Ou, S. Liu, H. Ma, X. Chen, T. Sridhar and H. Wang, *Green Chem. Eng.*, 2021, **2**, 122–131.
159. M. Safarpour, V. Vatanpour, A. Khataee and M. Esmaeili, *Sep. Purif. Technol.*, 2015, **154**, 96–107.
160. Y. Tao, Q. Xue, Z. Liu, M. Shan, C. Ling, T. Wu and X. Li, *ACS Appl. Mater. Interfaces*, 2014, **6**, 8048–8058.
161. Y. Wang, Q. Yang, C. Zhong and J. Li, *Appl. Surf. Sci.*, 2017, **407**, 532–539.
162. Z. Yuan, A. Govind Rajan, R. P. Misra, L. W. Drahushuk, K. V. Agrawal, M. S. Strano and D. Blankschtein, *ACS Nano*, 2017, **11**, 7974–7987.
163. D. Konatham, J. Yu, T. A. Ho and A. Striolo, *Langmuir*, 2013, **29**, 11884–11897.

164. K. Celebi, J. Buchheim, R. M. Wyss, A. Droudian, P. Gasser, I. Shorubalko, J.-I. Kye, C. Lee and H. G. Park, *Science*, 2014, **344**, 289–292.
165. G. Lei, C. Liu, H. Xie and F. Song, *Chem. Phys. Lett.*, 2014, **599**, 127–132.
166. D. Cohen-Tanugi and J. C. Grossman, *J. Chem. Phys.*, 2014, **141**, 74704.
167. Q. Chen and X. Yang, *J. Membr. Sci.*, 2015, **496**, 108–117.
168. L.-C. Lin and J. C. Grossman, *Nat. Commun.*, 2015, **6**, 8335.
169. N. Wei, X. Peng and Z. Xu, *ACS Appl. Mater. Interfaces*, 2014, **6**, 5877–5883.
170. N. Wei, X. Peng and Z. Xu, *Phys. Rev. E: Stat., Nonlinear, Soft Matter Phys.*, 2014, **89**, 12113.
171. W. L. Xu, C. Fang, F. Zhou, Z. Song, Q. Liu, R. Qiao and M. Yu, *Nano Lett.*, 2017, **17**, 2928–2933.
172. G. Algara-Siller, O. Lehtinen, F. C. Wang, R. R. Nair, U. Kaiser, H. A. Wu, A. K. Geim and I. V Grigorieva, *Nature*, 2015, **519**, 443–445.
173. M. Neek-Amal, F. M. Peeters, I. V Grigorieva and A. K. Geim, *ACS Nano*, 2016, **10**, 3685–3692.
174. A. Nicolaï, B. G. Sumpter and V. Meunier, *Phys. Chem. Chem. Phys.*, 2014, **16**, 8646–8654.
175. J. Yin, G. Zhu and B. Deng, *Desalination*, 2016, **379**, 93–101.
176. Y. Liu, Z. Yu, Y. Peng, L. Shao, X. Li and H. Zeng, *Chem. Phys. Lett.*, 2020, **749**, 137424.
177. R. Mukherjee, P. Bhunia and S. De, *Chem. Eng. J.*, 2016, **292**, 284–297.
178. A. Akbari, P. Sheath, S. T. Martin, D. B. Shinde, M. Shaibani, P. C. Banerjee, R. Tkacz, D. Bhattacharyya and M. Majumder, *Nat. Commun.*, 2016, **7**, 10891.
179. J. P. Ambre, K. B. Dhopte, P. R. Nemade and V. H. Dalvi, *J. Environ. Chem. Eng.*, 2019, **7**, 103300.
180. S. Zhang, M. H. Peh, Z. Thong and T.-S. Chung, *Ind. Eng. Chem. Res.*, 2015, **54**, 472–479.
181. S. Bano, A. Mahmood, S.-J. Kim and K.-H. Lee, *J. Mater. Chem. A*, 2015, **3**, 2065–2071.
182. J. Xue, J. Shen, R. Zhang, F. Wang, S. Liang, X. You, Q. Yu, Y. Hao, Y. Su and Z. Jiang, *J. Membr. Sci.*, 2020, **612**, 118465.
183. Z. Rao, K. Feng, B. Tang and P. Wu, *ACS Appl. Mater. Interfaces*, 2017, **9**, 2594–2605.
184. G. S. Lai, W. J. Lau, P. S. Goh, A. F. Ismail, N. Yusof and Y. H. Tan, *Desalination*, 2016, **387**, 14–24.
185. X. Kang, X. Liu, J. Liu, Y. Wen, J. Qi and X. Li, *Appl. Surf. Sci.*, 2020, **508**, 145198.
186. L. Chen, W. Wang, Q. Fang, K. Zuo, G. Hou, Q. Ai, Q. Li, L. Ci and J. Lou, *Appl. Mater. Today*, 2020, **20**, 100791.

CHAPTER 8

Responsive Polymeric Materials: Advances in Membrane-based Technologies for Water Treatment Processes

PERLA CRUZ-TATO[a,b*], ALONDRA S. RODRÍGUEZ-ROLÓN[a,b], JESSIKA PAZOL[a,b] AND EDUARDO NICOLAU[a,b]

[a]Department of Chemistry, University of Puerto Rico, Rio Piedras Campus, 17 Ave. Universidad Ste. 1701, San Juan, Puerto Rico, 00925-2537, USA; [b]Molecular Sciences Research Center, University of Puerto Rico, 1390 Ponce De León Ave, Suite 2, San Juan, Puerto Rico, 00931-3346, USA
*E-mail: perla.cruz@upr.edu

8.1 Introduction

8.1.1 Conventional Membranes and Intrinsic Challenges

Due to their water permeability and tunable selectivity, membrane-based technologies are considered state-of-the-art for water filtration processes. These membranes are composed of a polymeric or inorganic support and an active layer. Both components can be modified to modulate and enhance structural properties of the membrane, including the pore size, distribution and structure, crystallinity, wettability, and surface roughness.

Chemistry in the Environment Series No. 12
Novel Materials and Water Purification: Towards a Sustainable Future
Edited by Grigorios L. Kyriakopoulos and Miltiadis G. Zamparas

Published by the Royal Society of Chemistry, www.rsc.org

These properties affect the membrane performance, water flux, foulant rejection, and fouling.[1] Also, membranes can be tailored to incorporate specific and unique functionalities, such as a reactive or responsive layer.

Conventional membranes are selective to water molecules, yet their upscaling process still suffers from drawbacks, such as membrane fouling and high-energetic inputs as the separation driving force.[2] For this reason, researchers have been exploring the incorporation of novel materials to minimize the chemical and energetic input while extending the membrane lifespan and enhancing its performance.[3–5]

8.1.2 Responsive Materials—Next-generation Materials

The last decade has witnessed the development of "stimuli-responsive materials" and their application in water treatment processes.[6] These materials are unique and innovative since their physicochemical properties can be adapted in response to variations in their local environment. These include changes in *pH, temperature, ionic strength, electrical bias, light, and other chemical/physical interactions*. The response to these previously mentioned stimuli is switchable (*i.e.*, reversible), which means that the membrane can be "activated" and "deactivated" easily and reused, while conserving its intrinsic properties. The responsiveness of these materials occurs in two main steps: (1) a "stimulus" that triggers conformational transitions on a microscopic level; and (2) the amplification of these conformational transitions to a macroscopic level, which is reflected in the membrane's performance. The incorporation of these materials into membranes for water treatment has opened up a new research area with the potential for better *in situ* control over the filtration process. Moreover, responsive materials can promote anti-fouling and self-cleaning capabilities, enhance the membrane operational life and reduce the overall process cost.[7] Materials that can be employed to tailor the membrane's responsiveness include polymers,[8] metals,[9] semiconductors[10] and biomaterials.[11]

In terms of membrane fabrication, two common methods are typically employed: (1) the responsive material can be directly blended on the casting solution; or (2) it can be deposited (*i.e.*, chemically bonded or physically attached) over the porous or non-porous supports. In both cases, the membrane's reactivity is triggered by an external stimulus. The stimuli applied can be categorized as *direct*, *indirect* or *field responsive*. *Direct* stimulation is when the polymers are in direct contact with chemical agents (*e.g.*, pH), whereas *indirect* stimulation is a response towards a thermodynamic stimulus such as temperature or pressure.[12] In the third type of stimulus, a magnetic, electric or electromagnetic field is applied to produce a *field response*. Figure 8.1 shows the above-mentioned categories of stimuli and the types of membranes, which will be assessed in the following sections.

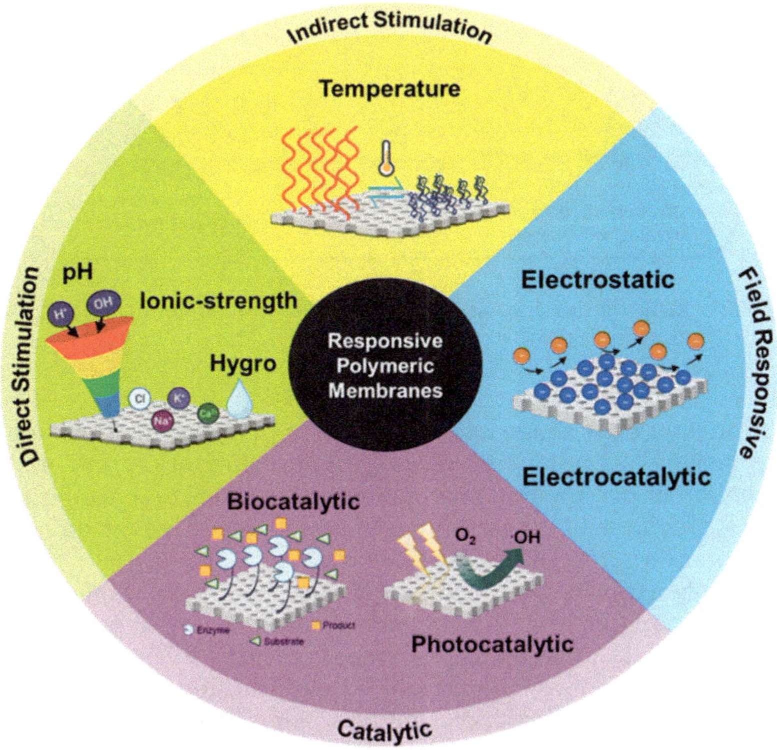

Figure 8.1 Responsive polymeric membranes classified by the different types of external stimuli: direct, indirect, electric field and catalytic.

8.2 Direct Stimulation of Polymeric Responsive Membranes

The membrane responsiveness can be triggered by different types of stimuli, with direct stimulation being one of three categories. In this case, the membrane is in direct contact with chemical or biological components that trigger the membrane response. The response is attributed to sites on the membrane that can interact with the added component and generate responses that modulate the membrane conformation (*e.g.* pore size) or performance (*e.g.* permeability/selectivity). The interaction can be non-specific, such as changes in pH or an electrostatic environment. Another type of interaction, which is considered more specific, is a biologically interfaced system, such as enzyme-based membranes.[12]

8.2.1 pH-responsive Membranes

The pH-responsive membranes are composed of a polymeric material that can undergo morphological and conformational changes when the pH of its environment is altered (illustrated in Figure 8.2). Moreover, the polymer

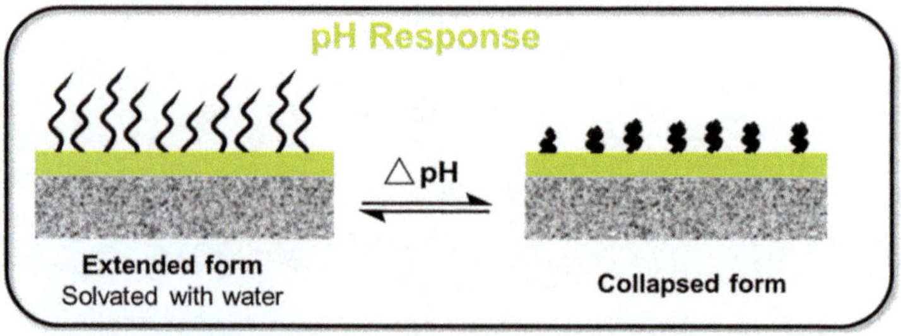

Figure 8.2 Representation of an example of the conformational changes in polymeric pH-sensitive chains.

chains can alter their shape upon a change in pH, affecting their solubility, surface activity/hydrophilicity, chain structure formation and swelling.[13] As a function of pH, the most notable mechanism of these conformational changes is alteration of the membrane pore diameter. This can directly affect the membrane's permeability and selectivity.[14]

These polymers should contain pH-sensitive or ionizable functional groups, such as carboxylic (COOH), sulfonate (HSO_4), pyridine, imidazole or amines (NH_2, NHR_2). The pH responsiveness occurs with protonation/deprotonation of these ionizable groups or by physical interactions in the polymer exposed to a basic or acidic solution. Some examples are acrylic acid (AAc)[15,16] and *N*,*N*-dimethylaminoethyl methacrylate (DMAEMA).[17,18]

Another example of a pH-responsive polymer and membrane is the vinyl pyridine, which is susceptible to the deprotonation/protonation equilibrium shown in Figure 8.3A. This equilibrium occurs between the pyridine and pyridinium salt below the pK_a value of 5.98. It has been reported that the appearance of the charged group (*i.e.*, pyridinium salt) influences the membrane performance, pores and selectivity.[19] Interestingly, this pH-dependent equilibrium (Figure 8.3A) could be affected by the position of the nitrogen in the polymer backbone. Furthermore, when P4VP and P2VP are compared (see Figure 8.3B) the main difference between them is the N position, which in P2VP is more sterically hindered than in P4VP. For this reason the nitrogen in P4VP is more sensitive to changes in pH.

A further aspect that must be considered in pH-responsive membranes is the interaction between dissolved ions and the ionizable groups in the polymer, and its effect on polymer conformation. Depending on the pH, the polymer could be present in an extended or collapsed conformation, which will ultimately reshape the membrane's pore size. An example of this is PAA-*g*-PVDF microfiltration membranes, where, at pH 1–6, the water permeation increases due to the conformation of the AA polymer.[20] At low pH, the polymer chains assume a collapsed conformation, interacting with each other and decreasing the pore steric obstruction. At high pH, the polymer chains adopt an extended conformation with a strong interaction with water and electrostatic repulsion among the negatively charged groups.[13]

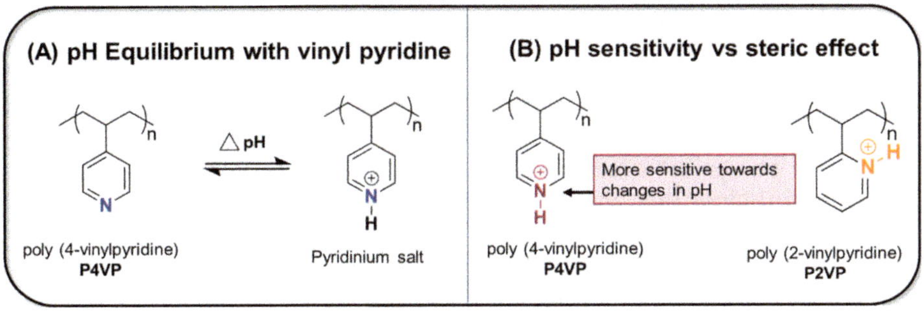

Figure 8.3 Scheme of pyridine-based polymer (A) pH equilibrium and (B) comparison between the P4VP and P2VP polymer chain.

8.2.2 Hygro-responsive Membranes

Polymeric materials can respond to changes in humidity due to the presence of hydrophilic moieties at the molecular level. These materials can convert the chemical potential of water into mechanical energy[21] and present simultaneous superhydrophilic/superoleophobic properties. This behavior is due to the favorable interactions of a membrane with polar liquids (*e.g.* water), leading to resistance for non-polar liquids. Once the membrane is in contact with water, the functional groups re-accommodate on the surface to interact with it, forming hydrogen bonds, decreasing the system's enthalpy and enhancing the water fluxes. Although hygro-responsive membranes are still in an early stage of development, they have had an impact on oil/water separation applications.[22,23]

Moreover, a well-studied class of moisture-responsive materials is hydrogels. Common hydrogels are poly(ethylene glycol) (PEG), 2-hydroxyethylmethacrylate (HEMA), polyacrylamide (pAAM), *N*-isopropylacrylamide (NIPAAM), chitosan and hyaluronic acid gels.[24]

8.2.3 Ionic Strength-responsive Membranes

The ionic strength in a solution can affect the interactions between charged groups and certain ions, such as K^+, Li^+, Na^+ and Cs^+, directly influencing the degree of swelling in the membrane. This dependence is observed in zwitterionic polymer membranes, as these molecules contain charged groups that can be easily affected by the ionic strength of the surrounding environment. Zwitterions with sulfonate and quaternary amine groups have shown an ionic strength-dependent flux in recent studies.[25] Other examples are poly-*N*-vinyl-lactams and bisacrylamide hydrogen, which show dependence on the presence of NaCl. In low concentrations, the polymer chains are extended, closing the pores and decreasing the water permeability. In contrast, at high NaCl concentrations, the polymeric chains collapse, leading to wider pores and higher permeability.

8.3 Indirect Stimulation of Responsive Membranes

Indirect stimulation is the second type of stimulus that can trigger a membrane response. Here, the membrane has responsive sites that are activated by changes in the thermodynamic environment. For example, temperature-responsive groups are susceptible to changes in temperature due to convective and conductive heat transfer.[12]

8.3.1 Temperature-responsive Membranes

Thermo-responsive materials change their properties or conformation at a specific temperature *via* phase transition. The specific temperatures are the lower (LCST) and upper critical solution temperature (UCST).[26] Below the LCST the polymer is soluble in a solvent, forming a homogeneous phase, and above the LCST phase separation occurs *via* an entropic process.[27] Due to this switchable solubility property, thermo-responsive membranes have been used in bio-separations, drug delivery and sensor applications.[28,29]

Well-known thermo-responsive polymers are pAAc, pAAM, pDEAAM and pNIPAAM, where the thermo-responsive material is the acrylamide (AAM).[30] The most common is poly(*N*-isopropylacrylamide) (pNIPAAM), which exhibits a phase transition from a hydrophilic to a hydrophobic state at temperatures higher than the LCST in an aqueous solution. Below the LCST, the NIPAAM polymer is soluble in water, forming hydrogen bonds between the amide groups and water molecules. The polymer chains are in an extended form and are solvated. Above the LCST of ~32 °C, the hydrogen bonds are broken, and the polymer chain collapses (desolvated) and the hydrophobic interactions within the isopropyl groups increase, promoting precipitation of the polymer.[31]

Other thermo-responsive polymers that have been used are poloxamers, pluronics, poly(*N*-vinylcaprolactam) and pNIPAAM derivatives.[32,33] The pNIPAAM polymers have been incorporated into organic polymeric membranes, such as polyvinylidene fluoride (PVDF),[34] polyvinyl alcohol (PVA),[35] polyamide membranes[36] and polysulfone (PS).[37] Due to the change in surface hydrophilicity/hydrophobicity as a function of temperature, these membranes present advantages in improving the anti-fouling properties and selectivity to water.

8.4 Field-responsive Polymeric Membranes

The third type of membrane stimulation can be classified as field responsive. This stimulation involves an external field, in which a magnetic, electric or electromagnetic field is employed.

In this type of membrane, metals or metallic structures (*e.g.*, Fe, Ni, Co) must be present within the polymer.[32] When a magnetic field is applied to these materials, changes in the polymer microstructure are induced, affecting the topography, viscosity and stiffness of the material.[38–41] This is considered a reversible process because the material returns to its original state when the magnetic field is removed.

Electro-responsive materials respond to an external electric field. These materials exhibit high electrical conductivity and are typically composed of metals, metal oxides, conductive polymers or ionic liquids.[42] Depending on the material's components, surfaces could be charged using an external electrical field or electric potential.[43]

Finally, electromagnetic-responsive materials have a specific response when irradiated with photons (*i.e.*, light) at a suitable wavelength (*e.g.*, UV, visible). These photo-responsive materials are "activated" when exposed to radiation. The activated material is then capable of carrying out redox reactions.[44]

8.4.1 Electro-responsive Membranes

The first examples of electro-responsive membranes (ERMs) were those employed in electrofiltration. Although such membranes were first studied in the 1970s,[45] there is a renewed interest in ERMs to remove contaminants, minimize fouling and selectively enhance water permeability. Also, it is known that ERMs may help in overcoming some of the limitations of conventional membranes *via* a "physical response", such as electrostatic adsorption or repulsion of pollutants, and a "chemical response", such as direct or indirect redox reactions of molecules of interest.

The fabrication of ERMs requires a highly electroconductive and porous surface. In the past decade, the most used materials for the ERM fabrication have been carbonaceous materials (~60%), metals and metal oxides (~32%), and conductive polymers (~8%).[46] These conductive materials are distinct in terms of conductivity, rate of electron transfer and the required applied potential. These affect the overall mechanism that can be used.

Carbonaceous materials are mainly studied because of their high conductivity, which is imparted by the sp^2 carbons hybridized with π–π electrons. Carbon nanotubes and graphene-based materials are the most common carbonaceous species incorporated in ERMs due to their facile fabrication and modification, high surface area and porosity.[47] Among the metals, steel mesh[48] substrates have served as cathodes for electrochemical reduction reactions and functionalization with electrocatalysts; for the metal oxides, numerous investigators have focused on Magnéli phase (Ti_4O_7),[49,50] the most conductive phase of the titanium oxides. Finally, conductive polymers usually possess a conjugated backbone with alternating single and double carbon bonds. The overlap of p orbitals in the π-bonds allows the electrons to be delocalized and to move freely between atoms.[51] Due to their high stability, polypyrrole[52] and polyaniline[53] are the most studied conductive polymers. However, these polymers are less commonly used due to their low electrical conductivity compared to other materials.

One of the main reasons that ERMs have been widely studied is their versatility in terms of the final application. Moreover, when an external potential is applied, these membranes can prevent and minimize all types of fouling, including inorganic scaling, organic fouling and biofouling, while increasing the overall membrane operational life. Also, through various electrocatalytic mechanisms, ERMs can reduce the concentration of pollutants in feed solutions.

8.4.1.1 Electrochemical Stimulus

Depending on the material and the applied potential, the ERM can act as the cathode or the anode and offer a particular response (*e.g.*, surface charge, selectivity to complete a redox reaction), translating into a unique decontamination mechanism. Some possible mechanisms to prevent or minimize all three types of fouling *via* electrochemical stimuli are electrostatic interactions and direct or indirect reduction/oxidation (redox) reactions (shown in Figure 8.4).

The electrostatic interactions occur due to a physicochemical process that does not involve an electron transfer. Depending on the applied potential, the membrane surface can be charged positively (*i.e.*, applying a positive voltage or anodic potential) or negatively (*i.e.*, applying a negative voltage or cathodic potential). This creates electrostatic interactions between the pollutants and the membrane surface, and it can be coulombic attraction or repulsion. The strength of these interactions depends mainly on the membrane properties (*e.g.*, specific surface area, pore size, electronic conductivity) and the ionization properties of the pollutant.[43,46]

The redox reactions involve an electron transfer process. In the direct redox reactions, the electron transfer occurs between the ERM and the pollutant (*e.g.*, organic molecules, inorganic ions or bacterial cells). In the anodic

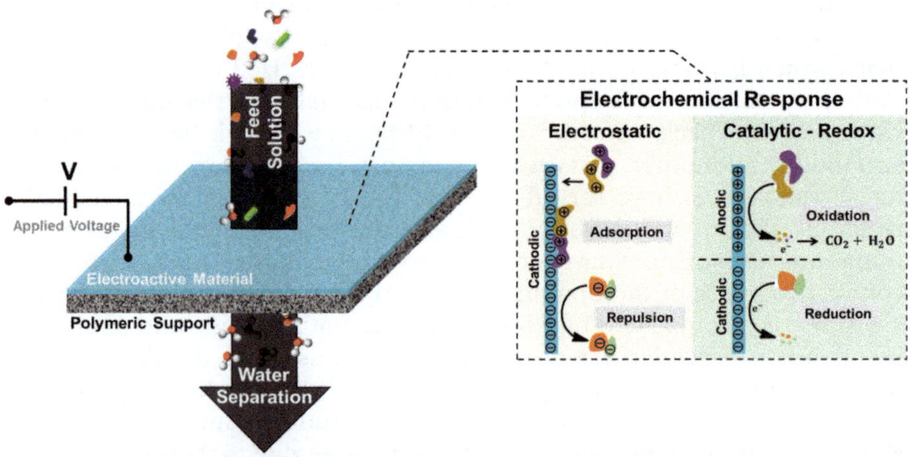

Figure 8.4 General representation of the electro-responsive membranes and the electrochemical responses to the applied stimuli: electrostatic interactions and catalytic reactions.

potential, the pollutants are oxidized, and the electrons are transferred to the ERM. In contrast, in the cathodic potential, the electrons transferred from the ERM reduce the pollutants. On the other hand, the indirect redox reactions rely on the formation of strong oxidizing species (*e.g.*, $^{\bullet}OH$, $Cl^{\bullet}$ and ClO^{-}) or the products from the bubbles from the water splitting reaction (hydrogen (or oxygen) evolution reactions). These species are responsible for completing the catalytic degradation or removal of the pollutants.

8.4.1.2 Anodic Potential Stimuli

The anodic potentials are those that create a highly positive surface and thus promote an oxidative environment. If the ERM surface is charged positively (*i.e.*, anode), it is prone to attract the negatively charged compounds electrostatically. This type of interaction has been employed as a pollutant removal mechanism.[54]

Anodic ERMs are known to reduce $CaSO_4$ fouling (*i.e.*, scaling) significantly. In this case, the anti-scaling mechanism is an electrokinetic reorganization of the electrical double layer (general representation in Figure 8.5D). Due to the membrane's positive charge, a layer of sulfate ions (SO_4^{2-}) is formed, which creates an anion/cation ratio imbalance at the membrane surface, reducing the nucleation rate of the $CaSO_4$ crystals.[55]

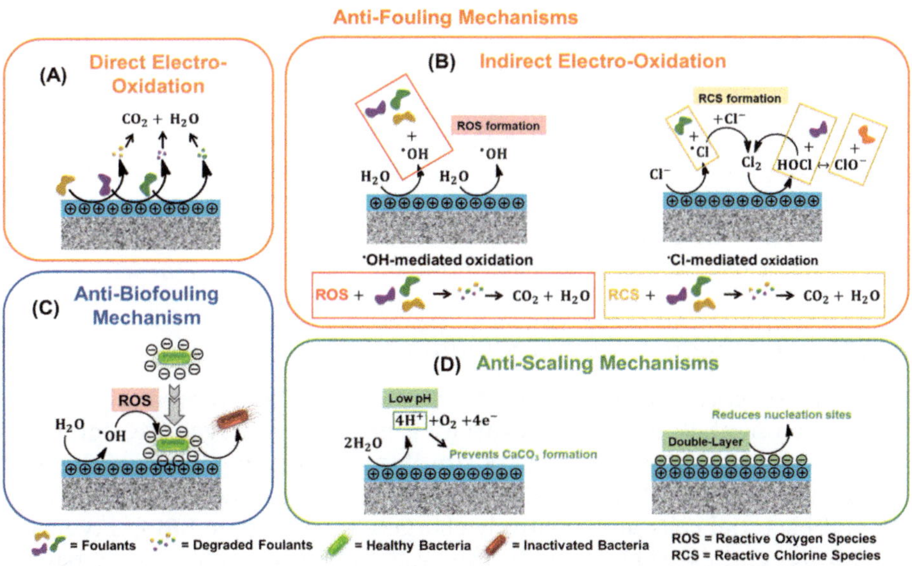

Figure 8.5 Anti-fouling mechanisms of ERMs under anodic stimuli. The degradation of organic pollutants can be carried out *via* (A) a direct electro-oxidation or (B) an indirect electro-oxidation, including the hydroxyl radical-mediated oxidation and the active chlorine-mediated oxidation. Representation of the (C) anti-biofouling mechanism and (D) anti-scaling mechanisms.

Direct electro-oxidation of pollutants (represented in Figure 8.5A) is one of the most used mechanisms with anodic ERMs. Adjusting the applied anodic potential allows the oxidation to be selective for specific molecules. The oxidation of organic molecules is of most interest because they are ultimately converted to CO_2 and H_2O, as is the case with perfluorooctanoic acid (PFOA).[56] Theoretically, this process can be completed successfully at low potentials (below the oxygen reduction reaction potential). However, due to sluggish kinetics, the reaction tends to be slow.

Anodic ERMs can also attract negatively charged bacterial cells and be inactivated *via* direct electro-oxidation, translating into an anti-biofouling approach. The electrons are transferred from the bacteria to the anodic ERMs, leading to cell inactivation by triggering cellular oxidative stress.[57–59]

On the other hand, indirect electro-oxidation (Figure 8.5B) is a rather complex process with regard to the mechanism by which the pollutants are oxidized. The most-accepted mechanisms agree on two main routes: (1) hydroxyl radical-mediated; and (2) active chlorine-mediated oxidation. In the $^{\bullet}OH$-mediated oxidation, the membrane oxidizes nearby water molecules without the addition of chemicals and produces the $^{\bullet}OH$ radical, which is a dominant reactive oxygen species (ROS). These radicals have a high standard potential (2.80 V *vs.* standard hydrogen electrode (SHE)) which enables non-selective and complete oxidation of most organic pollutants.[60–62] The anodic water oxidation also creates a local acidic environment, which prevents the formation of $CaCO_3$ crystals (*i.e.*, anti-scaling mechanism).[55,63] The $Cl^{\bullet}$-mediated oxidation is prompted to occur when chloride ions (Cl^-) are present in waste water. The oxidation of these chlorine ions generates reactive chlorine species (RCS), such as chlorine radicals, $Cl^{\bullet}$ (2.40 V *vs.* SHE), free chlorine, like Cl_2 (1.36 V *vs.* SHE), HClO (1.49 V *vs.* SHE) and ClO^- (0.89 V *vs.* SHE). These RCS can be formed with and without the involvement of $^{\bullet}OH$ and have been reported to remove various organic emerging pollutants, such as reactive dyes,[64] sulfamethazine[65] and paracetamol.[66] Also, as shown in Figure 8.5C, both indirect oxidation mechanisms can be used as bacterial inactivation mechanisms.[57–59]

8.4.1.3 Cathodic Potential Stimuli

The cathodic potential creates a negative interfacial charge, creating a reducing environment. When the ERM surface is negatively charged (*i.e.*, cathode), it can electrostatically attract the positively charged compounds of the feed solution (as a pollutant removal mechanism). As most natural organic matter and bacterial cells are negatively charged, the cathodic ERM is typically used to repel these pollutants. In this way, the electrostatic energy barrier is increased, and the charged pollutants migrate away from the membrane surface, thus inhibiting organic fouling and biofouling (Figure 8.6B). Some negative membrane/pollutant repulsion examples include natural organic matter (NOM),[67] per- and poly-fluoroalkyl substances (PFAS),[68] bovine serum albumin (BSA)[69] and bacterial cells,[70] which can be repelled from the cathodic membrane surface.

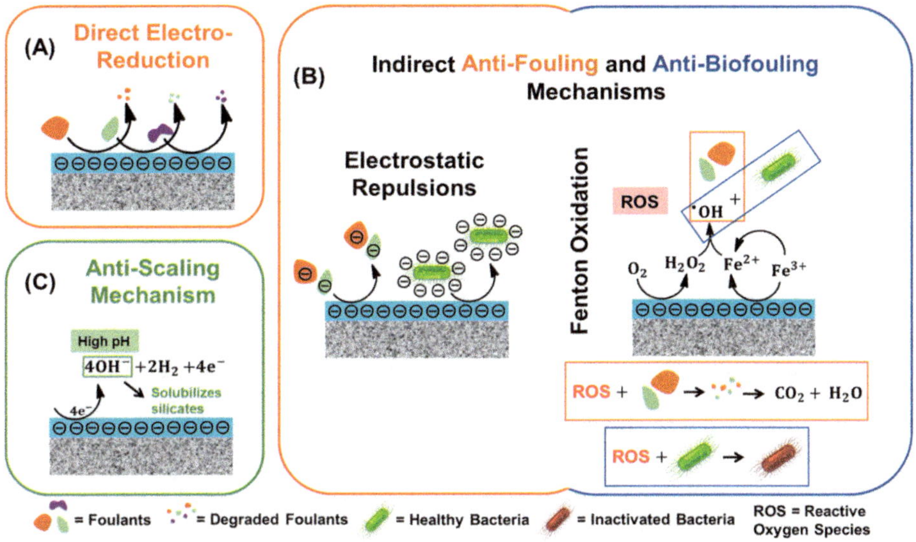

Figure 8.6 Anti-fouling mechanisms of the ERMs under cathodic stimuli. The degradation of organic pollutants can be carried out *via* (A) a direct electro-reduction. The indirect degradation of organic pollutant and the bacteria inactivation can be carried out *via* (B) electrostatic repulsions and Fenton oxidation. (C) Representation of the anti-scaling mechanism.

Cathodic ERMs have been studied for nitrate conversion *via* the nitrate reduction reaction (NO_3RR), which aims to obtain N_2 as the desired final product. However, depending on the catalyst and solution conditions (*e.g.*, pH, dissolved oxygen, ionic species and overpotential), the reaction pathway can be influenced, and NH_3 becomes a possible final product. This NO_3RR occurs *via* a direct electroreduction reaction, as represented in Figure 8.6A. The NO_3RR starts with the adsorption of nitrate onto the cathode surface, then it is reduced to nitrite (NO_2^-); this is the rate-determining step and controls the overall reaction kinetics. The adsorbed NO_2^- is rapidly converted to NO_{ad}, which is considered the intermediate to control the selectivity towards N_2 or NH_3 production. Several ERMs have been reported to produce NH_3 selectively as fertilizer.[9,71,72]

Although the electro-Fenton process is an indirect electro-oxidation, it is triggered by applying a cathodic potential at the ERM (Figure 8.6B). In general, hydroxyl radicals (•OH) can be generated indirectly *via* this process, the dissolved oxygen is reduced to hydrogen peroxide (H_2O_2) *via* a two-electron reduction and the produced H_2O_2 reacts with Fe^{2+} (the Fenton catalyst) to form •OH. Thus, this ROS is the one that degrades the pollutants, such as antibiotics[73–75] and dyes.[76,77] The Fenton catalyst (Fe^{2+}) is generated by three possible pathways: (1) directly adding Fe^{2+} to the feed solution; (2) electroreduction of

Fe^{3+} in the feed by the ERM; and (3) using a sacrificial anode to generate Fe^{3+}/Fe^{2+} in the feed solution.

The cathodic ERM can also be used as an anti-scaling process (shown in Figure 8.6C). The most common way to accomplish descaling using an electroconductive membrane is by modifying the local pH at the membrane surface. Cathodic water electrolysis can be employed to reach a high pH environment that prevents scale formation due to increased silica solubility. Silicates are an additional mineral that causes membrane scaling. The direct electrochemical reduction can achieve the descaling of silicate,[78] but also an indirect mechanism can be accessed, where the hydroxide ion from water electrolysis can dissolve the silicate scaling.[79]

In general, in ERM, combining anodic and cathodic potentials could enhance the overall result. This way, an electrostatic attraction can first occur, which increases the concentration of the pollutants at the membrane surface, and then be oxidized by switching the potential. If hydrogen evolution reaction or oxygen reduction reaction potentials are reached, all types of contaminants could be eliminated or removed from the membrane surface. An immense amount of research has been carried out on this topic; however, plenty of materials and combinations of materials need to be exploited. Thus, innovations in electro-responsive or electro-active membranes are necessary to develop sustainable and low-cost membranes. ERMs are extremely versatile membranes because, in theory, a membrane could have a wide range of working potentials, enabling its capacity to degrade different types of pollutants through various mechanisms.

8.5 Catalytic Membranes

Catalytic membranes are those that are capable of simultaneously completing the water separation process while catalyzing a redox reaction. In contrast to other responsive membranes, these incorporate a catalyst that increases the rate of a chemical reaction without undergoing permanent chemical change. Moreover, the reactions carried out by these catalysts are triggered by the applied stimuli. These membranes have been exploited in the last decade due to their complementary approach to reaching high water permeability and selectivity while mitigating or preventing membrane fouling. As stated before, membrane fouling and rejection of small and neutral molecules are still the main limitations for upscaling the membrane-based filtration process. Thus, incorporating a catalyst into the polymeric membranes brings us a step closer to overcoming these limitations. Different types of catalysts have been studied, including electrocatalysts (respond to an electrical bias, *e.g.*, voltage), photocatalysts (respond to photons, *e.g.*, light) or biocatalysts (respond to specific substrates, *e.g.*, enzymes). Incorporating these catalysts in the polymeric membranes will be discussed in detail in the following sections for photocatalytic and biocatalytic membranes. Electrocatalytic membranes have been discussed in the previous sections.

8.5.1 Photocatalytic Membranes

For a few years, the typical photocatalytic reactors were comprised of a membrane for the separation process and a photocatalyst in the feed solution for the catalytic conversion. Albeit photocatalysts have shown tremendous potential in treating waste water, while membrane technology has exhibited effective pollutant separation efficiency, both approaches have their shortcomings.[80] Hence, researchers have explored several possibilities to combine both methods into one hybrid technology: photocatalytic membranes (PCMs).[81,82] PCMs, as shown in Figure 8.7, combine the water separation process with a catalytic reaction triggered by photons as the main field-responsive stimulation.

These membranes are generally employed as an eco-friendly strategy to photo-degrade organic pollutants, inactivate bacterial cells and reduce heavy metals, which translates into an anti-fouling mechanism.[83,84] A photocatalyst (*e.g.*, a species that absorbs photons at a specific wavelength and utilize that energy to catalyze a reaction) is needed to achieve a good photoresponse. Due to the relatively small bandgap (energetic distance between the conduction (CB) and valence bands (VB)), semiconductors (SC) are the most used materials as photocatalysts. The small bandgap accounts for

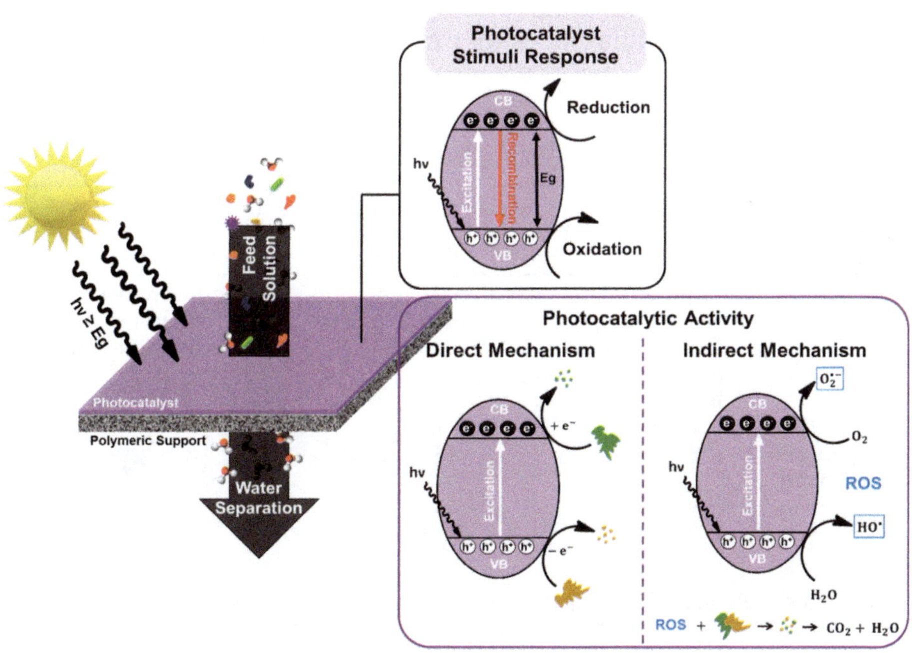

Figure 8.7 General representation of photocatalytic membranes (PCMs). Schematic representation of the photocatalyst response when irradiated with light and the two photocatalytic mechanisms, direct and indirect, for the degradation of pollutants.

the absorption of photons in the ultraviolet or visible light range. Briefly, the absorption of photons excites the electrons of the valence band to the photocatalyst's conduction band, which creates an electron/hole pair that, depending on the SC, can complete an oxidation or reduction reaction. However, there are different types of SCs, varying from intrinsic to extrinsic (when the material is doped).

8.5.1.1 Photocatalytic Principle

The primary feature of the photocatalyst is that it exhibits a high photocatalytic activity. Some of the SCs that have been extensively investigated are TiO_2, ZnO, Fe_2O_3, CuS, Bi_2WO_6, $BiVO_4$, g-C_3N_4, MoS_2 and WS_2, and varying from UV-responsive to visible light-responsive photocatalysts.[44] All of these SCs, have defined bandgaps, crystal structures and surface areas. However, some of the limitations of photocatalysts are wide bandgap energy, low light absorption capacity, low surface area, fast recombination rate of the electron/hole pair and low chemical stability. To address these limitations, many researchers have been studying modifications to photocatalysts, including introducing dopants,[85] forming heterojunctions[86] or incorporating a co-catalyst[87] into the bulk material. Table 8.1 shows the main differences between these multicomponent modifications.

Doping refers to the incorporation of foreign atoms into the lattice structure of the SC.[88,89] Adding dopants can create a lower bandgap than for the undoped SC, which moves the responsiveness towards the visible light

Table 8.1 Comparison between the different multicomponent photocatalysts.

	Multicomponent photocatalysts		
Type	Doped	Heterojunction	Co-catalyst
Definition	Incorporation of foreign atoms in the semiconductor crystal lattice	Combination of semiconductors to work synergistically	Addition of a second catalyst to work complementarily
Electron/hole separation and band alignment	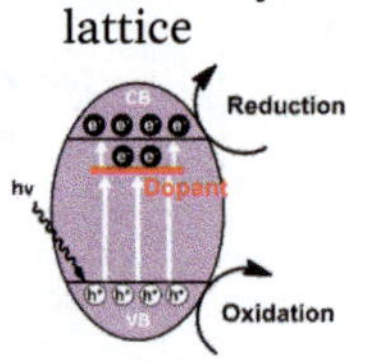	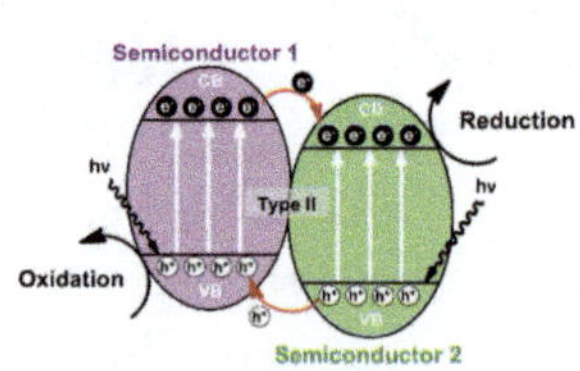	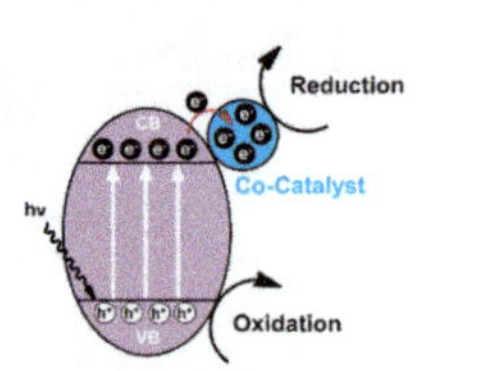
Advantages	Reduces bandgap (response in visible light) and enhances electron kinetics	Creates efficient electron/hole separation, reduces recombination rate, can extend the light absorption range	Creates efficient electron/hole separation, reduces recombination rate, and enhances reaction kinetics

range. These dopants can also generate a large dipole moment to change the electron transfer kinetics. This translates into a more favorable transition from the VB to the CB; thus, more electrons can be transferred.[90] Another method to enhance the SC performance is the formation of a heterojunction by combining SCs to work synergistically. In this case, the VB and CB of the SC have an energetic arrangement that allows the electron/hole photocarriers to flow between them. Usually, the bands are separated in different SCs, so the electron/hole separation is better.[91] Finally, use of a co-catalyst is when a second catalyst, which can be metallic, an SC or non-metallic, is added. Both catalysts will catalyze a complementary reaction (*e.g.*, oxidation occurs on the SC and reduction on the second catalyst). The recombination rate is improved because the photogenerated electrons or holes flow to the second catalyst, adding an energetic barrier to the recombination step.[92]

8.5.1.2 Photocatalytic Membrane Approaches

As mentioned in the previous sections, membrane fouling is a well-known concern in the filtration process. To address this, PCMs have been widely studied and are attracting more attention due to their low energy requirement. The main goal is to have a self-cleaning membrane that can harness energy from sunlight. To accomplish this, the photocatalyst bandgap (E_g) needs to be in the visible range (1.6–3.2 eV), and requires that the recombination rate is slow so that the electron/hole pair can be reactive for more extended periods.

These membranes can be fabricated by the blending method, where the photocatalyst is mixed in the casting solution.[10,93–96] Nonetheless, immobilizing the photocatalyst at the membrane surface is more effective by virtue of its higher surface area, exposure to the light source and accessibility to the pollutant. The common immobilization techniques include coatings (*e.g.*, spray, dip, spin), direct fabrication over the support (*e.g.*, hydrothermal) and self-assembly.[97] Studied polymeric supports include cellulose acetate (CA),[98–100] polysulfone (PSF),[101–103] polyvinylidene fluoride (PVDF),[104–106] polyethersulfone (PES),[107–109] polypropylene (PP),[110–112] polyamide (PA),[113–115] chitosan (CS)[116–118] and polyacrylonitrile (PAN).[84,119,120]

PCMs have been successfully applied in the degradation of organic pollutants and for disinfection of pathogenic bacteria, in many studies.[121] The photodegradation of the contaminants can be *via* a direct or indirect mechanism (see summary in Table 8.2 and Figure 8.7). In the direct mechanism, the pollutant is directly degraded by the photogenerated electron (*via* a reduction reaction) or the photogenerated hole (by an oxidation reaction). For this type of mechanism, band alignment is crucial. For reduction reactions, the CB of the SC needs to be energetically higher than the standard reduction potential of the foulant. On the other hand, for oxidation reactions, the VB of the SC needs to be lower than the oxidation potential of the foulant. The direct mechanism can be employed for the degradation of organic molecules;[122]

Table 8.2 Summary of the photocatalytic mechanisms: direct *vs.* indirect.

Step 1: Activation		Photocatalyst + $h\nu \rightarrow$ Photocatalyst*[h^+(VB) + e^-(CB)]	
Type of mechanism		Direct mechanism[125,126]	Indirect mechanism[127,128]
Step 2: Redox reactions	Reduction	Pollutant + $ne^- \rightarrow$ Red Product $Cr^{6+}_{(aq)} + 3e^- \rightarrow Cr^{3+}_{(aq)}$	$O_2 + e^- \rightarrow O_2^{\bullet -}$ −0.18 V *vs.* NHE
	Oxidation	Pollutant $\rightarrow$ Ox Product + ne^- $As^{3+}_{(aq)} \rightarrow As^{5+}_{(aq)} + 2e^-$	$H_2O + h^+ \rightarrow H^+ + HO^\bullet$ 2.8 V *vs.* NHE
Step 3: Secondary reactions		—	ROS + Pollutant $\rightarrow CO_{2(g)} + H_2O_{(l)}$

however, it is most commonly used for inorganic compounds, specifically heavy metals.[123–126]

The indirect mechanism, the most studied approach, is when the photocatalyst reacts with oxygen and hydroxyl groups in water, generating various ROS, such as hydroxyl radicals ($^\bullet OH$) and superoxide radical ions ($^\bullet O_2^-$). These ROS have been proven to successfully oxidize organic pollutants and ultimately liberate water and carbon dioxide; nonetheless, this is considered a non-selective degradation process.[44] In this mechanism, the energetic band alignment of the SC needs to be specific for the water and oxygen redox potentials (2.8 V and −0.18 V *vs.*, normal hydrogen electrode (NHE), respectively). These reactions can be promoted easily if the photocatalysts have a wide bandgap, specifically in the visible light range. For this reason, TiO_2 is the most studied SC for this application; it has a wide bandgap (3.0–3.2 eV) and its crystal lattice can be modified easily to enhance the photocatalytic activity. Common examples of foulants found in water are: (1) organic molecules, such as NOM,[127] dyes,[128] pharmaceuticals[95] and antibiotics;[129] (2) bacterial cells;[130] and (3) inorganic compounds, such as heavy metals.[131]

PCMs in the literature include a variety of polymeric supports and photocatalysts (single components, doped, heterojunctions and with co-catalysts). One important feature of the selected polymer support is its role in the photocatalytic process. In some cases, using a polymer that is oxygen-rich or amine-rich can reduce the recombination rate by assisting in the photocarrier separation process.[132] Likewise, the photogenerated electrons can combine with the polymer's surface oxygens and be converted into superoxide radicals by the photo-oxidation reaction, thus improving the ROS formation rate.[92]

Over the years and through extensive research, PCMs have displayed significant potential in combating water pollution. By virtue of combining photocatalysis and membrane technology, they overcome the limitations that both processes face when used individually. One of the most attractive features is the capacity to harvest the energy from sunlight to become a self-cleaning method. However, several aspects need to be addressed. For example, alternative photocatalyst deposition methods need to be studied because the current ones could produce leaching during operation. Polymeric membranes are exceptional for water separation, yet some polymer backbones are still prone to degradation when exposed to UV light. Another direction to focus on is ROS-based oxidation. The partial degradation of the pollutants can create intermediate compounds that can represent a greater environmental hazard and are rarely investigated. More studies are required to study the feasibility of upscaling the PCM reactions and their capabilities to treat different types of effluents. So far, all of the reported work uses simulated light sources; thus, there is a need to explore the feasibility of using direct sunlight as the photon source.

8.5.2 Biocatalytic Membranes

Another class of responsive membranes that has been proposed recently are those that exert a certain *in situ* reactivity against biological contaminants. Multiple natural contaminants will require some sort of catalysis to accelerate and achieve their biodegradation. It is well understood that not all catalysts are equally produced and each of them will provide a different response. Biocatalytic membranes more often embed enzymes as part of the polymer formulation and membrane fabrication. The study of this class of membranes is considered an emerging research topic to design novel sustainable materials for water treatment processes.[133] The biomaterial provides dual functionality for both mass transport performance and interfacial mechanisms that depend on an external stimulus (*i.e.*, direct stimulation). Natural biocatalysts, such as enzymes, offer significant advantages over metal, inorganic and chemically derived catalysts.[134,135] Treatment methods employing enzymes simplify the process without producing costly secondary by-products and post-reaction separation concerns.[136] Currently, many evolving biotechnologies include industrial biocatalysts with immobilized enzymes to catalyze reactions in the biosynthetic production of compounds, biomedical applications and bioremediation.[137,138] Immobilized biocatalysts have gained preference over the free enzyme processes to overcome further material handling constraints and for their reusability in the food and pharmaceutical manufacturing sectors. This industrial biocatalytic strategy has emerged in the water treatment field as well.

8.5.2.1 Biocatalytic Membrane Principles

In the broader context of responsive polymeric membranes, the materials act on detection of changes in variables such as pH, temperature, ionic strength, molecular structure, magnetic fields, light and electric fields.[139–142] Conversely, biomembranes that embed enzymes respond to substrate specificity instead. The functionalization of the interfaces of membranes using enzymes provides new horizon to catalyze slow reactions to degrade unwanted contaminants in water systems. Biocatalytic enzymatic membranes have a natural response to selective contaminants by means of the catalytic triad within the enzyme's active site.

Enzymes used as part of responsive materials design follow similar kinetics principles to catalyze reactions even in their immobilized form. As illustrated in Figure 8.8A, a homogeneous[143] catalysis occurs when free enzymes exist in the same phase as the reactants, while heterogeneous catalysis[144,145] occurs in different phases. Hence, biocatalytic membranes are heterogeneous catalysts that can easily regenerate for repeated reaction cycles. The enzymatic activity of a responsive membrane follows Michaelis–Menten kinetics (Figure 8.8B) and catalysis mechanisms (Figure 8.8C). The process efficiency is measured as the enzyme activity changes under different immobilization and physical conditions, leading to residual activity when compared to homogeneous catalysis. The extent of the residual catalysis can be expressed by the relationship in eqn (8.1).[146]

$$\%\ \mathrm{RA} = \frac{\mathrm{SA}_{\mathrm{Immobilized}}}{\mathrm{SA}_{\mathrm{Free}}} \times 100\% \tag{8.1}$$

where the percentage of residual activity (% RA) corresponds to the ratio of the specific activity for the immobilized enzyme ($\mathrm{SA}_{\mathrm{Immobilized}}$) on the membrane with respect to the specific activity of the free enzyme in solution ($\mathrm{SA}_{\mathrm{Free}}$).

The application of heterogeneous biocatalysts is notable and useful in the removal of industrial contaminants, such as pesticides, phenols, fatty acids and industrial dyes.[147,148] It is worth mentioning that there are seven categories for enzyme classifications in terms of the catalyzed substrate used by the International Union of Biochemists. The categories include oxidoreductases (EC 1), transferases (EC 2), hydrolases (EC 3), lyases (EC 4), isomerases

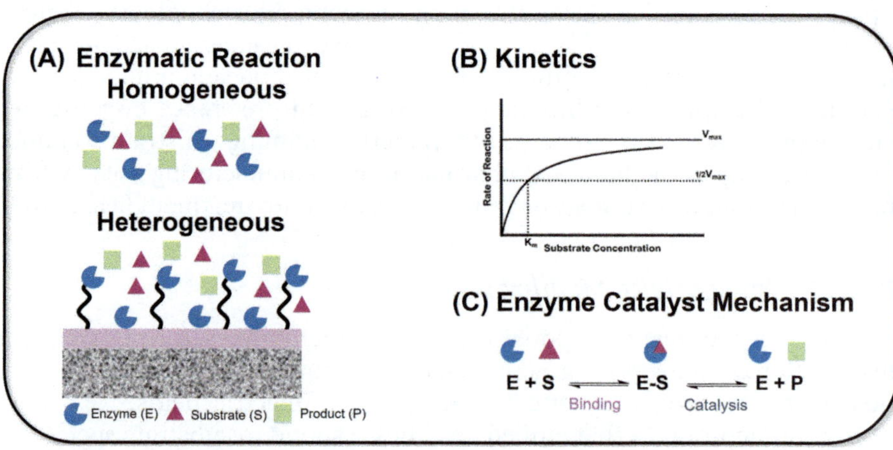

Figure 8.8 Theoretical aspects for biomembrane enzymatic reactions. (A) Classification for the different enzymatic reactions. (B) Michaelis–Menten kinetics applied to enzyme-catalyzed reactions of one substrate and one product. (C) General mechanism for enzyme catalysis where the enzyme (E) binds substrate (S) to produce product (P).

(EC 5), ligases (EC 6) and the most recently added translocases (EC 7).[149] Commonly used industrial enzymes are lipases (EC 3.1.1.3), which are ubiquitous in nature and have the ability to hydrolyze a wide range of substrates, including esters, alcohols, acids and lipids.[150] Indeed, lipases remain the dominant enzyme type for industrial applications.[151] Certain lipids, such as those shed away from bacterial cell walls (*i.e.*, endotoxins)[152] can be acutely toxic to humans. Other lipids are also known to be environmental pollutants.[153,154] Degrading such molecules represents a current challenge for many industrial processes in the field of water treatment, for which that biomembranes might provide a solution. Moreover, previous studies have demonstrated that lipases are able to fully expose their active center upon approaching a hydrophobic surface when immobilized.[151,155] This feature has been exploited to immobilize lipases on various hydrophobic surfaces.[133,156–158]

8.5.2.2 Enzyme Immobilization Approaches

Initial efforts towards enzyme immobilization over polymeric materials dates back to 1950s[159] when the principles were eventually translated to produce important products, such as high fructose corn syrup[160] during the 1970s and semi-synthetic antibiotics[161] (*i.e.*, penicillin). The characteristics of the support materials are important factors to consider. The polymeric support will eventually affect the performance of the immobilized catalyst. Physical parameters, such as contact area, mechanical properties and hydrophobicity, are key factors to achieve effective immobilization. Also, immobilization needs to provide the conditions that preserve enzymatic activity. Many efforts related to enzyme immobilization have been directed towards different hydrophobic supports (*e.g.*, nanoparticles,[162] activated carbon,[163] pore glasses,[156] di-block co-polymers, *etc.*). Investigators commonly employ one of four main immobilization mechanisms for the membrane polymeric materials, whereby immobilization may occur *via* covalent bonding, hydrophobic adsorption, enzyme entrapment or electrostatic attractions.[137,164]

As shown in Table 8.3, current trends in the polymeric support used for immobilization include metal–organic frameworks (MOFs), graphene oxide (GO), polyvinylidene fluoride (PVDF), poly(*N*-isopropylacrylamide)

Table 8.3 Examples of common supports recently used for enzyme immobilization for water treatment applications.

Enzyme	Support	Contaminant	Reference
Lac	TiO_2 sol–gel/PVDF	BPA	167
	pNIPAAM	BPA	168
	MOFs/carbon nanotubes	Phenols	169
Ly	GO/PES	Bacterial cell wall	170
CALB	PS-*b*-P4VP	LPS	133
HPR	MOFs	BPA	171
β-Gal	CA	Lactose	172

(pNIPAAM), cellulose acetate (CA), polyether sulfone (PES), among others. Moreover, recent investigations show that loading of lipases using polystyrene resins shows effective adsorption regardless of the size of the particles.[165] Other studies have applied poly (methyl methacrylate) resins (pMMA) as a support.[165,166]

In the quest to achieve the ultimate goal of providing safe water, additional efforts are still required regarding biocatalytic membrane configuration within a holistic treatment design. While researchers continue to make important progress in advancing the common knowledge for understanding the design of biofunctional interfaces as part of membrane materials with superior stability and reusability, much remains to be discovered. This relatively new research area would contribute additional novel ideas to enhance specific applications by combining both size-exclusion barriers along with reaction mechanisms, providing a sustainable solution for water treatment for generations to come.

8.6 Challenges and Future Directions

This chapter summarizes a variety of stimuli-responsive membranes that have been recently proposed to aid in water treatment. Undoubtedly, there is a tremendous growing demand for such stimuli-responsive membranes. This interest is mainly due to their tunable properties to gain *in situ* control over the filtration process, specifically for contaminant removal. A broad spectrum of materials has been explored to fabricate these responsive materials, including polymers, carbon-based materials, metals, non-metals, semiconductors, biological molecules and hybrids. All the reported materials enhance the membrane performance by means of water selectivity and permeability. This enhancement is achieved by modifying the membrane's structural architecture and its surface chemistry.

Moreover, responsive materials offer the opportunity to attain control of membrane pore size, anti-fouling and self-cleaning properties, reusability and contaminant removal *via* various mechanisms. The responsive polymeric membranes are a unique class of materials capable of addressing many of the current drawbacks in membrane-based technologies. The possibility to regenerate the membrane, degrade contaminants *in situ* or generate new molecules through biochemical or electrochemical reactions is intellectually and technologically stimulating. Indeed, this new class of membranes presents a unique opportunity to establish a new paradigm that focuses more on process sustainability.

Water treatment processes can be divided into many applications, varying in the complexity of the feed water, the separation driving force and the type of contaminants present. These pollutants include natural organic matter and emerging organic contaminants, such as pharmaceuticals, antibiotics, pesticides, dyes, heavy metals and microorganisms. For this reason, meaningful solutions to each class of pollutant are essential to sustain the growing demand for fresh and clean water. Different types of responsive polymeric membranes have been developed and tested for the removal of the above-mentioned foulants, as well as in the various membrane treatment processes, and the outcomes have been promising. Furthermore, the membranes can remove foulants and prevent fouling while exhibiting suitable water permeability and selectivity. However, despite the immense potential that these membranes have demonstrated, the number of studies that test their feasibility for the up-scaled processes is limited. Consequently, efforts need to be made to evaluate these membranes for an up-scaled approach.

Before envisioning the future of these responsive membranes, some crucial aspects and challenges must be addressed. In the case of pH-responsive membranes, the pore diameter can be tailored by changes in the acidity of the surrounding environment. However, in some cases, the resulting pore opening may be too wide, affecting the membrane's selectivity. To minimize this effect and take advantage of pore size expansion/contraction, materials with smaller pore constriction need to be explored. To up-scale pH-responsive membranes, alternatives to switch the membrane pH must be found; otherwise, vast amounts of chemicals are needed, which is not ideal.

Temperature-responsive membranes exhibit a controlled response to the stimulus of temperature; however, the need for a heat exchanger might affect the overall cost efficiency of the process. Implementing materials with stimulation at a lower temperature could be of benefit for upscaling because less energy input will be required.

ERMs have excellent anti-fouling properties, which can extend the membrane's lifespan, yet the applied potential defines the anti-fouling mechanisms. One of the limitations of these ERMs is the fact that they can mitigate only one type of fouling per applied voltage. For instance, if a cathodic potential is applied, electrostatic repulsions and electroreduction reactions can be carried out but positively charged molecules will be electrostatically attracted, thus causing fouling. One plausible alternative to minimize this effect is to switch periodically from anodic to cathodic potentials. Therefore, novel materials capable of acting as both cathode and anode will need to be implemented on membranes; this would broaden the electrochemical response towards foulants. Another downside of ERMs is the need for an applied potential that translates into a constant energy input. To reduce this, PCMs are of much interest because, ideally, their activation energy can be harvested from sunlight. So far, PCMs have the potential to overcome many of the drawbacks of other responsive membranes. The pore structure can be modified and defined by adding photocatalyst *via* the blending method, and the anti-fouling properties can be enhanced by depositing the photocatalyst at the surface. Even so, attention must be directed towards the photocatalyst/membrane surface interface because poor interaction could lead to leaching. This leaching can be minimized by exploring methods where the photocatalyst can be chemically bonded to the membrane instead of physically attached. Also, the possibility of generating intermediates during foulant photocatalysis is an alarming concern. In this matter, biocatalytic membranes are the most selective and specific approach to avoid intermediate species. Although this class of membranes is relatively new for water treatment, decent progress has been made in degrading target molecules. The main concern is their performance under a less controlled environment, in the presence of several types of foulants and under prolonged measurement.

In general, the field of responsive polymeric membranes has developed at a fast pace and will continue to grow as new polymers and novel approaches are employed. The combination of materials with different stimuli responses is a promising strategy. These hybrids could work synergistically and minimize the drawbacks that each counterpart exhibits when employed individually. In the future, after the above-mentioned drawbacks and concerns have been addressed, these membranes may replace current commercial candidates.

References

1. M. K. Sinha, M. K. Purkait, P. Mondal and R. Singh, *Stimuli Responsive Polymeric Membranes*, Elsevier, 2018.
2. F. Meng, S. R. Chae, A. Drews, M. Kraume, H. S. Shin and F. Yang, *Water Res.*, 2009, **43**, 1489.
3. X. Zhang, J. Ma, J. Zheng, R. Dai, X. Wang and Z. Wang, *Chem. Eng. J.*, 2022, **432**, 134425.
4. J. R. Werber, C. O. Osuji and M. Elimelech, *Nat. Rev. Mater.*, 2016, **1**, 16018.
5. W. J. Koros and C. Zhang, *Nat. Mater.*, 2017, **16**, 289.
6. K. Dutta and S. De, *J. Mater. Chem. A*, 2017, **5**, 22095.
7. D. Wandera, S. R. Wickramasinghe and S. M. Husson, *J. Membr. Sci.*, 2010, **357**, 6.
8. H. Musarurwa and N. Tawanda Tavengwa, *Microchem. J.*, 2022, **179**, 107503.
9. J. Zhao, L. Liu, Y. Yang, D. Liu, X. Peng, S. Liang and L. Jiang, *ACS Sustain. Chem. Eng.*, 2023, **11**, 2468.
10. I. H. Alsohaimi, M. S. Alhumaimess, M. Alzaid, A. A. Essawy, M. R. El-Aassar, R. M. K. Mohamed and H. M. A. Hassan, *J. Environ. Manage.*, 2023, **332**, 117351.
11. D. Yi, T. Bayer, C. P. S. Badenhorst, S. Wu, M. Doerr, M. Hohne and U. T. Bornscheuer, *Chem. Soc. Rev.*, 2021, **50**, 8003.
12. S. Darvishmanesh, X. Qian and S. R. Wickramasinghe, *Curr. Opin. Chem. Eng.*, 2015, **8**, 98.
13. G. Kocak, C. Tuncer and V. Bütün, *Polym. Chem.*, 2017, **8**, 144.
14. C. Zhao, S. Nie, M. Tang and S. Sun, *Prog. Polym. Sci.*, 2011, **36**, 1499.
15. S. Dai, P. Ravi and K. C. Tam, *Soft Matter*, 2009, **5**, 2513.
16. L. A. Connal, Q. F. Li, J. F. Quinn, E. Tjipto, F. Caruso and G. G. Qiao, *Macromolecules*, 2008, **41**, 2620.
17. J.-F. Gohy, B. G. G. Lohmeijer, S. K. Varshney, B. Décamps, E. Leroy, S. Boileau and U. S. Schubert, *Macromolecules*, 2002, **35**, 9748.
18. F. J. Xu, E. T. Kang and K. G. Neoh, *Biomaterials*, 2006, **27**, 2787.

19. G. Zhai, E. T. Kang and K. G. Neoh, *J. Membr. Sci.*, 2003, **217**, 243.
20. L. Ying, P. Wang, E. T. Kang and K. G. Neoh, *Macromolecules*, 2002, **35**, 673.
21. T. A. Saleh, G. Fadillah and E. Ciptawati, *J. Polym. Res.*, 2021, **28**, 197.
22. A. K. Kota, G. Kwon, W. Choi, J. M. Mabry and A. Tuteja, *Nat. Commun.*, 2012, **3**, 1025.
23. M. Tenjimbayashi, K. Sasaki, T. Matsubayashi, J. Abe, K. Manabe, S. Nishioka and S. Shiratori, *Nanoscale*, 2016, **8**, 10922.
24. S. H. Aswathy, U. Narendrakumar and I. Manjubala, *Heliyon*, 2020, **6**, e03719.
25. Q. Shi, Y. Su, W. Zhao, C. Li, Y. Hu, Z. Jiang and S. Zhu, *J. Membr. Sci.*, 2008, **319**, 271.
26. S. M. Husson, in *Responsive Membranes and Materials*, 2012, p. 73.
27. M. Wei, Y. Gao, X. Li and M. J. Serpe, *Polym. Chem.*, 2017, **8**, 127.
28. D. Wandera, S. R. Wickramasinghe and S. M. Husson, *J. Membr. Sci.*, 2010, **357**, 6.
29. M. A. C. Stuart, W. T. S. Huck, J. Genzer, M. Müller, C. Ober, M. Stamm, G. B. Sukhorukov, I. Szleifer, V. V. Tsukruk, M. Urban, F. Winnik, S. Zauscher, I. Luzinov and S. Minko, *Nat. Mater.*, 2010, **9**, 101.
30. S.-C. Low and Q.-H. Ng, in *Advanced Nanomaterials for Membrane Synthesis and Its Applications*, ed. W.-J. Lau, A. F. Ismail, A. Isloor, and A. Al-Ahmed, Elsevier, 2019, p. 69.
31. K. Pan, R. Ren, H. Li and B. Cao, *Polym. Adv. Technol.*, 2013, **24**, 22.
32. W. Zhou, Z. Qiao, E. Nazarzadeh Zare, J. Huang, X. Zheng, X. Sun, M. Shao, H. Wang, X. Wang, D. Chen, J. Zheng, S. Fang, Y. M. Li, X. Zhang, L. Yang, P. Makvandi and A. Wu, *J. Med. Chem.*, 2020, **63**, 8003.
33. F. Doberenz, K. Zeng, C. Willems, K. Zhang and T. Groth, *J. Mater. Chem. B*, 2020, **8**, 607.
34. L. Y. Chu, T. Niitsuma, T. Yamaguchi and S. i. Nakao, *AlChE J.*, 2003, **49**, 896.
35. Y.-M. Sun and T.-L. Huang, *J. Membr. Sci.*, 1996, **110**, 211.
36. S. Mondal and S. R. Wickramasinghe, *Sep. Purif. Technol.*, 2012, **90**, 231.
37. M. Hesampour, T. Huuhilo, K. Mäkinen, M. Mänttäri and M. Nyström, *J. Membr. Sci.*, 2008, **310**, 85.
38. J. D. Kiang, J. H. Wen, J. C. del Álamo and A. J. Engler, *J. Biomed. Mater. Res., Part A*, 2013, **101A**, 2313.
39. P. V. Testa, H. Morath, P. Goran, C. Carta and F. Ellinger, *2019 IEEE Asia-Pacific Conference on Applied Electromagnetics (APACE)*, 2019.
40. K. A. Tran, E. Kraus, A. T. Clark, A. Bennett, K. Pogoda, X. Cheng, A. Cēbers, P. A. Janmey and P. A. Galie, *ACS Appl. Mater. Interfaces*, 2021, **13**, 20947.
41. E. A. Corbin, A. Vite, E. G. Peyster, M. Bhoopalam, J. Brandimarto, X. Wang, A. I. Bennett, A. T. Clark, X. Cheng, K. T. Turner, K. Musunuru and K. B. Margulies, *ACS Appl. Mater. Interfaces*, 2019, **11**, 20603.
42. I. Carayon, A. Gaubert, Y. Mousli and B. Philippe, *Biomater. Sci.*, 2020, **8**, 5589.
43. A. Ronen, S. L. Walker and D. Jassby, *Rev. Chem. Eng.*, 2016, **32**, 533.

44. Y. Shi, J. Huang, G. Zeng, W. Cheng and J. Hu, *J. Membr. Sci.*, 2019, **584**, 364.
45. J. D. Henry, L. F. Lawler and C. H. A. Kuo, *AIChE J.*, 1977, **23**, 851.
46. M. Sun, X. Wang, L. R. Winter, Y. Zhao, W. Ma, T. Hedtke, J.-H. Kim and M. Elimelech, *ACS ES&T Engg*, 2021, **1**, 725.
47. X. Fan, G. Wei and X. Quan, *Environ. Sci.: Nano*, 2023, **10**, 11.
48. A. Mushtaq, H. Cho, H. Ryu, M. A. Ahmed, M. Saif Ur Rehman and J. I. Han, *Chemosphere*, 2022, **308**, 136160.
49. H. Lin, J. Niu, S. Liang, C. Wang, Y. Wang, F. Jin, Q. Luo and Q. Huang, *Chem. Eng. J.*, 2018, **354**, 1058.
50. S. Almassi, Z. Li, W. Xu, C. Pu, T. Zeng and B. P. Chaplin, *Environ. Sci. Technol.*, 2019, **53**, 928.
51. G. Wallace and G. Spinks, *Soft Matter*, 2007, **3**, 665.
52. J. Sun, G. Wang, H. Zhang, B. Zhang and C. Hu, *Chemosphere*, 2021, **270**, 129416.
53. P. Cruz-Tato, N. Rivera-Fuentes, M. Flynn and E. Nicolau, *ACS Appl. Polym. Mater.*, 2019, **1**, 1061.
54. K. Y. Foo and B. H. Hameed, *J. Hazard. Mater.*, 2009, **170**, 552.
55. W. Duan, A. Dudchenko, E. Mende, C. Flyer, X. Zhu and D. Jassby, *Environ. Sci.: Processes Impacts*, 2014, **16**, 1300.
56. D. Huang, K. Wang, J. Niu, C. Chu, S. Weon, Q. Zhu, J. Lu, E. Stavitski and J. H. Kim, *Environ. Sci. Technol.*, 2020, **54**, 10954.
57. M. S. Munna, *Am. J. BioSci.*, 2013, **1**, 59.
58. C. D. Vecitis, M. H. Schnoor, M. S. Rahaman, J. D. Schiffman and M. Elimelech, *Environ. Sci. Technol.*, 2011, **45**, 3672.
59. H. Liu, X. Y. Ni, Z. Y. Huo, L. Peng, G. Q. Li, C. Wang, Y. H. Wu and H. Y. Hu, *Environ. Sci. Technol.*, 2019, **53**, 3238.
60. X. Li, S. Shao, Y. Yang, Y. Mei, W. Qing, H. Guo, L. E. Peng, P. Wang and C. Y. Tang, *ACS Appl. Mater. Interfaces*, 2020, **12**, 21596.
61. M. Chen, L. Ren, K. Qi, Q. Li, M. Lai, Y. Li, X. Li and Z. Wang, *Bioresour. Technol.*, 2020, **311**, 123579.
62. Z. Wang, Y. Zhang, K. Li, J. Guo, C. Yang, H. Liu and J. Wang, *Sep. Purif. Technol.*, 2022, **290**, 120918.
63. W. Duan, A. Ronen, S. Walker and D. Jassby, *ACS Appl. Mater. Interfaces*, 2016, **8**, 22574.
64. D. Rajkumar and J. G. Kim, *J. Hazard. Mater.*, 2006, **136**, 203.
65. Y. Song, M. Xiao, Z. Li, Y. Luo, K. Zhang, X. Du, T. Zhang, Z. Wang and H. Liang, *Chemosphere*, 2022, **286**, 131680.
66. Q. Zhang, Y. L. Yang, Y. L. Zhou and J. M. Hong, *Chemosphere*, 2022, **287**, 132070.
67. Q. Zhang and C. D. Vecitis, *J. Membr. Sci.*, 2014, **459**, 143.
68. J. Radjenovic, N. Duinslaeger, S. S. Avval and B. P. Chaplin, *Environ. Sci. Technol.*, 2020, **54**, 14815.
69. J. Sun, C. Hu, B. Wu and J. Qu, *Chem. Eng. J.*, 2020, **393**, 124817.
70. X. Fan, Y. Liu, X. Wang, X. Quan and S. Chen, *Environ. Sci. Technol.*, 2019, **53**, 5292.

71. J. Gao, N. Shi, Y. Li, B. Jiang, T. Marhaba and W. Zhang, *Environ. Sci. Technol.*, 2022, **56**, 11602.
72. P. Gayen, J. Spataro, S. Avasarala, A. M. Ali, J. M. Cerrato and B. P. Chaplin, *Environ. Sci. Technol.*, 2018, **52**, 9370.
73. W. L. Jiang, X. Xia, J. L. Han, Y. C. Ding, M. R. Haider and A. J. Wang, *Environ. Sci. Technol.*, 2018, **52**, 9972.
74. J. Sun, Q. Wang, J. Zhang, Z. Wang and Z. Wu, *Electrochim. Acta*, 2018, **277**, 77.
75. Z. Li, C. Shen, Y. Liu, C. Ma, F. Li, B. Yang, M. Huang, Z. Wang, L. Dong and S. Wolfgang, *Appl. Catal., B*, 2020, **260**, 118204.
76. P. Liang, M. Rivallin, S. Cerneaux, S. Lacour, E. Petit and M. Cretin, *J. Membr. Sci.*, 2016, **510**, 182.
77. X. Li, L. Liu and F. Yang, *Sep. Purif. Technol.*, 2018, **193**, 220.
78. F. Han, J. Mao and K. Wang, *J. Environ. Chem. Eng.*, 2022, **10**, 108722.
79. L. Tang, A. Iddya, X. Zhu, A. V. Dudchenko, W. Duan, C. Turchi, J. Vanneste, T. Y. Cath and D. Jassby, *ACS Appl. Mater. Interfaces*, 2017, **9**, 38594.
80. A. W. M. S. Mozia, R. Molinari, L. Palmisano and V. Loddo, in *Handbook of Membrane Reactors*, ed. A. Basile, Woodhead Publishing, 2013, vol. 2, p. 236.
81. P. Argurio, E. Fontananova, R. Molinari and E. Drioli, *Processes*, 2018, **6**, 162.
82. M. R. Rahimpour and L. Mahmoodi, in *Current Trends and Future Developments on (Bio-) Membranes*, ed. S. Mozia, A. Basile, and R. Molinari, Elsevier, 2018, p. 173.
83. S. Mozia, *Sep. Purif. Technol.*, 2010, **73**, 71.
84. S. Salehian, M. H. Mehdipour, F. Fotovat and S. A. Mousavi, *Chemosphere*, 2022, **302**, 134893.
85. C. Hu, M.-S. Wang, C.-H. Chen, Y.-R. Chen, P.-H. Huang and K.-L. Tung, *J. Membr. Sci.*, 2019, **580**, 1.
86. W. Wang, Z. Wu, E. Eftekhari, Z. Huo, X. Li, M. O. Tade, C. Yan, Z. Yan, C. Li, Q. Li and D. Zhao, *Catal. Sci. Technol.*, 2018, **8**, 1704.
87. J. Yang, B. Liu and X. Zhao, *Chin. J. Catal.*, 2017, **38**, 2048.
88. O. Lisovski, S. Piskunov, Y. F. Zhukovskii and J. Ozolins, *IOP Conf. Ser.: Mater. Sci. Eng.*, 2012, **38**, 012057.
89. W. Shi, W. Yang, Q. Li, S. Gao, P. Shang and J. K. Shang, *Nanoscale Res. Lett.*, 2012, **7**, 590.
90. Y.-H. Chiu, T.-F. Chang, C.-Y. Chen, M. Sone and Y.-J. Hsu, *Catalysts*, 2019, **9**, 430.
91. L. Zhang and M. Jaroniec, *Appl. Surf. Sci.*, 2018, **430**, 2.
92. W. S. Koe, J. W. Lee, W. C. Chong, Y. L. Pang and L. C. Sim, *Environ. Sci. Pollut. Res. Int.*, 2020, **27**, 2522.
93. R. Ramezani Darabi, M. Jahanshahi and M. Peyravi, *Chem. Eng. Res. Des.*, 2018, **133**, 11.
94. D.-T. Tran, J. Mendret, J.-P. Méricq, C. Faur and S. Brosillon, *Chem. Eng. Process.*, 2020, **148**, 107781.

95. N. Rosman, W. N. W. Salleh, N. A. M. Razali, S. Z. N. Ahmad, N. H. Ismail, F. Aziz, Z. Harun, A. F. Ismail and N. Yusof, *Mater. Today: Proc.*, 2021, **42**, 69.
96. T. D. Kusworo, A. C. Kumoro, N. Aryanti, T. A. Kurniawan, F. Dalanta and N. H. Alias, *Chem. Eng. J.*, 2023, **451**, 138593.
97. Y. Zhang, Y. Wang, R. Xie, H. Huang, M. K. H. Leung, J. Li and D. Y. C. Leung, *Environ. Sci. Technol.*, 2022, **56**, 16582.
98. A. K. Nair and P. E. JagadeeshBabu, *J. Environ. Chem. Eng.*, 2017, **5**, 4128.
99. A. M. Asiri, V. Pugliese, F. Petrosino, S. B. Khan, K. A. Alamry, S. Y. Alfifi, H. M. Marwani, M. M. Alotaibi, D. Mukherjee and S. Chakraborty, *Polymers*, 2022, **14**, 636.
100. S. elbakry, M. E. A. Ali, M. Abouelfadl, N. A. Badway and K. M. M. Salam, *Surf. Interfaces*, 2022, **29**, 101748.
101. S. Salehian, H. Heydari, M. Khansanami, V. Vatanpour and S. A. Mousavi, *Sep. Purif. Technol.*, 2022, **285**, 120291.
102. M. A. Saflashkar, M. Homayoonfal and F. Davar, *Chem. Eng. J.*, 2022, **450**, 138150.
103. J. Kavitha, N. S. Naik, A. R. Phani, G. Arthanareeswaran, M. Rajalakshmi and M. Padaki, *Desalination*, 2023, **552**, 116444.
104. H. Rawindran, J.-W. Lim, P.-S. Goh, M. N. Subramaniam, A. F. Ismail, N. M. Radi bin Nik M Daud and M. Rezaei-Dasht Arzhandi, *J. Cleaner Prod.*, 2019, **221**, 490.
105. C. Li, T. Sun, G. Yi, D. Zhang, Y. Zhang, X. Lin, J. Liu, Z. Shi and Q. Lin, *J. Environ. Chem. Eng.*, 2022, **10**, 108488.
106. L. Valenzuela, M. Pedrosa, A. Bahamonde, R. Rosal, A. Torres-Pinto, C. G. Silva, J. L. Faria and A. M. T. Silva, *Catal. Today*, 2023, **418**, 114042.
107. N. E. Salim, N. A. M. Nor, J. Jaafar, A. F. Ismail, M. R. Qtaishat, T. Matsuura, M. H. D. Othman, M. A. Rahman, F. Aziz and N. Yusof, *Appl. Surf. Sci.*, 2019, **465**, 180.
108. Z. W. Heng, W. C. Chong, Y. L. Pang, L. C. Sim and C. H. Koo, *J. Environ. Chem. Eng.*, 2021, **9**, 105199.
109. C. Ding, X. Qin, Y. Tian and B. Cheng, *J. Membr. Sci.*, 2022, **659**, 120789.
110. S. Chaudhuri, C.-M. Wu and K. G. Motora, *J. Photochem. Photobiol., A*, 2023, **438**, 114525.
111. V. V. Pham, H. P. Phan La, H. V. Le, S. T. Truong, T. Q. Nguyen and T. M. Cao, *Ind. Eng. Chem. Res.*, 2022, **61**, 12427.
112. M. A. Hanif, H. Shin, D. Chun, H. G. Kim, L. K. Kwac and Y. S. Kim, *Membranes*, 2022, **13**, 50.
113. E. Cossich, R. Bergamasco, M. T. Pessoa de Amorim, P. M. Martins, J. Marques, C. J. Tavares, S. Lanceros-Méndez and V. Sencadas, *Mater. Chem. Phys.*, 2015, **164**, 91.
114. M. Amoli-Diva, E. Irani and K. Pourghazi, *Sep. Purif. Technol.*, 2020, **236**, 116257.
115. Y. Qing, Y. Li, Z. Guo, Y. Yang and W. Li, *J. Environ. Chem. Eng.*, 2022, **10**, 108648.
116. Y. Zhao, C. Tao, G. Xiao and H. Su, *RSC Adv.*, 2017, **7**, 11211.

117. H. Liang, C. Lv, H. Chen, L. Wu and X. Hou, *RSC Adv.*, 2020, **10**, 45171.
118. M. Kazemi, M. Jahanshahi and M. Peyravi, *Carbohydr. Polym.*, 2018, **198**, 164.
119. Y. Yue, K. Hou, J. Chen, W. Cheng, Q. Wu, J. Han and J. Jiang, *ACS Appl. Mater. Interfaces*, 2022, **14**, 24708.
120. R. Zhang, Y. Ma, W. Lan, D. E. Sameen, S. Ahmed, J. Dai, W. Qin, S. Li and Y. Liu, *Ultrason. Sonochem.*, 2021, **70**, 105343.
121. N. Nasrollahi, L. Ghalamchi, V. Vatanpour and A. Khataee, *J. Ind. Eng. Chem.*, 2021, **93**, 101.
122. R. Zhu, A. J. Diaz, Y. Shen, F. Qi, X. Chang, D. P. Durkin, Y. Sun, S. D. Solares and D. Shuai, *J. Membr. Sci.*, 2018, **563**, 531.
123. G. A. Mulungulungu, T. Mao and K. Han, *Colloids Surf., A*, 2021, **623**, 126714.
124. T. Rathna, J. PonnanEttiyappan and D. RubenSudhakar, *Environ. Technol. Innovation*, 2021, **24**, 102023.
125. Z. Lu, G. Zhou, B. Li, Y. Xu, P. Wang, H. Yan, M. Song, C. Ma, S. Han and X. Liu, *Appl. Catal., B*, 2022, **301**, 120787.
126. R. Molinari and P. Argurio, *Water Res.*, 2017, **109**, 327.
127. R. Yu, Y. Yang, Z. Zhou, X. Li, C. Liu, N. Wang and Y. Liu, *Sep. Purif. Technol.*, 2023, **307**, 122603.
128. R. Singh, M. K. Sinha and M. K. Purkait, *Sep. Purif. Technol.*, 2020, **250**, 117247.
129. J. C. Espíndola, R. O. Cristóvão, A. Mendes, R. A. R. Boaventura and V. J. P. Vilar, *Chem. Eng. J.*, 2019, **378**, 122114.
130. J. Song, X. Wu, M. Zhang, C. Liu, J. Yu, G. Sun, Y. Si and B. Ding, *Chem. Eng. J.*, 2020, **379**, 122269.
131. E. P. Ferreira-Neto, S. Ullah, T. C. A. da Silva, R. R. Domeneguetti, A. P. Perissinotto, F. S. de Vicente, U. P. Rodrigues-Filho and S. J. L. Ribeiro, *ACS Appl. Mater. Interfaces*, 2020, **12**, 41627.
132. N. Tavker and M. Sharma, *J. Cleaner Prod.*, 2020, **271**, 122510.
133. J. Pazol, A. Vázquez and E. Nicolau, *Colloids Surf., B*, 2019, **183**, 110418.
134. D. Yi, T. Bayer, C. P. Badenhorst, S. Wu, M. Doerr, M. Höhne and U. T. Bornscheuer, *Chem. Soc. Rev.*, 2021, **50**, 8003.
135. B. Hauer, *ACS Catal.*, 2020, **10**, 8418.
136. S. Chakraborty, H. Rusli, A. Nath, J. Sikder, C. Bhattacharjee, S. Curcio and E. Drioli, *Crit. Rev. Biotechnol.*, 2016, **36**, 43.
137. C. Spahn and S. D. Minteer, *Recent Pat. Eng.*, 2008, **2**, 195.
138. O. Kirk, T. V. Borchert and C. C. Fuglsang, *Curr. Opin. Biotechnol.*, 2002, **13**, 345.
139. I. Galaev and B. Mattiasson, *Smart Polymers: Applications in Biotechnology and Biomedicine*, CRC Press, 2007.
140. C. He, Z. Liu, J. Wu, X. Pan, Z. Fang, J. Li and B. A. Bryan, *Nat. Commun.*, 2021, **12**, 4667.
141. V. V. Khutoryanskiy and T. K. Georgiou, *Temperature-Responsive Polymers: Chemistry, Properties, and Applications*, John Wiley & Sons, 2018.

142. K. Dutta and S. De, *J. Mater. Chem. A*, 2017, **5**, 22095.
143. J. M. Berg, J. L. Tymoczko and L. Stryer, *Biochemistry: International Version*, Granite Hill Publishers, 2002.
144. Q. V. Ly, L. Cui, M. B. Asif, W. Khan, L. D. Nghiem, Y. Hwang and Z. Zhang, *Water Res.*, 2023, 119577.
145. A. S. Chouhan and A. K. Sarma, *Renewable Sustainable Energy Rev.*, 2011, **15**, 4378.
146. S. Serra, *Enzymes, Biocatalysis and Chemical Biology*, Mdpi AG, 2020.
147. C. S. Nunes and K. Malmlöf, in *Enzymes in Human and Animal Nutrition*, Elsevier, 2018, p. 331.
148. M. Narayanan, S. S. Ali and M. El-Sheekh, *J. Environ. Manage.*, 2023, **334**, 117532.
149. A. G. McDonald and K. F. Tipton, *FEBS J.*, 2023, **290**, 2214.
150. K. Blank, J. Morfill, H. Gumpp and H. E. Gaub, *J. Biotechnol.*, 2006, **125**, 474.
151. R. D. Schmid and R. Verger, *Angew. Chem., Int. Ed.*, 1998, **37**, 1608.
152. A. Steimle, I. B. Autenrieth and J.-S. Frick, *Int. J. Med. Microbiol.*, 2016, **306**, 290.
153. A. A. Mendes and H. F. d. Castro, *Braz. Arch. Biol. Technol.*, 2005, **48**, 135.
154. A. Valladão, A. Torres, D. Freire and M. Cammarota, *Bioresour. Technol.*, 2011, **102**, 7043.
155. C. Cambillau, S. Longhi, A. Nicolas and C. Martinez, *Curr. Opin. Struct. Biol.*, 1996, **6**, 449.
156. J. A. Bosley and J. C. Clayton, *Biotechnol. Bioeng.*, 1994, **43**, 934.
157. R. Fernandez-Lafuente, P. Armisén, P. Sabuquillo, G. Fernández-Lorente and J. M. Guisán, *Chem. Phys. Lipids*, 1998, **93**, 185.
158. A. M. Baron, N. Barouh, B. Barea, P. Villeneuve, D. A. Mitchell and N. Krieger, *Fuel*, 2014, **117**, 458.
159. N. Grubhofer and L. Schleith, *Naturwissenschaften*, 1953, **40**, 508.
160. C. M. Heckmann and F. Paradisi, *ChemCatChem*, 2020, **12**, 6082.
161. N. Kardos and A. L. Demain, *Appl. Microbiol. Biotechnol.*, 2011, **92**, 677.
162. N. Miletić, V. Abetz, K. Ebert and K. Loos, *Macromol. Rapid Commun.*, 2010, **31**, 71.
163. D. S. Rodrigues, G. P. Cavalcante, A. L. Ferreira and L. R. Gonçalves, *Chem. Biochem. Eng. Q.*, 2008, **22**, 125.
164. H. Zhang, J. Luo, S. Li, Y. Wei and Y. Wan, *Langmuir*, 2018, **34**, 2585.
165. B. Chen, M. E. Miller and R. A. Gross, *Langmuir*, 2007, **23**, 6467.
166. B. Chen, E. M. Miller, L. Miller, J. J. Maikner and R. A. Gross, *Langmuir*, 2007, **23**, 1381.
167. J. Hou, G. Dong, Y. Ye and V. Chen, *J. Membr. Sci.*, 2014, **469**, 19.
168. H. Zhang, J. Luo and Y. Wan, *iScience*, 2022, **25**, 103671.
169. D. Li, Y. Cheng, H. Zuo, W. Zhang, G. Pan, Y. Fu and Q. Wei, *J. Colloid Interface Sci.*, 2021, **603**, 771.
170. L. Duan, Y. Wang, Y. Zhang and J. Liu, *Appl. Surf. Sci.*, 2015, **355**, 436.
171. S. Xu, J. Liang, M. I. B. Mohammad, D. Lv, Y. Cao, J. Qi, K. Liang and J. Ma, *Chem. Eng. J.*, 2021, **426**, 131861.
172. H. A. Güleç, *Colloids Surf., B*, 2013, **104**, 83.

CHAPTER 9

Graphene-based Engineered Macrostructures for Water Purification

CHIRAG BATUKBHAI GODIYA*

Department of Chemical Engineering (BK21 FOUR Integrated Engineering Program), Kyung Hee University, Yongin 17104, South Korea
*E-mail: chiraggodiya@khu.ac.kr

9.1 Introduction

The swift pace of industrialization and human activities causes severe problems for the environment, thereby presenting a serious risk to all living organisms. Specifically, water pollution resulting from the direct or indirect release of various hazardous pollutants, such as heavy metal ions, drugs, conventional and emerging phenolic compounds, organic dyes, dioxins, fluorine-containing organics, and other emerging contaminants, has become a critical concern globally.[1] The detrimental effects of hazardous pollutants are widely recognized across various domains. One notable example is the prolonged consumption of water with high concentrations of fluoride, which can lead to the development of fluorosis, including dental and skeletal fluorosis of varying severity, which can cause serious harm to human health.[2] Heavy metal ions have a tendency to bioaccumulate in the body, leading to potential adverse effects when exposed to high concentrations, such as organ impairment, detrimental impacts on the central nervous system, increased cancer risk, irreversible brain damage, and, ultimately, mortality. Long-term

Chemistry in the Environment Series No. 12
Novel Materials and Water Purification: Towards a Sustainable Future
Edited by Grigorios L. Kyriakopoulos and Miltiadis G. Zamparas

Published by the Royal Society of Chemistry, www.rsc.org

exposure to mercury and lead has been associated with the development of autoimmunity, a condition where the immune system mistakenly targets healthy cells, resulting in significant health complications.[3] Likewise, exposure to phenolic compounds, including chlorophenols and phenols, can have detrimental effects on multiple organs in the human body. Prolonged exposure to bisphenol A has been associated with a range of health issues, including neurologic impairment, Reaven's syndrome, cardiovascular diseases, male sterility, inherent infirmity, and cancer.[4] Therefore, the efficient uptake of hazardous contaminants from industrial sewage is a crucial and pressing environmental objective.

Until now, numerous methods, including adsorption,[5] electrokinetics,[6] photocatalytic processes,[7] advanced oxidation processes,[8] reductive degradation,[9] decomposition,[10] coagulation–flocculation,[11] and membrane separation,[12] have been employed to remove hazardous contaminants from sewage. Given its numerous advantages, such as ease of operation, high efficiency, cost effectiveness, and minimal production of secondary pollutants, adsorption is a widely preferred method in wastewater treatment.[13] Accordingly, numerous functional materials have been investigated for their potential to take up wastewater pollutants, *i.e.*, molecular organic frameworks (MOFs),[14] membranes,[15] resins and fibers,[16] aerogels/hydrogels,[17] polymeric compounds,[18] and biochar-based nanomaterials.[19] Nevertheless, these adsorbents are not capable of effectively treating waste water that contains intricate mixtures of organic and inorganic pollutants. Therefore, there is a pressing need to create potent adsorbents that can effectively remove these challenging and complex substances in waste water.

In recent years, graphene (G) and its derivatives have attracted considerable interest as promising adsorbents due to their favorable physicochemical characteristics, including a large surface area, immense chemical functionalities, sp^2-conjugated carbon framework, ease of chemical functionalization to achieve appropriate surface architecture, and efficient elimination of diverse water contaminants.[20] However, graphene nanosheets (GNs) often suffer from notorious agglomeration issues, resulting in a substantial loss of their specific surface area and obstructing adsorption sites, thus impairing their intrinsic unique structural properties and negatively affecting their adsorption performance. Moreover, the direct use of nanoscale GNs can result in the release of GNs into the environment, causing secondary pollution in ecosystems and posing a potential hazard to living organisms.[21] Recent research has shown the remarkable potential of two-dimensional (2D) graphene-based membranes in selectively removing water contaminants due to their unique pore and channel structures. Furthermore, the utilization of three-dimensional (3D) graphene-based macrostructures (GBMs) facilitates the easy handling of GNs for water treatment, offering good recyclability and reusability.[22] Consequently, the development of multifunctional 2D/3D macrostructures through suitable physicochemical modification routes opens up new possibilities for broader applications of GBMs in addressing challenging environmental issues.

Extensive research has been conducted on GBMs, and insightful reviews discussing their potential as adsorbents have been published,[23] but a comprehensive evaluation of their effectiveness in treating different types of contaminants is necessary to expedite the development of GBMs in feasible sewage remediation applications. Hence, the objective of this chapter is to summarize the recent advances made in the development, fabrication, structural properties, and performance of GBMs in removing wastewater pollutants. Particular emphasis is placed on the construction of both 2D and 3D structures and their specific applications in sewage remediation. To ensure the safe and effective utilization of GBMs in sewage treatment, this chapter considers environmentally friendly fabrication approaches, functionalization, and recycling/regeneration. The adsorption mechanism, encompassing potential binding sites and affinity towards different pollutants, is summarized based on both experimental and theoretical studies. This comprehensive approach facilitates the uptake of pollutants from effluents. Furthermore, the chapter discusses the fundamental characteristics, future prospects, and challenges, and provides a critical overview of the advancement of GBMs for environmental applications. The recent advances presented in this chapter are expected to offer valuable insights for future investigations into GBM-based adsorbents, thereby paving the way for the synthesis of cost-effective, eco-friendly, effective, and long-lasting GBMs for wastewater treatment and other related fields.

9.2 Graphene-based Macrostructures and Synthetic Methods

In Figure 9.1, GBMs have been classified into different categories, including graphene nanosheets, graphene oxide (GO), reduced GO (rGO), chemically modified G, G-based 3D nanostructures, and composites of G (*i.e.*, polymer-based G, inorganic-based G, biomass-based G, MOF-G, and multicomponent-G). Subsequent sections provide a comprehensive exploration of the synthetic methodologies employed for the fabrication of these aforementioned GBMs and the potential of the above-specified GBMs.

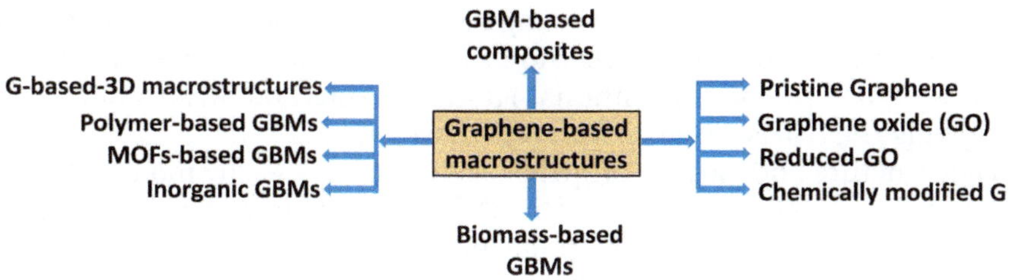

Figure 9.1 Overview of widely used graphene (G)-based adsorbents in wastewater treatment.

9.2.1 Pristine Graphene

Graphene is a 2D crystalline form of carbon with an ideal sp^2 aromatic structure and minimal structural defects, which has attracted considerable interest in diverse fields owing to its intriguing structure and surface properties. Currently, crystalline G is primarily fabricated through two distinct approaches: (1) mechanical or liquid-phase exfoliation of graphite; and (2) employing alternative carbon sources as starting materials in the methods of chemical vapor deposition, crystal growth on silicon carbide, or longitudinal shearing of carbon nanotubes.[20] However, the poor dispersibility of pristine G in a vast majority of organic solvents and aqueous media, coupled with its tendency to accumulate, hinders its use in adsorption applications. These limitations can be overcome by converting G into GO or rGO to enable its efficient use in various adsorption applications.

9.2.2 Graphene Oxide and Fabrication Methods

GO is an oxygenated derivative of graphene, which is generally synthesized by the oxidation of naturally occurring graphite. It is a 2D material comprised of carbon atoms organized in a hexagonal lattice. GO contains a significant number of randomly distributed oxygen (O) groups on both its basal planes (*i.e.*, epoxide and hydroxy) and edges (*i.e.*, carboxyl) moderately interrupted by sp^3-hybridized carbons.[24] The presence of O-containing groups in GO provides high dispersibility in water and polar solvents, as well as enabling the production of low-cost and scalable GBMs. This renders GOs as more favorable than crystalline G for environmental applications. Currently, GOs are typically produced using the Hummers' method, which involves oxidizing graphite with potent oxidizing agents (sulfuric acid (H_2SO_4), sodium nitrate ($NaNO_3$), and nitric acid (HNO_3)), which is subsequently purified and exfoliated in water.[25] Nevertheless, these conventional techniques exhibit a ubiquitous limitation of emitting hazardous gases, such as NO_x and ClO_2. In this context, Marcano and colleagues utilized an environmentally friendly and efficient approach to synthesize GO by substituting the corrosive HNO_3 with the less harmful oxidizing agent H_3PO_4. This approach necessitates lower energy input while concurrently resulting in a greater degree of oxidation without emitting hazardous gases.[26] Furthermore, Chen *et al.* have described a method for synthesizing GO utilizing $KMnO_4$ and H_2SO_4 as oxidizing agents. The GO synthesized with this method has equivalent properties and yield to those obtained by the conventional Hummers' method, while concurrently mitigating the release of toxic NO_2/N_2O_4 gases.[27]

The structure and surface properties of GO can be easily modified to control and alter the characteristics of the GBMs, so as to improve their efficiency in environmental remediation applications. For example, the O-containing moieties of GO can be readily tailored through chemical reactions or hybridization interactions, by adjusting the reaction conditions during the Hummers and Offeman oxidation processes (*i.e.*, oxidation degree, the proportion and positioning of O-containing moieties, and the number of remaining non-O-containing moieties). The content of O groups in GO plays a crucial role in determining its capacity for removing pollutants. More specifically, GO sheets with a higher O content demonstrate an increased affinity for positively charged contaminants, whereas those with lower O content are more effective in adsorbing negatively charged species. The preparation of GO possessing analogous O group content is challenging owing to the non-stoichiometric nature of GO. However, in comparison to crystalline G, GO possesses a more advantageous structure and broader applications in water treatment, which is elaborated in the subsequent section.

9.2.3 Reduced Graphene Oxide and Fabrication Methods

Reduced graphene oxide is a promising alternative to pristine G and GO owing to its balanced properties. During GO synthesis (in the majority of GO synthesis techniques), the use of strong oxidizing agents generally introduces defects in the G lattice, leading to diminishing of its structural properties. Reductive exfoliation treatments can potentially restore the graphene-like properties of GO, and convert it into rGO. Despite having a lower content of O-comprising moieties in comparison to GO, rGO possesses more repaired sp^2 domains, which leads to a richer aromatic structure than that of GO. rGO has been synthesized *via* chemical, thermal, biological, and electrochemical reduction paths. Among these, chemical reduction is the predominantly applied technique for rGO preparation, which uses hydrazine ($N_2H_4{\cdot}H_2O$), alkaline/acidic reducing agents, borohydrides, as well as eco-friendly alternatives. Among these, $N_2H_4{\cdot}H_2O$ is the most frequently employed reducing agent for GO reduction, yielding rGO with electronic and structural properties that closely resemble those of pristine G.[28] Under optimized conditions (95 °C for 3 h), GO reduction yields rGO with a significantly higher C/O ratio of ~15.1 compared to the ratio of ~3.1 for GO. In this process, the reaction of hydrazine with the epoxy groups of GO generates hydrazino alcohols and aminoaziridine groups, thermally eliminates di-imide, and ultimately forms C=C bonds. Additionally, the electroconductivity of the produced rGO is considerably enhanced, from 1.5×10^{-6} S cm^{-6} for GO to 5 S cm^{-6} for rGO, under these experimental conditions.[29] Sometimes, excessive reduction of GO can induce nano-aggregation of rGO, which in turn hinders its application efficiency. This can be prevented by incomplete reduction, pH and solvent regulation, or the fabrication of a core–shell structure. Sodium borohydride ($NaBH_4$) is also widely used as a reducing agent for GO. Despite being weaker than hydrazine, it avoids the additional integration of N in the rGO, as observed with hydrazine. Furthermore, borohydride-based reducing agents, such as NaBH(OAc), $NaBH_3(CN)$, and NH_3BH_3, have also been utilized for the reduction of GO.[30] Additionally, various acid and alkali reductants (HI, HBr, formic acid, NaOH, KOH) and green reductants (ascorbic acid, plant extracts, carbohydrates, microorganisms, amino acids) have been investigated for GO reduction.[31] The reduction of GO using HI as a reductant can achieve a C/O ratio of >12 in the resulting rGO when conducted at a temperature of 100 °C.[32] Conversely, in the case of alkali reductants, the reduction reaction occurs at temperatures below 100 °C and completes within a few minutes. The rGO obtained through the alkaline reduction method exhibits comparatively higher stability, which makes this approach more suitable in GO processing and applications. Given the potential toxicity linked to conventional chemicals, there is an increasing acceptance of utilizing green reductants for eco-friendly applications. For instance, the utilization of vitamin C as a reductant in reducing GO, under optimized operating conditions, has been shown to yield a C/O ratio of ~9.8 in the resulting rGO.[33] The thermal reduction of GO is commonly performed using two methods: (1) thermal annealing; and (2) solvothermal reduction. The former is often carried out in the presence of inert gases, and the resulting structure of rGO heavily relies on the annealing temperature employed. Notably, higher annealing temperatures yield rGO with a higher C/O ratio. In contrast, solvothermal reduction is performed in liquid environments, such as alkaline solutions, aqueous solutions, or organic solvents. The resulting structural properties of rGO can be finely adjusted by tuning the solvent conditions and operating temperature during the solvothermal process.[34] The solvothermal technique can be carried out in water as a solvent, representing a sustainable and eco-friendly process that significantly reduces the likelihood of impurity incorporation and simplifies operational aspects.

9.2.4 Chemically Modified Graphene and Fabrication Methods

In general, G is chemically modified to prevent the self-aggregation of its nanosheets, introduce additional functionalities, and ultimately enhance its adsorption capacity and rates. A recent trend pertains to the exploration of novel modifiers, for augmenting the adsorption efficiency of the resultant modified G (by tuning the types and parameters) towards specific water contaminants. Furthermore, novel methods, *i.e.*, gamma-ray irradiation, or use of laser for G modification, also perform a significant role in improving the adsorption property of the modified G and cost management. Accordingly, so far numerous GO modifiers have been reported, namely amino acids,[35] acids,[36] polyethyleneimine (PEI),[37] metals,[38] *etc.* By integrating these modifiers, a more versatile platform can be created for the G planes, allowing for effective removal of pollutants through synergetic effects within the modifier and G substrate. For example, Tadjenant *et al.* conducted a chemical reduction of GO using KOH while simultaneously functionalizing the rGO with PEI. The resulting rGO/PEI-KOH, demonstrated a remarkable Cr(VI) removal capacity, reaching 398.9 mg g^{-1}, which was consistent with the modified Langmuir isotherm and pseudo-second-order (PSO) kinetic model. The authors attributed the high effectiveness of the removal to the solid electrostatic interactions within the Cr(VI) and protonated $-NH_2$ moieties present on the rGO/PEI-KOH.[39]

In recent studies, various oxygen-containing modifiers have been identified as effective agents for modifying G and enhancing its adsorption capabilities.

Notably, Liu *et al.* introduced gallic acid (GA) to GO through π–π interactions, resulting in a remarkable selective capturing ability of GA-modified GO for Cr(III) ions *via* coordination complexation with the deprotonated carboxylic groups of GA.[40] Furthermore, modifiers containing both nitrogen and oxygen groups, such as EDTA,[41] dopamine,[42] sulfanilic acid,[43] *etc.*, have been extensively studied for their ability to establish favorable interactions, such as electrostatic, hydrogen bonding, or coordination interactions, between their functional groups and the contaminants. For instance, Bin *et al.* successfully integrated nitrogenous functional groups into GO by incorporating PEI and pyrrole, resulting in the formation of a PEI/polypyrrole/GO aerogel (PPGA). The amine-functionalized PPGA demonstrated outstanding Cr(VI) adsorbability, reaching 458.2 mg g^{-1} under optimized conditions, following the PSO rate kinetic model. Based on their experimental analysis, the authors proposed that the adsorption mechanism of Cr(VI) could be attributed to electrostatic attraction, redox reactions, and chelation phenomena.[44] Additionally, Kong *et al.* conducted a comparison of the Cr(VI) uptake capacity of GO modified with three distinct amine modifiers: diethylenetriamine (DETA), triethylenetetramine (TETA), and tetraethylenepentamine (TEPA). The nitrogen-containing moieties in DETA, TETA, and TEPA exhibited densities of ~7.2 wt%, ~10.2 wt%, and ~12.4 wt%, respectively, significantly influencing the adsorption performance of the modified GO. Specifically, the maximum monolayer Cr(VI) adsorption capacities for the three adsorbents were 245.0 mg g^{-1}, 257.2 mg g^{-1}, and 300.8 mg g^{-1}, respectively.[45]

Furthermore, the incorporation of sulfur functional groups onto GO through the process of sulfonation have reportedly shown effective enhancement in the adsorption capacity towards cationic contaminants, while also preventing the undesired aggregation of GO sheets. For instance, Diagboya *et al.* employed a novel approach to modify GO by incorporating $-NH_2$ and –SH groups, resulting in a dual-functionalized GO (GOSN). They conducted a comparative investigation on the uptake of Pb(II) by GOSN, GO modified solely with $-NH_2$ (GONH), and GO modified solely with –SH (GOSH) groups. They observed that GONH exhibited the highest Pb(II) uptake capacity (138.0 mg g^{-1}), followed by GOSH (101.5 mg g^{-1}), and GOSN (97.8 mg g^{-1}). Thermodynamic study indicated that the Pb(II) binding process was endothermic, spontaneous, with a rapid adsorption rate. The main adsorption mechanism was attributed to electrostatic interactions within the adsorptive sites and Pb(II). Interestingly, the adsorption isotherm of GO modified with O and N functionalities followed the Langmuir model, whereas GO modified with dual functionalities (GOSN) followed Freundlich or Brouers–Sotolongo–fractal models.[46]

9.2.5 Graphene-based 3D Nanostructures and Fabrication Methods

To overcome challenges associated with the powdered graphene NP-based adsorbents, significant efforts have been directed towards converting 2D graphene into macroscopic 3D graphene structures that enable easier handling and reuse of the material. Additionally, the development of intriguing 3D porous carbon network architectures enhances the adsorption process by facilitating rapid mass transfer and offering a large specific surface area, all while inhibiting the aggregation of G NPs.[47] The creation of 3D graphene structures can be achieved *via* self-assembly of 2D rGO nanosheets or by cross-linking them into 3D composite structures using biomaterials, polymers, and surfactants.[48] Recently, G-based gels are among the most extensively investigated adsorbents in wastewater treatment.[49–52] As a relevant example, Liu *et al.* fabricated methyltriethoxysilane (MTES)/G aerogels (MGAs) *via* a single-step solvothermal co-assembly method. The significantly low density, interconnected porous structure, excellent elasticity, large surface area, and grafting of $-CH_3$ moieties of MTES solidified the framework of the MGAs and a rendered superior selective oil/water separation performance. The MGA revealed superior elimination performance of various organic liquids with high capacities (ranging from 290 to 620 g/g) and remarkable recyclability (retention of 97% of its initial adsorbability after undergoing 20 cycles).[53] Guo *et al.* developed a G/hydroxypropyl methylcellulose (HPMC) adsorbent *via* a hydrothermal reduction coupled with an ice template method for the adsorption of crude oil. The incorporation of HPMC not only significantly improved the robustness and heat conductivity of the G aerogels, but also facilitated the formation of dense pore structures within the aerogel, enabling it to adsorb crude oil at a rate of 124.43 times its own mass. The G/HPMC aerogel exhibited an impressive adsorption capacity for crude oil, reaching 124.43 g/g within a 20 minute timeframe. Furthermore, even after undergoing five cycles of adsorption–extrusion desorption, it maintained 90% of its initial adsorbability.[54] Liu *et al.* prepared carboxymethyl cellulose (CMC)/G composite aerogel beads *via* a wet spinning-environmental pressure drying technique and utilized it for uptake of methylene blue (MB) in contaminated effluents. The maximum MB uptake capacity of CMC/G aerogel was reported to be 222.72 mg g^{-1}, following rate kinetics with the PSO model, with exceptional reusability, retaining about 90% of its adsorbability after 30 consecutive adsorption–desorption cycles.[55]

9.2.6 Polymer-based GBMs

Polymers are commonly used as key components in the fabrication of polymer-based G composites in separation and purification applications. In recent years, biopolymers have received significant interest for the development of high-performance polymer/G composites aimed at eliminating various wastewater pollutants due to their immense attainability, biodegradability, cost effectiveness, non-toxic nature, and availability of chemical moieties (N,O-comprising groups).[56] Zhong *et al.* used a catalytic membrane coated with a composite layer of GO and bacterial nanocelluloses. The membrane showed exceptional Congo red (CR) and basic blue dyes elimination efficacy, with removal rates of 99 and 96%, respectively, thanks to the surface-coated GO composite layer. Additionally, the researchers employed $NaBH_4$ to reduce the GO-coated composite membrane, leading to the catalytic conversion of various compounds, including methyl orange, MB, rhodamine B (RhB), and 4-nitrophenol. This successful conversion further substantiated the membrane's ability in treating complex water compositions.[57] Ahmad *et al.* have successfully fabricated GO-doped polyvinyl chloride inclusion membranes (PIMs) and applied them for the separation of aqueous ibuprofen (IBU) under varied pH conditions. The GO-doped PIM showed an extraction efficiency of ~84% for IBU at a pH of 10, with good chemical stability in acidic conditions.[58] Kong *et al.* have developed electrospun composite nanofibers consisting of MOF/polymer/GO for the efficient adsorption of arsenate (As(v)) in contaminated media. The MOF/polymer/GO composite exhibited a maximum monolayer As(v) adsorption capacity >180 mg g^{-1}. Based on a comprehensive experimental mechanistic investigation, the authors have indicated that the superior adsorption of As(v) onto the MOF/polymer/GO composite can be attributed to coordination and π–π interaction.[59]

9.2.7 Inorganic GBM Hybrids

Numerous investigations have been carried out to incorporate G with a diverse range of inorganic compounds, including $Al(OH)_3$, Fe_3O_4/Fe_2O_3, manganese dioxide (MnO_2), NiFe-layered double hydroxides (LDHs), montmorillonite (MMT), Fe_2O_3, and $NiFe_2O_4$/halloysite nanotubes. This has resulted in the creation of unique structures, morphologies, and additional functionalities, ultimately boosting the adsorption and recycling capabilities of GBMs. For instance, Akpotu *et al.* successfully synthesized a novel composite material by coupling MMT with GO/rGO and covalently bonding it to polyethylene glycol methyl ether (PEG) using an amidation method. This composite material demonstrated remarkable enrofloxacin adsorbability, achieving a maximum monolayer uptake capacity of 310.6 mg g^{-1}, which complies with the PSO rate kinetics model.[60] In another study, Alatalo *et al.* fabricated a novel adsorbent by incorporating zerovalent iron into layered G (Fe/G composite) and employed it for the uptake of tetracycline (TC) from effluents. The Fe/G composite exhibited an exceptional TC removal efficiency of 442 mg g^{-1}, which further increased to 660 mg g^{-1} after three consecutive adsorption/regeneration cycles, indicating the influence of iron oxide formation. The authors proposed two potential mechanisms for TC removal, namely, π–π interactions or carbonyl group interactions with the G sheets, as well as reduction catalysed by the α- and γ-Fe species.[61] Khataee *et al.* synthesized NiFe-LDH/rGO nanocomposite using a hydrothermal technique and applied it for the degradation of moxifloxacin (MOX). At optimized experimental conditions (catalyst dose: 1.0 g L^{-1}, MOX concentration: 20 mg L^{-1}, ultrasonic power: 150 W, pH: 8, time: 60 min) the sonophotocatalytic efficacy of the NiFe-LDH/rGO reached to 90.40% for the removal of MOX. Based on the experimental analysis, the authors indicated that the MOX degradation was primarily governed by the action of $^{\bullet}OH$ and $O_2^{\bullet-}$ radicals.[62] Overall, these investigations have shown that the combination of GBMs with various inorganic compounds can result in highly efficient adsorbents and catalysts for the elimination of various water contaminants.

9.2.8 Biomass-based GBMs

Biomass has become a promising material for creating 3D GBMs for environmental remediation due to its eco-friendliness, biocompatibility, plentifulness, and availability of active functionalities.[17] The surface functionalities of biomass, which primarily include O and N heteroatoms connected to C, can effectively bind both organic and inorganic pollutants. By hybridizing with G NPs, the resulting 3D GBMs exhibit even greater adsorption capacity and selectivity for various hazardous water pollutants. Recent studies have demonstrated the potential of these biomass-based GBMs for environmental remediation applications. For instance, Das *et al.* developed a chitosan (CS), PVA, and GO hybrid hydrogel, and applied it for the uptake of CR from contaminated solutions. By applying the GO/CS-PVA hydrogel (dose: 6 g L^{-1}) the researchers achieved impressive Cr removal rates of 88.1% and 81% at pH 2 and 7, respectively, for a 20 mg L^{-1} CR concentration.[63] Kaushik *et al.* fabricated a biomass (pear fruit)-derived G aerogel (f-GA) and utilized it for the uptake of organic dyes, *i.e.*, crystal violet (CV), MB, RhB, and their blends. The authors described that the good dissemination of O functionalities as surface defects on f-GA resulted in a greater number of adsorptive sites, promoting enhanced interactions with dye molecules compared to both pristine G and commercially available activated C. The maximum monolayer CV, MB, and RhB removal capacities were reported to be 334.4, 546.4, and 257.0 mg g^{-1}, respectively, complying with the PSO rate kinetics model.[64]

9.2.9 MOF-based GBMs

In recent years, MOFs have gained significant research attention in diverse fields, including catalysis, energy and gas storage, drug delivery, detectors, separation and purification, and photoelectric devices, thanks to their adaptable framework compositions, adjustable porosities, and accessible functionalities.[65] Incorporating MOFs into MOF/GO composites not only enhances the active adsorption sites and impedes the aggregation of graphene, but also may mitigate the inadequate selectivity issue of GO towards particular pollutants. As a consequence, the resultant MOF/GO composite may exhibit promising selectivity towards targeted contaminants (*i.e.*, dye, ammonia, benzene, pesticides, bisphenol A, and TCs). For instance, Eltaweil *et al.* successfully synthesized a Mn-doped UiO-66/GO-NH_2 (Mn-UiO-66/GO-NH_2) composite using a solvothermal technique, which revealed highly selective CR dye uptake capacity in the presence of MB and MO dyes. The maximum CR uptake capacity of the Mn-UiO-66/GO-NH_2 was reported to be ~1265.8 mg g^{-1}, abiding the adsorption process with PSO rate kinetics and adherence to both Langmuir and Freundlich isotherm models. Through empirical analysis, the authors indicated that the mechanism responsible for the uptake removal involved a synergistic interplay of electrostatic attraction, pore filling, π–π stacking, and hydrogen bonding.[66] In another contribution, Zuo *et al.* reported a hybrid MOF@rGO nanocomposite that demonstrated the ability of MOF@rGO to selectively and reversibly eliminate CrO_4^{2-} while also converting toxic Cr(VI) into less harmful Cr species. This dual functionality was achieved by harnessing the selective adsorption capabilities of Co-MOF and the electroconductivity of rGO. The targeted uptake of CrO_4^{2-} facilitated by Co-MOF played a crucial role in the electrocatalytic reductive transformation of Cr(VI) to less harmful Cr_{red} species, and subsequent regeneration of the adsorption capacity for the MOF@rGO. The researchers achieved effective Cr(VI) elimination from sewage and its conversion to less hazardous Cr species in brine *via* a cyclic electrochemical reactor process, which relies on the synergistic interplay of adsorption, electrocatalytic reduction, and desorption methods.[67]

9.3 Adsorption of Organic/Inorganic Pollutants on GBMs

GBMs have demonstrated efficacy in treating diverse types of water pollutants based on their surface charge (*i.e.*, positively or negatively charged, nonionic, or a mixture thereof), aqueous dissolvability (aqueous soluble, oily, or their mixture), and intricacy (multicomponent), as well as emerging contaminants. The following sections provide detailed information on the utilization of GBMs for the remediation of effluents polluted with these aforementioned pollutants.

9.3.1 GBMs for the Uptake of Cationic Contaminants

The diverse functionalities of GBMs offer several modes of interaction, including electrostatic attraction, complexation/coordination (with a particular emphasis on metallic ions), π–π stacking, and cation–π interactions (primarily for organic compounds) for the removal of cationic pollutants. These wide-ranging interaction modes make GBMs highly attractive functional materials for the adsorption of cationic species. Thus, substantial research efforts have been directed towards the elimination of cationic pollutants, including toxic metallic ions and cationic dyes, using GBMs as adsorbents. In the case of inorganic pollutants, numerous metal ions, *i.e.*, Cu^{2+}, Pb^{2+}, Hg^{2+}, Mn^{2+}, Cd^{2+}, Co^{2+}, Ni^{2+}, Zn^{2+}, as well as rare earth metals, have been widely investigated as contaminants using GBMs-based adsorbents. Regarding cationic dyes, MB has been extensively studied as a water contaminant using GBMs. For instance, Bao *et al.* developed PEI-*g*-amine-modified GO nanosheets for the selective uptake of Cr(VI) in effluents. The synthetic methodology involved heating of powdered GO (0.2 g) in deionized (DI) water (200 mL) at 50 °C for 1 h, followed by the addition of APTES under stirring. Subsequently, a 1 : 1 solution of PEI : glutaraldehyde (200 mL) was added to the

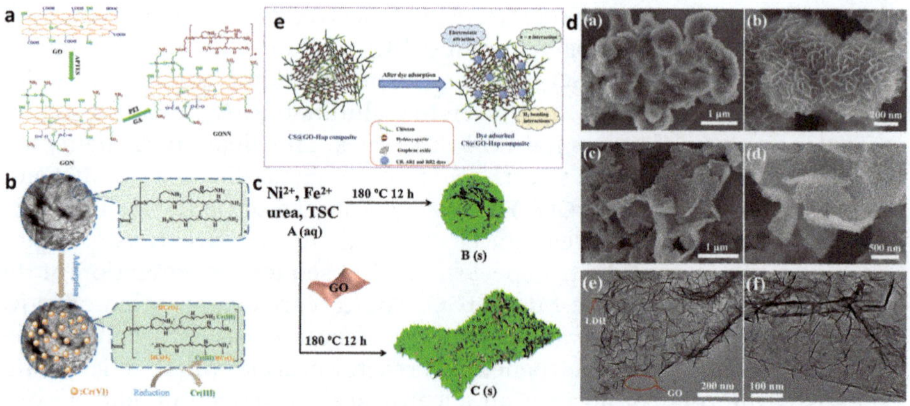

Figure 9.2 (a) Synthetic methodology of PEI-*g*-amine-modified GO nanosheets. (b) Proposed Cr(VI) binding mechanism with PEI-*g*-amine-modified GO.[68] Reproduced from ref. 68 with permission from Elsevier, Copyright 2020. (c) Fabrication and (d) SEM/TEM images of NiFe LDH microspheres and GO/NiFe LDH composite.[73] Reproduced from ref. 73 with permission from Elsevier, Copyright 2019. (e) Proposed CR, AR1, and RR2 dye-binding mechanisms on CS/GO-Hap composite.[78] Reproduced from ref. 78 with permission from Elsevier, Copyright 2020.

above solution for 12 h. The resulting mixture was then washed with DI water and subjected to freeze drying to obtain the final product (Figure 9.2a). The authors indicated that the GO-based adsorbent exhibited a maximum Cr(VI) adsorbability of 1185 mg g^{-1} and rapid rate abiding with the PSO kinetic model. Furthermore, the adsorbent demonstrated excellent recyclability, maintaining its performance over 40 cycles. Based on the experimental analysis, the authors suggested that the interaction between the negatively charged Cr(VI) species ($HCrO_4^-$ and $Cr_2O_7^{2-}$) and the protonated $-NH_2$ functional groups on the surface of the GO-based adsorbent dominated the Cr(VI) binding process. Moreover, the adsorbed Cr(VI) species can undergo reduction to Cr(III) by accepting protons from the electron-donating $-NH_2$ functional groups (Figure 9.2b).[68] Rostamian *et al.* investigated the utilization of poly(glycerol sebacate) (PGS)/GO, PGS/CS, and PGS/CS/GO nanocomposites as effective adsorbents for eliminating MB in effluents. The PGS/CS/GO nanocomposite exhibited an impressive MB elimination capacity of 178 mg g^{-1} and efficiency of 99% under optimized experimental conditions (contact time: 5 hours, MB concentration: 100 mg L^{-1}, pH: 7). Furthermore, even after seven repetitions, the MB adsorption capacity of the nanocomposite remained relatively stable at around 129 mg g^{-1}.[69] Abd-Elhamid *et al.* modified GO with sodium citrate, forming a GO nanosheet composite, and used it for uptake of MB, CV, and Cu^{2+} and Co^{2+} from effluents. The GO-based composite exhibited maximum monolayer MB, CV, Cu^{2+}, and Co^{2+} adsorbability of ~222.2, ~270.2, ~163.4, and ~145.3 mg g^{-1}, respectively.[70] Sarkar *et al.* developed a composite hydrogel comprising G quantum dots embedded in a matrix of Fe_3O_4/GO/PVA.

This innovative hydrogel demonstrated excellent performance in the uptake of MB and RhB dyes from contaminated solutions. The nanocomposite hydrogel exhibited a remarkable maximum monolayer adsorbability of 1000 mg g^{-1} for MB, conforming to adsorption kinetics with the PSO kinetic model.[71] Eltaweil *et al.* synthesized composite microbeads of CMC/carboxylated GO (CMC/GO-COOH) for effective uptake of MB from contaminated media. The CMC/GO-COOH showed maximum monolayer MB adsorbability of ~180.3 mg g^{-1}, abiding with the adsorption kinetics of the PSO model. On the basis of thermodynamic analysis, the authors indicated that the MB uptake process onto CMC/GO-COOH was exothermic.[72]

9.3.2 GBMs for the Uptake of Anionic Contaminants

To achieve effective interactions of GBMs with anionic pollutants, appropriate chemical modification, reduction, or hybridization of GO (alteration of surface characteristics and charge states of G) with specific compounds is imperative. This is particularly important for attaining targeted adsorption towards anionic pollutants, since, during the transformation of GO to rGO, the selective anionic dye uptake property of GO undergoes a transformation to omnidirectional adsorption towards both cationic and anionic dyes, regulated by π–π stacking interactions. For this purpose, $-NH_2$ group-containing compounds have been utilized extensively to modify G to make the GBMs suitable for anionic contaminant adsorption. As an example, Zheng *et al.* introduced a GO/NiFe LDH composite material for the uptake of CR and MO dyes from effluents. As depicted in Figure 9.2c, the synthesis of GO/NiFe LDH composite involved a single-step hydrothermal process, where in the GO dispersion (with a GO to water ratio of 1 : 1) $FeSO_4{\cdot}7H_2O$ (3 mmol), $NiSO_4{\cdot}6H_2O$ (1.5 mmol), urea (9 mmol), and $Na_3C_6H_5O_7{\cdot}2H_2O$ (0.45 mmol) were added, and the resulting solution was heated at 180 °C for 12 h. The obtained black GO/NiFe LDH composite was subsequently washed with DI water and ethanol, and finally dried through freeze drying. The GO/NiFe LDH composite exhibited a distinctive hierarchically structured 3D porous sandwich-like architecture. Transmission electron microscopy (TEM) images further revealed the loosely assembled LDH nanosheets on the surface of micrometer-sized GO nanosheets, with an LDH nanosheet thickness of 3 to 4 nm (Figure 9.2d). The inclusion of GO in the GO/NiFe LDH enhanced its dye adsorbability and resulted in a swift removal rate in comparison to pristine NiFe LDH microspheres.[73] In another investigation, Zhao *et al.* synthesized a GO/PEI (GP) aerogel through a sol–gel technique and applied it to the elimination of anionic MO and cationic MB dyes. The resulting GP aerogels demonstrated remarkable adsorption capacities, with an MO adsorbability of 331.0 mg g^{-1} at pH 2 and MB adsorbability of 249.6 mg g^{-1} at pH 10.5. The authors attributed the MO and MB binding mechanism with the GP aerogels to electrostatic and π–π interactions. Additionally, the GO/PEI aerogels displayed effective reusability for repetitive dye uptake applications.[74] Xiong *et al.* developed cellulose-mediated GO/Fe_3O_4 aerogel for the elimination of anionic CR and cationic MB dyes from sewage. The GO/Fe_3O_4 exhibited significant dye elimination capacities, with ~282 mg g^{-1} for CR and ~346 mg g^{-1} for MB. The adsorption kinetics followed the PSO model, while the adsorption isotherms were described well by both the Langmuir and Freundlich models. The magnetic nature of the GO/Fe_3O_4 aerogels allowed for easy separation from the solutions using a magnet, enabling their reuse for up to five repetitive cycles, with an adsorption capacity retention of over 68% for both dyes.[75] As an important contribution, Luna *et al.* reported CS/GO aerogels for the uptake of cationic MB and anionic indigo carmine (IC) dyes. The developed CS/GO adsorbents exhibited superior performance in eliminating MB and IC from contaminated media, with adsorption efficiencies of 168.6 and 376.8 mg g^{-1}, respectively. This exceptional performance was indicated to be attributed to the mixed characteristics of the biomaterials CS and GO nanosheets.[76] In another significant study, Chang *et al.* decorated the core–shell $MnFe_2O_4/TiO_2$ NPs onto the surface of rGO and utilized it as an adsorbent for the uptake of ciprofloxacin (CIP) and Cu^{2+} from sewage. Their experimental findings revealed that the maximum monolayer adsorbabilities for CIP and Cu^{2+} were ~22.8 and ~225.9 mg g^{-1}, respectively, following a PSO kinetic model within the investigated temperature ranges. According to the results of the thermodynamic experiments, the authors indicated that the binding process was spontaneous and involved endothermic chemisorption. Additionally, the $MnFe_2O_4/TiO_2$/rGO composite exhibited easy regeneration capability and demonstrated recycling CIP and Cu^{2+} adsorbabilities of ~76.5 and ~118.4 mg g^{-1}, respectively, after six repetitive adsorption/desorption cycles.[77] Sirajudheen *et al.* synthesized a CS-supported GO-hydroxyapatite composite (CS/GO-Hap) and evaluated its CR, Acid Red 1 (AR1), and Reactive Red 2 (RR2) dye elimination efficiency from effluents. Through batch-mode adsorption experiments, the CS/GO-Hap composite demonstrated effective elimination, with CR, AR1, and RR2 adsorbabilities of ~43.0, ~41.3, and ~40.0 mg g^{-1}, respectively, complying well with the adsorption isotherms and kinetics for the Freundlich and PSO models, respectively. Notably, the composite exhibited enhanced dye uptake at pH 2, while the efficiency decreased with increasing pH. Thermodynamic analysis indicated that the dye uptake process was spontaneous. Moreover, the CS/GO-Hap composite exhibited excellent regeneration ability, retaining over 65% of its initial dye uptake capacity. Based on the experimental analysis, the authors proposed that electrostatic attractions, chelation, hydrogen bonding, and π–π stacking interactions played dominant roles in the adsorption mechanism of dyes onto the CS/GO-Hap composite (Figure 9.2e).[78]

9.3.3 GBMs for the Removal of Non-ionic Pollutants

GBMs are also employed in the remediation of non-ionic contaminants, *i.e.*, polymers, microplastics (MPs), 2,4,6-trichlorophenol, and surfactants (*e.g.*, TX-100). In the case of non-ionic pollutant elimination, the absence of charged species leads to the dominance of hydrophobic interactions, hydrogen bonding, and π–π stacking interactions, rather than electrostatic attractions. As an illustrative example, Sun *et al.* developed a chitin/GO adsorbent for the uptake of various types of MPs, namely pristine polystyrene (PS), carboxylated polystyrene (CPS), and aminated polystyrene (APS). In the fabrication process of the chitin/GO adsorbent, epichlorohydrin created a cross-linked network with the –OH groups of two different chitin chains and the hydrogen bonds formed between chitin and GO, resulting in the immobilization of GO in the sponge. Subsequently, the chitin/GO sponge was immersed in ethanol, creating the second physical network (Figure 9.3a). The chitin/GO adsorbent demonstrated a maximum monolayer adsorbability of 5.8–8.4 for PS within temperature and pH ranges of 25–45 °C and 6–8, respectively. Moreover, the chitin/GO adsorbent exhibited excellent reusability, maintaining high adsorbabilities of 89.8%, 72.4%, and 88.9% for PS, CPS, and APS, respectively, even after undergoing three consecutive adsorption–desorption cycles. Based on

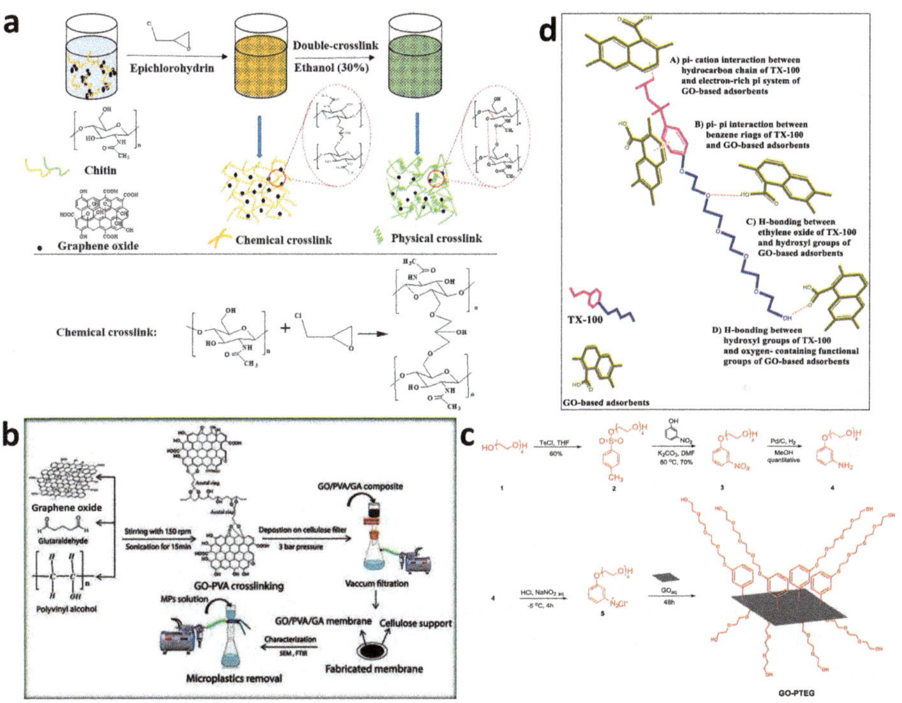

Figure 9.3 (a) An illustrative description of the fabrication of chitin/GO sponges.[79] Reproduced from ref. 79 with permission from Elsevier, Copyright 2020. (b) Schematic representation of the synthesis and microplastics separation performance of GO/PVA membrane.[80] Reproduced from ref. 80 with permission from Elsevier, Copyright 2023. (c) Diagram illustrating the synthetic process of GO-PTEG.[82] Reproduced from ref. 82 with permission from Elsevier, Copyright 2019. (d) Demonstration of the interactions between TX-100 molecules and GO adsorbent.[84] Reproduced from ref. 84 with permission from Elsevier, Copyright 2023.

the experimental analysis, the authors indicated that the key driving forces for MP elimination were electrostatic interactions, hydrogen bonding, and π–π interactions. Additionally, intraparticle diffusion played a significant role in the overall process of MP elimination. Furthermore, the authors suggested that the chitin/GO adsorbent could be biodegraded by microorganisms in the soil.[79] Dey *et al.* developed a GO/PVA functional membrane for the separation of high-density polyethylene (HDPE)-based MPs from effluents, utilizing electrostatic repulsion phenomena. As depicted in Figure 9.3b, the fabrication process involved stabilizing the GO sheets by bonding with PVA using glutaraldehyde, and subsequent deposition onto a cellulose substrate through vacuum filtration. The composite membrane demonstrated a remarkable HDPE-MP separation efficiency of 95% within 15 s at 3 bar transmembrane pressures and pH 8, which was attributed to the synergistic effects of electrostatic repulsion between the MPs and the membrane surface, as well as the sieving mechanism. The concentration of GO and the pH of the system played crucial roles in reducing surface defects and controlling the surface charge of MPs, leading to significant improvements in separation efficiency.[80] Jeong *et al.* fabricated a Fe_3O_4/laser-induced graphene (LIG) composite through a simple process involving laser irradiation on the Fe_3O_4, lignin, and polydimethylsiloxane surfaces. This composite was employed for the efficient uptake of MPs from domestic sewage. The Fe_3O_4/LIG adsorbent demonstrated a maximum multilayer adsorbability of 1400, 1250, 1050, and 775 mg g^{-1} for polyamide (particle size: 50 μm), PS (particle size: 10 μm), melamine (particle size: 10 μm), and melamine (particle size: 2 μm), respectively, complying well with the adsorption kinetics for both pseudo-first-order (PFO) and PSO models. Furthermore, the Fe_3O_4/LIG adsorbent could be reused seven times for each type of MP after undergoing heat treatment.[81] Cheminski *et al.* successfully synthesized a phenyl tetraethylene glycol (PTEG) derivative of GO through grafting of arenediazonium salt, resulting in enhanced stability of GO in water. Figure 9.3c illustrates the synthesis process of O-PTEG, which entails the grafting of the diazonium salt (5) onto GO. The GO-PTEG derivative exhibited remarkable adsorbability for TX-100 and dodecyltrimethylammonium bromide (DTAB). The maximum multilayer TX-100 and DTAB elimination capacities of GO-PTEG were reported as ~1689 and ~713 mg g^{-1}, respectively, with good agreement of experimental results with the PSO kinetic model. The presence of PTEG on GO facilitated micelle formation and binding on the surface through π-stacking and hydrophobic interactions.[82] de Figueiredo Neves *et al.* utilized GO for the effective uptake of cationic surfactants in contaminated solutions. Through optimization, GO exhibited remarkable elimination capacities for the pyridinium-containing surfactant (2083 mg g^{-1}), benzyl-containing surfactant (431 mg g^{-1}), and tetrakyl surfactant (329 mg g^{-1}). The synergistic effect of electrostatic, hydrophobic, and cation–π interactions play a major role in the surfactant binding process with GO, with pyridinium–π and π–π interactions being particularly strong.[83] Rezazadeh *et al.* investigated the binding mechanism of TX-100 molecules on a GO-based adsorbent. They indicated that TX-100 molecules adsorb flatly onto the adsorbent surface. Figure 9.3d demonstrates a schematic representation of the interactions between the top and bottom parts of TX-100 molecules and the adsorbent surface. The bottom part of TX-100 molecules interacts through hydrophobic bonds (π–cationic) with the hydrocarbon chain and the electron-rich π system of GO (Figure 9.3d, panel A), as well as through π–π interactions between the benzene rings of TX-100 and the adsorbents (Figure 9.3d, panel B). The top part of TX-100 molecules, which includes ethylene oxide and –OH groups, engages in hydrogen bonding with the –OH groups of the adsorbent (Figure 9.3d, panels C and D).[84]

9.3.4 GBMs for the Removal of Adsorbates of Varied Water Solubilities

9.3.4.1 GBMs for Oily Contaminants

GBMs with a low concentration of polar functional groups, *i.e.*, highly reduced GO or pristine G, are well suited for the separation of oily contaminants. Modifying the surface of G with hydrophobic compounds can effectively change the surface properties of GBMs, imparting hydrophobic and even oleophilic characteristics. As an illustrative example, Javadian *et al.* modified GO by incorporating Fe_3O_4/oleic acid (OA) core–shell nanospheres and evaluated its effectiveness for oil separation from oily waste water. As shown in Figure 9.4a, chemical covalent bonding is formed between Fe_3O_4 and OA in the composite nanospheres, with two O atoms of OA facilitating two coordination sites in the bond formation with Fe. In addition, physical bonding occurred between Fe_3O_4/OA and GO through interactions with the –COOH or –OH groups on the GN surface. Based on the demulsification experiments the authors revealed excellent demulsification performance of the Fe_3O_4/OA/GO composite, achieving a rapid separation efficiency of 99.99% and displaying consistent performance over six repeated cycles. Regarding the demulsification mechanism, the authors indicated that solid interactions occur within the Fe_3O_4/OA/GO and asphaltenes in oil–water interfaces, involving the polarized π orbitals of GO-based suspensions forming π–π and n–π interactions, which destabilize the oily waste water. The composite nanospheres exhibit enhanced hydrophobicity, resulting in a greater affinity for binding oil droplets and penetrating oil/water emulsion interfaces. These characteristics contribute to strong π–π interactions between the π-conjugated system of asphaltenes and the polarized π orbitals of Fe_3O_4/OA/GO nanosphere suspensions (Figure 9.4b).[85] In another contribution, Fan *et al.* successfully prepared an rGO incorporated with polyphenylene sulfide (PPS) fibrous membrane (rGO/PPS) with high chemical resistance and hydrophobicity for effective oil/water separation and rapid elimination of crude oil from water. The rGO/PPS membrane exhibited an excellent dichloromethane/water separation flux of 12 903 L m^{-2} h, with a removal efficiency of 99.99%. Even after undergoing ten cycles, the rGO/PPS membrane maintained a high separation flux and filtration efficiency. Notably, the membrane also maintains excellent conductivity, high filtration performance, and hydrophobicity after acid/alkali

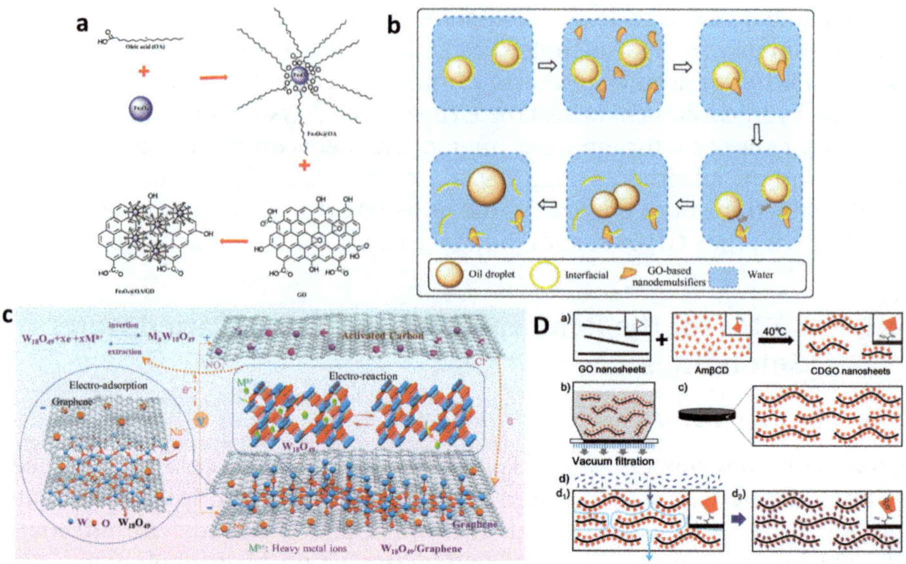

Figure 9.4 (a) Schematic depicting the fabrication process and (b) demonstrating the demulsification performance of Fe_3O_4/OA/GO nanodemulsifier.[85] Reproduced from ref. 85 with permission from Elsevier, Copyright 2019. (c) Illustrative electro-adsorption and electro-reaction coupling process using $W_{18}O_{49}$/G composite.[90] Reproduced from ref. 90 with permission from American Chemical Society, Copyright 2021. (D) Diagram of CDGO nanosheet fabrication.[91] Reproduced from ref. 91 with permission from Elsevier, Copyright 2020.

handling.[86] Also, Liu *et al.* fabricated a 1*H*,1*H*,2*H*,2*H*-perfluorodecanethiol (PFDT)-modified Au NP-loaded G sponge possessing high hydrophobicity and oleophilicity for efficient separation of oil from aqueous solution. The functionalized G foam exhibited exceptional oil adsorption capacity, reaching up to 25.8 g/g with high separation performance after up to ten consecutive cycles of use.[87] Kang *et al.* introduced a novel adsorbent, N-doped carbon nanospheres/G aerogel (MCNS/NGA), characterized by low density, high elasticity (up to ~95% compression), large surface area (~787.9 $m^2\ g^{-1}$), and excellent magnetic properties. The MCNS/NGA demonstrated efficient oil–water separation, exhibiting adsorbabilities ranging from 187 to 537 g/g for various organic solvents and oils. Additionally, the MCNS/NGA exhibited high durability and heat resistance, enabling regeneration through compression, condensation, and ignition processes.[88]

9.3.4.2 GBMs for Multicomponent Removal

In general, real industrial waste water comprises a significantly more intricate mixture of compounds beyond singular organic dyes, metal ions, antibiotics, and other emerging contaminants. Thus, developing GBMs for the elimination of multiple water-soluble contaminants simultaneously constitutes a noteworthy advancement towards the practical implementation of such adsorbents. Numerous investigations have showed the use of GBMs in effective elimination of multiple component from intricate waste waters. As a suitable example, Adeola *et al.* investigated the efficiency of G wool (GW) for the concurrent uptake of various polycyclic aromatic hydrocarbons (PAHs) from effluents. The GW showed superior uptake of PAHs, which was indicated to be primarily governed by hydrophobic interactions within the sorbate and the hydrophobic surface of the GW. Maximum PAH elimination on the GW was observed at an optimal dose of 2 g L^{-1}. On the basis of the thermodynamic analysis, the authors revealed that the uptake process of PAHs onto GW was spontaneous and endothermic. The GW adsorbent exhibited good regeneration capability, with high-performance efficiency up to six cycles.[89] Mao *et al.* investigated the efficiency of heavy metal ion elimination of a $W_{18}O_{49}$/G composite through the electro-adsorption and electro-reaction coupling process. The configuration of the asymmetric deionization system, which combines the processes of electro-adsorption and electro-reaction for the uptake of metal ions in mono, binary, and multicomponent systems, is shown in Figure 9.4c. The authors reported that the $W_{18}O_{49}$/G could uptake 99% of metal ions in binary (metal ions: 10 mg L^{-1} and NaCl: 100 mg L^{-1}) and multicomponent (seven metal ions: 10 mg L^{-1} each and NaCl: 100 mg L^{-1}) systems. Additionally, the $W_{18}O_{49}$/G composite exhibited excellent reusability, maintaining high-performance elimination after 20 repeat cycles. Through experimental analysis, the authors indicated that the binding mechanism of the heavy metal ions on the $W_{18}O_{49}$/G involved pseudocapacitive intercalation during the electro-adsorption and electro-reaction coupling processes.[90]

9.3.4.3 Emerging Contaminants

Recently, numerous emerging contaminants, *i.e.*, medications, phthalic acid compounds, insecticides, hygiene products, detergents, plasticizers, and phenols, which exhibit elevated toxicity to humans and living organisms have attracted significant research interest. Few attempts have been made to utilize GBMs for the purification of waste water containing emerging contaminants. For example, Chen *et al.* reported β-CD-functionalized GO membranes (CDGO), which demonstrated excellent capacity for eliminating BPA from sewage, along with high flux. The CDGO membrane was synthesized by grafting mono-(6-amino-6-deoxy)-β-CD (AmβCD) molecules onto GNs through a reaction between the $-NH_2$ groups of AmβCD and the epoxy groups of GO, forming CDGO nanosheets (Figure 9.4D, panel a). Subsequently, CDGO membranes were obtained using the suction filtration method (Figure 9.4D, panel b). The resulting CDGO membranes feature 2D nanochannels (Figure 9.4D, panel c), where the grafted β-CD molecules can function as active sites (Figure 9.4D, panel d). The CDGO membranes utilized host–guest interactions between β-CD and BPA molecules, resulting in a highly efficient elimination process for BPA. The water flux across the

CDGO membranes increased proportionally with the applied pressure, verifying the high stability of the CDGO membrane during operation. The CDGO membranes exhibited a BPA separation efficacy of ~100%, with fluxes one to two orders of magnitude higher than those achieved by reverse osmosis and nanofiltration (NF) membranes. This improvement in performance can be attributed to the high grafting density of β-CD molecules on both sides of the CDGO nanosheets, which significantly increased the specific surface area, as well as the presence of winding 2D nanochannels within the membrane structure, enhancing the adsorbability of BPA by the CDGO membranes. Furthermore, the CDGO membranes could be easily recycled by rinsing with ethanol and reused for numerous subsequent cycles, with a BPA elimination efficiency of ~100%.[91] In another investigation, Wang *et al.* synthesized a Fe_2O_3-GO composite and applied it for the efficient removal of BPA from effluents. Under optimized experimental conditions, the resulting Fe_2O_3-GO composite exhibited a BPA uptake capacity of 3293.9 mg g^{-1}, which was 1.9 and 1.2 times higher than that of GO and rGO, respectively. Additionally, the Fe_2O_3-GO composite demonstrated superior heat resistance, enhanced solid/liquid separation, and improved anti-fouling properties. When natural or effluent organic matter coexisted, the BPA adsorbability of Fe_2O_3-GO declined by 6.7–16.8%, which was lower than that observed for GO and rGO (11.8–39.4%). Based on the experimental analysis, the authors indicated that the enhanced BPA elimination property of Fe_2O_3-GO was attributed to the formation of Lewis acid–base (AB) interactions within the adsorptive sites of Fe_2O_3 (Lewis base) and BPA (Lewis acid). The presence of these AB interactions is advantageous in real-world applications, particularly when the environmental concentrations of BPA are low in effluents. Moreover, the reduced content of GO in the Fe_2O_3-GO composite resulted in a significant cost reduction of 77.2% for the adsorbent.[92]

9.4 Conclusions and Perspectives

Graphene-based materials have gained considerable global attention in the field of wastewater remediation, thanks to their exceptional physicochemical characteristics, including large specific surface area and porous structure, as well as their diverse chemical functionalities. This chapter provides an overview of recent advancements on GBMs for remediating effluents contaminated with diverse pollutants, *i.e.*, positively or negatively charged pollutants, non-ionic compounds, or mixtures thereof, aqueous soluble, oily, or mixed components, and emerging contaminants. The effects of surface modifications of GBMs with O-, N-, and S-containing functionalities on the elimination performance for pollutants has been reviewed. Significant emphasis has been placed on the development of innovative fabrication approaches for GBMs that exhibit enhanced capacity and selectivity in the elimination of water pollutants. Pristine G possesses numerous drawbacks, namely poor dispersibility in the majority of organic solvents and aqueous media, as well as a tendency to accumulate, which hinders its manufacturability and practical applications. The transformation of G into GO or rGO can address the limitations encountered with pristine G. Nevertheless, due to the non-stoichiometric nature of GO, achieving consistent O group contents remains challenging, and, therefore, future research should aim to develop improved fabrication methods for GO to enable precise control over O group contents, as this greatly affects the pollutant adsorbability. The creation of macroscopic 3D G structures facilitates rapid mass transfer, prevents the aggregation of G NPs, provides a large specific surface area, enables easier handling and material reuse, and overcomes the challenges associated with powdered G NP-based adsorbents. The utilization of magnetic GBMs offers an additional advantage of magnetic isolation and recyclability of adsorbents. While adsorption is a promising technique for wastewater treatment and possesses regeneration capacity for multiple cycles, the gradual loss of active sites in adsorbents after a few treatment cycles results in their disposal as waste, leading to secondary pollution. Therefore, it is imperative to address this issue in a sustainable manner. In order to facilitate the industrial-scale application of GBMs for environmental remediation, it is imperative to make additional efforts towards reducing costs through the utilization of low-cost functional components. With ongoing advancements in research, GBMs have shown considerable potential in environmental remediation, paving the way for exciting discoveries, such as the development of low-cost GBMs for large-scale effluent treatment.

References

1. C. B. Godiya, M. Liang, S. M. Sayed, D. Li and X. Lu, *J. Environ. Manage.*, 2019, **232**, 829–841.
2. G. Zhou, Q. Meng, S. Li, R. Song, Q. Wang, Z. Xu and Z. Xing, *J. Environ. Chem. Eng.*, 2022, **10**, 108421.
3. M. A. Shamim, H. Zia, M. Zeeshan, M. Y. Khan and M. Shahid, *J. Environ. Chem. Eng.*, 2022, **10**, 106991.
4. D. R. Rout and H. M. Jena, *Environ. Res.*, 2022, **214**, 114044.
5. Z. Mo, D. Tai, H. Zhang and A. Shahab, *Chem. Eng. J.*, 2022, 136320.
6. I. A. Phulpoto, Z. Yu, M. A. Qazi, F. Ndayisenga and J. Yang, *Chemosphere*, 2022, 136979.
7. J. Li, Y. Ding, K. Chen, Z. Li, H. Yang, S. Yue, Y. Tang and Q. Wang, *J. Alloys Compd.*, 2022, **903**, 163795.
8. S. Sathe, I. Chakraborty, B. Dubey and M. Ghangrekar, *Environ. Res.*, 2022, **204**, 112135.
9. C. B. Godiya, S. Kumar and B. J. Park, *Carbohydr. Polym. Technol. Appl.*, 2023, **5**, 100267.
10. M. M. Gharehveran, A. M. Walus, T. A. Anderson, S. Subbiah, J. Guelfo, M. Frigon, A. Longwell and J. G. Suski, *J. Environ. Chem. Eng.*, 2022, **10**, 108953.
11. M. Abu-Dalo, J. Abdelnabi, N. A. F. Al-Rawashdeh, B. Albiss and A. Al Bawab, *Environ. Nanotechnol., Monit. Manage.*, 2022, **17**, 100626.

12. R. Chen, L. Hu, H. Zhang, D. Lin, J. Wang, D. Xu, W. Gong and H. Liang, *J. Hazard. Mater.*, 2022, **438**, 129541.
13. C. B. Godiya, X. Cheng, D. Li, Z. Chen and X. Lu, *J. Hazard. Mater.*, 2019, **364**, 28–38.
14. T. Saeed, A. Naeem, I. U. Din, M. Farooq, I. W. Khan, M. Hamayun and T. Malik, *J. Hazard. Mater.*, 2022, **427**, 127902.
15. N. Nasrollahi, V. Vatanpour and A. Khataee, *Sci. Total Environ.*, 2022, 156010.
16. R. Zhao, D. Chen, N. Gao, L. Yuan, W. Hu, F. Cui, Y. Tian, W. Shi, S. Ma and G. Zhu, *Adv. Funct. Mater.*, 2022, **32**, 2200618.
17. C. B. Godiya, L. A. M. Ruotolo and W. Cai, *J. Mater. Chem. A*, 2020, **8**, 21585–21612.
18. J. Li, J. Wang, X. Yuan, Z. Wang, S. Sun, Q. Lyu and S. Hu, *Sep. Purif. Technol.*, 2023, **309**, 122913.
19. R. Kumar, P. Sharma, W. Yang, M. Sillanpää, J. Shang, P. Bhattacharya, M. Vithanage and J. P. Maity, *Environ. Res.*, 2022, 114043.
20. K. Yang, J. Wang, X. Chen, Q. Zhao, A. Ghaffar and B. Chen, *Environ. Sci.: Nano*, 2018, **5**, 1264–1297.
21. H. Hu, W. Wen and J. Z. Ou, *Environ. Sci.: Nano*, 2022, **9**, 3226–3276.
22. Q. Ma, J. Ming, X. Sun, N. Liu, G. Chen and Y. Yang, *Chemosphere*, 2022, **306**, 135512.
23. A. Saravanan, P. S. Kumar, S. Srinivasan, S. Jeevanantham, M. Vishnu, K. V. Amith, R. Sruthi, R. Saravanan and D.-V. N. Vo, *Chemosphere*, 2022, **298**, 134284.
24. S. Guo, S. Garaj, A. Bianco and C. Ménard-Moyon, *Nat. Rev. Phys.*, 2022, **4**, 247–262.
25. M. G. Azam, M. H. Kabir, M. A. A. Shaikh, S. Ahmed, M. Mahmud and S. Yasmin, *J. Water Process Eng.*, 2022, **46**, 102597.
26. D. C. Marcano, D. V. Kosynkin, J. M. Berlin, A. Sinitskii, Z. Sun, A. S. Slesarev, L. B. Alemany, W. Lu and J. M. Tour, *ACS Nano*, 2018, **12**, 2078.
27. J. Chen, B. Yao, C. Li and G. Shi, *Carbon*, 2013, **64**, 225–229.
28. M. Mahmudzadeh, H. Yari, B. Ramezanzadeh and M. Mahdavian, *J. Hazard. Mater.*, 2019, **371**, 609–624.
29. V. Agarwal and P. B. Zetterlund, *Chem. Eng. J.*, 2021, **405**, 127018.
30. C. K. Chua and M. Pumera, *J. Mater. Chem. A*, 2013, **1**, 1892–1898.
31. K. K. H. De Silva, H. H. Huang, R. K. Joshi and M. Yoshimura, *Carbon*, 2017, **119**, 190–199.
32. S. Pei, J. Zhao, J. Du, W. Ren and H.-M. Cheng, *Carbon*, 2010, **48**, 4466–4474.
33. S.-P. Xiao and H.-X. Huang, *Compos. Sci. Technol.*, 2022, **226**, 109533.
34. S. Dubin, S. Gilje, K. Wang, V. C. Tung, K. Cha, A. S. Hall, J. Farrar, R. Varshneya, Y. Yang and R. B. Kaner, *ACS Nano*, 2010, **4**, 3845–3852.
35. S. Mantovani, T. D. Marforio, S. Khaliha, A. Pintus, A. Kovtun, F. Tunioli, L. Favaretto, A. Bianchi, M. L. Navacchia, V. Palermo, M. Calvaresi and M. Melucci, *Environ. Sci.: Water Res. Technol.*, 2023, **9**, 1030–1040.

36. R. M. Firdaus, N. I. M. Rosli, J. Ghanbaja, B. Vigolo and A. R. Mohamed, *J. Nanopart. Res.*, 2019, **21**, 257.
37. F. Arshad, M. Selvaraj, J. Zain, F. Banat and M. A. Haija, *Sep. Purif. Technol.*, 2019, **209**, 870–880.
38. H. Wang, K. Zhang and M. Zhang, *J. Alloys Compd.*, 2019, **781**, 761–772.
39. Y. Tadjenant, N. Dokhan, A. Barras, A. Addad, R. Jijie, S. Szunerits and R. Boukherroub, *Chemosphere*, 2020, **258**, 127316.
40. G. Liu, R. Yu, T. Lan, Z. Liu, P. Zhang and R. Liang, *RSC Adv.*, 2019, **9**, 27060–27068.
41. C. Xing, A. Xia, L. Yu, L. Dong, Y. Hao and X. Qi, *Clean: Soil, Air, Water*, 2021, **49**, 2000272.
42. I. Kaminska, W. Qi, A. Barras, J. Sobczak, J. Niedziolka-Jonsson, P. Woisel, J. Lyskawa, W. Laure, M. Opallo, M. Li, R. Boukherroub and S. Szunerits, *Chem. –Eur. J.*, 2013, **19**, 8673–8678.
43. M.-S. Hosseini, M. Masteri-Farahani and S. Shahsavarifar, *J. Taiwan Inst. Chem. Eng.*, 2019, **102**, 34–43.
44. Y. Bin, Q. Liang, H. Luo, Y. Chen and T. Wang, *Environ. Sci. Pollut. Res.*, 2023, **30**, 6746–6757.
45. Q. Kong, J. Wei, Y. Hu and C. Wei, *J. Hazard. Mater.*, 2019, **363**, 161–169.
46. P. N. Diagboya, H. K. Mmako, E. D. Dikio and F. M. Mtunzi, *J. Environ. Chem. Eng.*, 2019, **7**, 103461.
47. Z. Sun, S. Fang and Y. H. Hu, *Chem. Rev.*, 2020, **120**, 10336–10453.
48. M. S. de Luna, Y. Wang, T. Zhai, L. Verdolotti, G. G. Buonocore, M. Lavorgna and H. Xia, *Prog. Polym. Sci.*, 2019, **89**, 213–249.
49. Y. Sun, X. Liu, X. Lv, T. Wang and B. Xue, *J. Cleaner Prod.*, 2021, **295**, 126406.
50. J. Ma, Y. Xiong, X. Dai and F. Yu, *Chem. Eng. J.*, 2020, **380**, 122387.
51. L. Jiang, Y. Wen, Z. Zhu, X. Liu and W. Shao, *Chemosphere*, 2021, **265**, 129169.
52. T. Shimizu, K. K. H. De Silva, M. Hara and M. Yoshimura, *Appl. Surf. Sci.*, 2022, **600**, 154098.
53. L. Liu, G. Kong, Y. Zhu, D. Lai, S. Zhang and C. Che, *Appl. Surf. Sci.*, 2022, **598**, 153694.
54. Q. Guo, S. Chen, Z. Liu, J. Yan and H. Liu, *J. Environ. Chem. Eng.*, 2022, **10**, 108312.
55. H. Liu, X. Tian, X. Xiang and S. Chen, *Int. J. Biol. Macromol.*, 2022, **202**, 632–643.
56. H. Rajabi, S. M. Jafari, J. Feizy, M. Ghorbani and S. A. Mohajeri, *Int. J. Biol. Macromol.*, 2020, **162**, 163–174.
57. Y. Zhong, S. Mahmud, Z. He, Y. Yang, Z. Zhang, F. Guo, Z. Chen, Z. Xiong and Y. Zhao, *J. Hazard. Mater.*, 2020, **397**, 122774.
58. A. L. Ahmad, O. I. Ebenezer, N. F. Shoparwe and S. Ismail, *Membranes*, 2022, **12**, 24.
59. L. Kong, Y. Wang, C. B. Andrews and C. Zheng, *Chem. Eng. J.*, 2022, **435**, 134830.

60. S. O. Akpotu, P. N. Diagboya, I. A. Lawal, S. O. Sanni, A. Pholosi, M. G. Peleyeju, F. M. Mtunzi and A. E. Ofomaja, *Chem. Eng. J.*, 2023, **453**, 139771.
61. S.-M. Alatalo, E. Daneshvar, N. Kinnunen, A. Meščeriakovas, S. K. Thangaraj, J. Jänis, D. C. W. Tsang, A. Bhatnagar and A. Lähde, *Chem. Eng. J.*, 2019, **373**, 821–830.
62. A. Khataee, T. S. Rad, S. Nikzat, A. Hassani, M. H. Aslan, M. Kobya and E. Demirbaş, *Chem. Eng. J.*, 2019, **375**, 122102.
63. L. Das, P. Das, A. Bhowal and C. Bhattachariee, *Environ. Technol. Innovation*, 2020, **18**, 100664.
64. J. Kaushik, V. Kumar, A. K. Garg, P. Dubey, K. M. Tripathi and S. K. Sonkar, *New J. Chem.*, 2021, **45**, 9073–9083.
65. Y. Shi, A.-F. Yang, C.-S. Cao and B. Zhao, *Coord. Chem. Rev.*, 2019, **390**, 50–75.
66. A. S. Eltaweil, H. M. Elshishini, Z. F. Ghatass and G. M. Elsubruiti, *Powder Technol.*, 2021, **379**, 407–416.
67. K. Zuo, X. Huang, X. Liu, E. M. G. Garcia, J. Kim, A. Jain, L. Chen, P. Liang, A. Zepeda, R. Verduzco, J. Lou and Q. Li, *Environ. Sci. Technol.*, 2020, **54**, 13322–13332.
68. S. Bao, W. Yang, Y. Wang, Y. Yu, Y. Sun and K. Li, *Chem. Eng. J.*, 2020, **399**, 125762.
69. M. Rostamian, H. Hosseini, V. Fakhri, P. Y. Talouki, M. Farahani, A. J. Gharehtzpeh, V. Goodarzi and C.-H. Su, *Chemosphere*, 2022, **289**, 133219.
70. A. I. Abd-Elhamid, E. M. A. Elgoud, S. S. Emam and H. F. Aly, *Sci. Rep.*, 2022, **12**, 9204.
71. N. Sarkar, G. Sahoo and S. K. Swain, *J. Mol. Liq.*, 2020, **302**, 112591.
72. A. S. Eltaweil, G. S. Elgarhy, G. M. El-Subruiti and A. M. Omer, *Int. J. Biol. Macromol.*, 2020, **154**, 307–318.
73. Y. Zheng, B. Cheng, W. You, J. Yu and W. Ho, *J. Hazard. Mater.*, 2019, **369**, 214–225.
74. Q. Zhao, X. Zhu and B. Chen, *Chem. Eng. J.*, 2018, **334**, 1119–1127.
75. J. Xiong, D. Zhang, H. Lin and Y. Chen, *Chem. Eng. J.*, 2020, **400**, 125890.
76. M. S. de Luna, C. Ascione, C. Santillo, L. Verdolotti, M. Lavorgna, G. G. Buonocore, R. Castaldo, G. Filippone, H. Xia and L. Ambrosio, *Carbohydr. Polym.*, 2019, **211**, 195–203.
77. L. Chang, Y. Pu, P. Jing, Y. Cui, G. Zhang, S. Xu, B. Cao, J. Guo, F. Chen and C. Qiao, *Appl. Surf. Sci.*, 2021, **541**, 148400.
78. P. Sirajudheen, P. Karthikeyan, K. Ramkumar and S. Meenakshi, *J. Mol. Liq.*, 2020, **318**, 114200.
79. C. Sun, Z. Wang, L. Chen and F. Li, *Chem. Eng. J.*, 2020, **393**, 124796.
80. T. K. Dey, M. Jamal and M. E. Uddin, *J. Water Process Eng.*, 2023, **52**, 103554.
81. S.-Y. Jeong, N. Sugita and B.-S. Shin, *Int. J. Precis. Eng. Manuf. – Green Technol.*, 2023, **10**, 807–818.
82. T. Cheminski, T. de F. Neves, P. M. Silva, C. H. Guimarães and P. Prediger, *J. Environ. Chem. Eng.*, 2019, **7**, 102976.

83. T. de F. Neves, P. K. Assano, L. R. Sabino, W. B. Nunes and P. Prediger, *Water, Air, Soil Pollut.*, 2020, **231**, 304.
84. N. Rezazadeh, S. Danesh and M. Eftekhari, *Environ. Nanotechnol., Monit. Manage.*, 2023, **20**, 100819.
85. S. Javadian, M. khalilifard and S. M. Sadrpoor, *J. Water Process Eng.*, 2019, **32**, 100961.
86. T. Fan, Y. Su, Q. Fan, Z. Li, W. Cui, M. Yu, X. Ning, S. Ramakrishna and Y. Long, *ACS Appl. Mater. Interfaces*, 2021, **13**, 19377–19386.
87. S. Liu, S. Wang, H. Wang, C. Lv, Y. Miao, L. Chen and S. Yang, *Sci. Total Environ.*, 2021, **758**, 143660.
88. W. Kang, Y. Cui, L. Qin, Y. Yang, Z. Zhao, X. Wang and X. Liu, *J. Hazard. Mater.*, 2020, **392**, 122499.
89. A. O. Adeola and P. B. C. Forbes, *Environ. Technol.*, 2022, **43**, 1255–1268.
90. M. Mao, T. Yan, J. Shen, J. Zhang and D. Zhang, *Environ. Sci. Technol.*, 2021, **55**, 3333–3340.
91. Z.-H. Chen, Z. Liu, J.-Q. Hu, Q.-W. Cai, X.-Y. Li, W. Wang, Y. Faraj, X.-J. Ju, R. Xie and L.-Y. Chu, *J. Membr. Sci.*, 2020, **595**, 117510.
92. Y. Wang, X. Wei, Y. Qi and H. Huang, *Chemosphere*, 2021, **263**, 127563.

CHAPTER 10

Textile Azo Dye Removal Using a Quaternary Ammonium Anion Exchanger Prepared From Sugarcane Bagasse

NATÁLIA ROBERTA MARQUES TANURE[a], LILIANE CATONE SOARES[a], LUISA CARDOSO MAIA[a], MEGG MADONYK COTA ELIAS[a], AMÁLIA LUÍSA PEDROSA XAVIER[b], OSCAR FERNANDO HERRERA ADARME[c], GUILHERME MAX DIAS FERREIRA[d], EDUARDO RIBEIRO DE AZEVEDO[e], LUIS HENRIQUE MENDES DA SILVA[f] AND LEANDRO VINÍCIUS ALVES GURGEL[a*]

[a]Department of Chemistry, Institute of Exact and Biological Sciences, Federal University of Ouro Preto, Campus Universitário Morro do Cruzeiro, Rua Quatro, 786, Bauxita, 35402-136, Ouro Preto, Minas Gerais, Brazil; [b]Federation of Cargo Transport Companies of the State of Minas Gerais, Avenida Antônio Abrahão Caram, 728, Pampulha, 31275-000, Belo Horizonte, Minas Gerais, Brazil; [c]School of Agricultural Engineering, University of Campinas, Avenida Candido Rondon, 501, 13083-875, Cidade Universitária, Campinas, São Paulo, Brazil; [d]Department of Chemistry, Institute of Natural Sciences, Federal University of Lavras, Campus Universitário, 37200-000, Lavras, Minas Gerais, Brazil; [e]Department of Physics and Interdisciplinary Science, São Carlos Institute of Physics, University of São Paulo, Avenida Trabalhador São-carlense, 400, Parque Arnold Schimidt 13566-590, São Carlos, São Paulo, Brazil; [f]Department of Chemistry, Federal University of Viçosa, Avenida Peter Henry Rolfs, s/n, Campus Universitário, 36570-900, Viçosa, Minas Gerais, Brazil
*E-mail: legurgel@ufop.edu.br

Chemistry in the Environment Series No. 12
Novel Materials and Water Purification: Towards a Sustainable Future
Edited by Grigorios L. Kyriakopoulos and Miltiadis G. Zamparas

Published by the Royal Society of Chemistry, www.rsc.org

10.1 Introduction

The use of colour to customize clothing dates back many thousands of years, and until the advent of synthetic dyes, thanks to the young chemist W. H. Perkin, dyeing was a relatively rare and expensive process.[1] Unfortunately, synthetic dyes are toxic and cause adverse effects on living beings. In fact, even natural dyes are rarely low impact due to their use with mordants.[2]

Among all classes of synthetic dyes, azo dyes, which mainly provide the colours yellow, orange, and red, are the most used by the textile industry and, therefore, the most commonly released into the environment.[3] Many azo dyes can be mutagenic and carcinogenic, and can cause allergic reactions.[4] The azo bond can be cleaved through enzymatic, thermal, or photochemical degradation forming aromatic amines whose toxicity depends on the position, number, and nature of other substituents in the aromatic ring.[3]

Different chemical, physical, physical–chemical, and biological methods have been developed to remove dyes and pigments from textile waste water, such as coagulation/flocculation, reverse osmosis, aerobic and anaerobic degradation, advanced oxidative processes, filtering membranes (nano- and ultrafiltration), adsorption, among others. These methods are often combined for best results due to the recalcitrant nature of synthetic dyes.[5]

Adsorption has been successfully used to remove dyes from textile waste water, with high selectivity and efficiency, simplicity of operation, non-generation of toxic by-products, possibility of dye and adsorbent recovery, and adsorbent reuse in various cycles.[5,6] Furthermore, the concept of a bio-based circular economy can be applied to wastewater treatment using adsorbents produced from biomass residues, especially when they can be reused in multiple adsorption–desorption cycles, making the process more sustainable and lower cost.

In this work, a quaternized bioadsorbent (STEA) was produced from sugarcane bagasse (SB), one of the main agro-industrial residues produced in Brazil. For this, SB was chemically modified with epichlorohydrin (EPI) and triethylamine (TEA) using *N*,*N*-dimethylformamide (DMF) as solvent, using a new method based on that developed by Gao *et al.*[7] In this new method, a two-step one-pot reaction is performed instead of a three-step reaction as described by Gao *et al.*,[7] and no catalyst is needed. Chemical modification of SB with EPI and TEA is greener and more sustainable because: (1) bio-based EPI is commercially available;[8] and (2) TEA can be synthesized from a greener route using ammonia[9] and bioethanol.[10] STEA bioadsorbent was used to remove the anionic azo dye remazol golden yellow (RGY) RNL (or Reactive Orange 107) from spiked aqueous solutions in batch and fixed-bed column modes. The reuse and recovery of STEA was also assessed to meet the circular bioeconomy concept.

10.2 Materials and Methods

10.2.1 Bioadsorbent and Dye

Remazol golden yellow RNL ($C_{16}H_{16}N_4O_{10}S_3Na_2$, 566 g mol^{-1}) was donated by a textile industry from Itabirito, Minas Gerais, Brazil. The dimensions and geometry of the RGY molecule (Scheme 10.1) were estimated using the Universal Force Field.[11] The Avogadro software (SourceForge, version 1.2.0) was the tool used to build, edit, visualize, and analyse the molecular and chemical data of the RGY molecule.

Raw SB was provided by the Jatiboca sugar and ethanol mill (Urucânia, Minas Gerais, Brazil). It was dried to ~10% moisture, ground (De Leo, Model EDB-5), and sieved (Bertel, Model 2713). The raw SB particles retained on the 100 mesh (149 μm) were suspended in distilled water at 70 °C under stirring (Corning®, Model PC-420) to remove water-soluble compounds. After 1 h, the raw SB particles were separated by filtration under reduced pressure on a Büchner funnel and washed with ethanol (95%) and dried at 90 °C (FANEM, Model 515C). Afterwards, the raw SB particles were washed in a Soxhlet system with an *n*-hexane/ethanol mixture (1 : 1 v/v) to eliminate extractives and phenolic compounds. After 4 h of extraction, the raw SB particles were dried at 90 °C for 60 min and then stored in a desiccator at room temperature.

10.2.2 Synthesis of STEA

STEA was synthesized in a two-step one-pot reaction, as shown in Scheme 10.2. In this synthesis, 10.5 g SB, 211 mL anhydrous DMF, and 135 mL EPI were added in that order to a 1 L double-neck round-bottom flask equipped with an Allihn condenser connected to a glass bulb packed with a drying agent (powdered $CaCl_2$). Before use, DMF was dehydrated with a previously activated 3 Å molecular sieve at 150 °C. The suspension was

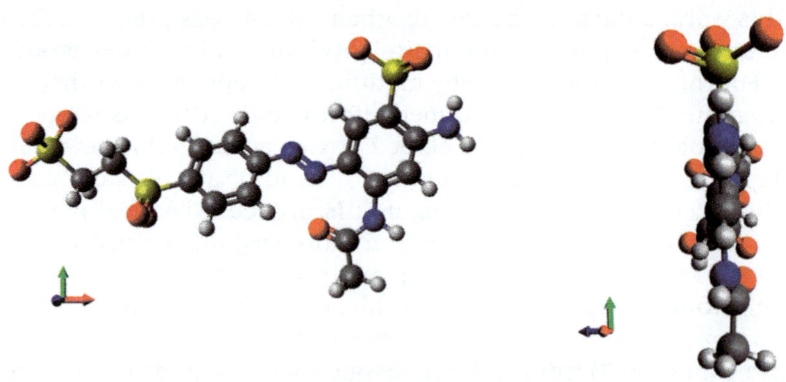

Scheme 10.1 Three-dimensional molecular structure of RGY dye, with carbon, nitrogen, oxygen, sulfur, and hydrogen atoms presented in gray, blue, red, yellow, and white colours, respectively.

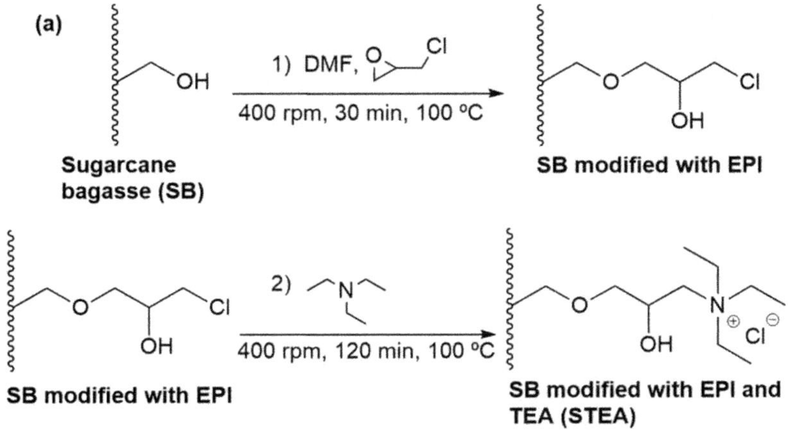

Scheme 10.2 Synthesis route of STEA from raw SB using EPI and TEA.

maintained under reflux for 30 min under stirring (400 rpm) in a canola oil bath at 100 °C, and then 232 mL TEA (fresh distilled) were gradually added to the mixture through a pressure-equalizing dropping funnel. The suspension was stirred at 500 rpm and 100 °C for another 2 h. Prior to use, TEA was purified by refluxing with NaOH pellets for 3 h and then distilled. STEA was separated by filtration under reduced pressure on a glass Büchner funnel (porosity no. 2) and washed with DMF, ethanol (95%), distilled water, 0.1 mol L^{-1} HCl, distilled water, ethanol (95%), and diethyl ether. STEA was dried under reduced pressure at 35 °C (Marconi, Model MA030/12) for 36 h and stored in a desiccator.

10.2.3 Characterization of SB and STEA

The characterization of SB and STEA was conducted by elemental C, H, and N analysis, Fourier-transform infrared (FTIR) spectrometry (ABB Bomen, model MB 3000), and ^{13}C solid-state nuclear magnetic resonance (^{13}C SS NMR) spectroscopy (Bruker, Model Avance 400). Specific surface area (Brunauer–Emmett–Teller (BET)) and pore size distribution (Barrett–Joyner–Halenda (BJH)) were determined by N_2 adsorption/desorption isotherms (Quantachrome, Model Nova Station), as described by Teodoro *et al.*[12] Chlorine content in SB and STEA was determined by the Schöniger method, which involves the combustion of organic samples to quantify the chlorine content through precipitation volumetry.[13]

10.2.4 Adsorption Studies

The adsorption studies with STEA were initially performed in batch to evaluate the performance of the STEA and assess the properties of the STEA-RGY adsorption system. Based on the batch results, fixed-bed column experiments were conducted. This continuous process is widely used in various base industries due to its simplicity of operation and effectiveness in adsorption/desorption cycles, making it more suitable for large-scale applications.

Dye solutions were prepared on the day of the experiment to avoid RGY degradation and using ultrapure water (≥18 MΩ cm). The initial pH of the solutions was adjusted by adding drops of aqueous HCl (0.1 mol L^{-1}) or NaOH (0.1 mol L^{-1}) solutions. RGY quantification was performed on a UV-visible spectrophotometer (Biospectro, Model SP-220) at the maximum absorption wavelength (λ_{max} = 411 nm). The adsorption capacity of RGY at time t (q_t) or equilibrium (q_e) was calculated using eqn (10.1):

$$q_x = (C_i - C_x)\frac{V}{w} \tag{10.1}$$

where q_x (mmol g^{-1}) is the STEA adsorption capacity at time t ($x = t$) or equilibrium ($x = e$), C_i and C_x (mmol L^{-1}) are the initial RGY concentration and the RGY concentration at time t or equilibrium, V (L) is the volume of RGY solution, and w (g) is the weight of STEA.

10.2.4.1 Batch Adsorption Studies

All batch adsorption experiments were performed with 20.0 ± 0.1 mg STEA and at 25.0 ± 0.1 °C. Masses of STEA were weighed into cylindrical glass jars on an analytical balance (Shimadzu, Model Libror AEG-45SM). The glass jars were then placed into conical flasks (250 mL) containing 100.0 mL RGY solution, under stirring (50 rpm), in an orbital shaker-incubator (Tecnal, Model TE-424). The separation of solid and liquid fractions was performed by centrifugation at 3000 rpm for 10 min. The supernatant was collected to determine the concentration of the RGY solution at time t or equilibrium. Batch experiments to investigate the effect of solution pH, kinetics, and initial RGY concentration on STEA adsorption capacity, as well as regeneration and reuse of STEA and competitive adsorption of RGY with sulfate, were performed as described below.

10.2.4.1.1 Exploratory Study. A preliminary exploratory study to evaluate the adsorption of RGY on raw SB was conducted. For this, 100.0 mL of an aqueous 0.58 mmol L^{-1} (300.0 mg L^{-1}) RGY solution was added to 20.0 ± 0.1 mg of raw SB in a conical flask (250 mL) under agitation for 24 h. The test was performed in triplicate. The same conditions were applied to STEA to evaluate its applicability as an adsorbent for RGY.

10.2.4.1.2 Effect of RGY Solution pH. The experiments to evaluate the adsorption capacity of RGY as a function of solution pH were performed with an aqueous 0.58 mmol L^{-1} RGY solution adjusted to different pH values (2.00, 3.00, 4.00, 5.00, 6.00, 7.00, 8.00, 9.00, and 10.00) and agitation for 24 h, as described in Section 10.2.4.1.1.

10.2.4.1.3 Kinetics. The adsorption capacity of RGY as a function of contact time was investigated with an aqueous 0.58 mmol L^{-1} RGY solution at pH 7.00 at times of 10, 17, 25, 35, 40, 55, 70, 90, 120, 150, 180, 240, 300, 360, 540, 720, 840, 960, 1080, 1200, 1300, and 1440 min. Equilibrium time was determined through graphical analysis. Kinetic parameters were obtained by fitting the intraparticle diffusion (IPD),[14] Boyd,[15] and homogeneous surface diffusion model (HSDM)[16] to experimental data.

10.2.4.1.4 Isotherm. The equilibrium experiments were performed by varying the adsorbate to adsorbent ratio, in which the STEA mass was kept constant (20.0 ± 0.1 mg) with increasing RGY concentration. The following RGY concentrations were used 0.01, 0.03, 0.05, 0.08, 0.10, 0.13, 0.18, 0.21, 0.26, 0.36, 0.46, 0.54, 0.64, 0.72, 0.80, 0.90, 1.03, 1.16, 1.21, and 1.29 mmol L^{-1}. The initial pH was adjusted to 7.00 and the contact time was 12 h. The isotherm of RGY on STEA was modelled with the Langmuir,[17] Redlich–Peterson (R–P),[18] and Dubinin–Radushkevich (D–R)[19] models.

10.2.4.1.5 Regeneration and Reuse Studies. The STEA reuse potential was evaluated through desorption/re-adsorption experiments. Initially, adsorption experiments were performed to load STEA samples with RGY (C_i = 1.63 mmol L^{-1}, pH 7.00, 840 min, 50 rpm, 25.0 ± 0.1 °C, and 0.2 g STEA L^{-1}). The RGY-loaded STEA samples were separated by single filtration through a JP-41 filter paper (black ribbon, JProlab, Brazil) and washed with excess ultrapure water. The desorption experiment was performed by adding RGY-loaded STEA samples (20.0 ± 0.1 mg) into conical flasks (250 mL) containing 100.0 mL of an aqueous 0.1 mol L^{-1} HNO_3 solution. The flasks were kept under agitation for 24 h (Section 10.2.4.1.1). After quantification of the supernatant, the desorbed STEA samples were washed with ultrapure water to remove excess HNO_3 solution. Finally, STEA recovered from desorption experiments was reused for subsequent adsorption experiments (C_i = 0.54 mmol L^{-1}, V = 100.0 mL, pH 7.00, 840 min, 50 rpm, and 0.2 g STEA L^{-1}). The desorption efficiency (E_{des}) and re-adsorption efficiency ($E_{re\text{-}ads}$) were determined as reported by Fideles *et al.*[20]

10.2.4.1.6 Competitive Adsorption Study. The competition for surface sites of STEA between RGY and sulfate (SO_4^{2-}) was investigated. Three RGY to SO_4^{2-} molar ratios were evaluated (3 : 1, 1 : 3, and 1 : 1), in which 100% of the species in solution was equivalent to an amount of matter of 2.00 mmol. A control experiment with only the RGY dye was also performed for comparison. The procedures used in this study were the same as those reported in Section 10.2.4.1 and 10.2.4.1.1. A pH of 7.00 and a contact time of 14 h were used in this study. The source of sulfate was anhydrous Na_2SO_4. The experiments were performed in triplicate.

Sulfate was quantified on an ion chromatograph (Methrom) equipped with an autosampler and a conductivity detector was used. The chromatographic separation was performed in a Metrosep A Supp 5 – 150 × 4.0 mm column (Methrom), equipped with a guard column at 30 °C. The sample injection volume was 10 μL. The mobile phase was composed of a solution containing 3.20 mmol L^{-1} sodium carbonate (Na_2CO_3) and 1.00 mmol L^{-1} sodium bicarbonate ($NaHCO_3$), which was pumped in isocratic mode at a flow rate of 0.7 mL min^{-1}. A multi-element standard was used to obtain the analytical curve, which was performed at seven concentration levels, from 1 to 20 mg L^{-1}, diluted in ultrapure water.

10.2.4.2 Fixed-bed Column Studies

The continuous adsorption experiments were performed as described by Xavier *et al.*[21] A jacketed glass column (1.0 cm internal diameter × 10.0 cm height) was packed with STEA (500.0 ± 0.1 mg). The bed height was 2.2 cm. The temperature (25.0 ± 0.1 °C) was controlled by a thermostatic circulating water bath (Marconi, Model MA470). An aqueous RGY solution (198 mg L^{-1}; pH 7.00) was pumped through the column in upward flow at a rate of 2.0 mL min^{-1}. Effluent samples were collected at specific time intervals: every 40 min during the first 240 min, every 25 min from 240 to 540 min, and every 30 min from 540 to 810 min. The total operating time of the column was 810 min. The desorption step was performed with an aqueous 0.1 mol L^{-1} HNO_3 solution pumped in upward flow at a rate of 10 mL min^{-1}. During 240 min of operation, the effluent was collected every 5 min during the first 30 min, every 10 min from 30 to 60 min, and every 15 min for the remaining time. Subsequently, ultrapure water was percolated through the column until the pH of the effluent and influent were equal. Three complete and successive adsorption/desorption cycles in a fixed-bed column were performed. Breakthrough curves were obtained, and the breakpoint and exhaustion point were considered as 5% and 95% of the influent RGY concentration.

The amount of RGY adsorbed on STEA was determined by integrating the area under the breakthrough curves, as described by Teodoro *et al.*[12] The desorption efficiency (E_{des}) and the length of the mass transfer zone (MTZ) were also calculated as described by Teodoro *et al.*[12] The breakthrough curves were adjusted with the Thomas[22] and Bohart–Adams[23] original models using MatLab® software (MathWorks, version 2018). The generic algorithm function was used for convergence (global minimum solution) and data optimization.

10.2.5 Isothermal Titration Calorimetry

The thermal effects resulting from the interaction between RGY and STEA were monitored by isothermal titration calorimetry (ITC). The measurements were performed in a nanocalorimeter (TA Instruments®, Model TAM III), equipped with two 4 mL stainless steel reaction cells, one for sample and the another for reference, controlled by TAM assistant software. The experiments were performed in duplicate with the temperature controlled at 25.0000 ± 0.0001 °C. In each titration, 5.00 mg STEA and 2.70 mL solvent (water at pH 7.00) were added to the sample cell. The reference cell was filled with solvent only. Consecutive injections of 10 μL RGY solution (9.00 g L^{-1}; pH 7.00), at intervals of 180 min, were performed inside the sample cell using a 500 μL Hamilton syringe, which was controlled by a piston pump.

For each system, titration of dye solution over water (pH 7.00) without STEA and titration of water (pH 7.00) over water plus STEA were performed as control experiments.

The adsorption enthalpy change ($\Delta_{ads}H$) and the standard adsorption enthalpy change ($\Delta_{ads}H^{\circ}$) were calculated as described by Pereira *et al.*[24] The change in standard free energy ($\Delta_{ads}G^{\circ}$) and entropy ($\Delta_{ads}S^{\circ}$) for adsorption of RGY on STEA were calculated as described by Fideles *et al.*[20]

10.3 Results and Discussion

10.3.1 Chemical and Structural Characterization of Raw SB and STEA

The contents of C, H, N, and Cl in raw SB and STEA obtained by elemental analysis are presented in Table 10.1. The N content increased from 0.23% (0.1642 ± 0.0001 mmol g^{-1}) in SB to 1.57% (1.12 ± 0.02 mmol g^{-1}) in STEA. The increase of 0.96 ± 0.05 mmol N g^{-1} after chemical modification of raw SB with EPI and TEA confirmed the incorporation of quaternary ammonium groups ($R–N^+(CH_2CH_3)_3$) into the STEA structure. The Cl content increased from 0% in SB to 3.81% (1.08 ± 0.01 mmol g^{-1}) in STEA, also confirming the introduction of EPI and TEA into the STEA structure. Assuming that all Cl atoms were replaced by N atoms, the nucleophilic substitution reaction in step 2 of the STEA synthesis is expected to produce a chemically modified material with a Cl/N ratio of 1 : 1. The Cl/N ratio (Cl/N = 0.96 ± 0.02) indicated that almost all Cl atoms in EPI that were introduced in the first step of the reaction were replaced by N atoms in TEA. As expected, the Cl atoms that remained in the STEA structure are in the form of counterions of the $R–N^+(CH_2CH_3)_3$ group. These results indicated the effectiveness of the proposed synthesis method.

The ^{13}C SS NMR spectrum of STEA showed some differences when compared to the SB spectrum. The signal at 7.7 ppm in the STEA spectrum is attributed to the CH_3 groups attached to the quaternary nitrogen atom ($R–N^+(CH_2CH_3)_3$) in STEA. The presence of this sign confirms the successful synthesis of STEA. The slight shift in signal from 63.6 ppm (raw SB) to 64.1 ppm (STEA) can be attributed to the change in the group attached to C-6 carbon atom of the β-D-anhydroglucose unit (AGU) in cellulose ($R–CH_2–OH$ in raw SB and $R–CH_2–O–CH_2–R$ in STEA). This indicates that the chemical modification occurred at the OH group attached to the C-6 carbon atom, since no chemical shifts were observed at the C-2 and C-3 carbon atoms in the AGU of cellulose. The OH group at the C-6 position is more susceptible to chemical

Table 10.1 Elemental C, H, N, and Cl analysis for raw SB and STEA.

Material	C (%)	H (%)	N (%)	Cl (%)
Raw SB	44.4 ± 0.1	6.1 ± 0.2	0.23 ± 0.01	—
STEA	44.8 ± 0.1	6.7 ± 0.2	1.57 ± 0.03	3.81 ± 0.04

modification due to its lower steric hindrance, although the primary hydroxyl at C-6 is a worse nucleophile compared to the secondary hydroxyl at C-2 or C-3. Another difference is the absence of signals at 20.5 and 171.9 ppm in the STEA spectrum. These signals corresponded to the CH_3 of acetyl groups (CH_3–C(=O)–O–R) in hemicelluloses and carbonyl groups of carboxylic acid derivatives (R–C(=O)–O–R, CH_3–C(=O)–O–R) in hemicelluloses. The suppression of these signals related to hemicelluloses indicates that the chemical modification resulted in the solubilization of the hemicelluloses.

BET specific surface area and BJH pore size distribution are presented in Table 10.2. The mean pore diameter of raw SB and STEA is within the range of 20–500 Å; therefore, both are classified as mesoporous materials.[25] Mesoporous and microporous materials are important in the adsorption processes of larger molecules, such as dyes. There was no significant difference in specific surface area and mean pore diameter when comparing raw SB and STEA. A slight decrease in density was observed, confirming the structural chemical modifications observed in STEA in relation to raw SB.

10.3.2 Adsorption Studies

10.3.2.1 Batch Adsorption Studies

10.3.2.1.1 Exploratory Study. Preliminary investigation revealed that raw SB cannot adsorb RGY dye. This result proves that the functional groups already present in raw SB, such as R–OH (primary, secondary, phenolic hydroxyl), R–COOH, R–C(=O)–O–R, and R–C(=O)–R groups, are unable to remove RGY from the aqueous phase by adsorption. Therefore, a chemical modification of raw SB, such as the introduction of R–$N^+(CH_2CH_3)_3$ into the SB structure, is essential to promote RGY adsorption. The exploratory study with STEA resulted in an adsorption capacity for RGY of 0.73 ± 0.02 mmol g^{-1}. Therefore, STEA proved to be a potential adsorbent for RGY dye.

10.3.2.1.2 Effect of RGY Solution pH. The influence of solution pH on the adsorption of RGY on STEA was evaluated over a pH range of 2.00–10.00 (Figure 10.1a). The equilibrium adsorption capacity (q_e) of RGY increased with increasing solution pH, with higher q_e values observed under alkaline conditions (pH 9–10). This result confirms the influence and effectiveness of

Table 10.2 Textural properties of raw SB and STEA.

Parameter	Material	
	Raw SB	STEA
Density (g cm^{-3})	1.52 ± 0.01	1.33 ± 0.03
Specific surface area (m^2 g^{-1})	1	2
Pore volume (cm^3 g^{-1})	0.003	0.003
Maximum pore diameter (Å)	370	342
Mean pore diameter (Å)	37	36

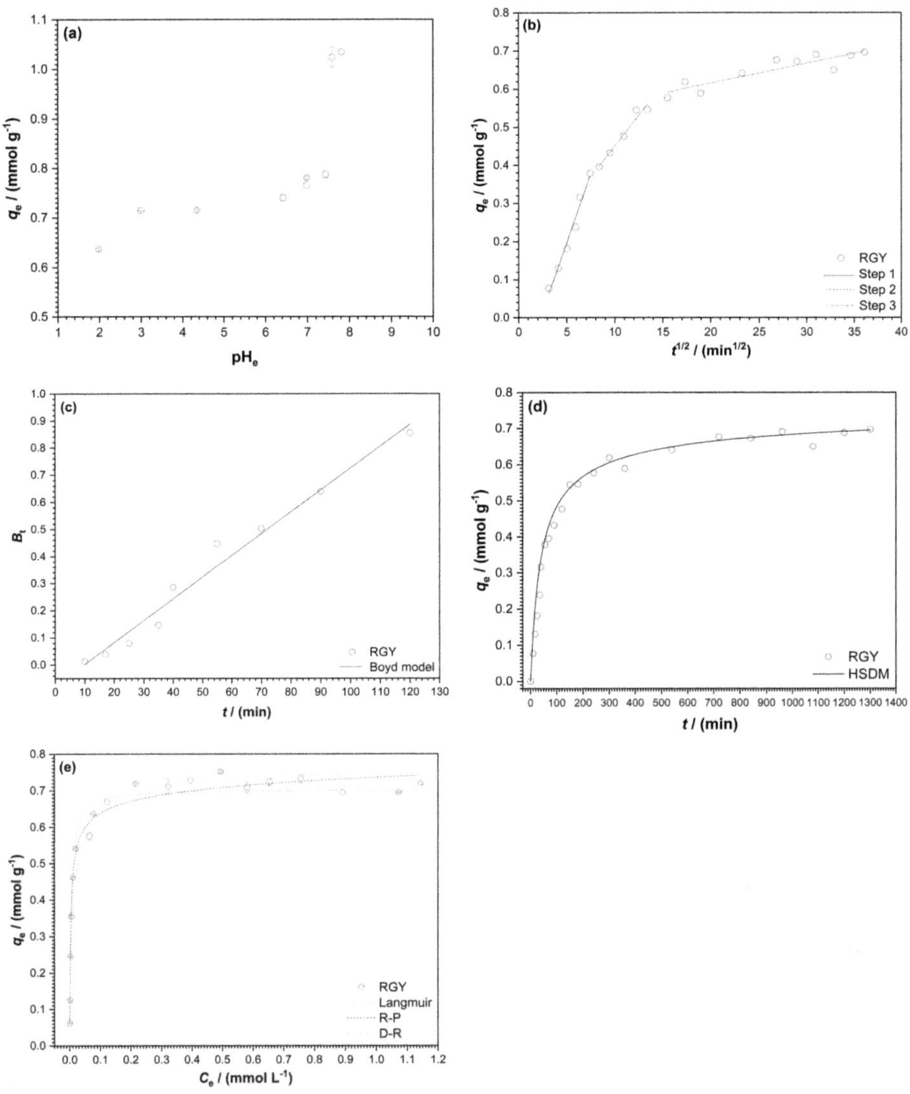

Figure 10.1 (a) Effect of solution pH on the adsorption of RGY on STEA. (b) IPD plot for adsorption of RGY on STEA. (c) Boyd plot for RGY adsorption on STEA. (d) HSDM model fitted to experimental kinetic data. (e) Adsorption isotherm of RGY on STEA.

the positive formal charge of the R–$N^+(CH_2CH_3)_3$ group introduced into the STEA structure.

The pK_a values of RGY are 3.00, 3.50, and 6.00, corresponding to the deprotonation of the sulfate, sulfonic, and amide groups, respectively.[26] At pH 2.00, the lowest q_e value of RGY on STEA was observed among all the pH values evaluated (q_e = 0.64 mmol g^{-1}). For pH < pK_{a1} (3.00), RGY is fully protonated (neutral) and this result is probably attributed to unfavourable electrostatic interactions between the STEA sites and the neutral form of RGY, resulting in its limited adsorption. Therefore, it is likely that STEA removed RGY from the aqueous phase through different interactions, including hydrogen bonding, dipole–dipole, ion–dipole, π-stacking, and London dispersion forces. For pH above pK_{a1} (3.00), there was an increase in the negative charge of the RGY due the successive deprotonation of the sulfonic and sulfate groups. For pH above 6.00 (pH > pK_{a2}), RGY has the highest density of negative charges due to deprotonation of the amide group. Therefore, the increase in the q_e value is attributed to the electrostatic attraction between the negative charges of RGY and the positive charges of the $R{-}N^+(CH_2CH_3)_3$ group in STEA.

The significant increase in the q_e value observed from pH 8.00 onwards indicates that some structural modification in the RGY structure may have occurred. This may involve conversion of the RGY dye to its hydrazone form following deprotonation of the acetamido group, or its hydrolysis due to contact with the alkaline solution (0.1 mol L^{-1} NaOH) used to adjust the pH of the RGY solution.[27] A nucleophilic addition reaction can occur between RGY and cellulose under alkaline conditions,[28] as indicated by eqn (10.2) and (10.3):

$$RGY-SO_2CH_2CH_2OSO_3Na+NaOH \rightarrow RGY-SO_2CH{=}CH_2+Na_2SO_4+H_2O \quad (10.2)$$

$$RGY{-}SO_2CH{=}CH_2+HO{-}Cellulose \rightarrow RGY{-}SO_2CH_2{-}CH_2{-}O{-}Cellulose \quad (10.3)$$

It is possible that the adsorption of RGY above pH 9.00 occurred not only by electrostatic attraction with the surface sites of STEA, but also by chemical reaction with the R–OH groups in STEA, which were not chemically modified in its synthesis (chemisorption). A pH of 7.00 was chosen for further studies in order to: (1) avoid chemisorption of RGY on STEA and make its reuse more feasible; and (2) enable the large-scale application of STEA, since the dye effluent from a textile wastewater treatment plant normally has a pH close to 7.

10.3.2.1.3 Kinetics. The phenomenological model of IPD was used to investigate the possible existence of more than one RGY adsorption step on STEA and to identify the rate-limiting step of the mass transfer of RGY from the solution bulk to the surface, and the porosity of the STEA.[29] The values of the kinetic parameters estimated for each model evaluated are presented in Table 10.3. The multilinear behavior exhibited by the IPD plot (Figure 10.1b) indicates that the adsorption occurred in three steps.[30,31] The first step is associated with the mass transfer of RGY through the boundary layer surrounding the STEA particles (external diffusion). The second step is associated with the intraparticle diffusion of RGY molecules on the surface and in the porosity of the STEA.[12] The last step is associated with the saturation of STEA sites on the surface and in the porosity, when intraparticle diffusion decreases due to a decrease in the concentration of RGY in solution and reduced availability of STEA sites.

Table 10.3 Kinetic parameters estimated by the IPD, Boyd, and HSDM models for adsorption of RGY on STEA.

Intraparticle diffusion model (IPD)		
Step 1	$k_{d,1}$ (mmol g^{-1} min$^{-1/2}$)	0.07 ± 0.01
	C (mmol g^{-1})	−0.17 ± 0.01
	R^2_{adj}	0.9683
Step 2	$k_{d,2}$ (mmol g^{-1} min$^{-1/2}$)	0.03 ± 0.01
	C (mmol g^{-1})	0.13 ± 0.04
	R^2_{adj}	0.9475
Step 3	$k_{d,3}$ (mmol g^{-1} min$^{-1/2}$)	0.005 ± 0.001
	C (mmol g^{-1})	0.51 ± 0.03
	R^2_{adj}	0.7824
Boyd		
B	0.008 ± 0.001	
R^2	0.9746	
D_i (m^2 min^{-1})	(4.5 ± 0.6) × 10^{-12}	
Homogeneous surface diffusion model (HSDM)		
k_c (m min^{-1})	5.71 × 10^{-3}	
D_S (m^2 min^{-1})	5.28 × 10^{-13}	
Biot number	4572.2	
R^2	0.9705	
RMSE	0.0469	

The initial step exhibited the highest diffusion constant (k_d) value, attributed to the intense initial electrostatic attraction between negatively charged RGY molecules and positively charged surface sites of STEA. As expected, the adsorption rate becomes slower with increasing time as $k_{d,1} > k_{d,2} > k_{d,3}$. For the first step, the C value was not zero, indicating that the rate of RGY adsorption on STEA was initially governed by film diffusion and later by intraparticle diffusion. The mean and maximum pore sizes of STEA are 36.41 and 342.0 Å, respectively, and the dimensions of the RGY molecule calculated by molecular mechanics are 18.0, 9.3, and 6.4 Å. Therefore, RGY adsorption can occur through three types of diffusion: molecular, Knudsen, and surface.[32,33] Knudsen and surface diffusions are considered slower diffusions due to the ratio between the pore size of the STEA and the RGY molecule. The ratio between mean and maximum pore sizes of STEA and RGY size indicates that diffusion is more restricted in pores of medium size (ratio $\ll 10$). However, the limitation is much lower in the larger pores of STEA (ratio > 10).

The parameters estimated by fitting the Boyd model to the experimental data are shown in Figure 10.1c and Table 10.3. The plot is linear ($R^2 = 0.9746$) up to 120 min and did not intersect the origin, implying that both mass transfer across the boundary layer and diffusion within the porosity control the rate of RGY adsorption on STEA.[34] These findings corroborate the results obtained by the IPD model.

The HSDM considers both the resistance to mass transfer through the boundary layer of water molecules surrounding the adsorbent particles and the resistance to intraparticle diffusion (surface and pore diffusion).[35] Therefore, the relationship between the convective mass transfer coefficient (k_c) and the surface diffusion coefficient (D_s) can be used to elucidate the dominant mechanism in the adsorption process. The fit of the HSDM to experimental data is shown in Figure 10.1d. The steeper initial slope of the RGY adsorption curve as a function of time (<100 min) indicates a rapid adsorption rate, and from 100 min onwards the adsorption rate decreases, suggesting the occurrence of mass transfer limitation.[36] The k_c and D_s values are presented in Table 10.3, as well as the Biot number, which is the ratio between the resistance to convective mass transfer outside the particle and the resistance to intraparticle diffusion. The D_s value (5.28×10^{-13} m^2 min^{-1}) estimated in this study agrees with the D_s value reported by Atun *et al.*,[37] who studied the adsorption of anionic dyes on Red Mud. A very high Biot number (4572) was estimated for adsorption of RGY on STEA. For Biot numbers higher than 100, the overall adsorption rate is controlled by intraparticle diffusion, and for Biot numbers lower than 1, the overall adsorption rate is controlled by film diffusion. Therefore, the adsorption of RGY on STEA was a process dominated by intraparticle diffusion.[38]

10.3.2.1.4 Isotherm. The isotherm for RGY adsorption on STEA is shown in Figure 10.1e and the parameters estimated by fitting the isotherm models to equilibrium data are presented in Table 10.4.

Table 10.4 Isotherm model parameters estimated by the Langmuir, R–P, and D–R models for adsorption of RGY on STEA.

Experimental data	
$q_{e,exp}$ (mmol g^{-1})	0.71 ± 0.01
Langmuir	
$Q_{max,est}$ (mmol g^{-1})	0.70 ± 0.01
b (L mmol^{-1})	221 ± 24
R^2_{adj}	0.9864
χ^2_{red}	0.0031
Redlich–Peterson (R–P)	
$Q_{max,est}$ (mmol g^{-1})	0.74 ± 0.01
K_R (L g^{-1})	211 ± 24
a_R (L mmol^{-1})	286 ± 31
β	0.95 ± 0.01
R^2_{adj}	0.9924
χ^2_{red}	0.0017
Dubinin–Radushkevich (D–R)	
$Q_{max,est}$ (mmol g^{-1})	0.76 ± 0.02
β (mmol^2 kJ^{-2})	0.0053 ± 0.0003
E (kJ mol^{-1})	9.778 ± 0.005
R^2_{adj}	0.9835
χ^2_{red}	0.0038

The box-type isotherm indicates a very favourable adsorption of RGY on STEA, as substantial amounts of RGY dye were adsorbed on STEA, even at low concentrations.[39] All models showed high values of R^2_{adj} and low values of χ^2_{red}. The Langmuir model better predicts the experimental adsorption capacity ($Q_{max,exp}$), compared to the R–P and D–R models. The Langmuir constant (b) is related to the affinity between adsorbate/adsorbent, and the high value of b obtained (221 L $mmol^{-1}$) suggests a very high affinity between the STEA surface sites and the RGY molecules. The β value of the R–P model was very close to 1, therefore, the R–P model simplifies to the Langmuir model. In addition, the β value (0.95 ± 0.01) suggests that the adsorption system exhibited a certain degree of homogeneity. However, this does not mean that the adsorption system under study followed the assumptions of the Langmuir model, since the STEA surface consists of different adsorption sites with different affinities for RGY. The D–R model provides information about the characteristic adsorption energy (E). Based on a series of adsorption data involving different organic molecules on various microporous adsorbents obtained from carbonaceous raw materials, the E value can be used to distinguish between physisorption ($E < 8$ kJ mol^{-1}) and chemisorption (8–16 kJ mol^{-1}).[40] For RGY adsorption on STEA, $E = 9.778 \pm 0.005$ kJ mol^{-1} and, therefore, it can be classified as chemisorption. The ITC study (see Section 10.3.3) will provide a more comprehensive analysis of the energy associated with the adsorption of RGY on STEA.

The $Q_{max,exp}$ value was 0.71 mmol g^{-1}. When this value is compared with the amount of nitrogen incorporated into the STEA structure (0.96 mmol g^{-1}), it suggests that part of the $R\text{–}N^+(CH_2CH_3)_3$ groups may not be involved in the adsorption of RGY, or the RGY dye may be adsorbed through electrostatic interaction with more than one quaternary ammonium group. The sulfonate and/or sulfate groups in the RGY structure probably interact with the positive charges of $R\text{–}N^+(CH_2CH_3)_3$ on the STEA surface. The $n_{R\text{-}N^+(CH_2CH_3)_3}/Q_{max,RGY}$ ratio suggests that 1.4 $R\text{–}N^+(CH_2CH_3)_3$ sites may be involved in the adsorption of each dye molecule. Another possibility is that steric effects prevent other RGY molecules from approaching vacant adsorption sites.

10.3.2.1.5 Regeneration and Reuse Studies. The HNO_3 solution has a pH lower than the pK_{a1} of RGY and, under this condition, the sulfonic and sulfate groups of RGY are protonated. Consequently, electrostatic interactions between RGY and $R\text{–}N^+(CH_2CH_3)_3$ are disrupted, leading to desorption of RGY. Despite this, RGY desorption was partial ($E_{des} = 56 \pm 1\%$), which suggests that other types of interactions, in addition to electrostatic interaction, may have occurred between STEA sites and RGY functional groups. It is difficult to state which type of binding/interaction remained unaltered and, consequently, prevented a higher desorption efficiency being obtained.[41]

A new adsorption study was conducted with STEA recovered after desorption to evaluate its reuse potential, even with partially regenerated adsorption sites. The $E_{re\text{-}ads}$ value of 45.7 ± 0.5% was expected considering that approximately half of the STEA active sites were available after desorption. The $E_{re\text{-}ads}$ value was similar to the E_{des} value, indicating that almost the same amount of RGY that was desorbed from STEA was subsequently re-adsorbed on STEA. It can be inferred that the RGY molecules adsorbed on STEA did not interfere with the re-adsorption. Therefore, RGY molecules did not act as chemically modified adsorption sites for new RGY molecules.

10.3.2.1.6 Competitive Adsorption Study. Sulfuric acid is often used to adjust the pH of textile dye effluent in the wastewater treatment plant and therefore SO_4^{2-} can be a competitor for STEA adsorption sites. The results of the competitive study performed to examine the effectiveness of STEA in removing RGY in the presence of SO_4^{2-} for RGY : SO_4^{2-} molar ratios of 3 : 1, 1 : 3, and 1 : 1, and 100% RGY were, respectively, 0.68 ± 0.03, 0.73 ± 0.02, 0.69 ± 0.02, and 0.71 ± 0.02 mmol g^{-1}. The presence of both components in solution did not result in any statistically significant difference in RGY removal, even at higher sulfate ratios. The bicomponent mixture had no significant effect on the adsorption capacity of RGY when compared to the monocomponent test. This suggests that the bicomponent mixtures did not exhibit any synergistic or antagonistic interactions.[42]

10.3.2.2 Fixed-bed Column Studies

The images of the fixed-bed column throughout the adsorption experiment are shown in Figure 10.2. The breakthrough curves obtained from three adsorption cycles of RGY on STEA are presented in Figure 10.3d. The parameters estimated by fitting the Thomas and Bohart–Adams models to experimental data are presented in Table 10.5. Both models showed determination coefficients (R^2) higher than 0.99 and low root mean squared errors (*RMSE*). The maximum adsorption capacities estimated ($Q_{max,est}$) by the models for the three cycles were close to those obtained experimentally ($Q_{max,exp}$), with differences of 3.0–6.5%. The excellent fits of the Thomas model to

Figure 10.2 Images showing the adsorption of RGY on STEA in a fixed-bed column, at times of (a) 0.0 h, (b) 0.5 h, (c) 3.0 h, (d) 6.0 h, and (e) 12.0 h.

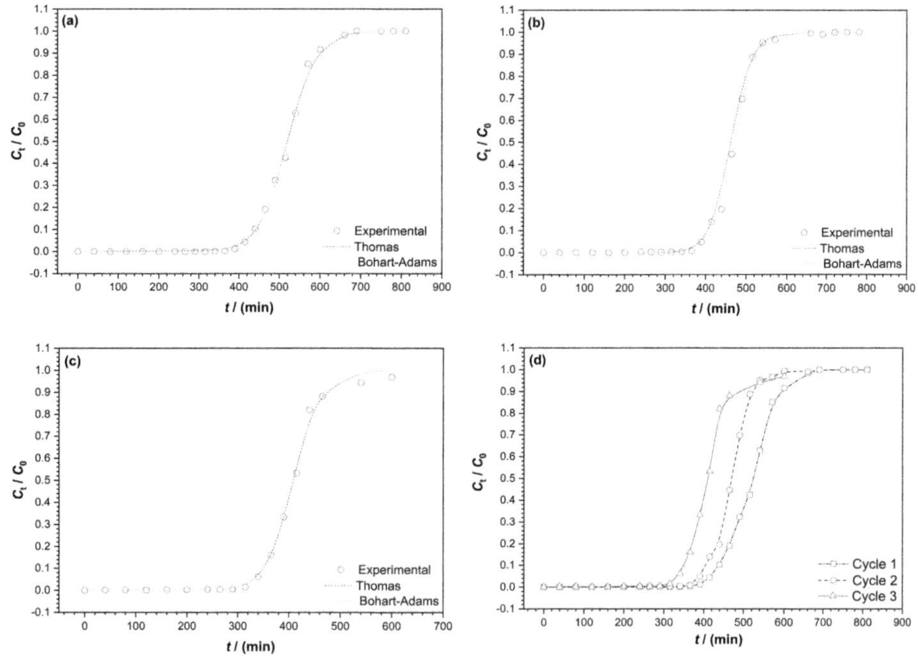

Figure 10.3 Breakthrough curves for RGY adsorption on STEA in a fixed-bed column: (a) first cycle, (b) second cycle, (c) third cycle, and (d) comparison between the breakthrough curves.

Table 10.5 Fixed-bed column parameters estimated by the Thomas and Bohart–Adams models for the three RGY adsorption cycles on STEA.

	Experimental		
Parameter	1st cycle	2nd cycle	3rd cycle
$Q_{max,exp}$ (mg g^{-1})	422.3	371.0	331.8
Thomas			
k_{Th} (mL min^{-1} mg^{-1})	0.16	0.22	0.23
$Q_{max,est}$ (mg g^{-1})	415.3	355.4	313.7
RMSE	0.0177	0.0323	0.0207
R^2	0.9982	0.9947	0.9969
Bohart–Adams			
k_{B-A} (mL min^{-1} mg^{-1})	0.16	0.23	0.23
$Q_{max,est}$ (mg g^{-1})	410.9	356.7	310.0
RMSE	0.0141	0.0141	0.0202
R^2	0.9989	0.9990	0.9970

the experimental data can be explained by the fact that the Thomas model assumes that the adsorption process follows the Langmuir isotherm.[43] As noted in Section 10.3.2.1.4, this model described very well the equilibrium data for adsorption of RGY on STEA. The Thomas model also assumes an adsorption process in a fixed-bed column without axial dispersion and obeying second-order reversible kinetics.[44] The Bohart–Adams model assumes that the adsorption rate is proportional to the adsorption sites still available and the concentration of adsorbate species in solution (quasi-chemical kinetics).[45,46] This model does not consider axial dispersion and is adequate to describe adsorption systems with high affinity equilibrium behavior (box-type isotherm), L2 classification according to Giles *et al.*[47]

The values of the Thomas (k_{Th}) and Bohart–Adams ($k_{B\text{-}A}$) adsorption rate constants were almost identical. These constants include the effects of mass transfer of RGY molecules from the fluid phase to the STEA surface (external diffusion), intraparticle diffusion (surface and pore diffusion), and the adsorption of RGY itself.[21,48] The Langmuir model approaches a rectangular isotherm when $bC_0 \gg 1$ and, under this condition, the Thomas and Bohart–Adams models provide similar predictions.[49] Considering the Langmuir constant value (221 L mmol^{-1}) and the concentration of RGY solution used in the fixed-bed column experiments (0.350 mmol L^{-1}), the estimated value for bC_0 was 77.4, therefore, the condition of $bC_0 \gg 1$ is satisfied. In this condition, the second-order reversible kinetic constant of the Thomas model approaches the quasi-chemical kinetic constant of the Bohart–Adams model. This explains the similarity between the k_{Th} and $k_{B\text{-}A}$ values, resulting in the Thomas and Bohart–Adams models providing nearly identical predictions for the breakthrough curves of the three adsorption cycles.[21] Therefore, it can be concluded that both models are able to predict the RGY breakthrough curves.

To evaluate the reuse potential of STEA, three complete adsorption/desorption cycles were conducted in a fixed-bed column. This study is important for industrial applications to reduce cost and minimize waste generation. Despite a slight decrease in adsorption capacity in subsequent cycles due to incomplete desorption of RGY, the benefits of reusing STEA outweigh this limitation. The breakthrough curves for the three cycles are shown in Figure 10.3d. The adsorption/desorption/re-adsorption efficiency results are presented in Table 10.6. STEA exhibited good performance for RGY adsorption in a fixed-bed column. The breakthrough curves for the three cycles showed similar behavior, with a slight increase in the slope of the curve and a decrease in the breakthrough time as the number of cycles increased (Figure 10.3a–c). The

Table 10.6 Results for the three RGY adsorption/desorption cycles in a fixed-bed column packed with STEA.

	Adsorption/desorption cycles		
Parameter	1st cycle	2nd cycle	3rd cycle
q_A (mg)	211.1	185.5	165.9
$q_{A,des}$ (mg)	124.3	132.6	113.4
Q_{max} (mg g^{-1})	422.3	371.0	331.8
$E_{re\text{-}ads}$ (%)	—	87.86	78.56
E_{des} (%)	58.87	71.50	68.39
MTZ (cm)	0.82	0.90	0.99

breakthrough curve for the first cycle indicated that the system operated for 415 min without the need for regeneration. During all this time, the effluent solution remained colourless. Following the breakthrough point, the curve showed a steep slope, reaching exhaustion after 660 min of operation. The length of the *MTZ* was determined to be 0.82 cm. The $Q_{max,exp}$ value was 422.3 mg g^{-1} (0.81 mmol g^{-1}). This value is higher than that determined using the batch isotherm (0.71 mmol g^{-1} or 369.2 mg g^{-1}). The increase in the $Q_{max,exp}$ value for the fixed-bed column adsorption compared to the batch adsorption can be attributed to the continuous increase in the concentration gradient at the MTZ interface traveling the column. In contrast, the batch concentration gradient decreases over time.[50] For the second cycle, the system operated for 390 min, reaching the exhaustion time at 660 min also. The Q_{max} value was 371.0 mg g^{-1}, with an *MTZ* length of 0.90 cm. For the third cycle, the operating and exhaustion times were 315 and 570 min, respectively, with a Q_{max} value of 331.8 mg g^{-1} and an *MTZ* length of 0.99 cm. Therefore, a 12.1% reduction in the Q_{max} value was observed from the first cycle to the second cycle, and an 10.6% reduction between the second and third cycles. However, STEA maintained a high adsorption capacity. The increase in the *MTZ* length from 0.82 to 0.99 cm was accompanied by a decrease in adsorption rate and bed utilization over the cycles. A decrease in breakthrough time and Q_{max} value was expected due to incomplete desorption of RGY from STEA. This decrease can be attributed to the progressive saturation of the active adsorption sites by RGY, as the amount of RGY molecules not desorbed from STEA gradually increased over the cycles.

When comparing adsorption data obtained in continuous and batch modes, in addition to the increase in Q_{max}, there was a significant increase in the E_{des} value, obtaining almost 60% efficiency in the first cycle. The increase in the E_{des} value can be attributed to the higher concentration gradient between the desorption solution and the RGY-saturated STEA surface in continuous mode compared to batch mode.

The results presented in Table 10.6 show that the amounts of RGY re-adsorbed in the second and third cycles were higher than the amounts of RGY desorbed in the previous cycles. A hypothesis is the presence of active adsorption sites that remained unused in the first and second cycles. These STEA sites were then occupied in the subsequent cycles, leading to an increase in the amount of RGY adsorbed on STEA. In addition, there was an increase in the E_{des} value in the second and third cycles, compared to the first cycle. These strong and varied interactions between STEA and RGY, which were previously formed, probably occurred at the beginning of the first cycle and were not disrupted by the desorption agent (HNO_3). In subsequent cycles, these sites are no longer available and thus the desorption and re-adsorption processes occurred more efficiently due to weaker interactions between adsorbate and adsorbent. As reported by Moghaddam *et al.*,[51] the reversibility of adsorption depends on the existence of strong or weak bonds/interactions between the adsorption sites and the adsorbate species. Despite the reduction in the Q_{max} value over the three cycles, it remains significantly high, and most of the

RGY molecules desorbed from STEA in the first cycle were re-adsorbed in subsequent cycles. The results of three consecutive adsorption/desorption cycles in a fixed-bed column confirmed the effectiveness of STEA for large-scale applications.

10.3.3 Enthalpy of RGY Adsorption on STEA

The plot of the integral adsorption enthalpy change ($\Delta_{ads}H_{int}$) as a function of the amount of RGY adsorbed (q_e) on the STEA surface is shown in Figure 10.4a. The plot of $\Delta_{ads}H_{int}$ as a function of STEA surface coverage is shown in Figure 10.4b.

The $\Delta_{ads}H$ values were negative over the entire range of q_e values studied, indicating that the adsorption of RGY on STEA was an exothermic process at a given concentration. The $\Delta_{ads}H$ values became more negative at lower q_e values, reaching approximately -7 ± 1 kJ mol^{-1} for a q_e of 17.3 mg g^{-1}. This indicates that the higher energy adsorption sites were occupied first, releasing more energy with the formation of the RGY–STEA site interaction. As the

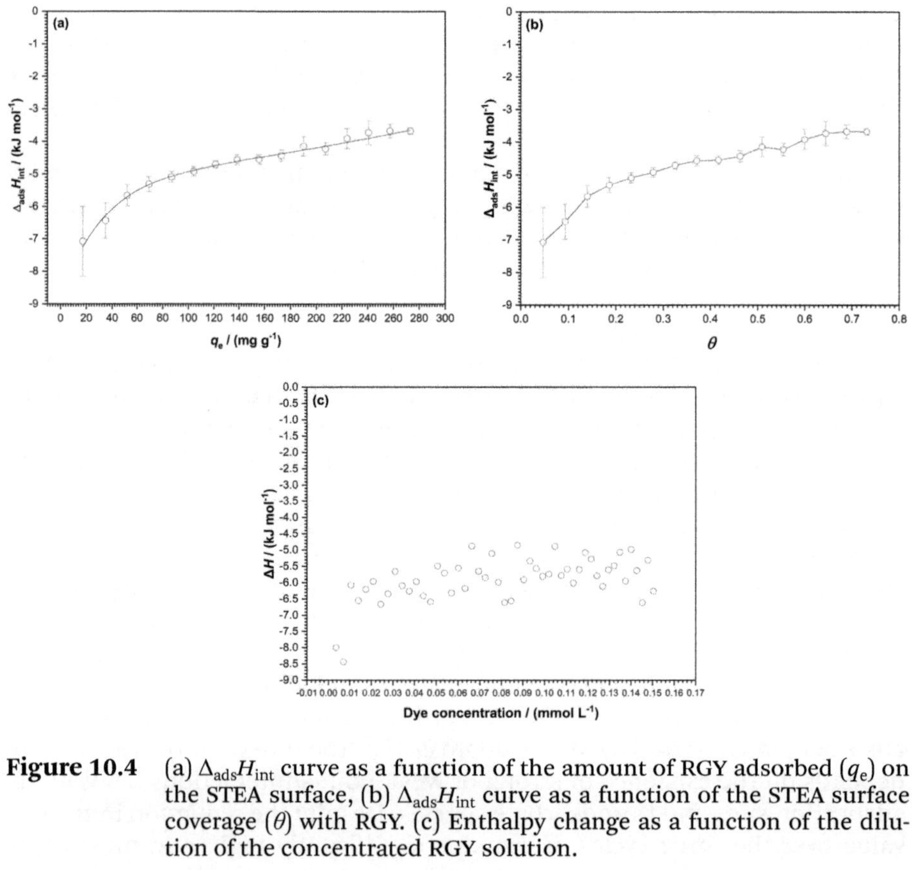

Figure 10.4 (a) $\Delta_{ads}H_{int}$ curve as a function of the amount of RGY adsorbed (q_e) on the STEA surface, (b) $\Delta_{ads}H_{int}$ curve as a function of the STEA surface coverage (θ) with RGY. (c) Enthalpy change as a function of the dilution of the concentrated RGY solution.

q_e value gradually increased, it was observed that the $\Delta_{ads}H$ values became less negative. The highest $\Delta_{ads}H$ value observed was −3.7 ± 0.1 kJ mol^{-1} for a q_e value of 273.1 mg g^{-1}. This indicates that the $\Delta_{ads}H$ values showed a minimal variation with increasing q_e, with a minimum $\Delta_{ads}H$ value of −7 ± 1 kJ mol^{-1} and a maximum $\Delta_{ads}H$ value of −3.7 ± 0.1 kJ mol^{-1} observed. This behavior probably indicates the presence of a wide range of surface coverages with adsorption sites of similar energy levels.

The $\Delta_{ads}H$ value can be interpreted as the sum of terms related to four independent subprocesses that occur during RGY adsorption on STEA. Eqn (10.4) shows the distinct enthalpic contributions to the $\Delta_{ads}H$ value.

$$\Delta_{ads}H = \Delta_{ads}H^{des} + \Delta_{ads}H^{RGY-STEA} + \Delta_{ads}H^{(RGY-RGY)sur} + \Delta_{ads}H^{(RGY-RGY)sol} \quad (10.4)$$

The first term, $\Delta_{ads}H^{des}$, represents the enthalpy change related to the desolvation subprocess of the RGY molecules and STEA sites. This subprocess includes: (1) the disruption of water–RGY and water–STEA interactions; (2) the formation of water–water interactions; and (3) the change in the structure of the solvation layer surrounding STEA sites caused by RGY adsorption.[52] The value of the $\Delta_{ads}H^{des}$ term can be positive or negative, depending on the energy balance involved in the formation and disruption of interactions during the adsorption process. The second term, $\Delta_{ads}H^{RGY-STEA}$, corresponds to the establishment of interactions between the RGY molecules and the STEA sites, and is always negative. The third and fourth terms result from the enthalpy change associated with the interactions formed and disrupted between the RGY molecules on the STEA surface ($\Delta_{ads}H^{(RGY-RGY)sur}$) and in the solution bulk ($\Delta_{ads}H^{(RGY-RGY)sol}$), respectively.[52] The $\Delta_{ads}H^{(RGY-RGY)sur}$ term is normally positive due to electrostatic repulsion between negatively charged RGY molecules on the STEA surface. The $\Delta_{ads}H^{(RGY-RGY)sol}$ term can be positive or negative, depending on the concentration of the RGY solution. As the concentration increases, there is a corresponding increase in electrostatic repulsion, leading to a positive contribution of the $\Delta_{ads}H^{(RGY-RGY)sol}$ term for the process. The $\Delta_{ads}H^{(RGY-RGY)sur}$ term increases its contribution to the process with increase in the degree of coverage of the STEA surface, as the enthalpy change becomes less negative during the adsorption process.

Figure 10.4a shows that in the initial stage of adsorption of RGY on STEA there is a sharp increase that overcomes the slope in the remaining curve. This behavior is probably attributed to a stronger interaction between RGY molecules and STEA sites at the beginning of the adsorption process, resulting in a more negative $\Delta_{ads}H$ value. This observation can be correlated with the results obtained in the STEA reuse test, where the desorption was incomplete (56%). This implies that a part of the RGY remained adsorbed on the material, probably due to the formation of stronger interactions during the process. The weakly adsorbed RGY molecules were desorbed and then re-adsorbed during subsequent adsorption/desorption cycles. Therefore, this indicates that the strongest interactions occurred specifically at low STEA surface coverages, while another type of interaction likely occurred at higher surface coverages. In addition, RGY dimerization in solution can also occur, which can influence the value of the $\Delta_{ads}H^{(RGY-RGY)sol}$ term. According to Walker and Weatherley,[53] dyes with ionic characteristics tend to self-associate or aggregate in aqueous solutions. At very low concentrations (10^{-4} to 10^{-6} mol L^{-1}), dyes can form dimers. Subsequently, the dimeric units also aggregate with each other, and so on. As the RGY concentration range employed in the ITC measurements was 10^{-5}–10^{-3} mol L^{-1}, there is a possibility of dimer formation. In this scenario, when RGY molecules dimerize inside the calorimetric cell due to the dilution of concentrated RGY solution in STEA suspension, the adsorption process requires the disruption of dimeric interactions before adsorption. This requires an energy consumption, which may or may not offset the contribution of the $\Delta_{ads}H^{(RGY-RGY)sur}$ term, depending on the algebraic sign of this term. Consequently, it can make the $\Delta_{ads}H$ values constant again.[52]

In order to examine the RGY dimerization potential, a specific investigation was conducted using the ITC technique in the concentration range of 10^{-4}–10^{-6} mol L^{-1} reported in the literature,[53] where dimerization is expected. Figure 10.4c shows the enthalpy change as the concentrated RGY solution is diluted, without the presence of STEA. No energy associated with the RGY aggregation process was observed. The dilution energies practically oscillated around a constant value, with the RGY concentration varying from 3.5×10^{-6} to 1.5×10^{-4} mol L^{-1}. In aggregation processes, in general, curves with sigmoid shapes are observed.[52]

Finally, by extrapolating the $\Delta_{ads}H_{int}$ *versus* C_e curve to $C_e \rightarrow 0$ (infinite dilution), the standard adsorption enthalpy change ($\Delta_{ads}H^{\circ}$) was determined to be -12 ± 5 kJ mol^{-1}. Under this condition, RGY–RGY interactions can be disregarded.

10.3.4 Adsorption Thermodynamics

The thermodynamic parameters estimated for the RGY adsorption process on STEA were $\Delta_{ads}H^{\circ} = -12 \pm 5$ kJ mol^{-1}, $\Delta_{ads}G^{\circ} = -31 \pm 3$ kJ mol^{-1}, and $T\Delta_{ads}S^{\circ} = 19 \pm 7$ kJ mol^{-1}. The adsorption of RGY on STEA was an exothermic process ($\Delta_{ads}H^{\circ} < 0$). The adsorption reaction equation is usually represented as a reaction between a solute entity (S) with a surface site (A) to form an adsorbed entity ($S_{(aq)} + A_{(s)} \rightleftharpoons AS_{(s)}$). The $\Delta_{ads}G^{\circ}$ value was negative, indicating a strong displacement of the reaction equilibrium towards the products. This implies that RGY is preferentially adsorbed on the STEA surface.[54] The positive algebraic sign of the entropic term ($T\Delta_{ads}S^{\circ}$) and its greater magnitude than the enthalpic term ($\Delta_{ads}H^{\circ}$) show that the process was entropically directed. The increase in entropy is mainly attributed to the desolvation of STEA sites and RGY molecules during adsorption. This process involves the release (migration) of water molecules with lower degrees of freedom from the solvation layer of RGY molecules and STEA sites to the solution bulk, resulting in an increase in the degrees of freedom of the system.[20]

Compared to the dye adsorption studies reported in the literature, the system under study showed a relatively high entropy increase. Usually, high values of $T\Delta_{ads}S^{\circ}$ are associated with chemisorption, which involves a significant displacement of water molecules from the solvation layer of adsorption sites and adsorbate molecules to the solution bulk.[20,52] As $\Delta_{ads}S^{\circ}$ is positive and $\Delta_{ads}G^{\circ}$ is negative, it can be stated that the adsorption process is spontaneous, following the principle of minimum energy and maximum entropy, with $\Delta_{ads}H^{\circ} \neq 0$ due to deviations from ideality.[54]

To obtain more information about the interactions between STEA and RGY, FTIR analyses were performed on STEA samples containing RGY adsorbed at distinct steps of the adsorption isotherm: beginning, slope, and plateau. The corresponding spectra are shown in Figure 10.5. When analysing the FTIR spectra for STEA–RGY samples at the slope (q_e = 0.36 mmol g^{-1}) and plateau (q_e = 0.71 mmol g^{-1}) of the adsorption isotherm, two bands at 1259–1230 cm^{-1} were observed (Figure 10.5b). These bands correspond to the vibrations of the sulfonate and sulfate groups, due to the higher concentration of RGY on STEA in these samples. In contrast, at the beginning of the isotherm (q_e = 0.06 mmol g^{-1}), only the band at 1250 cm^{-1} can be observed. The absence of the other IR band at low surface coverages may suggest the presence of a chemical bond between RGY and STEA. RGY can form covalent bonds with hydroxyl groups in AGU of cellulosic fibres due to its electrophilic vinyl sulfone group. This property is one of the main reasons why RGY is widely used in the textile industry, as it guarantees strong fixation of the colour in the fabric. For the reaction with the fibre to occur, the sulfate group of the sulfate ethyl sulfone is eliminated, resulting in the formation of the vinyl sulfone compound. Consequently, the absence of this band may be attributed to the loss of this group. However, at higher surface coverages, the appearance of the IR band related to the sulfate group indicates that chemosorption is not the dominant process when higher amounts of RGY are present on the STEA surface. Instead, RGY physisorption appears to occur.

The band at 1316 cm^{-1} in the spectra of STEA and STEA samples loaded with RGY at the beginning of the isotherm, associated with vibration of the C–N bond, appears at 1300 cm^{-1} for STEA samples loaded with RGY at the slope and plateau of the isotherm. This band shift suggests the occurrence of electrostatic interactions between the quaternary ammonium group of STEA and the negatively charged RGY dye molecule. Therefore, at low surface coverage (8.5% of Q_{max}), the FTIR spectra (Figure 10.5) suggest the occurrence of chemosorption, while at high surface coverages (50–100% of Q_{max}), it indicates the occurrence of electrostatic interaction (physisorption). Therefore, the results suggest the presence of mixed interactions between STEA and RGY. The occurrence of chemosorption is also suggested by the incomplete desorption of RGY molecules from STEA and the high value of *b* (Langmuir affinity constant), which indicates a high affinity of the STEA sites for RGY molecules, in addition to the high value of the entropic term ($T\Delta_{ads}S^{\circ}$). The occurrence of physisorption, in addition to chemisorption, is supported by the following observations: (1) the low magnitude of $\Delta_{ads}H$ values obtained

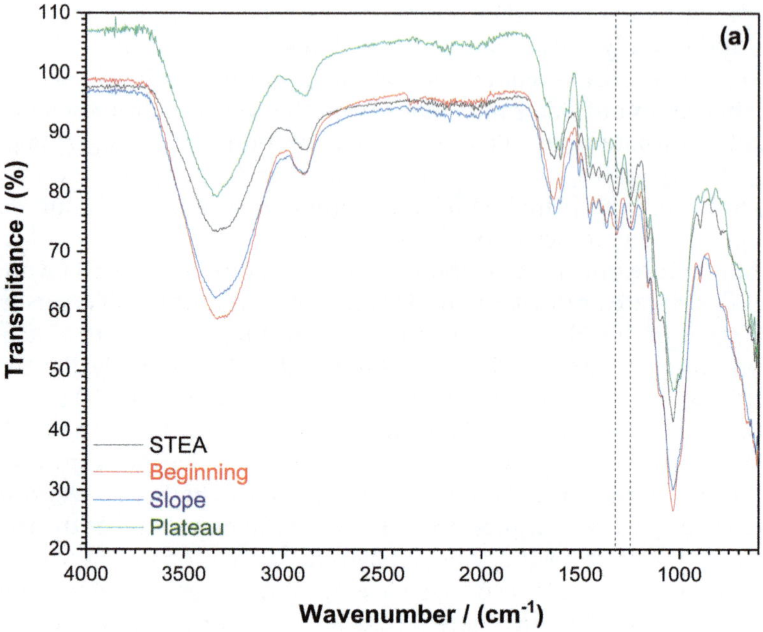

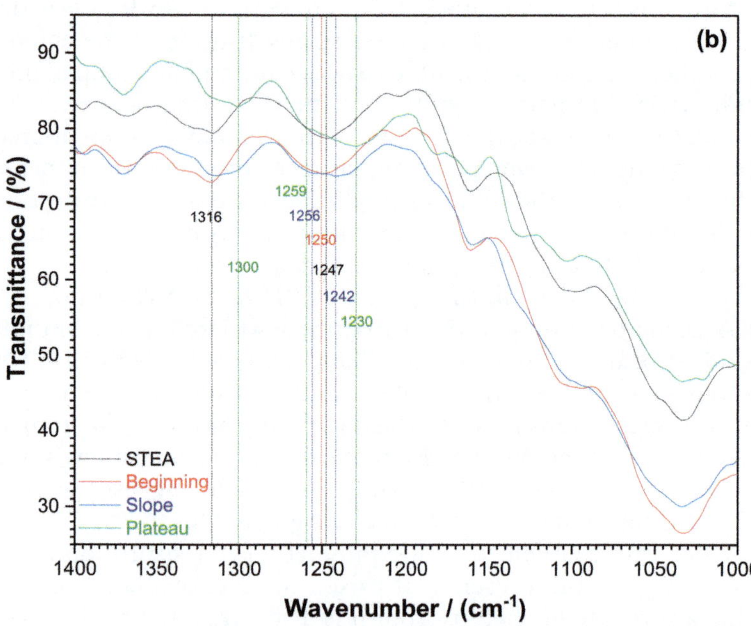

Figure 10.5 FTIR spectra of STEA samples loaded with RGY at the beginning (C_i = 0.01 mmol L^{-1}, C_e = 0.0004 mmol L^{-1}, q_e = 0.06 mmol g^{-1}), slope (C_i = 0.08 mmol L^{-1}, C_e = 0.006 mmol L^{-1}, q_e = 0.36 mmol g^{-1}), and plateau (C_i = 0.54 mmol L^{-1}, C_e = 0.39 mmol L^{-1}, q_e = 0.71 mmol g^{-1}) of the adsorption isotherm at (a) 4000–600 cm^{-1} and (b) 1500–1000 cm^{-1}.

in ITC measurements (Figure 10.4b); (2) at least 50% of the RGY molecules can be desorbed, and the STEA sites can be reused for up to three cycles; and (3) the adsorption study as a function of solution pH revealed a significant variation in the adsorption capacity of RGY as the pH of the solution changed.

10.4 Conclusions

A quaternary ammonium anion exchanger (STEA) was successfully prepared from sugarcane bagasse by a two-step one-pot reaction, and it was effective in removing the remazol golden yellow (RGY) RNL dye from spiked aqueous solutions in batch and continuous mode. The intraparticle diffusion model revealed that the adsorption occurred in three steps, with the adsorption of RGY on STEA initially controlled by film diffusion and later by intraparticle diffusion, which was corroborated by the Boyd model. However, the HSDM model and Biot number showed that the rate of adsorption of RGY on STEA was more influenced by intraparticle diffusion than by film diffusion. Competitive studies of RGY and sulfate showed that sulfate did not interfere with the adsorption of RGY on STEA. The high adsorption capacity of RGY over the three cycles of adsorption/desorption in a fixed-bed column demonstrated the potential of STEA for large-scale applications. The isothermal titration calorimetric measurements suggested a compensatory effect between the enthalpic contributions related to the desolvation of STEA sites and RGY molecules, the formation of STEA–RGY interactions, and the formation of attractive or repulsive interactions between RGY molecules on the STEA surface and in the solution bulk, or even indicated the existence of adsorption sites with similar energy. The combination of macroscopic and microscopic experiments suggested mixed interactions involving physisorption and chemisorption between STEA and RGY. STEA proved to be an efficient alternative for removing RGY from water, reinforcing its potential to contribute to the circular bioeconomy in the sugar energy industry.

Acknowledgements

The authors are grateful to the Federal University of Ouro Preto (UFOP, grant number 23109.014957/2022-44) and National Council for Scientific and Technological Development (CNPq, grant numbers 307445/2019-4 and 307844/2022-6) for funding this research.

References

1. A. Abel, *Colour Design*, Elsevier, 2012, pp. 557–587.
2. R. Kant, *Nat. Sci.*, 2012, **04**, 22.
3. V. Pande, S. C. Pandey, T. Joshi, D. Sati, S. Gangola, S. Kumar and M. Samant, *Smart Bioremediation Technologies*, Elsevier, 2019, pp. 255–287.
4. A. Gičević, L. Hindija and A. Karačić, *Toxicity of Azo Dyes in Pharmaceutical Industry*, Springer, Cham, 2020.

5. U. Ewuzie, O. D. Saliu, K. Dulta, S. Ogunniyi, A. O. Bajeh, K. O. Iwuozor and J. O. Ighalo, *J. Water Process Eng.*, 2022, **50**, 103273.
6. F. Amalina, A. S. A. Razak, S. Krishnan, A. W. Zularisam and M. Nasrullah, *Cleaner Waste Syst.*, 2022, **3**, 100051.
7. B.-Y. Gao, X. Xu, Y. Wang, Q.-Y. Yue and X.-M. Xu, *J. Hazard. Mater.*, 2009, **165**, 461.
8. Solvay, Solvay Epicerol® Earns Roundtable on Sustainable Biomaterials Certification, https://www.solvay.com/en/press-release/solvayepicerol-earns-roundtable-sustainable-biomaterials-certification.
9. M. H. Hasan, T. M. I. Mahlia, M. Mofijur, I. M. R. Fattah, F. Handayani, H. C. Ong and A. S. Silitonga, *Energies*, 2021, **14**, 3732.
10. H. Zabed, J. N. Sahu, A. Suely, A. N. Boyce and G. Faruq, *Renewable Sustainable Energy Rev.*, 2017, **71**, 475.
11. A. K. Rappe, C. J. Casewit, K. S. Colwell, W. A. Goddard and W. M. Skiff, *J. Am. Chem. Soc.*, 1992, **114**, 10024.
12. F. S. Teodoro, L. C. Soares, J. G. Filgueiras, E. R. de Azevedo, Á. J. Patiño-Agudelo, O. F. H. Adarme, L. H. M. da Silva and L. V. A. Gurgel, *Environ. Sci. Pollut. Res.*, 2022, **29**, 26425.
13. W. Schoniger, *Mikrochim. Acta*, 1956, **44**, 869.
14. W. J. Weber and J. C. Morris, *J. Sanit. Eng. Div.*, 1963, **89**, 31.
15. G. E. Boyd, A. W. Adamson and L. S. Myers, *J. Am. Chem. Soc.*, 1947, **69**, 2836.
16. E. Worch, *Adsorption Technology in Water Treatment*, De Gruyter, 2012.
17. I. Langmuir, *J. Am. Chem. Soc.*, 1916, **38**, 1145.
18. O. Redlich and D. L. Peterson, *J. Phys. Chem.*, 1959, **63**, 1024.
19. M. M. Dubinin and L. V. Radushkevich, *Proc. USSR Acad. Sci.*, 1947, **55**, 327.
20. R. A. Fideles, G. M. D. Ferreira, F. S. Teodoro, O. F. H. Adarme, L. H. M. da Silva, L. F. Gil and L. V. A. Gurgel, *J. Colloid Interface Sci.*, 2018, **515**, 172.
21. A. L. P. Xavier, O. F. H. Adarme, L. M. Furtado, G. M. D. Ferreira, L. H. M. da Silva, L. F. Gil and L. V. A. Gurgel, *J. Colloid Interface Sci.*, 2018, **516**, 431.
22. H. C. Thomas, *J. Am. Chem. Soc.*, 1944, **66**, 1664.
23. G. S. Bohart and E. Q. Adams, *J. Am. Chem. Soc.*, 1920, **42**, 523.
24. A. R. Pereira, L. C. Soares, F. S. Teodoro, M. M. C. Elias, G. M. D. Ferreira, R. M. L. Savedra, M. F. Siqueira, C. Martineau-Corcos, L. H. M. da Silva, D. Prim and L. V. A. Gurgel, *J. Colloid Interface Sci.*, 2020, **576**, 158.
25. J. Rouquerol, D. Avnir, C. W. Fairbridge, D. H. Everett, J. M. Haynes, N. Pernicone, J. D. F. Ramsay, K. S. W. Sing and K. K. Unger, *Pure Appl. Chem.*, 1994, **66**, 1739.
26. T. P. F. Teixeira, S. I. Pereira, S. F. Aquino and A. Dias, *Environ. Eng. Sci.*, 2012, **29**, 685.
27. W. Smyth, S. McClean, E. O'Kane, I. Banat and G. McMullan, *J. Chromatogr. A*, 1999, **854**, 259.
28. A. G. Tamirat, A. Sendek and S. Libsu, *Int. J. Adv. Res.*, 2014, **2**, 234.
29. A. N. Módenes, A. P. de Oliveira, F. R. Espinoza-Quiñones, D. E. G. Trigueros, A. D. Kroumov, C. E. Borba, C. L. Hinterholz and R. Bergamasco, *Environ. Technol.*, 2017, **38**, 1474.

30. K. Bhattacharyya and A. Sharma, *Dyes Pigm.*, 2005, **65**, 51.
31. M. Doğan, Y. Özdemir and M. Alkan, *Dyes Pigm.*, 2007, **75**, 701.
32. D. M. Ruthven, *Principles of Adsorption and Adsorption Processes*, John Wiley & Sons, 1984.
33. M. Schmal, *Chemical Reaction Engineering: Essentials, Exercises and Examples*, CRC Press, 2014.
34. L. Yu and Y. Luo, *J. Environ. Chem. Eng.*, 2014, **2**, 220.
35. A. Bonilla-Petriciolet, D. I. Mendoza-castillo and H. E. Reynel-ávila, *Adsorption Processes for Water Treatment and Purication*, Springer International Publishing, 1st edn, 2017.
36. N. Cherkasov, *J. Mol. Liq.*, 2020, **301**, 112378.
37. G. Atun, S. Ortaboy, E. Tüzün and E. T. Acar, *J. Dispersion Sci. Technol.*, 2023, **44**, 144.
38. A. Muthukkumaran and K. Aravamudan, *J. Environ. Manage.*, 2017, **204**, 424.
39. A. M. M. Vargas, A. L. Cazetta, A. C. Martins, J. C. G. Moraes, E. E. Garcia, G. F. Gauze, W. F. Costa and V. C. Almeida, *Chem. Eng. J.*, 2012, **181–182**, 243.
40. Q. Hu and Z. Zhang, *J. Mol. Liq.*, 2019, **277**, 646.
41. I. D. Mall, V. C. Srivastava and N. K. Agarwal, *Dyes Pigm.*, 2006, **69**, 210.
42. V. Hernández-Montoya, M. A. Pérez-Cruz, D. I. Mendoza-Castillo, M. R. Moreno-Virgen and A. Bonilla-Petriciolet, *J. Environ. Manage.*, 2013, **116**, 213.
43. S. Kizito, S. Wu, S. M. Wandera, L. Guo and R. Dong, *Sci. Total Environ.*, 2016, **563–564**, 1095.
44. V. C. Srivastava, B. Prasad, I. M. Mishra, I. D. Mall and M. M. Swamy, *Ind. Eng. Chem. Res.*, 2008, **47**, 1603.
45. K. H. Chu, *J. Hazard. Mater.*, 2010, **177**, 1006.
46. H. C. Kang and W. H. Weinberg, *Surf. Sci.*, 1994, **299–300**, 755.
47. H. C. Giles, D. Smith and A. Huitson, *J. Colloid Interface Sci.*, 1974, **47**, 755.
48. R. A. Ogwada and D. L. Sparks, *Soil Sci. Soc. Am. J.*, 1986, **50**, 1162.
49. O. M. Couto, I. Matos, I. M. da Fonseca, P. A. Arroyo, E. A. da Silva and M. A. S. D. de Barros, *Can. J. Chem. Eng.*, 2015, **93**, 68.
50. V. Gupta, S. K. Srivastava and R. Tyagi, *Water Res.*, 2000, **34**, 1543.
51. S. S. Moghaddam, M. R. A. Moghaddam and M. Arami, *Water Sci. Technol.*, 2010, **61**, 1673.
52. G. M. D. Ferreira, G. M. D. Ferreira, M. C. Hespanhol, J. de Paula Rezende, A. C. dos Santos Pires, L. V. A. Gurgel and L. H. M. da Silva, *Colloids Surf., A*, 2017, **529**, 531.
53. G. M. Walker and L. R. Weatherley, *Chem. Eng. J.*, 2001, **83**, 201.
54. T. Shahwan, *Colloid Interface Sci. Commun.*, 2021, **41**, 100369.

CHAPTER 11

Iron Oxide Nanomaterials for Water Purification

ATEEQ RAHMAN[a*], LIKIUS SHIPWIISHO DANIEL[b] AND VEIKKO UAHENGO[a]

[a]Department of Physics, Chemistry and Material Sciences, School of Science, Faculty of Agriculture, Natural Science and Engineering; [b]Science and Technology Division, Multidisciplinary Research Services, Centre for Research Service, University of Namibia, Private Bag 4331, Mandumedumefayo, Windhoek, Namibia
*E-mail: arahman@unam.na, ateeqr786@yahoo.com

11.1 Introduction

Water resources are the most fundamental component of our planet, and water quality is declining continuously because of human activities. Around the world about 780 million people do not have access to clean drinking water.[1] This is more prominent in developing countries where wastewater management is usually non-existent, and swift action is required. Existing wastewater management and processes are improving their ability to provide suitable clean water to meet human needs and other environmental requirements. Clean water is vital for all living organisms to thrive. The toxic contaminants present in water have increased globally due to rapid industrialization and large increases in population.[2] In agriculture, the demand for clean water has intensified. The consumption of fresh and clean water with above-average pollutants is about 70%, 22%, and 8% in industry, domestic sectors, and other forms of consumption, respectively.[3] Municipal liquid

Chemistry in the Environment Series No. 12
Novel Materials and Water Purification: Towards a Sustainable Future
Edited by Grigorios L. Kyriakopoulos and Miltiadis G. Zamparas

Published by the Royal Society of Chemistry, www.rsc.org

Table 11.1 Different sources of waste water. Data from ref. 90

Agricultural	Domestic	Industrial
Poultry waste	Washing laundry	Canneries
Slage liquor	Toilet	Milk dairies
Piggery waste	Septic tank	Breweries
Vegetable waste	Shower	Beverages
Dairy farming waste	Kitchen	Abattoir
Fire water	Academic institutions	Fertilizer
Sediment runoff	Hotels/restaurants	Pesticide
Slaughtering waste	Hostels	Pharmaceutical
Nutrient runoff	Hospitals	Paper and pulp
Farming	Small business activities	Tanneries

waste is waste water that contains contaminants such as organic materials, microorganisms, inorganic soluble compounds, and toxic heavy metals. These contaminants change the chemical, biological, and physical characteristics of clean water.[4] Waste water is divided into two groups, municipal and industrial waste water, municipal wastewater sources normally contain faeces and urine, industrial and agricultural wastewater sources, as well as domestic, contain organic and inorganic chemicals.[5] Table 11.1 shows different sources of waste water. Waste water usually contains microorganisms, including viruses, bacteria, and protozoa, and toxic heavy metals, for example, heavy metals, radionuclides, and trace elements. Waste water is one of the major causes of waterborne diseases, including deadly diseases such as typhoid, malaria, and cholera. Polluted water caused the death of more than 1.5 million children under the age of five in 2004.[6] Over the decades wastewater treatment has become necessary due to the toxic effects of pathogens and the hazards of wastewater pollution for humans, agriculture, and animals. Wastewater treatment at the individual level and government level is vital to conserve and protect the environment from pollution. This chapter emphasizes the development of iron oxide NPs by the biosynthesis method, the characterization of iron oxide NPs, wastewater treatment of iron oxide NPs, and its limitations, which will form a new route for wastewater treatment in water-scarce countries. The future outlook for developments of various methods of iron oxide NP synthesis and their application in diverse areas of water treatment is discussed.

11.2 Wastewater Treatment Conventional Methods

The treatment of waste water usually involves physical, chemical, and biological processes for water purification from various contaminants.[7] Methods such as boiling, filtration, distillation, and chlorination are common procedures to treat waste water. However, the treatment of industrial effluent water based on the pollutants in the waste water generally uses adsorption, membrane separation, activated sludge methods, *etc.* Amongst several chemicals found in industrial waste water, those from dyes, pharmaceuticals, the paper industry, cotton, and slaughter houses are considered as important pollutants. Dyes can remain in the environment for an extended period of time, because of their high thermal and photostability. Several dyes and their breakdown products are carcinogenic, mutagenic, and toxic to life. Dyes increase the biochemical oxygen demand of sewage, thus affecting aquatic life. Dyes are not readily degradable from water by wastewater treatment systems using conventional techniques, such as adsorption, ultrafiltration, chemical, and electrochemical methods which are cumbersome and require more energy, longer process times, and are costly. These limitations of conventional methods are summarized in Table 11.2. Although many organic

Table 11.2 Limitations of wastewater treatment by conventional methods.

Method	Limitations
Flotation	High initial capital cost, pH-dependent selectivity and high maintenance and operational cost
Chemical precipitation	Excess of reagent is necessary. Product can be a low-quality mixture that can limit use. Not very selective method
Flocculation and coagulation	Needs alkaline additives to achieve optimum pH and is complex and inefficient
Biological treatment	Microorganisms are environmentally sensitive. Intermediates can destroy microbial cells. Method is expensive and time consuming
Ion exchange	High initial capital cost and high maintenance and operational cost
Incineration/thermal oxidation	High initial capital cost and high running cost
Electrochemistry	High initial cost of equipment, high maintenance cost
Membrane filtration	High investment costs for small- and medium-sized industries. High energy needs. Design of membrane filtration systems can vary considerably
Evaporation	Expensive for high volumes of waste water. High investment costs for small- and medium-sized industries
Liquid–liquid (solvent) extraction	High investment (equipment)
Advanced oxidation processes (AOP)	Economically non-viable for medium- and small-sized industries, technical constraints, and low throughput
Nanofiltration	Photolysis economically non-viable for medium- and small-sized industries, technical constraints and low throughput
Ultrafiltration	This method requires pretreatment and high water-cleaning energy. Salt and univalent ions have limited retention
Microfiltration	This method requires high energy and fails to remove dissolved inorganics
Carbon filter	Fails to remove metals, fluoride, sodium, nitrates, volatile organics, colors and so on. Method requires regular cleaning and membrane fouling can also occur
	This method fails to remove fluoride, nitrates, sodium, metals and so on. This is susceptible to mold and clogging occurs with undissolved solids

pollutants originate in industrial effluents, methylene blue is a pollutant that disturbs the aesthetics of the environment.[8,9]

11.3 Nanotechnology

The basic properties of nanoparticles (NPs) and nanomaterials (NMs) have been classified on the basis of size, excellent shape, and attractive physicochemical properties.[10] Owing to their interesting characteristics, NPs/NMs are promising contenders for applications in advanced technologies. NPs have attracted the attention of the scientific community due to their exceptional properties, such as optoelectrical, magnetic, and thermal stability. A variety of NPs of diverse metal oxides, such as CuO, ZnO, NiO, SnO_2, and Fe_2O_3, have been successfully biosynthesized using plant extracts.[11–14] The recent expansion in nanoscience and nanotechnology indicate the opportunities that exist for the development of improved water resources and arrangements using these methods.[15] NPs are in great demand due to their applications in various branches of science, engineering, and medicine, including gene therapy, drug delivery, antibacterial agents, catalysts, biosensors, clean-up, and recycling technologies.[16] In recent years, scientists and industries have pursued sustainable possibilities to produce NPs and NMs from renewable waste resources to meet global demand. The application of inorganic waste, for example, mine waste, electronic (e-waste), and industrial waste, as well as lignocellulosic waste, for the production and recovery of NPs provides a future vision for environmental research and nanotechnology.[17] Hence, recycling these secondary-metal and carbon-rich wastes provides a promising strategic plan to promote a sustainable circular economy and environmental remediation.[18,19] Therefore, the effectiveness of organic and inorganic wastes to synthesize NPs can summon a new scientific era. NP production methods primarily involve physical, chemical, and hybrid means. The extremely effective, integrated, and multifunctional progress enabled by nanoscience and nanotechnology is expected to offer high rates of presentations and reasonable and affordable wastewater treatment solutions compared to existing large infrastructure.[20] The application of metal and metal oxide nanoparticles is expanding in several areas of science and engineering, such as agriculture, environment, engineering, electronics, pharmaceuticals, biosensors, catalysis, medicine, textile, food industries, environmental remediation, diagnostics, and medicine.[21] Nanoparticles of Fe_2O_3 (iron oxide), TiO_2 (titanium dioxide), and CuO (copper oxide) have received much attention owing to their simple method of preparation, non-toxicity, and low cost of manufacture.[22] Among them, iron oxide nanoparticles are regarded as the most important owing to their exceptional chemical and physical properties, which increases their opportunities for application.[23] However, metal contamination in water is of extreme concern because of the toxic effect on plants, animals, and humans. Thus nanoparticles are employed for wastewater treatment in the transformation of contaminants into less toxic species and iron oxide can be exploited as a photocatalyst and as an effective immobilization and adsorptive technology, which facilitates the removal of contaminants.[23]

11.4 General Method for the Preparation of Iron Oxide NPs by Biosynthesis Route

Although the biosynthesis of iron oxide NPs using plant extracts is a harmless and simple method, some difficulties are still encountered, such as preparation of morphologically well-defined iron oxide NPs in good yield.[24] The extensive use of plants can be attributed to the presence of many phytochemicals, which act as reducing and stabilizing agents during the synthesis reaction.[25] The presence of flavonoids,[26] amines, terpenoids,[26] carbohydrates,[27] and phenols[28] boosts the growth of iron oxide NPs (Tables 11.3 and 11.4). These biomolecules are expected to provide diverse chemical functional groups for the stabilization, oxidation, and reduction of metal oxide precursors.[24,29] These extracts can reduce and diffuse the metal ions, resulting in homogeneous particle size distribution. Flavonoids dissolve secondary metabolites in plants and have scavenging capability for oxygen molecules, increasing the electron-donating ability. Phenolic compounds, such as caffeic acid, protocatechuic acid, and gallic acid, comprise hydroxyl groups that can form bonds with metallic ions (Table 11.4). For example, the reduction potential of polyphenolic compounds is adequate to reduce metallic iron to create zerovalent iron, yielding magnetic iron oxide NPs (Table 11.5).[29]

11.4.1 Biosynthesis of Iron Oxide Nanoparticles

Iron oxide NPs possess a remarkable sorption potential owing to their high porosity, specific surface area, and strong magnetic response, which result in an extraordinary sorption capacity.[30,31] In recent years, iron oxide NPs have gained attention because of their high Brunauer–Emmett–Teller (BET) surface area, super magnetic properties, and high pore volume.[32]

Iron(III) oxide (haematite) one of the main three iron oxides and frequently is known as rust, is reddish brown in color, and has paramagnetic properties.[35] Among these, iron oxide NPs have been found to be fairly interesting due to their various applications in diverse fields.[21] Several plant extracts to date, such as green tea,[33] pomegranate,[34] banana,[35] *Skimmia*,[36] daisy[37,39] and soya bean[39] have been employed for the assembly of Fe_2O_3NPs of discrete morphologies (Tables 11.3, 11.5 and 11.6). The green-synthesized iron oxide NPs have been employed in diverse fields of applications.[34,38–40] Several methods, such as microwave irradiation, photochemical, and chemical reduction, are being used for the physical and chemical synthesis of iron oxide NPs, which require hazardous and expensive chemicals.[41] The nanoparticles formed by physical or chemical synthesis also show high amounts of agglomeration, which increases the volume to surface area ratio.[42] However, in the case of green-synthesized iron oxide

Table 11.3 Different plant extracts procured from various plant parts for the synthesis of iron oxide nanoparticles (α-Fe_2O_3). Data from ref. 90

Pant source	Iron precursor	Conc. in leaves	Extraction	Stirring	Size (nm)	Morphology
Cassia auriculata (flowers)	$FeCl_3$	0.1 M	50 g in 100 ml	1 h	15–35 mm	Spherical
Ocimum tenuiflorum (extract)	$FeCl_3{\cdot}6H_2O$ $FeCl_2{\cdot}4H_2O$	1 M + 1 M	10 g in 100 ml	2 h	9–21 mm	Spherical
Corymbia citriodora (hook)	$FeSO_4{\cdot}7H_2O$	0.1 M	20 g in 500 ml		–80 mm	Rhombic shapes
Leucas aspera Jatropha podagrica	$FeCl_3$	1 M		30 min	30	Spherical
Hibiscus rosa-sinensis flower	$FeCl_3$	1 M	5 g in 100 ml		51	Spherical
Beta vulgaris and *Punica granatum*	$FeCl_3$	1 M	10 g in 200 ml	30 min	20–21	Semi-spherical
Ficus carica leaf extract	$FeCl_3{\cdot}6H_2O$	0.01 M	10 g in 200 ml	1 h	43–57	—
Simarouba glauca leaf extract	$FeCl_3$	0.1 M	20 g in 200 ml	15 min	70–122	Sphere
Bauhinia tomentosa leaf extract	$FeCl_3$	0.01 M			70	—
Aegle marmelos extracts	$Fe(NO_3)_3$		5 g in 100 ml	60 min	—	—
Tamarix aphylla extract	$FeCl_3$		30 g in 250 ml			Spherical
Wedelia urticifolia	$FeCl_3$	0.1 M	10 g in 100 ml		15–20	Rod
Carica papaya leaf extract	$FeCl_3{\cdot}6H_2O$	0.1 M	20 g in 1 L	30 min	21.59	—
Withania coagulans and $FeCl_2{\cdot}4H_2$	$FeCl_3{\cdot}6H_2O$	1 : 2 M	30 g in 250 ml		15–20	Nanorods
Punica granatum seed extract	$FeCl_3$	1 M	20 g in 250 ml		25–55	—
Ruellia tuberosa	$FeSO_4$	1 M	5 g in 100 ml		20–80	Nanorod
Peltophorum pterocarpum	$FeSO_4{\cdot}7H_2O$	0.1 M		60 min	—	Nanorod
Psidium guajava	$FeCl_3$	0.5 M	10 g in 100 ml	2 h	—	Rod
Moringa oleifera						
Rhizophora mucronata Lam	$FeCl_3$	0.1 M	5 g		65	Sphere
Amaranthus spinosus	$FeCl_3$	0.5 M	40 ml	90 min	—	Spherical
Cocos nucifera	$Fe(NO_3)_2{\cdot}6H_2O$	0.1 M	50 ml	1 h–30 min	20–30	Crystallite
Syzygium aromaticum (CE)	$FeCl_3$	2 : 1 M	Calcination	1 h	16, 25 nm	Small-sized BNPs
Manglicolous fungi	$FeCl_3$	1 : 1 M	$FeCl_2 : FeCl_3$ 1 : 1	72 h		

Table 11.4 Antioxidant activity of different plant extracts for iron oxide NPs. Data from ref. 90 [a]

Plant name	Size (nm)	Material	[Conc.]	DPPH	TAC/ORAC
Achyranthes aspera	20–40	Fe_2O_3	70 μg ml^{-1}	84.98%	NA
Ficus carica	43–57	γ-Fe_2O_3	12.118 mg ml^{-1}	1 g to 5.14 mg 1 g–7.36 mg	Troloks
Stevia rebaudiana Bertoni	25	SPIONS	0–200 mg ml^{-1}	IC_{50} = 65 mg ml^{-1}	NA
Phoenix dactylifera L.	12.61	Fe_2O_3	0.01 M	IC_{50} = 26.4	180 mg g^{-1} mg ml^{-1}
Grewia optiva	15–60	Fe_2O_3	15 μM	32.3%	NA
Prunus persica	13–70	Fe_2O_3	15 μM	43.2%	NA
Terminalia bellirica	21.32	Fe–O	NA	70%	28 μg ml^{-1}
Moringa oleifera	45	Fe–O Na		70%	49 μg ml^{-1}

[a]SPIONS: superparamagnetic iron oxide nanoparticles; DPPH: 2,2-diphenyl-1-picrylhydrazyl; TAC: total antioxidant capacity; ORAC: oxygen radical absorbance capacity.

Table 11.5 Synthesis of iron oxide NPs from extracts of different plant parts. Data from ref. 90

Sample no.	Plant source	Shape	Size (nm)	Material
1.	*Alchornea laxiflora*	Crystallite size	14.7	Chitosan–iron@silver (CS–Fe@Ag)
2.	*Aloe vera*, mulberry leaves, *Gundelia* tournefortii	Crystallite	35.5	M–CuO@Ag/PAN/ZIF-67 ACT
3.	*Nephrolepis auriculata*	Spheroidal	40–70	NA-Fe-NPs
4.	Green tea	Spherical	60–120	GT-nZVI
5.	Green tea	Core–shell	50–80	nFe Cr(VI)
6.	Tangerine peel	200–300	T-Fe_3O_4 Pb(II)	
7.	Eucalyptus	30–50	EL-FeNP As(V)	

Table 11.6 Different plant extracts for the synthesis of iron oxide for dye degradation. Data from ref. 90

Sample no.	Plant name	Dye	Degradation	Fe(NPs) (mg)	Dye conc. (ppm)	Time (min)
1.	Pomegranate seeds (*Punica grananum*)	Reactive blue	95.08	15	20	56
2.	*Wedelia urticifolia*	Methylene blue	97.72	40	10	30
3.	*Peltophorum pterocarpum*	Methylene blue	92	15	20	27
4.	*Psidium guajava Moringa oleifera*	Methylene blue	Nil	Nil	Nil	60
5.	*Kappaphcus alvarezil*	Textile waste	75	50	10	180

NPs, expensive and hazardous chemicals are not used, and agglomeration of the nanoparticles is not observed.[43] Various materials, such as bacteria, plants, and fungi,[44] are employed for the green synthesis of nanoparticles. Iron oxide nanoparticles are characterized by a magnetic core, which simplifies their recovery, a shell that provides stability and prevents oxidation, and the surface of the ligand is attached to a particular pollutant. It possesses various advantages such as low-cost chemicals and magnets, implementation in continuous manner, and target specificity.[45]

11.4.2 Removal of Contaminants from Wastewater Iron Oxide NPs

Numerous researchers, including Prasad *et al.*,[45] have reported the application of green-synthesized nanoparticles in the removal of heavy metals. Currently, green-synthesized iron nanoparticles (iron oxide NPs) have received increased attention due to their ability to eliminate heavy metals from the fluid phase.[46] One of the essential roles of iron-based NPs is in the adsorptive removal of contaminants from waste water. At field-scale and laboratory-scale, iron oxide NMs play significant roles in contaminant removal.[47] In the treatment of waste water, NMs of iron oxides and of metals such as Fe, Al, Ni, and Zn use zerovalent nanoparticles. Nanozerovalent Al has a very high reduction ability and is thermodynamically active in the presence of water, utilizing oxides/hydroxides above the surface and complete electron transfer from the surface of the pollutants.[48] Conversely, Fe and Ni nanoparticles have a low ability towards reduction. And, in comparison to many redox-labile pollutants, nanoparticles of Zn or Fe have excellent potential as reducing agents. Despite the low ability towards reduction, in wastewater treatment that incorporates properties like adsorption, oxidation, cost effectiveness, and precipitation, Fe nanoparticles are beneficial compared to Zn. In fact, among all the zerovalent nanoparticles of metals, Fe nanoparticles are the most widely studied and used in the treatment of waste water to remove contaminants.[49] Parekh[50] reported the removal of heavy metals from polluted effluent, such as mercury. Magnetic nanoparticles were prepared by functionalization with dithiocarbamate and the adsorption of mercury at variable pH and different ionic strengths in the presence of salinity and competing ions was investigated. Asrarian *et al.*[51] reported the adsorption of aluminium from industrial waste water by magnetic NPs (MNPs). The methodology was rapid, easy, inexpensive, effective, and safe for the removal of aluminium from contaminated water. The adsorbent preparation was simple and the duration needed was low. Variation in pH and high electrolyte concentration (NaCl) did not have a marked effect on the aluminium remediation efficiency.[51] In a different method, reported by Xu *et al.*,[52] poly-allylamine-hydrochloride (PAAH) stabilized MNPs were used to remove bacterial pathogens, such as *Escherichia*, *Acinetobacter*, *Pseudomonas*, and *Bacillus*, with over 99.5% of bacteria removed when the bacterial count was observed to be less 10^5 colony-forming units (CFU) mL^{-1}. It was observed that the MNPs exhibited different adhesion effects on cells based on the bacterial species, which was greater for *Acinetobacter* and *Pseudomonas*, and this was attributed to the external cell structure and ion exchange capacity, but not the zeta potential, of the bacterial cell surface. However, from this data it is evident that MNPs showed high efficiency for removal of bacterial from drinking water reservoirs in Sheffield and Leeds, and it was observed that the total bacteria residual counts were as low as 78 CFU mL^{-1}, which is identical to WHO drinking water standard.[51,52] Benkhaya *et al.*[53] reported the use of iron oxide NPs for the removal of azo dyes in the textile industry. The different plant extracts used in the synthesis of iron oxide nanoparticles for dye degradation are presented in Table 11.6. The dyes possess compounds such as benzene rings, naphthalenes, aromatic heterocycles, or enolizable aliphatic groups, which are bonded to the azo group and are characteristic of the dye color. Dyes are classified as carcinogenic and affect aquatic life exposed to dye-containing effluents.[53] However, dye-rich effluents can be treated using nanosized iron oxide by the photocatalytic degradation method. Rather *et al.*[54] reported the synthesis of magnetite iron oxide from the extract of *Wedelia urticifolia* was more than 99% effective in the degradation of methylene blue. This is attributed to the total number of active sites of oxygen and hydroxyl radicals.[54] For the extract of *Peltophorum pterocarpum* the highly mesoporous structure showed a surface area of 66.44 m^2 g^{-1}, which is 11 times higher than that of commercial hematite iron oxide. The higher surface area contributed to the faster degradation of methylene blue, with 90% of the dye being degraded in 220 min.[55] *Moringa oleifera* extract was used for the synthesis of iron oxide NPs that showed degradation of dye within 30 min.[56] The degradation efficiency was ascribed to the high surface properties observed during the synthesis. It is interesting to note that the degradation mechanism of iron oxide NPs shows that electrons were transferred from the valence band to the conduction band, initiating the formation of holes in the valence band during activation of the iron oxide photocatalyst. The activity of the iron oxide NP photocatalysts was determined using the interface transfer mechanism, which includes electrons, holes, and their deactivation by recombination.[57,58] When the iron oxide NPs are at the photoexcited surface, a redox reaction occurs. The oxygen present scavenges electrons to produce superoxide, its protonated form, the hydroperoxyl radical, and hydrogen peroxide, and recombination between the electrons and holes occurs. The activation of the entire catalytic system and the photonic stimulation of the iron oxide NP catalyst is the first step as a result of this chain of events.[59] The mechanism is ascribed to the oxygen in the aqueous solution and chemisorbed oxygen on the catalyst surface being combined, with intermediate superoxide radicals produced due to the photogenerated electrons of the iron oxide NPs. Hydroperoxyl radicals are produced when the superoxide radicals combine with dissociated water (H^+), to produce H_2O_2. On the surface of the catalysts the hydroxyl radicals (OH^-) are formed due to the photogenerated holes (H^+) with adsorbed water (OH_2). Organic molecules are formed from the bond cleavage of the contaminant molecules, which are not harmful, and this is attributed to the reaction of the reactive radicals and intermediate species with the contaminant molecules. The redox reaction serves as the basis for the entire methodology. The iron oxide NPs possess photocatalytic properties with high-intensity white light-emitting diodes for the degradation of water. The photocatalytic adsorption effect of iron oxide NPs under visible light conditions greatly degrades the water contaminants compared to the absence of light and was monitored by gas chromatography-mass spectrometry (GC-MS), high-pressure liquid chromatography (HLPC), and UV–visible absorption for the pollutant degradation.[59] Owing to its extremely high reductive ability, nanozerovalent Al is thermodynamically unstable in the presence of water, and favors the formation of oxides/hydroxides on the surface, hindering (completely) the transfer of electrons from the metal surface to contaminants.[60] In comparison to Fe, Ni possesses a less negative standard reduction potential, which indicates a lower reducing ability. With a moderate standard reduction potential, nanozerovalent Fe or Zn possess good potentials to act as reducing agents in comparison to a number of redox-labile contaminants. In spite of a weaker reduction ability, Fe possesses many advantages over Zn for applications in water pollution treatment, with excellent properties for adsorption, precipitation, and oxidation (in the presence of dissolved oxygen), and low cost. Hence, zerovalent iron (nZVI) nanoparticles have been broadly studied. Due to the extremely small size and thus large specific surface area, nZVI exhibits good adsorption properties and shows strong reducing ability.[61] These characteristics contribute enormously to its excellent activity in the removal of contaminants. Under anaerobic conditions, FeO is oxidized by H_2O or H^+ and generates Fe^{2+} and H_2, both of which are potential reducing agents for contaminants. In the oxidation/reduction reaction between nZVI and contaminants, Fe^{2+} will be oxidized to Fe^{3+}, which forms $Fe(OH)_3$ with increase in pH. $Fe(OH)_3$ is an effective flocculant and enables the removal of contaminants, for example, Cr(VI).[62] However, nZVI can reduce and oxidize a diverse range of organic compounds in the company of dissolved oxygen and subsequently ZVI transfers two electrons to O_2 to yield H_2O_2. The resulting H_2O_2 can be reduced to H_2O by nZVI. Furthermore, the combination of H_2O_2 and Fe^{2+} (known as the Fenton reaction) can produce hydroxyl radicals ($HO^•$), which possess strong oxidizing ability towards a wide variety of organic compounds. With the effects of adsorption, reduction, precipitation, and oxidation (in the presence of dissolved oxygen), nZVI has been effectively applied in the removal of a large number of contaminants, including halogenated organic compounds,[63] nitroaromatic compounds,[64] organic dyes,[65] phenols,[66] heavy metals,[67] inorganic anions, such as phosphates[68] and nitrates,[69] metalloids,[70] and radio elements.[70] In recent years, nZVI has also been applied in soil remediation[71] and pilot-scale and full-scale applications at real-water-contaminated field spots have already been accomplished.[72] In spite of its many advantages, nZVI also has particular disadvantages, such as aggregation, oxidation, and separation. However, various modification methods have been proposed to improve the performance of nZVI in water and wastewater treatment. Common modifications normally include doping with other metals, surface coating, conjugation with supports, encapsulation in a matrix, and emulsification.[73] Impregnation with other metals could hypothetically enhance the reactivity of nZVI.[74] Surface coating and conjugation with supports can avoid aggregation and improves the dispersibility of nZVI.[75] Additionally, both conjugation with supports and encapsulation in a matrix enable the separation of nZVI from the degraded system.[76] Furthermore, the emulsification of nZVI is designed to solve the delivery problem of nZVI in dense non-aqueous phase liquid.[77]

11.5 Characterization of Iron Oxide NPs

Iron oxide NPs have been characterized by scanning electron microscopy (SEM), transmission electron microscopy (TEM), X-ray diffraction (XRD), Fourier-transform infrared (FTIR) spectroscopy, differential/thermogravimetric analysis (DGA/TGA), N_2 adsorption–desorption, and determination of Brunauer–Emmett–Teller (BET) surface area.

11.5.1 Characterization of Fe NPs by FTIR Spectroscopy

Iron oxide NPs from clove extract were characterized by FTIR spectroscopy. The peaks obtained at 3272, 1733, 1573, 1322, 1125 and 1000 cm^{-1} were ascribed to the presence of several biomolecules in clove extract, which were coated on the surface of the iron oxide NPs. A broad peak was observed at 3272 cm^{-1} corresponding to the O–H stretching vibration of phenolic moieties in clove extract.[78] Additionally, peaks were observed at 1733, 1322, and 1573 cm^{-1} illustrating the C–O stretching frequency of the carbonyl group and aromatic rings. The presence of O–H stretching vibration was designated peaks at 1000 cm^{-1}. These peaks indicated the presence of β-caryophyllene and α-humulene as the chief constituents of clove extract capped on the surface of iron oxide NPs. The FTIR spectrum of calcined (c) iron oxide NPs of clove extract indicated that the absence of peaks was due to the elimination of biomolecules during the calcination at high temperatures. A band at 1125 cm^{-1} is due to the OH^- bending of water molecules. An additional peak was observed due to the Fe–O bond at 517 cm^{-1}.[79]

11.5.2 Characterization of Fe NPs by XRD

The XRD of c-Fe_2O_3 NPs displayed major peaks at 2θ values of 24.2°, 33.4°, 35.8°, 49.2°, 54.3°, and 64°. These peaks relate to the (111), (121), (210), (212), (103), and (331) crystal planes. The strongest peak, at 35.8°, is attributed due to the α or γ phases of haematite. The relative intensities of the supplementary peaks clearly indicated that α-iron oxide (ref. 01-089-664) is the main phase in the as-prepared iron oxide NPs. The inset shows a broad peak at 20–30°, which is characteristic of amorphous materials. The absence of diffraction lines in the XRD pattern of iron oxide NPs again signifies their amorphous nature, which was reported by Ebrahiminezhad *et al.*[80] Finally, the amorphous nature of synthesized iron oxide NPs can be attributed to the rich coating of biological moieties in the clove extract on the surface of the iron oxide NPs.

11.5.3 Characterization of Iron Oxide NPs by Electron Microscopy

Morphological structures of green-synthesized α-iron oxide NPs were characterized by field-emission (FE) SEM (Figure 11.1a and b).[81] From the FESEM analysis the irregular and uneven shape of iron oxide NPs can be seen, which are common characteristics of plant extract-synthesized nanostructures.[81] Karpagavinayagam *et al.*[82] reported the synthesis of iron oxide NPs with *Avicennia marina* flower extract. The elemental composition of iron oxide NPs were determined by energy-dispersive X-ray spectroscopy (Figure 11.2b) and showed similar morphologies, which indicated the presence of C and O atoms, along with Fe. As reported by Karpagavinayagam *et al.*,[91] Figure 11.2b shows the presence of peaks due to C and O and this is attributed to the presence of organic molecules from clove extract on the NPs. From the TEM images in Figure 11.3(a–c) it is evident that the iron oxide and calcined iron oxide NPs were found to be irregular in shape and 16.5 nm (Figure 11.3d) in size.[91] These irregular shapes are very common when plant extract is used for the synthesis of iron oxide NPs, which is due to aggregation. Justus *et al.*[83] reported that the TEM images of c-Fe_2O_3 can be clearly attributed to the

Figure 11.1 (a and b) From the SEM images of the iron oxide NPs it can be seen that they are in nanoscale range and show uniform distribution with flake-like structures. This can be attributed to the occurrence of diverse groups of phytochemicals, which enables reduction and stabilization of the nanoparticles during the synthesis procedure. Reproduced from ref. 81 with permission from Elsevier, Copyright 2022.

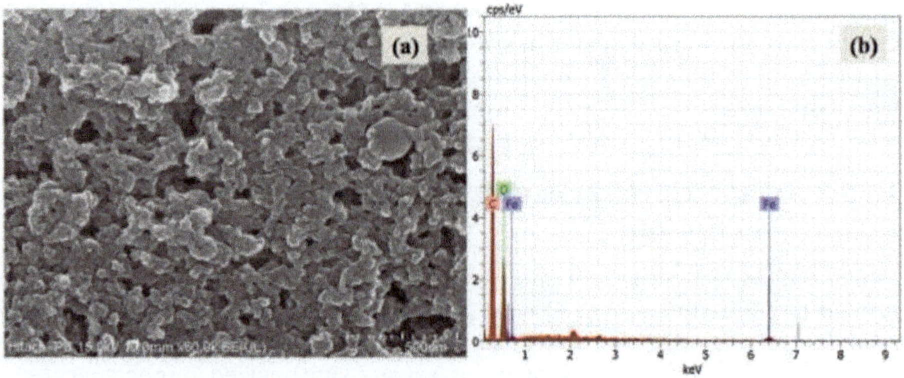

Figure 11.2 (a and b) SEM images of iron oxide NPs. Reproduced from ref. 91 with permission from Elsevier, Copyright 2021.

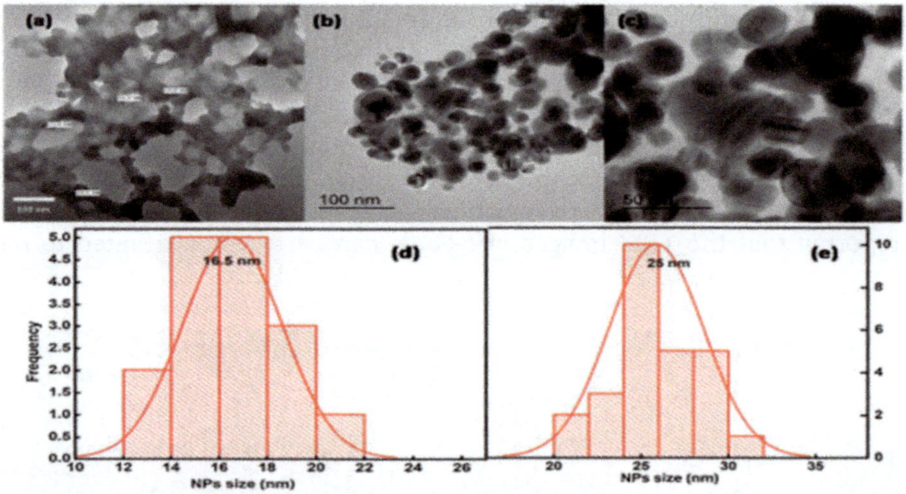

Figure 11.3 (a–e) TEM images of iron oxide NPs. Reproduced from ref. 91 with permission from Elsevier, Copyright 2021.

absence of plant extract coating.[83] The capping of plant extract plays a crucial role in stabilization of small-sized NPs, which is attributed to the size of c-Fe_2O_3, which was found to be 25 nm (Figure 11.3e).[91]

11.5.4 Characterization of Fe NPs by Thermogravimetric Analysis

The thermal stability of synthesized iron oxide and c-iron oxide NPs was evaluated by TGA. The TGA thermogram (Figure 11.4a) displayed a three-step weight loss.[91] The first weight loss of 15.3% was observed at around 110 °C,

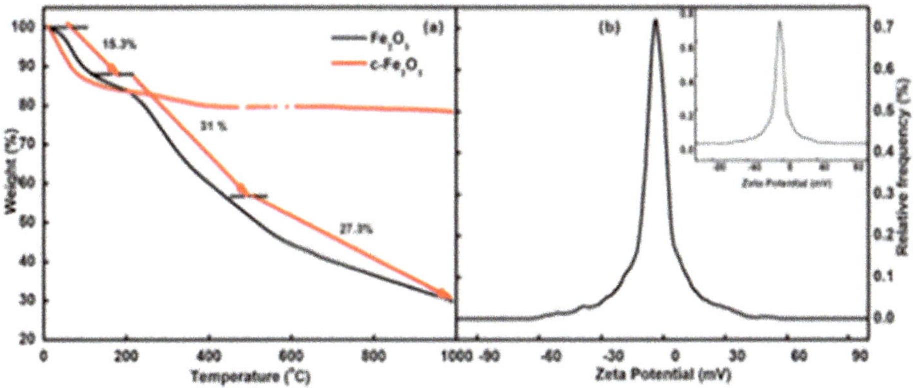

Figure 11.4 (a) TGA thermograms and (b) zeta potential of iron oxide and c-Fe_2O_3. Reproduced from ref. 91 with permission from Elsevier, Copyright 2021.

which may be due to the loss of adsorbed water and other volatile components on the surface of the iron oxide. The second weight loss, of 31% at around 290 °C, was ascribed due to the decomposition of biomolecules present in the capping agent on the surface of the iron oxide NPs. The third weight loss, of 26.3%, might be due to the adsorbed oxygen species.[84] For the thermogram of calcined iron oxide a gradual loss of mass upon heating up to 820 °C showed the conversion of undesirable phases, *i.e.*, FeO to Fe_2O_3,[85] and this was also indicated in the XRD analysis. The colloidal stability and charge on the surface of the iron oxide NPs was characterized by zeta potential measurements, with the charge on the surface of iron oxide and c-Fe_2O_3 observed to be −10 and −15 mV, respectively, demonstrating the moderate stability and negatively charged surface of the NPs in aqueous solution (Figure 11.4b). The existence of a negative charge on the NP surface provided stability to the system by avoiding the aggregation of NPs *via* electrostatic repulsions.[86]

11.5.5 Characterization of Iron Oxide NPs by Nitrogen Adsorption/Desorption Isotherm

The nitrogen adsorption/desorption isotherm of Fe_2O_3 and c-Fe_2O_3 showed that the synthesized iron oxide NPs displayed a type IV isotherm with a H3 hysteresis loop, illustrating the presence of a mesoporous metal oxide structure.[87] The pore size distribution of iron oxide NPs was evaluated using a Barrett–Joyner–Halenda (BJH) plot from the desorption edge of the isotherm. The average pore diameter and surface area for iron oxide NPs and c-Fe_2O_3 was found to be 3.5 nm (mesoporous), 82.8 $m^2\ g^{-1}$ and 2.5 nm (mesoporous), 59.1 $m^2\ g^{-1}$, respectively. The higher surface area of the biosynthesized iron oxide NPs was attributed to the stabilization and capping of iron oxide nuclei due to the growth from plant extract, resulting in higher surface area. The

Table 11.7 BET surface area of iron oxide NPs from plant extract. Data from ref. 89

Synthesis method	BET surface area (m^2/g)
Green synthesis-*Aloe vera* leaf extract	18.8
Green synthesis-*Camellia sinensis* leaf extract	22.5
Green synthesis-*Cynometra ramiflora* fruit extract	107.97
Green synthesis-*Spondias dulcis* leaf extract	197.84
Chemical synthesis	79.2
Chemical synthesis	90.71
Chemical synthesis	105.1

high surface area can be compared to the doping effect of carbonaceous substances from the polyphenolics present in clove extract.[88]

11.5.6 BET Surface Area of Iron Oxide NPs

BET surface area analysis[89] was performed in order to gain knowledge regarding the surface properties, such as surface area, pore size, and volume of the green-synthesized iron oxide NPs. The surface area was 12.83 mg^2 g, with a pore size of 35.67 nm, which is in the range 2–50 nm, implying that the sorbent is mesoporous as reported by Vinayagam *et al.*[89] The green synthesis of iron oxide NPs using plantain leaf extract and the surface area of the synthesized iron oxide NPs was 11.31 mg^2 g. The structure of the iron oxide NPs was assigned as mesoporous based on the BET surface area and pore volume results. This data indicates that the iron oxide NPs can be studied as a photocatalyst for the degradation of various pollutants from waste water.[89] Table 11.7 presents the BET surface areas for *Aloe vera*, *Camellia sinensis* leaf extract, *Cynometra ramiflora* fruit extract, and *Spondias dulcis* leaf extract, *etc.*[89]

11.6 Conclusions and Future Outlook

Nanotechnology is playing major role in our everyday lives by enhancing the effectiveness of everyday objects. It delivers a clean environment by providing safer air and water, and clean water and renewable energy for a sustainable future. Nanotechnology has gained wide interest with more research and development being a top priority for institutions, industries, and other organizations to solve the problems of access to fresh water at a global level. Nanotechnology has become established as an advanced scientific field where extensive research is carried out to implement the technology. For more than 15 years, inorganic metal NPs have been at the forefront of research, and they have proved their capability in various fields, with the synthesis and application of versatile metallic NPs, such as Pt, Ag, Cu, Au, Fe, and Pd. The iron oxide NPs exhibit distinct qualities that are superior to their bulk metals, such as optoelectronic and dimensional properties; increased surface-to-volume ratio; reactivity; efficiency; and functional alterations, which can affect their potential as multifunctional technical tools for broad-spectrum applications. This chapter has discussed the synthesis of iron oxide NPs from various plant extracts and their characterization by XRD, FTIR, SEM, TEM, TGA/DTA, N_2 adsorption and desorption studies, and BET surface area, and finally the application of iron oxide NPs for waste water. Advancement of iron oxide NP applications will provide a productive area of research, and further research is required to investigate the synthesis, characterization and advanced applications. The application of plant extract as a green solution for the synthesis of iron oxide NPs is simple, with different sizes, shapes, and properties exhibited, and the rate of reaction can be regulated by studying different reaction parameters, such as time, temperature, pH, substrate and concentration. The future of using plant extract is pivotal for the synthesis of iron oxide nanoparticles, and is strongly recommended for the requirements of safe, biocompatible agents. In general, the scientific community must address different tasks to develop new and efficient nanotechnology for society, which will benefit the development of green routes for economic and ecofriendly processes to reproduce and recycle iron oxide NPs, and, finally, to avoid the use of hazardous materials for the synthesis of iron oxide NPs in the area of wastewater treatment.

Conflict of Interest

There are no conflicts of interests among the authors.

Acknowledgements

The authors acknowledge the Department of Physics, Chemistry and Material Science, University of Namibia, for supporting this project.

References

1. H. Y. A. E. Ali, *Chiral Pollutants: Distribution, Toxicity and Analysis by Chromatography and Capillary Electrophoresis*, 2004.
2. C. Fant, R. Srinivasan, B. Boehlert, L. Rennels, S. C. Chapra, K. M. Strzepek, J. Corona, A. Allen and J. Martinich, Climate Change Impacts on US Water Quality Using Two Models: HAWQS and US Basins, *Water*, 2017, **9**, 118–121.
3. R. R. W. Herschy, Water quality for drinking: WHO guidelines, *Encycl. Earth Sci. Ser.*, 2012, pp. 876–883
4. M. M. Abou El-Nour, A. Eftaiha, A. Al-Warthan and R. A. A. Ammar, Synthesis and applications of silver nanoparticles, *Arabian J. Chem.*, 2010, **3**(3), 135–140.
5. L. K. Adams, D. Y. Lyon and P. J. J. Alvarez, Comparative eco-toxicity of nanoscale TiO_2, SiO_2, and ZnO water suspensions, *Water Res.*, 2006, **40**(19), 3527–3532.

6. S. A. AL-Thabaiti, F. M. Al-Nowaiser, A. Y. Obaid, A. O. Al-Youbi and Z. Khan, Formation and characterization of surfactant stabilized silver nanoparticles: A kinetic study, *Colloids Surf., B*, 2008, **67**(2), 230–237.
7. N. P. Cheremisinoff, *Handbook of Water and Wastewater Treatment Technologies*, Butterworth-Heinemann, 2002.
8. A. K. Mittal, Y. Chisti and U. C. Banerjee, Synthesis of metallic nanoparticles using plant extracts, *Biotechnol. Adv.*, 2013, **31**, 346–356.
9. J. V. Lakshmi, R. Sharath, M. N. Chandraprabha, E. Neelufar, A. Hazra and M. Patra, Synthesis, characterization and evaluation of antimicrobial activity of zinc oxide nanoparticles, *J. Biochem. Technol.*, 2012, **3**, 151–154.
10. D. Vollath, Nanomaterials an introduction to synthesis, properties and application, *Environ. Eng. Manage. J.*, 2008, **7**(6), 865–870.
11. M. Nasrollahzadeh, S. M. Sajadi, A. R. Vartooni and S. M. Hussin, Green synthesis of CuO NPs using aqueous extract of Thymus vulgaris leaves and their catalytic performance for N-arylation of indoles and amines, *J. Colloid Interface Sci.*, 2016, **466**, 113–119.
12. D. Suresh, P. C. Nethravathi, R. H. Udayabhanu, H. Nagabhushanaa and S. C. Sharma, Green synthesis of multifunctional zinc oxide (ZnO) NPs using Cassia fistula plant extract and their photodegradative, antioxidant and antibacterial activities, *Mater. Sci. Semicond. Process.*, 2015, **31**, 446–454.
13. A. A. Ezhilarasi, J. J. Vijaya, K. Kaviyarasu, M. Maaza, A. A. Yeshamariame and L. J. Kennedy, Green synthesis of NiO NPs using Moringa oleifera extract and their biomedical applications: cytotoxicity effect of NPs against HT-29 cancer cells, *J. Photochem. Photobiol., B*, 2016, **164**, 352–360.
14. A. Diallo, E. Manikandan, V. Rajendran and M. Maaza, Physical and enhanced photocatalytic properties of green synthesized SnO_2 NPs via Aspalathus linearis, *J. Alloys Compd.*, 2016, **681**, 561–570.
15. G. Vilardi, M. Stoller, N. Verdone and L. Di Palma, Production of nano zero valent iron particles by means of a spinning disk reactor, *Chem. Eng. Trans.*, 2017, **57**, 751–756.
16. X. Li, H. Xu, Z. S. Chen and G. Chen, Biosynthesis of NPs by Microorganisms and Their Applications, *J. Nanomater.*, 2011, 270974.
17. A. Zamani, A. P. Marjani and Z. Mousavi, Agricultural waste biomass-assisted nanostructures: synthesis and application, *Green Process. Synth.*, 2019, **8**(1), 421–429.
18. T. Dutta, K. H. Kim, A. Deep, J. E. Szulejko, K. Vellingiri, S. Kumar, E. E. Kwon and S. T. Yun, Recovery of nanomaterials from battery and electronic wastes: a new paradigm of environmental waste management, *Renewable Sustainable Energy Rev.*, 2018, **82**, 3694–3704.
19. K. K. Brar, D. L. Martinez, M. Naghdi, S. K. Brar, B. S. Chadha, P. Singh and R. Y. Surampalli, Industrial practices in sustainability. *Sustainability: Fundamentals and Applications*, 2020, pp. 313–338.
20. X. Qu, J. Brame, Q. Li and P. J. J. Alvarez, Nanotechnology for a safe and sustainable water supply: enabling integrated water treatment and reuse, *Acc. Chem. Res.*, 2013, **46**(3), 834–843.

21. J. Braunschweig, J. Bosch and R. U. Meckenstock, Iron oxide NPs in geomicrobiology from biogeochemistry to bioremediation, *N. Biotechnol.*, 2013, **30**, 793–802.
22. H. S. Rai, M. S. Bhattacharya, J. Singh, T. K. Bansal, P. Vats and U. C. Banerjee, Removal of dyes from the effluent of textile and dyestuff manufacturing industry: a review of emerging techniques with reference to biological treatment, *Crit. Rev. Environ. Sci. Technol.*, 2005, **35**(3), 219–238.
23. I. Bavasso, G. Vilardi, M. Stoller, A. Chianese and L. Di Palma., Perspectives in nanotechnology based innovative applications for the environment, *Chem. Eng. Trans.*, 2016, **47**, 55–60.
24. A. U. Mirza, *et al.*, Biogenic synthesis of iron oxide nanoparticles using Agrewia optiva and Prunus persica phyto species: Characterization, antibacterial and antioxidant activity, *J. Photochem. Photobiol., B*, 2018, **185**, 262–274.
25. S. Vasantharaj, Biosynthesis of iron oxide nanoparticles using leaf extract of Ruellia tuberosa: Antimicrobial properties and their applications in photocatalytic degradation, *J. Photochem. Photobiol., B*, 2019, **192**, 74–82.
26. S. M. Sajadi, M. Nasrollahzadeh and M. Maham, Aqueous extract from seeds of Silybum marianum L. as a green material for preparation of the Cu/Fe_3O_4 nanoparticles: A magnetically recoverable and reusable catalyst for the reduction of nitroarenes, *J. Colloid Interface Sci.*, 2016, **469**, 93–98.
27. D. A. Demirezen, Y. S. Yıldız and D. D. Yılmaz, Amoxicillin degradation using green synthesized iron oxide nanoparticles: Kinetics and mechanism analysis, *Environ. Nanotechnol., Monit. Manage.*, 2019, **11**, 100219.
28. M. S. Al-Ruqeishi, T. Mohiuddin and L. K. Al-Saadi, Green synthesis of iron oxide nanorods from deciduous Omani mango tree leaves for heavy oil viscosity treatment, *Arabian J. Chem.*, 2019, **12**(8), 4084–4090.
29. S. Bishnol, A. Kumar and R. Selvaraj, Facile synthesis of magnetic iron oxide nanoparticles using inedible Cynometra ramiflora fruit extract waste and their photocatalytic degradation of methylene blue dye, *Mater. Res. Bull.*, 2018, **97**, 121–127.
30. S.-H. Huang and D.-H. Chen, Rapid removal of heavy metal cations and anions from aqueous solutions by an amino-functionalized magnetic nano-adsorbent, *J. Hazard. Mater.*, 2009, **163**(1), 174–179.
31. Y. Pang, *et al.*, Preparation and application of stability enhanced magnetic nanoparticles for rapid removal of Cr(VI), *Chem. Eng. J.*, 2011, **175**, 222–227.
32. Y. Chang and H. C. Zeng, Controlled synthesis and self-assembly of single-crystalline CuO Nanorods and Nanoribbons, *Cryst. Growth Des.*, 2004, **4**(2), 397–402.
33. T. Shahwan, S. A. Sirriah and M. Nairat, Green synthesis of iron NPs and their application as a fenton-like catalyst for the degradation of aqueous cationic and anionic dyes, *Chem. Eng. J.*, 2011, **172**(1), 258–266.
34. A. Rao, A. Bankar, A. R. Kumar, S. Gosavi and S. Zinjarde, Removal of hexavalent chromium ions by Yarrowia lipolytica cells modified with phyto-inspired Fe°/Fe_3O_4 NPs, *J. Contam. Hydrol.*, 2013, **146**, 63–73.

35. S. Thakur and N. Karak, One-step approach to prepare magnetic iron oxide/reduced graphene oxide nanohybrid for efficient organic and inorganic pollutants removal, *Mater. Chem. Phys.*, 2014, **144**(3), 425–432.
36. T. Alam, R. A. A. Khan, A. Ali, H. Sher, Z. Ullah and M. Ali, Biogenic synthesis of iron oxide NPs via Skimmia laureola and their antibacterial efficacy against bacterial wilt pathogen Ralstonia solanacearum, *Mater. Sci. Eng., C*, 2019, **98**, 101–108.
37. M. A. Nazir, N. A. Khan, C. Cheng, S. Shoaib, A. Shah, T. Najam, M. Arshad, A. Sharif, S. Akhtar and A. Rehman, Surface induced growth of ZIF-67 at Co-layered double hydroxide: removal of methylene blue and methyl orange from water, *Appl. Clay Sci.*, 2020, **190**, 105564.
38. M. Senthil and C. Ramesh, Biogenic synthesis of Fe_3O_4 NPs using Tridax procumbens leaf extract and its antibacterial activity on Pseudomonas aeruginosa, *Dig. J. Nanomater. Biostruct.*, 2012, 7(4), 1655–1661.
39. Y. Cai, Y. Shen, A. Xie, S. Li and X. Wang, Green synthesis of soya bean sproutsmediated superparamagnetic Fe_3O_4 NPs, *J. Magn. Magn. Mater.*, 2010, **322**(19), 2938–2943.
40. T. Shahwan, S. A. Sirriah and M. Nairat, Green synthesis of iron NPs and their application as a fenton-like catalyst for the degradation of aqueous cationic and anionic dyes, *Chem. Eng. J.*, 2011, **172**(1), 258–266.
41. P. Banerjee, S. R. Barman, A. Mukhopadhayay and P. Das, Ultrasound assisted mixed azo dye adsorption by chitosan–graphene oxide nanocomposite, *Chem. Eng. Res. Des.*, 2017, **117**, 43–56.
42. T. Hyeon, Chemical synthesis of magnetic nanoparticles, *Chem. Commun.*, 2003, **8**, 927–934.
43. A. Bharde, D. Rautaray, V. Bansal, A. Ahmad, I. Sarkar, S. M. Yusuf and M. Sastry, Extracellular biosynthesis of magnetite using fungi, *Small*, 2006, **2**, 135–141.
44. H. Hashimoto, S. Yokoyama, H. Asaoka, Y. Kusano, Y. Ikeda, M. Seno and R. Murakami, Characteristics of hollow microtubes consisting of amorphous iron oxide nanoparticles produced by iron oxidizing bacteria, Leptothrix ochracea, *J. Magn. Magn. Mater.*, 2007, **310**, 2405–2407.
45. K. S. Prasad, P. Gandhi and K. Selvaraj, Synthesis of green nano iron particles (GnIP) and their application in adsorptive removal of As (III) and As (V) from aqueous solution, *Appl. Surf. Sci.*, 2014, **317**, 1052–1059.
46. V. Madhavi, T. N. V. K. V. Prasad, A. V. B. Reddy, B. R. Reddy and G. Madhavi, Application of phytogenic zerovalent iron nanoparticles in the adsorption of hexavalent chromium, *Spectrochim. Acta, Part A*, 2013, **116**, 17–25.
47. A. T. Nguyen, C. T. Hsieh and R. S. Juang, Substituent effects on photodegradation of phenols in binary mixtures by hybrid H_2O_2 and TiO_2 suspensions under UV irradiation, *J. Taiwan Inst. Chem. Eng.*, 2016, **62**, 68–75.
48. M. Ozmen, K. Can, G. Arslan, A. Tor, Y. Cengeloglu and M. Ersoz, Adsorption of Cu (II) from aqueous solution by using modified Fe_3O_4 magnetic nanoparticles, *Desalination*, 2010, **254**(1–3), 162–169.

49. D. C. Reynolds, D. C. Look, B. Jogai, C. W. Litton, G. Cantwell and W. C. Harsch, Valence-band ordering in ZnO, *Phys. Rev. B: Condens. Matter Mater. Phys.*, **60**, 1999, (4), 2340.
50. A. Parekh, *Use of Magnetic Nanoparticles for Wastewater Treatment*, Massachusetts Institute of Technology, Cambridge, USA, 2013.
51. R. Asrarian, R. Jadidian, H. Parham and S. Haghtalab, Removal of Aluminum from water and wastewater using magnetic iron oxide nanoparticles, *Adv. Mater. Res.*, 2014, **829**, 752–756.
52. Y. Xu, C. Li, X. Zhu, W. E. Huang and D. Zhang, Application of magnetic nanoparticles in drinking water purification, *Environ. Eng. Manage. J.*, 2014, **13**, 2023–2029.
53. S. Benkhaya, S. M. Rabet and A. El Harfi, Classifications, properties, recent synthesis and applications of azo dyes, *Heliyon*, 2020, **6**(1), 03271.
54. M. Y. Rather and S. Sundarapandian, Magnetic iron oxide nanorod synthesis by Wedelia urticifolia (Blume) DC. leaf extract for methylene blue dye degradation, *Appl. Nanosci.*, 2020, **10**(7), 2219–2227.
55. S. Anchan, *et al.*, Biogenic synthesis of ferric oxide nanoparticles using the leaf extract of Peltophorum pterocarpum and their catalytic dye degradation potential, *Biocatal. Agric. Biotechnol.*, 2019, **20**, 101251.
56. N. Madubuonu, *et al.*, Biosynthesis of iron oxide nanoparticles via a composite of Psidium guavaja-Moringa oleifera and their antibacterial and photocatalytic study, *J. Photochem. Photobiol., B*, 2019, **199**, 111601.
57. J. Li, *et al.*, Application of α-Fe_2O_3-based heterogeneous photo-Fenton catalyst in wastewater treatment: A review of recent advances, *J. Environ. Chem. Eng.*, 2022, **10**(5), 108329.
58. J. Niu, *et al.*, Microwave-based preparation of γ-Fe_2O_3/$SrTiO_3$ photocatalyst for efficient degradation of organic pollutants in water, *Mater. Chem. Phys.*, 2022, **288**, 126357.
59. M. F. Al-Hakkani, *et al.*, Cefotaxime removal enhancement via bionanophotocatalyst α-Fe_2O_3 using photocatalytic degradation technique and its echo-biomedical applications, *Sci. Rep.*, 2022, **12**(1), 11881.
60. P. Xu, *et al.*, Use of iron oxide nanomaterials in wastewater treatment: a review, *Sci. Total Environ.*, 2012, **424**, 1–10.
61. F. Fu, D. D. Dionysiou and H. Liu, The use of zero-valent iron for groundwater remediation and wastewater treatment: a review, *J. Hazard. Mater.*, 2014, **267**, 194–205.
62. G. Yang, Z. Cheng, H. Bao, L. Zhang, H. Zhang, H. Jia and J. Wang, Mechanistic insight of weak magnetic field trigger transformation of amorphous Fe(III)-(oxy)hydroxide for enhanced ferrate (VI) towards selective removal of natural organic matter, *Chemosphere*, 2022, **303**(2), 134967.
63. D.-W. Liang, Y.-H. Yang, W.-W. Xu, S.-K. Peng, S.-F. Lu and Y. Xiang, Nonionic surfactant greatly enhances the reductive debromination of polybrominated diphenyl ethers by nanoscale zero-valent iron: mechanism and kinetics, *J. Hazard. Mater.*, 2014, **278**, 592–596.
64. Z. Xiong, B. Lai, P. Yang, Y. Zhou, J. Wang and S. Fang, Comparative study on the reactivity of Fe/Cu bimetallic particles and zero valent iron

(ZVI) under different conditions of N_2, air or without aeration, *J. Hazard. Mater.*, 2015, **297**, 261–268.
65. G. E. Hoag, J. B. Collins, J. L. Holcomb, J. R. Hoag, M. N. Nadagouda and R. S. Varma, Degradation of bromothymol blue by 'greener' nanoscale zero-valent iron synthesized using tea polyphenols, *J. Mater. Chem.*, 2009, **19**(45), 8671–8677.
66. X. Y. Wang, M. P. Zhu, H. L. Liu, J. Ma and F. Li, Modification of Pd–Fe nanoparticles for catalytic dechlorination of 2,4-dichlorophenol, *Sci. Total Environ.*, 2013, **449**, 157–167.
67. M. N. Arancibia, S. E. Baltazar and A. García, *et al.*, Nanoscale zero valent supported by Zeolite and Montmorillonite: template effect of the removal of lead ion from an aqueous solution, *J. Hazard. Mater.*, 2016, **301**, 371–380.
68. Z. Markova, K. M. Šišková and J. Filip, *et al.*, Air stable magnetic bimetallic Fe-Ag nanoparticles for advanced antimicrobial treatment and phosphorus removal, *Environ. Sci. Technol.*, 2013, **47**(10), 5285–5293.
69. G. G. Muradova, S. R. Gadjieva, L. Di Palma and G. Vilardi, Nitrates removal by bimetallic nanoparticles in water, *Chem. Eng. Trans.*, **47**, 2016, 205–210.
70. L. Ling, B. Pan and W.-X. Zhang, Removal of selenium from water with nanoscale zero-valent iron: mechanisms of intraparticle reduction of Se(IV), *Water Res.*, 2015, **71**, 274–281.
71. M. T. Gueye, L. D. Palma and G. Allahverdeyeva, Hexavalent chromium reduction by nano zero valent iron in soil, *Chem. Eng. Trans.*, 2016, **47**, 289–294.
72. W.-X. Zhang, Nanoscale iron particles for environmental remediation: an overview, *J. Nanopart. Res.*, 2003, **5**(3–4), 323–332.
73. M. Stefaniuk, P. Oleszczuk and S. Yok, Review on nano zerovalent iron (nZVI): From synthesis to environmental applications, *Chem. Eng. J.*, 2016, **287**, 618–632.
74. Y. Liou, S.-L. Lo, C.-J. Lin, W. H. Kuan and S. C. Weng, Chemical reduction of an H. unbuffered nitrate solution using catalyzed and uncatalyzed nanoscale iron particles, *J. Hazard. Mater.*, 2005, **127**(1–3), 102–110.
75. R. Singh and V. Misra, Stabilization of zero-valent iron nanoparticles: role of polymers and surfactants, in *Handbook of Nanoparticles*, Springer, New York, NY, USA, 2015, pp. 1–19.
76. Y. Liu, L. Zhang, Q. Song and Z. Xu, Recovery of palladium as NPs from waste multilayer ceramic capacitors by potential-controlled electrodeposition, *J. Cleaner Prod.*, 2020, **257**, 120370.
77. N. D. Berge and C. A. Ramsburg, Oil-in-water emulsions for encapsulated delivery of reactive iron particles, *Environ. Sci. Technol.*, 2009, **43**(13), 5060–5066.
78. V. Kumar, A. Jain, S. Wadhawan and S. K. Mehta, Synthesis of biosurfactant coated magnesium oxide NPs for methylene blue removal and selective Pb2+ sensing, *IET Nanobiotechnol.*, 2018, **12**(3), 241–253.

79. N. F. Jaafar, A. A. Jalil, S. Triwhyono, M. N. M. Muhid, N. Sapawae, M. A. H. Satar and H. Asaari, Photodecolorization of methyl orange over α-Fe_2O_3 supported HY catalysts: the effects of catalyst preparation and dealumination, *Chem. Eng. J.*, 2012, **191**, 112–122.
80. A. Ebrahiminezhad, A. Z. Hoseinabadi, A. Berenjian and Y. Ghasemi, Green synthesis and characterization of zerovalent iron NPs using stinging nettle (Urtica dioica) leaf extract, *Green Process. Synth.*, 2017, **6**(5), 1.
81. N. S. Mohan, R. Gokulkumar, J. Shankar, R. Sridharan, B. Logesh, R. Sasikumar, K. Vallarasu and V. Vijayalakshmi, A facile green approach of Fe_2O_3, Fe_2O_3@Ag, Fe_2O_3@AC and Fe_2O_3@Ag@AC NPs synthesized via Cocos nucifera L for waste water treatment applications, *Results Chem.*, 2022, **4**, 100626.
82. P. Karpagavinayagam and C. Vedhi, Green synthesis of iron oxide nanoparticles using Avicennia marina flower extract, *Vacuum*, 2019, **160**, 286–292.
83. J. S. Justus, S. D. D. Roy and A. M. E. Raj, Synthesis and characterization of hematite nanopowders, *Mater. Res. Express*, 2016, **3**, 105037–105046.
84. B. Purnama, A. T. Wijayanta, and S. Suharyana, Effect of calcination temperature on structural and magnetic properties in cobalt ferrite nano particles, *J. King Saud Univ., Sci.*, 2019, **31**(4), 956–960.
85. K. Shimizu, T. Kubo, A. Satsuma, T. Kamachi and K. Yoshizawa, Surface oxygen atom as a cooperative ligand in Pd NPs catalysis for selective hydration of nitriles to amides in water: experimental and theoretical studies, *ACS Catal.*, 2012, **2**(12), 2467–2474.
86. H. M. Fahmy, Oxidative impact of carob leaf extract–synthesized Iron oxide magnetic nanoparticles on the kidney, liver, testis, and spleen of wistar rats, *BioNanoScience*, 2020, **10**, 54–61.
87. G. T. Adithya and C. Sivasankari, Stable and microcrystalline Ce-Fe Bi-metal oxide nano particles: synthesis, characterization and fluoride adsorption performance in drinking water, *Indian J. Chem. Technol.*, 2019, **26**, 122–130.
88. J. O. Tijani, O. Ugochukwu, L. A. Fadipe, M. T. Bankole, A. S. Abdulkareem and W. D. Roos, One-step green synthesis of WO_3 nanoparticles using Spondias mombin aqueous extract: effect of solution pH and calcination temperature, *Appl. Phys. A*, 2019, **125**, 162.
89. R. Vinayagam, S. Pai, T. Varadavenkatesan, M. K. Narasimhan, S. Narayanasamy and R. Selvaraj, Structural characterization of green synthesized α-Fe_2O_3 nanoparticles using the leaf extract of Spondias dulcis, *Surf. Interfaces*, 2020, **20**, 100618.
90. D. D. Suppiah, N. M. Julkapli, S. Sagadevan and M. R. Johan, Eco-friendly green synthesis approach and evaluation of environmental and biological applications of iron oxide nanoparticles, *Inorg. Chem. Commun.*, 2023, **152**(2023), 110700.
91. J. Ayushi, W. Shweta and S. K. Mehta, Biogenic synthesis of non-toxic iron oxide NPs via Syzygium aromaticum for the removal of methylene blue, *Environ. Nanotechnol., Monit. Manage.*, 2021, **16**, 100464.

CHAPTER 12

Factors Influencing TiO_2-based Composites for Water Decolorization: A Systematic Review

LIKIUS SHIPWIISHO DANIEL[a,b*], VEIKKO UAHENGO[b] AND ATEEQ RAHMAN[b]

[a]Science and Technology Division, Multidisciplinary Research Services, Centre for Research Service, University of Namibia, Private bag X Windhoek, Namibia; [b]Department of Physics, Chemistry and Material Sciences, Faculty of Agriculture, Natural Science and Engineering, University of Namibia
*E-mail: daniels@unam.na

12.1 Introduction

12.1.1 Dye As an Aquatic Pollutant

Any biodegradable waste material that comes from plants or animals is referred to as organic waste. Pollution from agriculture,[1] soil,[2] leftover food, and various types of waste can contaminate water through nutrient runoff.[3] Sewage, urban runoff, hazardous waste, and industrial non-hazardous waste,[4] construction sites, and landfills can also contribute to water pollution in many ways, such as through nutrient enrichment, oxygen depletion, contamination by pathogens, sedimentation, and altered pH and toxic substances.[5] Harmful substances, such as pesticides, synthetic chemicals, and

Chemistry in the Environment Series No. 12
Novel Materials and Water Purification: Towards a Sustainable Future
Edited by Grigorios L. Kyriakopoulos and Miltiadis G. Zamparas

Published by the Royal Society of Chemistry, www.rsc.org

heavy metals, enter the water and can gather in the food chain. This leads to an increase in toxin concentration as they go higher up in the food chain. Dyes are a prevalent type of aquatic pollutant. They can cause significant negative impacts to aquatic ecosystems by altering the water's color and properties.[6] Dyes are chemicals used in many industries to add color to products, such as textiles, printing, papers, rubber, paints, cosmetics, plastics, leather, and food. According to their origin, chemical structure, and solubility, dyes are typically categorized into numerous categories. In essence, dyes can be categorized as either natural or synthetic.[7] In the food industry, natural dyes are frequently used and are divided into three primary categories based on their origins: plants, animals, and minerals.[8] While synthetic dyes are used extensively in the textile industry and are categorized by chromosphere groups in their chemical structures as azo dyes, anthraquinone dyes, phthalein dyes, phthalocyanine dyes, *etc.*[9] Dye contaminants decrease photosynthesis, cause eutrophication, and disrupt aquatic life by preventing sunlight from penetrating the water.[10] Furthermore, these harmful dyes can also find their way into our food and water supplies. This can lead to serious health complications for humans and animals alike. It is essential that the industrial use of toxic dyes be reduced or, better yet, phased out altogether, in favor of safer, more sustainable alternatives. Otherwise, we risk leaving a lasting negative impact on both our environment and our health. Therefore, there is a pressing requirement to develop practical dye effluent purification methods prior to their release into water bodies. Dyes are hard to get rid of from industrial waste water because they do not break down easily and can resist digestion. They are also very stable when exposed to light and do not break down naturally.[6]

To reduce the negative effects of organic waste on water pollution, it is essential to implement effective waste management procedures. To protect the environment, we must implement measures such as sewage treatment, improved agricultural practices, composting and recycling, and the promotion of responsible refuse management. We should also educate individuals and businesses about the significance of these measures. For the decolorization of water, diverse physical, chemical, and biological remediation methods for decolorizing dye effluents have been studied to date.[11] Typical methods for treating water pollution, such as activated carbon,[12] adsorption,[13] membrane filtration,[14,15] and reverse osmosis,[16] have several issues. They take a long time to process, cost a lot to run, need special equipment, and can produce more pollution. These techniques merely transport the pollutant to a different phase and do not degrade or break down the dyes.[17] Unstable metabolites of chemical processes such as electrochemical,[18] chlorination,[19] and ozonation[20] have negative effects on human health. Biological methods, such as microbiological[21] and enzymatic[22] treatments, result in insufficient degradation and inadequate mineralization of by-products.[23] To surmount such obstacles, researchers have concentrated on developing a method for the degradation of dyes that is effective, economical, and environmentally friendly.

12.1.2 Advanced Oxidation Process As Solution to the Degradation of Dye

Advanced oxidation processes (AOPs) based on the production of reactive oxygen species (ROS) (*e.g.*, H_2O_2, $HO^•$, $O_2^•$, and O_3) are highly capable of oxidizing a broad variety of pollutants.[24] AOPs and their combinations, such as ozonation, the Fenton and photo-Fenton processes, ultraviolet (UV) irradiation, hydrogen peroxide (H_2O_2), TiO_2-mediated photocatalysis, sonolysis, and sonophotocatalysis, have been used to achieve this objective under varying operational conditions.[25] Photocatalysis is one of the most recently developed AOPs. Photocatalysis has been extensively studied for its ability to remove organic pollutants from waste water. This method offers many advantages over traditional treatment and has attracted a lot of attention from researchers.[26,27] Recently, Kumari and his team[28] studied the processes to produce rare earth-doped metal oxides and nanocomposites, as well as their effectiveness in reducing organic contaminants. According to their review, new photocatalysts made of titanium oxide have a lot of promise for removing organic contaminants from waste water. The photocatalysts have several outstanding qualities, including excellent stability, non-toxicity, and photocatalytic degradation potential.

Photocatalysis can be a useful method for decolorizing water. When exposed to light, photocatalysis entails the use of a catalyst, typically a semiconductor, to initiate a chemical reaction. Photocatalysis is frequently used in the context of water treatment to degrade and remove organic pigments and pollutants that cause water discoloration. Titania, also known as titanium(IV) oxide, or TiO_2, is a well-known and extensively researched photocatalyst. In nature, titanium(IV) oxide is known as titania or titanium dioxide. It can be found in many different forms, including anatase, rutile, brookite, TiO_2(B), TiO_2(H), and TiO_2(R), as well as the newly discovered and incredibly uncommon mineral akaogiite. Titania mixes, such as ilmenite, a black iron-titanium oxide with the chemical formula $FeTiO_3$ that is widely used to make titania *via* the sulfate method, are more frequently found in nature. Everyday items that we use, such as food, medications, cosmetics, clothing, paper, and photocatalysts, include titanium. Titanium dioxide (TiO_2) is a good photocatalyst because it can soak up light and create ROS molecules when exposed to light. This photocatalyst has benefits, such light stability, high activity, non-toxicity, environmental friendliness, and low cost, but it also has drawbacks that restrict its use. The low quantum performance, wide bandgap energy, and ineffective use of visible light of TiO_2 has prompted substantial study of other photocatalysts. The energy bandgap of TiO_2 can only be activated by ultraviolet (UV) light, which falls within a narrow spectrum of sunlight. Therefore, the objective is to discover a substitute photocatalyst to attain the greatest efficiency under ideal circumstances. The progress of research to implement decolorization of dyes is aided by an understanding of the fundamental principles governing photocatalyst performance.

12.1.3 Fundamental Principles Governing Photocatalyst Performance

Bearing in mind that in 1972, when Fujishima and Honda first reported the splitting of water into hydrogen under UV light by titanium dioxide (TiO_2) as a photocatalyst,[29] TiO_2-based materials have attracted growing interest in the materials field. Today it is well known as the most efficient and environmentally benign photocatalyst and has been used widely for the photodegradation of a broad range of pollutants in both water and air. The water contaminated with organic dyes is exposed to light, typically UV or visible light, depending on the catalyst's absorption properties. TiO_2 is a semiconducting material. Semiconductors are characterized by their distinctive electrical conductivity. Semiconductors absorb light with a threshold wavelength that provides sufficient photon energy to bridge the distance between the valence and conduction bands.[30] This required threshold wavelength corresponds to the minimum photon energy and is dependent on the band-gap energy. Figure 12.1 illustrates how a semiconductor can be excited with minimal photon energy. As light illuminates the photocatalyst, the incident photon energy is absorbed by the valence band (VB) electrons of the semiconductor, causing a transition. When there is enough energy, an electron is generated in the conduction band (CB) and a hole is generated in the VB, hence the electron moves from the CB to the VB. Eqn (12.1) demonstrates this phenomenon.

$$\text{Semiconductor} + h\nu \Rightarrow e^-_{cb} + h^+_{vb} \tag{12.1}$$

where e^-_{cb} and h^+_{vb} are the electrons in the conduction band and the electron vacancy in the valence band, respectively.

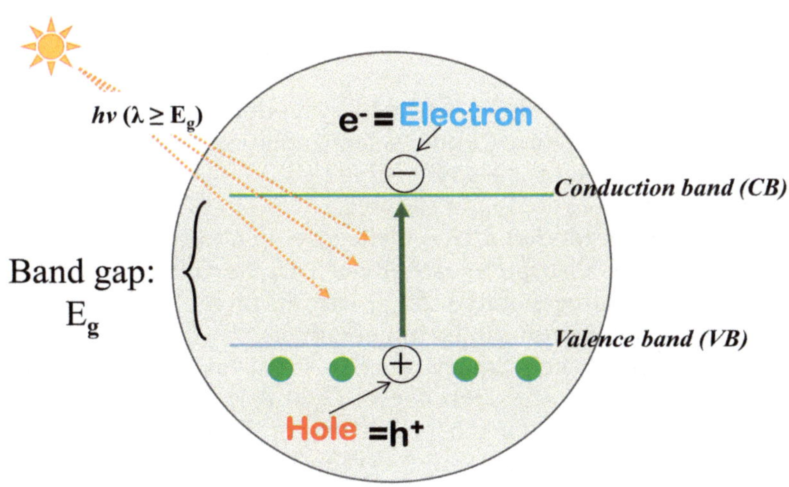

Figure 12.1 Excitation of semiconductor: e^-–h+ pair formation by illuminated metal oxide semiconductor particle.

The excess energy of photogenerated electrons is converted to thermal energy, and the electrons then appear steadily at the base of the CB. Alternatively, if the energy is insufficient to make the VB electron transition to the bottom of the CB, the photogenerated electrons will relinquish their energy and return to the VB without a valid transition.

Due to their peculiar catalytic and optoelectronic properties, metal oxide semiconductor materials, such as TiO_2 (bandgap of, 3.0 eV), ZnO (3.2 eV), WoO_3 (3.2 eV), SnO_3 (3.8 eV), Fe_2O_3 (2.3 eV), and other metals, have been the subject of extensive research.[31] These metal oxide semiconductors have numerous applications in the photocatalysis process. However, light-harvesting and energy conversion efficiencies are quite low, necessitating modification of the material to improve performance. Moreover, most binary metal chalcogenides, including PbS, PbSe, CdS, CdSe, and Ag_2S, can also be utilized in photocatalysis due to their ability to absorb within a bandgap of 1.1–1.4 eV; however, they can be toxic and only a few of them have the appropriate energy levels for creating light-harvesting systems and facilitating photo-induced charge separation, which is required for redox reactions to take place.[32] This restriction, along with the weak photo-corrosion stability of many semiconductors, severely restricts the number of possible light-harvesting materials. For a semiconductor to produce hydroxyl radicals ($OH^{\bullet}$), the valence band potential must be greater than the $OH^{\bullet}$ potential. ZnO, TiO_2, WO_3, and SnO_2 can produce hydroxyl radicals.[33,34] Semiconductors with large bandgaps are effective at separating charges when exposed to light because their conductance and valence bands contain the requisite negative and positive potentials. The requirement for significant energy input is a drawback of using semiconductors with a wide bandgap. As a result, the photo-responsiveness of these semiconductors under sunlight is low, as the portion of the solar spectrum in the UV region is small (4%), *e.g.*, the bandgap of TiO_2 anatase is 3.2 eV and of rutile is 3.0 eV, corresponding to an absorbance threshold of 388 and 415 nm, respectively.[35]

Titanium dioxide can only be activated by a photon of light of 400 nm in wavelength,[34,35] limiting its use under solar irradiation. Solar light irradiation containing 2–3% UV light, 40–43% visible light, and 50–55% near-infrared light, as well as indoor room light (fluorescent, contains a small quantity of UV light) irradiation, rendered TiO_2 ineffective. In addition, the low surface area and inadequate adsorption capacity of TiO_2, as well as the high rate of recombination of photogenerated charge carriers on the surface of TiO_2, prevent its practical applications to its full potential.[36,37]

Figure 12.2 depicts the primary processes that take place when a semiconductor is excited in the presence of O_2 and H_2O, which corresponds to an aqueous phase or an atmospheric environment. The photo-formed electrons and holes produce active oxygen species, such as O_2 and OH radicals. This process is referred to as a photocatalytic reaction, and semiconductors are now used as photocatalysts.[38,39] During this period, the majority of photogenerated electrons and holes generated by eqn (12.1)

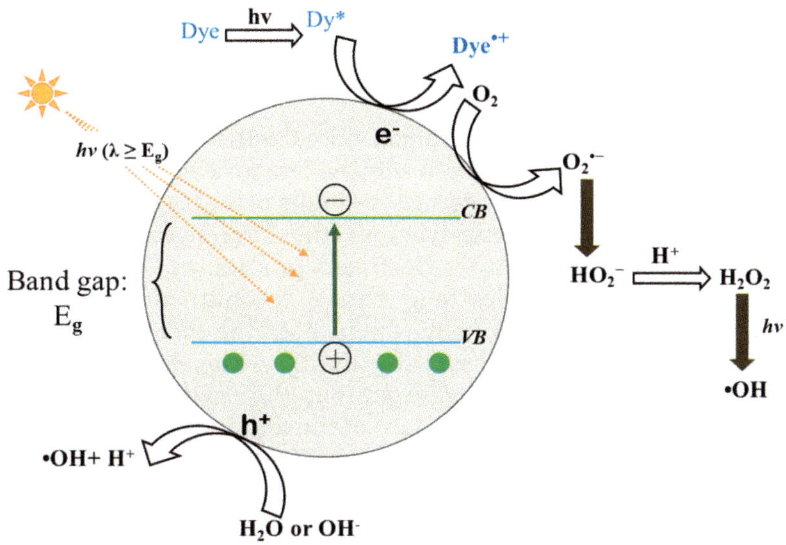

Figure 12.2 Mechanical generation of ROS in a photocatalyst for the decolorization of water.

migrate to the surface of the semiconductor material to reduce or oxidize, respectively, the dye and water. However, if the electron and hole meet during migration, they will recombine and lose their reduction or oxidation capabilities.[40]

As shown in eqn (12.2) and (12.3), h^+_{vb} can readily react with surface-bound H_2O to produce $^{\bullet}OH$ radicals, whereas e^-_{cb} can react with O_2 to produce superoxide radical anion of oxygen.[41,42]

$$h^+_{vb} + H_2O_{(ads)} \rightarrow HO^{\bullet} + H^+ \quad (12.2)$$

$$h^+_{vb} + OH^-_{(ads)} \rightarrow HO^{\bullet} \quad (12.3)$$

$$e^-_{cb} + O_{2(ads)} \rightarrow O_2^{\bullet -} \quad (12.4)$$

$$2O_2^{\bullet -} + H_2O \rightarrow 2HO^{\bullet} + 2OH^- + O_2 \quad (12.5)$$

This reaction precludes the electron and hole produced in the previous step from combining. The $^{\bullet}OH$ and O_2 produced in this manner can then react with the dye to form other species, resulting in the discoloration of the dye. Eqn (12.6)–(12.10) demonstrate how this occurs.

$$O_2^{\bullet -} + H^+ \Rightarrow HO_2^{\bullet} + 2OH^- + O_2 \quad (12.6)$$

$$H_2O_2 \Rightarrow 2HO^{\bullet} \quad (12.7)$$

$$2HO^{\bullet} + dye \Rightarrow dye_{ox} \quad (12.8)$$

$$\text{Dye}+e^-_{cb} \Rightarrow \text{Dye}^\bullet\left(\text{reduction products}\right) \tag{12.9}$$

$$2\text{HO}^\bullet+\text{dye} \Rightarrow \text{intermediates} \Rightarrow CO_2+H_2O \tag{12.10}$$

The generated ROS react with the organic dyes in the water, resulting in their degradation and transmutation into simpler, less toxic compounds. This process degrades the pigmented molecules, resulting in the decolorization of water.

Following the photocatalytic reaction, the catalyst can be regenerated and reused in subsequent water treatment procedures. The ability of the catalyst to regenerate enables long-term and continuous water remediation processes. These equations represent both the direct approaches, the Langmuir–Hinshelwood process[43] and the Eley–Rideal process,[44] and the indirect approach.[42] It can be indicated that photogenerated electrons and holes must have a considerable amount of consumption channels, otherwise one of them will accumulate in the catalyst, and the two will recombine in large quantities, resulting in the catalytic reaction being stopped.

12.1.4 TiO_2-based Composite As a Promising Photocatalytic Material for Water Decolorization

Titania is a promising and environmentally benign photocatalytic semiconductor for water decolorization. Titania is an eco-friendly method of water purification that utilizes light energy and does not require the use of chemicals. It can be utilized for an extended period, making it a lasting and sustainable solution. While it offers several advantages, such as broad applicability on a variety of organic dyes, long-term operation, and versatility as a catalyst, successful implementation in real-world water treatment scenarios requires careful consideration of matrix effects, reactor design, cost effectiveness, and efficiency. The effectiveness of TiO_2 is dependent on light intensity and wavelength. Improving the efficacy and light absorption of titania is an ongoing research objective. Doping can improve the photocatalytic properties of TiO_2, making it more efficient and effective for water purification, air purification, and solar energy conversion. Other components can be added to TiO_2 to decrease its bandgap size and enable absorption of visible light, in addition to its main property of absorbing UV light. In materials science, we use the techniques of composites, alloys, incorporates, dopes, deposits, and impregnates to add foreign elements to the parent photocatalyst without creating new crystallographic phases or structures. The addition of components to TiO_2 reduces the activation energy, thereby accelerating the reaction and increasing the overall efficacy of the process. Although this technique was discovered in the early 1980s,[45] composite semiconductors are commonly used in designing visible-light-driven TiO_2 structures. Several decades of research on composites based on TiO_2 have been reviewed.[46] In fact, comparable visible-light-sensitive photocatalysts were reported as early as 1984.[47] Many researchers have tried to improve the material used to collect light by adding a foreign substance to TiO_2, which can shift its photo-response into the visible spectrum. This can enhance its photocatalytic properties. The literature review revealed that previous authors had conducted bibliometric studies in the related areas. Sasidharan and Priya,[48] Kanan *et al.*,[49] Dharma *et al.*,[50] Musial *et al.*,[51] Shoneye *et al.*,[52] Al-Nuaim *et al.*,[53] and Madkhali *et al.*[54] all reviewed TiO_2-based composites for water decolorization. Akhter *et al.*[55] and Zeshan *et al.*[56] report the recent development of non-metal-doped titanium dioxide photocatalysts for degradation of different dyes and the study of their strategic factors. Peiris *et al.*,[39] Al Zoubi *et al.*,[57] and Perović *et al.*[58] reviewed the recent development and future prospects of TiO_2 photocatalysis. Groeneveld's group[59] summarized parameters that affect the photodegradation of dyes and pigments in solution and on substrates. The parameters reviewed include operative (such as catalyst dosage, reaction time, light intensity), environmental conditions (such as temperature), and water characteristics (such as initial dye concentration).

As we can see above, the current state of knowledge regarding TiO_2-based composites for water decolorization is relatively advanced, but there are still uncertainties and areas of ongoing research. It is well established that TiO_2-based composites can significantly enhance the photocatalytic degradation of organic dyes in water compared to pure TiO_2. The enhanced photocatalysis in TiO_2-based composites is often attributed to synergistic effects between TiO_2 and the incorporated materials. These effects can include improved charge separation, increased surface area, modified band structure, and reduced electron–hole recombination. However, the influence of composite components on TiO_2 still has uncertainties. Uncertainties exist regarding the catalyst design (optimal fabrication method and supportive substrate), operational parameters (such as catalyst dosage, reaction time, light intensity), environmental conditions (such as pH, temperature), and water characteristics (such as initial dye concentration, organic matter content), dye removal efficiency and degradation rate of dyes, reactor configuration, and cost estimation.[59] These factors influence the performance of TiO_2-based composites for the decolorization of water and require further investigation.[51–60] Hence understanding of the influence of the above-mentioned factors on the current state of TiO_2-based composites for water decolorization through systematic review is crucial for their practical application. The systematic review in this chapter aims to summarize the potential factors and challenges associated with the use of TiO_2-based composites for water decolorization. This will be done through research and development efforts to understand how one can enhance the effectiveness and practicality of TiO_2-based composites for water decolorization applications.

12.1.5 Specific Research Questions

By incorporating the above-mentioned elements, the explicit research questions were formulated as follows:

1. How do factors such as catalyst design, operational parameters, environmental conditions, and water characteristics influence the performance of TiO_2-based composites in water decolorization?

2. What are the variations in these factors observed in the literature, and how do they affect parameters such as dye removal efficiency, degradation rate of dyes, and cost estimation of the use of TiO_2-based composites in wastewater treatment?

Based on the these two research equations and the question formulation frameworks proposed by Booth *et al.*[61] and Munn *et al.*[62] the following explicit statements of questions were derived to address the effectiveness and factors influencing TiO_2-based composites in the decolorization of water: the formulation considers the population of interest (TiO_2-based composites), the interventions (composite composition, fabrication methods, dye decolorization), the comparison (potential control groups or reference standards), the outcome measures (color removal efficiency, degradation rate, cost), and the relevant context (environmental conditions: pH, time, water composition, temperature).

12.2 Study Methodology

The study was conducted in five stages including: design and search strategy, study selection and data extraction, review of entry and exit criteria, qualitative assessment, and, finally, statistical analysis of data. To avoid bias in the study, each of the above steps was done by two researchers independently, and finally the agreement of the results obtained was examined by a third researcher.

12.2.1 Search Design: Eligibility Inclusion and Exclusion Criteria and Grouping

12.2.1.1 Inclusion Criteria

Study Design. Only peer-reviewed primary and secondary research studies were included. Experimental studies, laboratory-scale investigations, pilot-scale studies, book chapters, and reviewed articles were considered.

Intervention. Studies that focused on the use of TiO_2-based composites as the primary method for water decolorization were included. The composites include TiO_2 nanoparticles with various compositions, morphologies, surface modifications, or combinations with other materials.

Outcome Measures. Studies that report quantitative or qualitative data on the effectiveness of TiO_2-based composites in decolorizing water are included. Key outcome measures may include color removal efficiency, degradation rate of dyes or organic pollutants, energy consumption, or up-scaling.

Factors of Interest. Studies that investigate the influence of various factors on the performance of TiO_2-based composites in water decolorization were included. This includes studies examining the effect of composite composition, synthesis methods, operational parameters (such as catalyst dosage, reaction time, light intensity), environmental conditions (such as pH, temperature), and water characteristics (such as initial dye concentration, organic matter content).

Language and Publication Date. We only considered articles written in English and published from 2017 to 2023.

12.2.1.2 Exclusion Criteria

Non-English Studies. Studies published in languages other than English were excluded due to limitations in language proficiency.

Review Articles and Conference Abstracts. Conference abstracts, encyclopaedias, editorials, and opinion pieces were excluded.

Irrelevant Studies. Studies that focus on topics other than TiO_2-based composites for water decolorization were excluded. This includes studies on other photocatalysts, adsorbents, or water treatment methods that are not primarily centered around TiO_2-based composites.

12.2.1.3 Grouping for Synthesis

This study was grouped based on its research focus, methodology, and outcomes. The grouping was done as follows:

Composite Composition. Studies investigating the influence of different composite compositions (*e.g.*, doped TiO_2, mixed metal oxides) on the photocatalytic activity of TiO_2-based composites for water decolorization performance based on their classification.

Operational Parameters. Studies focusing on the optimization and influence of operational parameters (*e.g.*, reaction time, light intensity) on the decolorization efficiency of TiO_2-based composites.

Environmental Conditions. Studies investigating the effect of different environmental conditions (*e.g.*, pH, temperature, coexisting substances) on the performance and stability of TiO_2-based composites in water decolorization.

Water Characteristics. Studies examining the influence of water characteristics (*e.g.*, initial dye concentration) on the decolorization efficiency and degradation kinetics of TiO_2-based composites.

These groupings allowed a comprehensive synthesis of the literature, enabling analysis and interpretation of the effectiveness and influencing factors of TiO_2-based composites for water decolorization.

12.2.2 Search Strategy

The search strategy was prepared based on the PRISMA 2020 checklist and was adjusted based on the specific databases and resources available, as well as feedback from preliminary searches and expert input.[63] The search strategy was then constructed based on the PICO-style approach, which allows a comprehensive exploration of relevant literature related to the effectiveness and factors influencing TiO_2-based composites for water decolorization.[64]

We first identify the main concepts, which are: TiO_2-based composites, water decolorization, color removal efficiency, and influencing factors. According to PICO-style, these components act as our population or participants. A list of keywords and synonyms for each concept was then generated and are listed as: "titanium dioxide composites," "TiO_2-based composites" OR "titania composites" OR "TiO_2 nanocomposites" OR "TiO_2 hybrids" OR "TiO_2 photocatalysts," "composite composition," "water treatment," "dye degradation," "dye decomposition," OR "color removal," OR "dye removal," OR "water" OR "wastewater decolorization" OR "dye-contaminated water". The keywords were combined using Boolean operators (AND, OR) to form search strings: "TiO_2 composites AND decolorization AND efficiency," "TiO_2 composites AND factors influencing," "TiO_2-based composites AND water decolorization AND effectiveness," "TiO_2-based composites AND water decolorization AND factors influencing," (TiO_2* water decolori* OR dye degradati* OR color remov* effectiv*), factor* influen*. This was also done considering truncation to account for variations in spelling or word endings. Specific terms or phrases relevant to our overall research question were also included: synthesis methods OR surface modifications OR operational parameters OR environmental conditions OR composite composition.

12.2.3 Study Selection and Data Extraction

The Scopus database and Google Scholar were both systematically searched. Scopus was used to retrieve the bibliographic data for this study because it not only satisfies the stringent quality criteria for indexation,[65,66] but also because the databases indexed by Scopus are more comprehensive. Scopus and Google Scholar are the most significant citation and abstract databases, as well as the most popular search databases.[67] Their data encompass a wide range of topics.

One of us (D.L.S.) scanned the Scopus database on 15 March 2023, using the specified search terms and combinations. He contemplated the use of appropriate filters, whose names and date coverage are listed above. He searched for more studies by doing a "snowball" search. He looked at the lists of references in other studies and used Google Scholar to find more articles that were related to our topic. Using predefined inclusion and exclusion criteria, he then assessed the quality and relevance of the retrieved articles. Then, he analyzed information from the studies and summarized the results, considering how well TiO_2-based materials worked, what influenced their performance, and any commonalities seen across studies. On 15 April 2023, we searched Google Scholar and additional supplementary search engines for online publications. We examined the article titles and abstracts to determine their relevance to the research question. The database search was updated on 18 May 2023, and the avalanche and additional searches were updated on 30 May 2023. We checked if our study was good for the review on TiO_2-based composites for water decolorization by looking at the titles and abstracts of the articles we found and seeing whether they were related to our question and criteria. At this juncture, studies that were manifestly unrelated or did not meet the inclusion criteria were excluded. The full text of the remaining articles was then retrieved and exhaustively evaluated against the predetermined inclusion criteria. This phase entailed a thorough reading of the methods, results, and discussion sections to ensure that the study satisfied all applicable criteria. For the period 2017–2023, a total of 662 publications were returned. We employed narrative synthesis, which entails qualitative analysis and interpretation of the findings from individual studies. We then summarized and analyzed the key findings, trends, and patterns of TiO_2-based composites for water decolorization, providing a comprehensive understanding of their efficacy and factors that influence them. Relevant findings and data from the included studies were summarized in narrative form, highlighting similarities, differences, and significant contextual factors. This method of synthesis enables an all-encompassing interpretation of the evidence and the identification of potential voids or limitations in the existing literature. Meta-analysis is appropriate for quantitative research when a substantial number of comparable studies with quantitative data are available. It combines individual studies to get a more accurate estimate of the overall effectiveness and factors influencing TiO_2-based composites for water decolorization. We gathered information on important measurements, such as how well colors are removed and how fast the process happens, from the studies that we included.

12.3 Results and Discussion

Based on the search strategy, a total 662 articles were collected from Scopus and 178 articles were removed after duplication checks. The titles, abstracts, and texts of the articles were studied, and 353 articles were deleted based on the exclusion criteria. Finally, 131 articles were included in the study. The composition of the TiO_2-based composites, which may include doping with metals or non-metals, fabrication methods or surface modifications (such as calcination temperature, doping concentration, or coating materials), the surface morphologies, operational conditions (such as pH, temperature, light source), and the presence of other water constituents, all influence the application, effectiveness, stability, catalyst recovery, and practical implementation in water purification and are discussed in this systematic review.

12.3.1 TiO_2-based Composites for Water Decolorization

To start with, Sasidharan and team discussed how photocatalysis works and explained how TiO_2 and its derivatives are used in photocatalysis.[68] The modifications of TiO_2 are classified under various headings:[69] metal/metal oxide-based composites (Pd-TiO_2[70]), non-metal-based composites (*e.g.*, N-TiO_2,[71] carbon derivative-based composites (rGO-TiO_2[72]), semiconductor-based composites (Cu_2O/TiO_2), support structure-based composites (polysulfone/N-Pd-TiO_2[73]), and co-modified composites (N-TiO_2/rGO[72])). This detailed concise classification is based on the classification done by Nasr *et al.*[74] in their review. Fang *et al.*[75] developed a generalized classification for visible-light-active reduced TiO_2 that consisted of non-metal-doped TiO_2, metal-based reduced TiO_2, carbon-based TiO_{2-x}, and TiO_{2-x} with distinctive morphology. In addition, Higashimoto[76] categorized visible-light-active photocatalysts as follows: bulk impurities, heterojunction in metal, and interfacial surface complexes.

12.3.2 Current Drawbacks Identified in the Literature

Recently, several reviewers have extensively presented the degradation mechanism of modified titania. Sasidharan's group examined titania-based composites containing metals, non-metals, semiconductors, carbon derivatives, and other support structures designed to overcome the limitations of photocatalysis using pure TiO_2.[68] Chen *et al.*[77] reviewed TiO_2-based photocatalysts for the photocatalytic destruction of organic contaminants. Kanan *et al.*[49] highlighted recent advances in photocatalytic degradation of pesticides and major organic pollutants using TiO_2-based photocatalysts. In analyzing numerous literature sources, Al-Nuaim and colleagues[53] concluded that TiO_2-based photocatalysts degrade dye, pesticides, microorganisms, anti-inflammatory agents, and petroleum hydrocarbons more effectively than other types of wastewater treatment methods. Dharma *et al.*[50] explained different ways to modify TiO_2 and discussed how its ability to remove oilfield-produced water through photocatalysis can be improved. Peiris *et al.*[39] provided a summary of the most recent advances and prospects for TiO_2 photocatalysis. Charge carrier recombination was one of the key challenges they found in the photocatalytic process. By preserving charge carrier separation, metal and non-metal doping can reduce photogenerated electron and hole recombination. Then, it was suggested that doping could be used as a technique to enable TiO_2 to absorb more light energy by narrowing its band gap. Al Zoubi *et al.*[57] reviewed the literature on ways to boost the capacity of TiO_2 for photocatalytic degradation of organic pollutants. They highlighted the most recent advancements in the field of modified TiO_2 for degrading organic pollutants. Investigations into additional methods of modification, as well as doping with organic and inorganic substances, were conducted. Humayun *et al.*[78] found that TiO_2 can be modified by adding metals and non-metals, forming junctions with noble metals and semiconductor oxides, and changes with inorganic acids. To summarize recent work on TiO_2-based nanoparticles (NPs) and nanocrystals (NCs) for the photocatalytic mineralization of different pesticides, Zeshan *et al.*[56] presented a review paper. They talked about the fundamentals of TiO_2 photocatalysis, several types of photocatalytic reactors, and ideal experimental conditions for using TiO_2 to mineralize pesticides. In their review, Shoneye *et al.*[52] discussed innovative techniques for enhancing the performance, durability, charge separation, and water contaminant removal of TiO_2 photocatalysts. These adjustments are crucial to making the technology more useful. Semiconductor photocatalysis was extensively studied both experimentally and theoretically. Defect engineering can improve photocatalytic performance.

All these reviewers investigated numerous methods to improve the performance of TiO_2 and counter some of its drawbacks. The impact of composite components on TiO_2 is still a subject of some concern, notwithstanding these reviews. All these researchers discovered six distinct obstacles that limit the efficiency of TiO_2-based composites in organic effluents. The challenges of using TiO_2 for organic effluents are particle aggregation, limitations in mass transfer, low attraction, high band energy, scattering, and problems with recovery. The treatment settings need to be designed and optimized further to attain higher removal efficiency at reduced costs. They advise scientists to look at the structure, support substrate, light intensity, dopant, and particle size of TiO_2-based photocatalysts.

By looking at structure, Jabbar *et al.*[79] reviewed different techniques to study the properties of materials, including X-ray diffraction, Fourier-transform infrared spectroscopy, X-ray photoelectron spectroscopy, energy-dispersive X-ray spectroscopy, Raman spectra, scanning, transmission and high-resolution transmission electron microscopy, Brunauer–Emmett–Teller surface area, diffuse reflectance spectroscopy, electrochemical impedance spectroscopy, photoluminescence spectroscopy, and transient photocurrent responses. We recommend reading their work to gain a better understanding of the characterization techniques for heterojunction photocatalysts. The primary processing factors that have not been thoroughly investigated are the choice of dopant and a TiO_2 supportive substrate.[60] Similarly, Akpan and Hameed[80] demonstrated that several variables, including the initial pH of the solution to be degraded, oxidizing agents, the temperature at which the catalysts must be calcined, the presence of any dopant(s), and the loading of the catalysts, have an impact on the photocatalytic degradation of any given dye in waste water. The design of the catalyst, the operating conditions, the environmental considerations, and the water factors all need to be clarified. There are worries about removing dye effectively, the speed at which dyes break down, the slurry reactor's design, and estimating the cost of scaling up. To maximize the sustainability of TiO_2-based composites for long-term industrial applications and overcome all these drawbacks, this systematic review investigates some of these factors.

12.3.3 Factors Influencing Catalyst Design

To understand how photocatalysis is designed, we reviewed TiO_2-based composites and their supportive substrates. Many photocatalysts have been tested to remove dye from water and can be grouped into three categories.

The first class of photocatalysts includes single-component materials, such as TiO_2, ZnO, CuO, Cu_2O, and Fe_2O_3.[6] Due to their 3–5 eV wide bandgap, single-material photocatalysts can only be activated by UV light. For a semiconductor material to absorb light efficiently, its bandgap must be small.[81] This is because the bandgap of the semiconductor material fundamentally determines the putative upper limit of the energy conversion efficiency.

Doping is an effective way to get over this obstacle and advance towards the development of the second class of photocatalysts.[54,82] Doping TiO_2 with metal/metal oxide-based composites (Pd-TiO_2[90]), non-metal-based composites (*e.g.* N-TiO_2,[71] C-TiO_2,[83–87] S-TiO_2[88]), carbon derivative-based composites (rGO-TiO_2,[72] CN-TiO_2,[89,90] CAC-TiO_2[91]), semiconductor-based composites (Cu_2O/TiO_2,[92] WO_3–TiO_2,[93] ZnO–TiO_2,[94] SnO_2–TiO_2,[95,96] MoS_2–TiO_2,[97,98] CdS–TiO_2[99]), support structure-based composites (polysulfone/N-Pd-TiO_2,[73] conducting polymer polypyrrole-TiO_2,[100] PDMS-TiO_2,[101] cellulose-TiO_2,[102] kaolin-TiO_2,[103] eggshell-TiO_2,[104] xylan/PVA/TiO_2,[105] fiber-TiO_2,[106] PAE-TiO_2[107]), and co-modified composites (N-TiO_2/rGO, S-TiO_2/rGO,[72] rGO-TiO_2-B,[108] erbium and MWCNT-TiO_2,[109] rGO–ZnO–TiO_2,[70] TiO_2/ZnO/Bi_2O_3-graphene[110]) can solve the wide bandgap problem, and is a route to design a suitable photocatalyst. Dila and Türkten[82] synthesized doped TiO_2 photocatalysts and studied their decolorization efficiencies under solar illumination. Using the wet impregnation method, they demonstrated that undoped TiO_2 can be successfully doped with metals and non-metals. They compared the effectiveness of doped photocatalysts in reducing color and found that the new ones worked better than TiO_2 when exposed to simulated solar light. Perović *et al.*[58] reported the latest achievements in photocatalytic water treatment using TiO_2-based heterojunction composites. They reviewed photocatalytic degradation mechanism principles and demonstrated that TiO_2 and narrow bandgap semiconductors with visible light sensitivity are essential for photocatalytic applications. All the above-mentioned researchers wanted to design a synergistic catalysis between TiO_2 and dopants to enable efficient charge separation, sufficient visible light sensitivity, and excellent photocatalytic activity. Hence, these are primary factors that influence the design of TiO_2-based composite. However, the effectiveness of a TiO_2-based composite to remove dye from water can also be influenced by factors such as the type and amount of dye used, the duration of the reaction, as well as the absence of agglomeration of the photocatalyst.

12.3.4 Supportive Substrate As Factor Influencing Catalyst Design

Photocatalysts are typically available in particle form and can be used as a suspension or an immobilized structure.[14] In suspended mode, photocatalysts are more exposed to the solution and light,[111] but photocatalyst extraction and recovery may be difficult. TiO_2-based composite photocatalyst agglomeration occurs when used in a suspension, making it necessary to use more equipment and energy to separate and reuse these photocatalysts. Several researchers have investigated the use of TiO_2-based composite photocatalyst powder to enhance their performance and solve this issue. TiO_2/GO composite powder improves the ability to adsorb organic pollutants due to the large specific surface area and effective charge carrier separation.[112] Romero-Moran and team[113] have demonstrated that the fabrication of a suspension of TiO_2/SiO_2 composite photocatalyst is effective for the breakdown of organic contaminants in water. Titania composite powders with varying carbon content have been demonstrated to be effective for air filtration, but agglomeration of suspended particles in a liquid medium can restrict their efficiency.[114] This brings us to photocatalyst immobility as another factor to consider during catalytic design.

Photocatalysts immobilized on solid support materials comprise the third class of TiO_2-based composite photocatalysts. Up-scaling photocatalytic systems can be difficult because of the lack of proper disposal methods for the photocatalysts and the toxicity of nanoparticles.[115] Reusing and recycling photocatalysts is important. Immobilizing them on a substrate helps with this.[116] Once the photocatalysts are immobilized on a support, it is significantly easier to separate them.[116,117] It has been widely recognized that immobilized photocatalysts on inert substrates can overcome agglomeration, and hence remove costly phase separation methods.[117,118] To improve the efficiency of TiO_2-based photocatalytic systems, researchers have developed new photocatalytic materials with diverse functions. According to studies, immobilized photocatalysts demonstrated improved catalytic activity, consequently increasing the rate of organic pollutant degradation.[116] Ali and colleagues[119] reported on the use of solid-supported photocatalysts for wastewater treatment, focusing on the role played by the supports in this process. As a result of research into several solid supports, stability, reusability, and affordability have improved, but they still lack efficient involvement in the optical, electrical, and chemical stages of photocatalysis. Opto- and electro-active solid supports are promising for solid-supported photocatalysts.

In an ideal situation, the photocatalysts and the support should have a strong affinity for one another, and the photocatalysts should be stable on the support surface even when exposed to strong oxidizing agents.[120] Moreover, the coating procedure should not inhibit photocatalysis activity[111,121] and the support should have a high surface area.[122,123] Various immobilization techniques, such as the sol–gel technique, hydrothermal treatment method, dip-coating method, solvent-casting method, chemical vapor deposition (CVD), electrophoretic deposition, molecular precursor method (MPM), and photoetching, have been utilized to date for the preparation of TiO_2-based composites. However, due to the complex procedures, expensive equipment, and high calcination temperature requirements, most of these methods are not commonly employed.[124] Alhaji *et al.*[125] state that the type of contaminant, the medium (air or liquid) through which degradation occurs, and the type of support determine the most suitable immobilization techniques. Ali *et al.*[119] summarized some of the most well-known immobilization techniques used for the fabrication of TiO_2-based photocatalysts. Also, Atmianlu and colleagues[126] provided a comprehensive review of the immobilization of nanomaterials on substrates of various materials and morphologies. They provided an exhaustive summary of the status of various substrates used to immobilize nanomaterials, as well as their characteristics, advantages, and disadvantages. To immobilize nanomaterials, numerous substrates, including beads, membranes, fibers, three-dimensional porous structures, and organic and inorganic supports, have been studied. Considering the benefits and drawbacks of beads, spherical carbon beads have superior mechanical, chemical, thermal, and mechanical resistance. Consequently, they may be the optimal option for these treatment procedures.

Much recent research has demonstrated TiO_2-based composite immobilization on supporting substrates such as glass plates, porous glasses, glass beads, glass tubes, glass fiber, quartz,[127,128] silica,[83] activated carbon,[129–133] aluminium,[134] pumice stone, perlite granules, fabric, plastic, polymers,[135–137] clays,[116] zeolite, ceramic membranes,[138–141] Teflon tubes, plant fibers, metals, reactor walls, fly ash, Vycor glass, optical fibers, Raschig rings, *etc.*[116] Nanophotocatalysts can also be fixed in the reactor using materials such as stainless steel, metal wire mesh, or metal foams, or in easily separable materials such as quartz sand, calcium alginate hydrogel beads, black volcanic ash, *etc.*[117,142] In water treatment applications, Joseph and Vijayanandan[117] discussed the support materials used to immobilize nanophotocatalysts. Carbonaceous materials were identified as the most used support material due to their thermal and mechanical stability, and increased surface area.[129] For the immobilization of nanophotocatalyst on carbonaceous substrates, various immobilization methods, such as spraying,[130,131] hydrothermal method, and dip coating,[128,132] are used.

Polymeric membranes are commonly used substrates.[15,143–145] Lotfi *et al.*[15] investigated this novel method. They demonstrated that TiO_2-coated polyethersulfone (PES-TiO_2) membranes in a flow-through photocatalytic membrane reactor under UV light (365 nm) can degrade steroid hormone at 100 ng L^{-1}. The residual β-oestradiol in the purified water was close to that for the drinking-water guideline concentration and the limit of quantitation of the assay system. During the immobilization of photocatalysts on membrane surfaces, membrane dimension, porosity, membrane material, and photocatalyst loading rate are the most important parameters to consider.[138] Due to particle agglomeration, excessive accumulation of photocatalyst on the support membrane may reduce the surface area.[146] Few polymers can sustain N-TiO_2 and photocatalyze ozonation. Ozone degradation of the polymer may destroy the catalyst and contaminate the water. Thus a support based on catalytic application technology should be chosen.[116] Supportive substrates vary by supply, function, structure, thermal behavior, polymerization process, and preparation.[115,116,130] Before the final immobilization, therefore, photocatalyst dosage must be optimized.[117,147] UV light can degrade polymeric membranes if they are exposed for too long. To avoid this issue, it is essential to select the most stable membrane possible. Polyethersulfone and polystyrene membranes are more susceptible to UV degradation than polyacrylonitrile membranes[145] because they contain sulfur-containing compounds. In addition to UV resistance, the stability of the membranes influences how well photocatalysts adhere to the polymer membrane surface. Ceramic membranes demonstrate superior mechanical and chemical stability compared to polymeric membranes.[140] However, ceramic membranes have limited applications owing to their high cost.[138] To make photocatalytic systems affordable for commercial use, we need to focus on creating cheap polymeric membranes that work well and are economically feasible. The advantages of adsorptive membranes include their high efficiency, mechanical strength, chemical resistance, and high flexibility. Consequently, they are the best option for membrane substrates. In comparison to other three-dimensional porous structures, aerogel has superior nanoparticle immobilization properties. Organic supports, which are regarded as a vulnerable support for nanoparticles, have several drawbacks, including limited mechanical, thermal, and chemical stability. Due to their numerous benefits and few drawbacks, inorganic supports are the best substrate.[126]

Nevertheless, the use of a supportive substrate is based on the fabrication method employed during the photocatalyst design. Many methods can be used to incorporate dopants in TiO_2 on glass, such as sol dip coating, atmospheric plasma spraying, gas phase, MPM,[73] CVD, *etc.* Various gas-phase methods have been found to be capable of depositing TiO_2-based composite thin films of good quality and are well established across the coatings industry.[126] However, they are associated with ultrahigh vacuum systems and complicated and expensive instrumentation, and may involve toxic or corrosive chemical precursors. Alternative fabrication methods, such as the electrospray deposition, the sol–gel method, and MPM have been devised and represent active research areas. The MPM method is relatively new.[127,148] However, it has been found to be capable of effectively fabricating thin films of various metal oxides and of metals. Dip coating was identified as the most frequently used immobilization approach.[116]

Immobilized photocatalysts have been examined for their crystallinity, adhesion, and regeneration. A good supportive substrate should adhere to the following standards:[120] photocatalyst-to-support affinity is essential for the stable anchoring of the catalyst; the technique of attachment adopted must not have an impact on the catalytic activity. Additionally, the substrate must have a considerable affinity for the adsorption of contaminants and a large specific surface area. In order to stop the photocatalyst from being washed away from the support surface, a range of reaction circumstances should be avoided. Immobilized TiO_2 photocatalysis depends on the strength of adherence of the TiO_2-based composite coating toward the support.[116] Leaching immobilized TiO_2 can be utilized to evaluate TiO_2-based composite coating adhesion. Given the variety of immobilization technologies, leaching of TiO_2 immobilized on glass should be extensively researched.[120] The composite of photocatalyst and substrate must be stable over the long term, because it is subjected to high quantities of oxidative radicals during the photocatalytic reaction.[149] TiO_2 can have its photo-reactivity increased by transition metal doping, which can increase its effectiveness in photocatalysis and other uses.

Due to their low negative impurity formation energies under O-rich growth conditions, Co-, Cr-, and Ni-doped TiO_2 are rather simple to prepare experimentally, suggesting that these doping systems are accessible and stable.[150]

During this systematic review, it was determined that additional research should be conducted to design new supporting substrates that circumvent the innate mass transfer limitations of immobilized catalysis. Replication of laboratory- and bench-scale test models of immobilized photocatalysts requires further investigation. In other words, a laboratory-to-plant strategy is required. The emphasis must be placed on developing new immobilized photocatalysts that are not only capable of photocatalytic activity under visible and/or solar light spectrum, but also can be recycled repeatedly without significant loss of photocatalytic activity. Nonetheless, it is important to note that large-scale implementation of these immobilized catalysts can aid in reducing the environmental stress caused by toxic effluents from industries, toxic organic compounds, and pharmaceuticals from domestic drainage, *etc.*

12.3.5 Influence of Light Intensity and Irradiation Time

The ability of TiO_2-based composites to absorb photon energy is influenced by the supportive substrate, as we have seen above, as well as by the wavelength, power of the light source, and irradiation time. Therefore, a key element in effective photocatalysis is the choice dopant for the energy (light) source with the proper light intensity and wavelength, such as solar, visible, and irradiation.[151–153] Most photocatalysts are incapable of exploiting visible and infrared light, which account for 43 and 52% of solar energy, respectively.[88] TiO_2-based composites are typically built to be excited by a broad light spectrum to be activated by natural solar energy. This feature would surely help for industrial-scale photocatalytic reactor for economic and environmental sustainability.[154] Noble metal modification of TiO_2 appears to be a potential step in this regard. Jahdi *et al.*[155] examined photocatalytic degradation under natural solar irradiation utilizing noble metal-doped titania-based photocatalysts and fabricated F-Pd co-doped TiO_2 nanocomposites. When they switched from simulated solar light to direct natural sunshine, they noticed a significant rise in degradation rate: from 57% within 360 minutes to 93% within 90 minutes. This result is explained by the authors as a wide wavelength range absorbed by the TiO_2-based nanocomposites and low irradiation intensity when utilizing a solar simulator. This wide range can also include infrared. According to Li and team,[151] the hybrid nanomaterial Pt-rGO-TiO_2, which has a wide light wavelength absorption range (800–2500 nm), can be employed as a highly active photo-thermal-responsive catalyst for effective volatile organic compound degradation when exposed to infrared radiation. With a considerable toluene conversion of 95% and a CO_2 production of 72% under infrared irradiation intensity of 116 mW cm^{-2}, the maximum photo-thermal conversion efficiency could reach 14.1% and maintain its stability for about 50 hours. This study demonstrated the significant potential for redox processes that TiO_2-based composites as infrared light photon absorbers show through the photo-thermal effect, offering an alternative and workable method for converting solar energy.

Natural sunshine is important, but not enough is available at all latitudes. The effects of solar light, ultraviolet-light-emitting diodes (UV-LEDs), and traditional blacklight fluorescent lamps (UV-BL) on photocatalytic dye degradation have been studied. UV-LEDs power photocatalytic devices and are environmentally friendly. LEDs are cheaper and last longer than traditional light sources. LEDs are growing more and more popular.[156] LEDs are semiconductor chips that convert direct current into light of a certain wavelength based on the semiconductor's bandgap.[157] Chen *et al.* (2017) investigated LED manufacturing and photocatalytic reactor designs.[157] UV-LEDs have several advantages over conventional UV lamps, including higher quantum yields, lower power requirements, no warm-up time, a long lifespan, and fewer disposal issues. They eliminate the warming of mercury lamps (causing long-term exposure instability), high energy consumption (due to low photonic efficiency), and short lifespan. LEDs are mercury-free, which is beneficial because mercury is persistent, bioaccumulative, and dangerous.

The development of ultraviolet light-emitting diodes (UV-LED) opens new possibilities for water treatment and photoreactor design. Photocatalytic dye degradation was assessed using a multi-chip array of UV-LEDs of various wavelengths.[156] Even at high intensity (UV-AH), UV-C (260 nm) outperformed UV-B (300 nm) and UV-A (365 nm). Shorter wavelengths have higher energy, allowing direct photolysis of pollutant compounds. Photocatalytic studies using xenon and low-pressure mercury lamps (black lamps) were also investigated.[139] Black bulbs shine at 340–400 nm UV. Their strength diminishes at 375 nm, while solar radiation intensity increases. Xenon lights approximate solar energy distribution at the Earth's surface in the wavelength range of 290–700 nm. They also have peaks in the non-sunlight range (800–1000 nm), which can be reduced with cut-off filters. LEDs, on the other hand, can use bulbs that emit light of the desired wavelength. Mercury and xenon lamps can be used to test the performance of the newly built photocatalyst, even though LEDs and natural sun irradiation seem to be the most promising and sustainable light sources for modern photocatalytic remediation systems. Artificial UV sources are more reproducible than solar light, allowing for greater degradation efficiency of textile dyes.[158] Due to its non-hazardous nature and abundance, solar light is expected to emerge as an alternative, cost-effective source of illumination.

In the photodegradation process, the light intensity has a significant impact on the catalyst's ability to adsorb molecules and the photocatalytic activity of the e^-–h^+ pairs.[156,157,159] Naturally, increased UV intensity leads to more electron excitation from the catalyst and higher OH radical generation, increasing photocatalytic oxidation.[156] The photodegradation rate is independent of the irradiation intensity at higher light intensities, and increasing the irradiation intensity only slightly improves the photocatalytic activity.[59] This process results from the competitiveness of e^-–h^+ recombination on the catalyst surface with variable light intensity, and lowers the photocatalytic efficiency.[159,160] According to research, charge carrier recombination is minimal at low irradiation intensities and predominates in the photocatalytic performance of e^-–h^+ couples, although this is not the case at higher intensities.[158] It was discovered that as the light intensity increased, the rate of photodegradation also increased. This indicated that the enhanced extent of degradation was caused by the increased probability of hydroxyl radical formation on the catalyst surface, which was made possible by the increased light intensity. The light absorption range of the photocatalyst employed dictates the selection of the irradiation source for photodegradation purposes, but other light source parameters, such as power, intensity, or device type, also play a role and can have a big impact on the photocatalytic process.[6] Optimizing the catalyst dose is best done through laboratory experiments. Factors to consider include the type and amount of dye, reactor layout, light intensity, and catalyst properties. Furthermore, the price and sustainability of the energy source should be evaluated.

Collazzo *et al.*[161] compared the ability of nitrogen-doped TiO_2 to decolorize direct black 38 dye using sunlight and visible light sources. They observed that nitrogen-doped TiO_2 degraded more rapidly due to its greater absorption in the visible region when exposed to visible light. The influence of irradiation time on the photo-decolorization of methyl orange (MO) with titania and Co- and La-modified titania was investigated by Kumar and Pandey.[162] In photocatalytic reactions, the greatest MO photo-decolorization was observed after three hours of exposure to visible light. Thus, it can be concluded that photodegradation accelerates with increasing exposure time. Using a Fe_2O_3-TiO_2-zeolite nanocomposite photocatalyst, Davari *et al.*[163] investigated the effect of irradiation time on diphenhydramine photodegradation. It was discovered that the degradation of diphenhydramine is time dependent. For instance, the rate of diphenhydramine degradation increased from approximately 60% after 60 minutes of irradiation to 90% after 180 minutes. Some medications, like diphenhydramine, need more time to break down completely due to their complex structure.[164] Using an *ex situ* technique, Sharma *et al.*[165] created TiO_2-Fe_2O_3 nanocomposites with variable molar ratios. They used the fabricated catalyst to deactivate the dangerous pathogenic bacterium *Escherichia coli* when exposed to natural solar radiation. In 120 minutes of direct sunlight (600–900 watts per square meter), 99.28% of *E. coli* was destroyed. In broad terms, the photochemical degradation efficiency correlates to reaction time, and the degree of color decolorization increases as reaction time increases. Increased irradiation time improves dye degradation. This is because the dye molecules are exposed to light irradiation, producing more reactive radicals that remove the dye more effectively. In addition, extending the irradiation time enhances photocatalytic performance by increasing the sensitivity of the photocatalyst nanoparticles. Longer reaction times lead to more electrons moving from the VB to the CB, resulting in an increase in active species produced by light.[6,165] Overall, the quantum efficiency and kinetics of a photocatalytic degradation process can be affected by the intensity, wavelength, and duration of irradiation of the light source. The energy required for electron excitation from the VB to the CB of the catalyst is provided by the light irradiation. The brightness and color of the light are both important for photocatalytic systems to work properly.

12.3.6 Influence of Initial Dye Concentration

The concentration of dye at the beginning affects the decomposition rate in photocatalysis.[49] The decomposition rate increases until it reaches a maximum value with the rise in dye concentration. Recombination of ˙OH radicals in the cavity is more significant at low dye concentrations than ˙OH radical interaction with contaminants in the solution. More dye concentration leads to less hydroxyl radicals that can participate in recombination reactions. This increases the probability of collision between dye molecules and hydroxyl radicals, which results in a higher efficiency of decolorization. If there is too much dye, then decolorization will not work well because there are too many dye molecules for the ˙OH radicals to interact with. At higher dye concentrations, there is more competition for a limited number of active sites due to increased production of chemical intermediates.

Additionally, at high dye loading, dye molecules absorb most of the light, rather than the catalyst surface, lowering the photocatalytic activity of TiO_2-based composites. Dye molecules act like filters. When there is a lot of dye, they stop light from spreading and only a few photons can get to the surface. Numerous studies looking into how initial dye loading affects photocatalytic degradation efficiency have been published. For instance, Ahmedchekkat *et al.*[166] found that increasing the rhodamine B loading from 10 mg L^{-1} to 30 mg L^{-1} resulted in a drop in the decolorization percentage from 94% to 60%. TiO_2–ZnO–CuO nanoparticles were created using the sol–gel process by Mansouri *et al.*[167] as a catalyst for the photocatalytic degradation of methyl *tert*-butyl ether (MTBE) solution. The goal of the study was to determine how conditions of 1, 2, and 3 g L^{-1} of catalyst loading affected the photocatalytic degradation of MTBE. It was evident that as the catalyst loading increased from 1 to 3 g L^{-1}, the percentage of degradation efficiency increased. The percentage of MTBE degradation, however, decreases as catalyst dosage is raised from 1 to 2 g L^{-1}, reversing the trend that had been seen. The growth in the number of TiO_2–ZnO–CuO nanoparticles caused an increase in the number of absorbed photons and molecules as the catalyst concentration rises. The particles become denser in the light area as a result.[168] Some photocatalyst particles might not have sufficient energy to produce a hydroxyl radical and start oxidizing MTBE, which could explain this phenomenon. It might also be the result of TiO_2–ZnO–CuO aggregation, which would make it more opaque, reduce the surface active spots for adsorbing organic chemicals and UV, and lessen the amount of e^-–h^+ and OH free radicals, which would impair degradation.[168,169] By looking at how different initial concentrations of dyes affect the rate of degradation, it is obvious that raising the initial concentration of dyes results in less degradation. Similar outcomes have been found for the photocatalytic oxidation of various organic substance interfaces.[52,56,78] As a result, at low dye concentrations, dye degrades quickly because more water molecules will adsorb onto the available TiO_2-based composites, generating hydroxyl radicals. Less water is available for active locations at high dye concentrations. As a result, the rivalry between the molecules of dye and water to adsorb on the catalyst is increased, which lowers the rate of breakdown. Pilot-scale investigations must optimize dye loading, since the dye structure determines the optimal concentration. Numerous experiments have shown that an optimal initial dye loading of 0.1–100 μM maximizes decolorization efficiency.[170,171] The initial concentration of the reactant is considered a crucial factor in the photocatalytic degradation of pollutants. Utilizing the adsorption of material on the photocatalyst surface, the relationship between the photocatalytic activity of titania and the initial pollutant concentration was reported by Wu *et al.*[170] The photocatalytic activity of TiO_2 typically decreases as the concentration of the initial reactant increases. This is due to the limited number of activities accessible on the nanoparticle surface of photocatalysts.

12.3.7 Influence of Solution pH and the Water Matrix

The pH of the solution affects the rate and efficiency of TiO_2-based photocatalytic dye degradation.[172–175] Initial solution pH is crucial for the photocatalytic properties, ionization state of contaminants, activity, and solubility of oxidizing agents.[176] The pH effect on photocatalysis of TiO_2-based composites is mainly due to the attraction between the semiconductor surface, organic molecules, and reactive radicals.[163] The surface charge properties and particle aggregation of photocatalysts are dependent on the pH of a given solution.[6] Yuan *et al.*[176] and Długosz *et al.*[177] studied the surface charge of sulfamethoxazole as a model pollutant in two separate investigations. At acidic to weakly alkaline conditions (pH 4 to 8) the decomposition rate of sulfamethoxazole decreased as pH increased.[176] At pH > 5.6, sulfamethoxazole may exist primarily in anionic form, and in neutral form between 1.85 and 5.50.[177] At pH 4, sulfamethoxazole was able to absorb light, which corresponded to greater photochemical reactivity,[173] resulting in a higher decomposition efficiency. Kader *et al.*[178] synthesized $Ag/MoO_3/TiO_2$ nanocomposites using the sol–gel method and studied the UV-induced photodegradation of methyl orange dye. The efficiency of the photocatalyst ($Ag/MoO_3/TiO_2$) diminished as pH and initial pigment concentrations increased. The pH greatly affected the breakdown of MO dye with the order of degradation for pH being 11.0 < 7.0 < 3.0. Because the surface of TiO_2 is positively charged under acidic conditions, it is stimulated to adsorb negatively charged organic dyes. The highest removal was observed at the lowest pH. Due to increased surface adsorption at a lower pH, it is anticipated that the photocatalytic efficiency will increase.[49] The effect of pH on the surface charge of a catalyst can be described using the point of zero charge (PZC).[175] Below pH_{pzc}, the surface of the catalyst is positively charged, while above pH_{pzc}, it is negatively charged.[6] For example, the point of zero charge for TiO_2 particles is approximately 6.8;

therefore, below pH_{pzc}, (pH 6.8) or in an acidic medium, the catalyst surface becomes positively charged, whereas above that, it becomes negatively charged, as shown in eqn (12.11) and (12.12).[175]

$$TiOH_{(surface)} + H^+ \rightarrow TiOH^+_{2(surface)} \quad (12.11)$$

$$TiOH_{(surface)} + OH^- \rightarrow TIO^-_{(surface)} + OH_2 \quad (12.12)$$

Additionally, the initial solution pH impacts the generation rate of hydroxyl radicals during degradation, and thus the decolorization efficiency. Acidic pH increases the formation rate of ˙OH radicals by preventing their recombination.[175] In contrast, a relatively weaker coulombic repulsion can occur between negatively charged TiO_2 and organic compound anions in an alkaline environment, indicating a decrease in photocatalytic reaction due to less dye adsorption.[6] Furthermore, pH may alter the surface charge of photocatalysts for the degradation of organic compounds. Due to the abundance of negatively charged sulfonate groups,[111] the adsorption of MO on the TiO_2 surface under neutral or basic conditions is negligible (3%). The result was in fair agreement with the report of González and Martínez.[179] A pH of 7 was found to be the best for removing color from methyl blue (MB) dye.[180]

The water matrix is a crucial factor that can influence the photocatalytic degradation kinetics of TiO_2-based composites. The removal of dye from water may be affected by water components, such as inorganic anions and natural organic matter. This effect may have an influence on both the rate of removal and the formation of transformation products. Both natural water and effluent typically contain a mixture of various inorganic anions and organic compounds. Depending on the concentration and characteristics of inorganic species, the photocatalysis of TiO_2-based composites can follow various routes.[6,181] Inorganic matrices can influence the photocatalytic degradation of pigment molecules in composites containing TiO_2. They can either accelerate or retard the process. Matrices can transport organic compounds to cavitation bubbles, prevent reactive radicals, reduce catalyst activity by obstructing light, and occupy active sites on the photocatalyst surface. Thus, dye molecules migrate from the bulk solution to the bubble–bulk interface, where degradation is accelerated.[182] Anions can stop photocatalytic reactions in TiO_2-based materials by scavenging radicals and competing with dye molecules to react with oxidative radicals. Because of the scavenging reaction between inorganic anions and oxidative radicals, less reactive radicals are produced, decreasing the overall degradation efficiency.[168] Anions can also deactivate the photocatalyst towards dye molecules by adsorbing to the surface of the catalyst and obstructing numerous active sites.[6]

12.3.8 Influence of Temperature

Temperature affects chemical reactions, but we do not know much about how it affects the breakdown of pollutants in water using photocatalysis. Previous studies have demonstrated that the rate of photocatalytic decolorization of TiO_2-based composites is significantly temperature dependent.[183–185]

In broad terms, the temperature of the solution influences crucial photocatalytic factors, such as the amount of dissolved gas and vapor pressure. Both the viscosity and surface tension of a bulk liquid decrease as its temperature increases. Higher temperature increases the transfer of free radicals and speeds up the breakdown of dye molecules, leading to more degradation. It has been observed that increasing the temperature slows down the e^-–h^+ recombination. Increasing the reaction temperature usually makes the photocatalytic activity better. But if the temperature is over 80 °C, it might stop the adsorption of organic compounds on titania and cause the charge carriers to recombine. Lowering the temperature below 80 °C is better for adsorption but lowering it further to 0 °C makes the activation energy increase.[53] So, it is best to keep the reaction temperature between 20 °C and 80 °C to remove organic matter effectively by photocatalysis.[183–185]

Many researchers have examined the influence of operational temperature on the photocatalytic degradation of dyes. Karimi *et al.*[186] assessed the influence of temperature on photocatalytic degradation using an electromagnetic heater. The temperature of the solution was altered between 30 and 50 °C while the other parameters remained constant. As the temperature rose, the degradation rate of dyes progressively increased. As the temperature rose, the solution bubble ascended, resulting in the production of free radicals. In addition, the rise in temperature made the reaction more competitive with the electron–hole recombination. Likewise, the rising temperature may accelerate the rate of dye oxidation at the interface. Mozia *et al.*[187] investigated photodegradation using Acid Red 18 and discovered that increasing the reaction temperature increased the photodegradation rate constant. Soares *et al.*[188] also reported that temperature affected the reaction. The optimal operational temperature range was determined to be between 40 and 50 °C. Desorption of the generated products, which is slower than degradation on the surface and adsorption of the reactants, limits the reaction at low temperature. By increasing the temperature, the adsorption of the pigment onto the photocatalyst becomes the limiting step. Zhou and Ray[189] hypothesized that at elevated temperatures, a decrease in the adsorptive capacity associated with dissolved oxygen and organics decreases the rate constantly. Therefore, the optimal temperature is typically between 293 and 353 K.[158] Low temperature facilitates the spontaneous exothermic adsorption of the reactants.[190] When the reaction temperature exceeds 80 °C, charge carrier recombination is also substantially accelerated.[191] The optimal reaction temperature for effective photodegradation of dye molecules is between 20 and 80 °C according to Hashimoto *et al.*[192] Using various photocatalysts, Kumar and Pandey[162] demonstrated the effect of temperature on the photodegradation of methyl red and MO. Maximum photo-decolorization was observed at elevated temperatures, while low photo-decolorization efficiency was observed at low temperatures.[193]

In general, it has been found that temperature has a negligible impact on the photofading process of organic dyes and pigments. As anticipated, an increase in temperature increased the rate of fading, but to a lesser extent than other parameters.[194] Temperature differences have not been found to alter the pathways of the photodegradation process. This is a surprising finding when one considers that a rise in temperature results in an increase in energy; consequently, reactions with higher activation energies may become preferable. However, because the thermal energies involved are negligible in comparison to the photon energies absorbed, the alterations are relatively insignificant. Unfortunately, there are no recent studies (2017–2023) that have evaluated the effect of temperature on the photodegradation of dyes.

12.4 Challenges and Future Development

Although advanced TiO_2-based composites have been the subject of considerable and rigorous research, many aspects of the photodegradation process remain to be fully understood. Improving the ability of photodegradation to break down pollutants is not well studied. Operational factors like pH level, temperature, dye concentration, organic types, and light source and intensity are not well considered. A life cycle assessment of the technologies is required for widespread use, particularly the treatment of industrial organic pollutants. One of the most crucial methods for determining the viability and costs of a process, as well as its environmental effects, is a life cycle assessment. Although photodegradation has been the subject of many studies, nothing is known about how long photocatalysts can last. Using photocatalytic processes for organic degradation on an industrial scale is currently difficult. To attain good continuous performance under solar radiation in a larger variety of operating situations, sustainability should be improved. To improve TiO_2 photocatalysis, we need to find the best doping elements that enhance its performance and stability, while being compatible with its physical and chemical properties, such as surface sites, shape, optics, and electron transport. Using organic pollutants, like phenol, in laboratory studies can aid in the application of photodegradation at an industrial level. We suggest that in order to considerably lessen the impact of organic contaminants in waste water, it is time for scientists to consider studying photocatalysis through scale-up of photoreactors.

12.5 Conclusion

According to this systematic evaluation conducted between 2017 and 2023, photocatalytic technologies have achieved substantial advances in the treatment of severely contaminated industrial waste water. According to a survey of the published articles, TiO_2-based photocatalysts degrade organic pollutants more effectively than alternative photocatalytic systems for wastewater treatment. Indeed, photocatalytic composites for water decolorization based on TiO_2 appear promising. Their photocatalytic performance is influenced by several factors. The photocatalytic activity of the TiO_2-based composite can be influenced by the substrate it is coated or immobilized on. The substrate should support effective charge transfer and light absorption processes and offer good adherence. In photocatalysis, the reaction or irradiation time is critical. A successful dye degradation is more likely the longer the TiO_2-based composite is exposed to light. There is, however, a time window after which efficiency may not rise appreciably. The kinetics of photocatalytic processes can be impacted by temperature. In general, higher temperatures can speed up reactions, but going too high can cause the catalyst to degrade thermally or cause other unfavorable side effects. The TiO_2-based composite and the dye being processed determine the ideal photocatalytic temperature. A crucial factor is the pH of the fluid being treated. This may have an impact on the chemical composition of the dye as well as on the surface charge of the photocatalyst. The specific composite and the dye that is being destroyed determine the ideal pH range for photocatalysis. While certain photocatalysts work better in an alkaline environment, others are more active in an acidic one. The photocatalytic effectiveness is influenced by the dye's initial concentration in the aqueous matrix. Increased light absorption and more effective dye breakdown may result from higher dye concentrations. The catalyst active sites could become saturated at very high concentrations, which would diminish its effectiveness. Depending on the particular TiO_2-based composite and dye, there may be different ideal dye concentrations. The water matrix properties and composition can also affect how well the photocatalytic process works. The photocatalytic activity of TiO_2-based composites can be influenced by elements such as the presence of organic or inorganic impurities, dissolved oxygen, and other water constituents. Variations in the water matrix can affect the availability of reactive species and the adsorption of the dye on the catalyst surface. Variables can interact, so the best photocatalytic conditions may be different based on the specific composite, dye, and equipment used. To attain the best photocatalytic performance, it is imperative to tune these parameters.

Acknowledgements

The authors are indebted to the UNAM and the Centre for Research Service (CRS) for their assistance in conducting this research and for providing the necessary resources. This research was not supported by a grant.

References

1. L. Lin, H. Yang and X. Xu, Effects of water pollution on human health and disease heterogeneity: a review, *Front. Environ. Sci.*, 2022, 975.
2. I. Manisalidis, E. Stavropoulou, A. Stavropoulos and E. Bezirtzoglou, Environmental and health impacts of air pollution: a review, *Front. Public Health*, 2020, 14.
3. H. K. Jeswani, G. Figueroa-Torres and A. Azapagic, The extent of food waste generation in the UK and its environmental impacts, *Sustainable Prod. Consumption*, 2021, **26**, 532–547.

4. D. M. Chen, B. L. Bodirsky, T. Krueger, A. Mishra and A. Popp, The world's growing municipal solid waste: Trends and impacts, *Environ. Res. Lett.*, 2020, **15**(7), 074021.
5. X. Zhao, Y. Zheng, F. Quan, S. Hu, Q. Wu, M. Luo, Y. Gu, S. Tang and J. Jiang, Road runoff as a significant nonpoint source of parabens and their metabolites in urban rivers, *Chemosphere*, 2022, **301**, 134632.
6. M. Pirsaheb and N. Moradi, A systematic review of the sonophotocatalytic process for the decolorization of dyes in aqueous solution: Synergistic mechanisms, degradation pathways, and process optimization, *J. Water Process Eng.*, 2021, **44**, 102314.
7. K. D. Mahato and U. Kumar, A review of organic dye based nanoparticles: preparation, properties, and engineering/technical applications, *Mini-Rev. Org. Chem.*, 2023, **20**(7), 655–674.
8. A. L. Ahmad and S. W. Puasa, Reactive dyes decolourization from an aqueous solution by combined coagulation/micellar-enhanced ultrafiltration process, *Chem. Eng. J.*, 2007, **132**(1–3), 257–265.
9. K. Shakir, A. F. Elkafrawy, H. F. Ghoneimy, S. G. Beheir and M. Refaat, Removal of rhodamine B (a basic dye) and thoron (an acidic dye) from dilute aqueous solutions and wastewater simulants by ion flotation, *Water Res.*, 2010, **44**(5), 1449–1461.
10. S. Garg, Z. Z. Chowdhury, A. N. Faisal, N. P. Rumjit and P. Thomas, Impact of industrial wastewater on environment and human health. *Advanced Industrial Wastewater Treatment and Reclamation of Water: Comparative Study of Water Pollution Index during Pre-industrial, Industrial Period and Prospect of Wastewater Treatment for Water Resource Conservation*, 2022, pp. 197–209.
11. R. Monsef, M. Ghiyasiyan-Arani, O. Amiri and M. Salavati-Niasari, Sonochemical synthesis, characterization and application of $PrVO_4$ nanostructures as an effective photocatalyst for discoloration of organic dye contaminants in wastewater, *Ultrason. Sonochem.*, 2020, **61**, 104822.
12. A. Rahman, H. J. Hango, L. S. Daniel, V. Uahengo, S. J. Jaime, S. V. Bhaskaruni and S. B. Jonnalagadda, Chemical preparation of activated carbon from Acacia erioloba seed pods using H_2SO_4 as impregnating agent for water treatment: an environmentally benevolent approach, *J. Cleaner Prod.*, 2019, **237**, 117689.
13. M. T. Yagub, T. K. Sen, S. Afroze and H. M. Ang, Dye and its removal from aqueous solution by adsorption: a review, *Adv. Colloid Interface Sci.*, 2014, **209**, 172–184.
14. V. Loddo, L. Palmisano, T. Marino and R. Molinari, Membranes for photocatalysis in water and wastewater treatment, in *Advanced Membrane Science and Technology for Sustainable Energy and Environmental Applications*, Woodhead Publishing, 1 January 2011, pp. 746–768.
15. S. Lotfi, K. Fischer, A. Schulze and A. I. Schäfer, Photocatalytic degradation of steroid hormone micropollutants by TiO_2-coated polyethersulfone membranes in a continuous flow-through process, *Nat. Nanotechnol.*, 2022, **17**(4), 417–423.

16. S. K. Nataraj, K. M. Hosamani and T. M. Aminabhavi, Nanofiltration and reverse osmosis thin film composite membrane module for the removal of dye and salts from the simulated mixtures, *Desalination*, 2009, **249**(1), 12–17.
17. R. Shokoohi, M. Salari, A. Shabanloo, N. Shabanloo, S. Marofi, H. Faraji, M. V. Tabar and M. Moradnia, Catalytic activation of persulphate with Mn_3O_4 nanoparticles for degradation of acid blue 113: process optimisation and degradation pathway, *Int. J. Environ. Anal. Chem.*, 2022, **102**(16), 3786–3805.
18. N. Nippatla and L. Philip, Electrochemical process employing scrap metal waste as electrodes for dye removal, *J. Environ. Manage.*, 2020, **273**, 111039.
19. G. A. R. Oliveira, E. R. A. Ferraz, F. M. D. Chequer, M. D. Grando, J. P. F. Angeli, M. S. Tsuboy, J. C. Marcarini, M. S. Mantovani, M. E. Osugi, T. M. Lizier, M. V. B. Zanoni and D. P. Oliveira, Chlorination treatment of aqueous samples reduces, but does not eliminate, the mutagenic effect of the azo dyes Disperse Red 1, Disperse Red 13 and Disperse Orange 1, *Mutat. Res., Genet. Toxicol. Environ. Mutagen.*, 2010, **703**(2), 200–208.
20. P. Colindres, H. Yee-Madeira and E. Reguera, Removal of Reactive Black 5 from aqueous solution by ozone for water reuse in textile dyeing processes, *Desalination*, 2010, **258**(1–3), 154–158.
21. N. D. Parmar and S. R. Shukla, Decolourization of dye wastewater by microbial methods-A review, *Indian J. Chem. Technol.*, 2018, **25**(4), 315–323.
22. L. Y. Jun, L. S. Yon, N. M. Mubarak, C. H. Bing, S. Pan, M. K. Danquah, E. C. Abdullah and M. Khalid, An overview of immobilized enzyme technologies for dye and phenolic removal from wastewater, *J. Environ. Chem. Eng.*, 2019, 7(2), 102961.
23. R. Shokoohi, M. Salari, R. Safari, H. Z. Nasab and S. Shanehsaz, Modelling and optimisation of catalytic ozonation process assisted by ZrO_2-pumice/H_2O_2 in the degradation of Rhodamine B dye from aqueous environment, *Int. J. Environ. Anal. Chem.*, 2021, **101**(15), 2629–2653.
24. Y. Deng and R. Zhao, Advanced oxidation processes (AOPs) in wastewater treatment, *Curr. Pollut. Rep.*, 2015, **1**, 167–176.
25. M. Rahimi, N. Moradi, M. Faryadi and S. Safari, Removal of ammonia by high-frequency ultrasound wave (1.7 MHz) combined with TiO_2 photocatalyst under UV radiation, *Desalin. Water Treat.*, 2016, **57**(34), 15999–16007.
26. M. P. Rayaroth, C. T. Aravindakumar, N. S. Shah and G. Boczkaj, Advanced oxidation processes (AOPs) based wastewater treatment-unexpected nitration side reactions-a serious environmental issue: A review, *Chem. Eng. J.*, 2022, **430**, 133002.
27. G. A. Ismail and H. Sakai, Review on effect of different type of dyes on advanced oxidation processes (AOPs) for textile color removal, *Chemosphere*, 2022, **291**, 132906.

28. H. Kumari, S. Suman, R. Ranga, S. Chahal, S. Devi, S. Sharma, S. Kumar, P. Kumar, S. Kumar, A. Kumar and R. Parmar, A Review on Photocatalysis Used For Wastewater Treatment: Dye Degradation, *Water, Air, Soil Pollut.*, 2023, **234**(6), 349.
29. A. Fujishima and K. Honda, Electrochemical photolysis of water at a semiconductor electrode, *Nature*, 1972, **238**(5358), 37–38.
30. H. Yoo, I. S. Lee, S. Jung, S. M. Rho, B. H. Kang and H. J. Kim, A review of phototransistors using metal oxide semiconductors: Research progress and future directions, *Adv. Mater.*, 2021, **33**(47), 2006091.
31. X. Yang, Y. Deng, H. Yang, Y. Liao, X. Cheng, Y. Zou, L. Wu and Y. Deng, Functionalization of Mesoporous Semiconductor Metal Oxides for Gas Sensing: Recent Advances and Emerging Challenges, *Adv. Sci.*, 2023, **10**(1), 2204810.
32. M. Dadkhah and J. M. Tulliani, Green synthesis of metal oxides semiconductors for gas sensing applications, *Sensors*, 2022, **22**(13), 4669.
33. Z. A. M. Hir and R. Ali, Photodegradation of benzene, toluene, p-xylene in aqueous phase using zno, tio2, sno2, wo3 and fe2o3 photocatalysts prepared via sol-gel method, *Gading J. Sci. Technol.*, 2021, **4**(1), 25–32.
34. L. S. Daniel, H. Nagai, N. Yoshida and M. Sato, Photocatalytic activity of vis-responsive Ag-nanoparticles/TiO_2 composite thin films fabricated by molecular precursor method (MPM), *Catalysts*, 2013, **3**(3), 625–645.
35. A. Di Paola, M. Bellardita and L. Palmisano, Brookite, the least known TiO_2 photocatalyst, *Catalysts*, 2013, **3**(1), 36–73.
36. *Inorganic Pollutants in Wastewater: Methods of Analysis, Removal and Treatment*, ed. A. Mohammad, and A. M. Asiri, Materials Research Forum LLC, October 2017.
37. R. Asahi, T. Morikawa, H. Irie and T. Ohwaki, Nitrogen-doped titanium dioxide as visible-light-sensitive photocatalyst: designs, developments, and prospects, *Chem. Rev.*, 2014, **114**(19), 9824–9852.
38. L. Xiang and X. Zhao, Wet-chemical preparation of TiO_2-based composites with different morphologies and photocatalytic properties, *Nanomaterials*, 2017, 7(10), 310.
39. S. Peiris, H. B. de Silva, K. N. Ranasinghe, S. V. Bandara and I. R. Perera, Recent development and future prospects of TiO_2 photocatalysis, *J. Chin. Chem. Soc.*, 2021, **68**(5), 738–769.
40. J. Luo, S. Zhang, M. Sun, L. Yang, S. Luo and J. C. Crittenden, A critical review on energy conversion and environmental remediation of photocatalysts with remodeling crystal lattice, surface, and interface, *ACS Nano*, 2019, **13**(9), 9811–9840.
41. B. Sun, Q. Li, M. Zheng, G. Su, S. Lin, M. Wu, C. Li, Q. Wang, Y. Tao, L. Dai, Y. Qin and B. Meng, Recent advances in the removal of persistent organic pollutants (POPs) using multifunctional materials: a review, *Environ. Pollut.*, 2020, **265**, 114908.
42. W. S. Koe, J. W. Lee, W. C. Chong, Y. L. Pang and L. C. Sim, An overview of photocatalytic degradation: photocatalysts, mechanisms, and development of photocatalytic membrane, *Environ. Sci. Pollut. Res.*, 2020, **27**, 2522–2565.

43. K. V. Kumar, K. Porkodi and A. F. Rocha, Langmuir–Hinshelwood kinetics–a theoretical study, *Catal. Commun.*, 2008, **9**(1), 82–84.
44. A. Fernandez-Perez and G. Marban, Titanium dioxide: A heterogeneous catalyst for dark peroxidation superior to iron oxide, *J. Environ. Chem. Eng.*, 2020, **8**(5), 104254.
45. W. Jaegermann, The semiconductor/electrolyte interface: a surface science approach, *Mod. Aspects Electrochem.*, 1996, **30**, 1–85.
46. I. M. S. Pillai and K. L. Priya, TiO_2-Based Composites for Water Decolorization, *Novel Materials for Dye-Containing Wastewater Treatment*, 2021, pp. 103–142.
47. A. Zada, Y. Qu, S. Ali, N. Sun, H. Lu, R. Yan, X. Zhang and L. Jing, Improved visible- light activities for degrading pollutants on TiO_2/g-C_3N_4 nanocomposites by decorating SPR Au nanoparticles and 2, 4-dichlorophenol decomposition path, *J. Hazard. Mater.*, 2018, **342**, 715–723.
48. I. M. S. Pillai and K. L. Priya, TiO_2-Based Composites for Water Decolorization, *Novel Materials for Dye-Containing Wastewater Treatment*, 2021, pp. 103–142.
49. S. Kanan, M. A. Moyet, R. B. Arthur and H. H. Patterson, Recent advances on TiO_2-based photocatalysts toward the degradation of pesticides and major organic pollutants from water bodies, *Catal. Rev.*, 2020, **62**(1), 1–65.
50. H. N. C. Dharma, J. Jaafar, N. Widiastuti, H. Matsuyama, S. Rajabsadeh, M. H. D. Othman, M. A. Rahman, N. N. M. Jafri, N. S. Suhaimin, A. M. Nasir and N. H. Alias, A review of titanium dioxide (TiO_2)-based photocatalyst for oilfield-produced water treatment, *Membranes*, 2022, **12**(3), 345.
51. J. Musial, D. T. Mlynarczyk and B. J. Stanisz, Photocatalytic degradation of sulfamethoxazole using TiO_2-based materials–Perspectives for the development of a sustainable water treatment technology, *Sci. Total Environ.*, 2023, **856**, 159122.
52. A. Shoneye, J. S. Chang, M. N. Chong and J. Tang, Recent progress in photocatalytic degradation of chlorinated phenols and reduction of heavy metal ions in water by TiO_2-based catalysts, *Int. Mater. Rev.*, 2022, **67**(1), 47–64.
53. M. A. Al-Nuaim, A. A. Alwasiti and Z. Y. Shnain, The photocatalytic process in the treatment of polluted water, *Chem. Pap.*, 2023, **77**(2), 677–701.
54. N. Madkhali, C. Prasad, K. Malkappa, H. Y. Choi, V. Govinda, I. Bahadur and R. A. Abumousa, Recent update on photocatalytic degradation of pollutants in waste water using TiO_2-based heterostructured materials, *Results Eng.*, 2023, 100920.
55. P. Akhter, A. Arshad, A. Saleem and M. Hussain, Recent Development in Non-Metal-Doped Titanium Dioxide Photocatalysts for Different Dyes Degradation and the Study of Their Strategic Factors: A Review, *Catalysts*, 2022, **12**(11), 1331.
56. M. Zeshan, I. A. Bhatti, M. Mohsin, M. Iqbal, N. Amjed, J. Nisar, N. Almasoud and T. S. Alomar, Remediation of pesticides using TiO_2 based photocatalytic strategies: A review, *Chemosphere*, 2022, 134525.

57. W. Al Zoubi, A. A. S. Al-Hamdani, B. Sunghun and Y. G. Ko, A review on TiO_2-based composites for superior photocatalytic activity, *Rev. Inorg. Chem.*, 2021, **41**(4), 213–222.
58. K. Perović, F. M. dela Rosa, M. Kovačić, H. Kušić, U. L. Štangar, F. Fresno, D. D. Dionysious and A. L. Bozic, Recent achievements in development of TiO_2-based composite photocatalytic materials for solar driven water purification and water splitting, *Materials*, 2020, **13**(6), 1338.
59. I. Groeneveld, M. Kanelli, F. Ariese and M. R. van Bommel, Parameters that affect the photodegradation of dyes and pigments in solution and on substrate–An overview, *Dyes Pigm.*, 2022, 110999.
60. A. Akhoondi, A. Sharma, D. Pathak, M. Yusuf, T. B. Demissie, R. T. Guo and A. Ali, Hydrogen evolution via noble metals based photocatalysts: A review, *Synth. Sintering*, 2021, **1**(4), 223–241.
61. P. Booth, S. A. Chaperon, J. S. Kennell and A. M. Morrison, Entrepreneurship in island contexts: A systematic review of the tourism and hospitality literature, *Int. J. Hospitality Manage.*, 2020, **85**, 102438.
62. C. Munn-Giddings, A. McVicar and L. Smith, Systematic review of the uptake and design of action research in published nursing research, 2000–2005, *J. Res. Nurs.*, 2008, **13**(6), 465–477.
63. M. J. Page, J. E. McKenzie, P. M. Bossuyt, I. Boutron, T. C. Hoffmann, C. D. Mulrow, L. Shamseer, J. M. Tetzlaff and D. Moher, Updating guidance for reporting systematic reviews: development of the PRISMA 2020 statement, *J. Clin. Epidemiol.*, 2021, **134**, 103–112.
64. R. McDougall, Systematic reviews in bioethics: types, challenges, and value, *J. Med. Philos.*, 2014, **39**(1), 89–97.
65. N. Donthu, S. Kumar, D. Mukherjee, N. Pandey and W. M. Lim, How to conduct a bibliometric analysis: An overview and guidelines, *J. Bus. Res.*, 2021, **133**, 285–296.
66. J. Paul, W. M. Lim, A. O'Cass, A. W. Hao and S. Bresciani, Scientific procedures and rationales for systematic literature reviews (SPAR-4-SLR), *Int. J. Consum. Stud.*, 2021, **45**(4), O1–O6.
67. V. N. Amrutha and S. N. Geetha, A systematic review on green human resource management: Implications for social sustainability, *J. Cleaner Prod.*, 2020, **247**, 119131.
68. S. Pillai, M. Indu and K. L. Priya, TiO_2-Based Composites for Water Decolorization, *Novel Materials for Dye-Containing Wastewater Treatment*, 2021, pp. 103–142.
69. W. Fang, M. Xing and J. Zhang, Modifications on reduced titanium dioxide photocatalysts: A review, *J. Photochem. Photobiol., C*, 2017, **32**, 21–39.
70. C. H. Nguyen, C.-C. Fu and R.-S. Juang, Degradation of methylene blue and methyl orange by palladium-doped TiO_2 photocatalysis for water reuse: efficiency and degradation pathways, *J. Cleaner Prod.*, 2018, **202**, 413–427.
71. P. Le, L. Hieu and T.-N. Lam, *et al.*, Enhanced photocatalytic performance of nitrogen-doped TiO_2 nanotube arrays using a simple annealing process, *Micromachines*, 2018, **9**, 618.

72. A. Brindha and T. Sivakumar, Visible active N, S co-doped TiO_2/graphene photocatalysts for the degradation of hazardous dyes, *J. Photochem. Photobiol., A*, 2017, **340**, 146–156.
73. A. T. Kuvarega, N. Khumalo, D. Dlamini and B. B. Mamba, Polysulfone/N, Pd co-doped TiO_2 composite membranes for photocatalytic dye degradation, *Sep. Purif. Technol.*, 2018, **191**, 122–133.
74. M. Nasr, C. Eid, R. Habchi, P. Miele and M. Bechelany, Recent progress on titanium dioxide nanomaterials for photocatalytic applications, *ChemSusChem*, 2018, **11**(18), 3023–3047.
75. W. Fang, M. Xing and J. Zhang, Modifications on reduced titanium dioxide photocatalysts: A review, *J. Photochem. Photobiol., C*, 2017, **32**, 21–39.
76. S. Higashimoto, Titanium-dioxide-based visible-light-sensitive photocatalysis: Mechanistic insight and applications, *Catalysts*, 2019, **9**(2), 201.
77. D. Chen, Y. Cheng, N. Zhou, P. Chen, Y. Wang, K. Li, S. Huo, P. Cheng, P. Peng, R. Zhang, L. Wang, H. Liu, Y. Liu and R. Ruan, Photocatalytic degradation of organic pollutants using TiO_2- based photocatalysts: A review, *J. Cleaner Prod.*, 2020, **268**, 121725.
78. M. Humayun, F. Raziq, A. Khan and W. Luo, Modification strategies of TiO_2 for potential applications in photocatalysis: a critical review, *Green Chem. Lett. Rev.*, 2018, **11**(2), 86–102.
79. Z. H. Jabbar, B. H. Graimed, M. A. Issa, H. J. Khadim, S. H. Ammar, A. A. Okab and Y. A. Shafiq, A review study summarizes the main characterization techniques of nano-composite photocatalysts and their applications in photodegradation of organic pollutants, *Environ. Nanotechnol., Monit. Manage.*, 2022, 100765.
80. U. G. Akpan and B. H. Hameed, Parameters affecting the photocatalytic degradation of dyes using TiO_2-based photocatalysts: a review, *J. Hazard. Mater.*, 2009, **170**(2–3), 520–529.
81. H. Zhang, J. Liu, T. Xu, W. Ji and X. Zong, Recent Advances on Small Band Gap Semiconductor Materials ($\leq$ 2.1 eV) for Solar Water Splitting, *Catalysts*, 2023, **13**(4), 728.
82. K. A. Y. A. Dila and N. Türkten, Preparation of doped TiO_2 photocatalysts and their decolorization efficiencies under solar light, *Mühendislik Bilimleri ve Tasarım Dergisi*, 2020, **8**(3), 655–663.
83. H. Gunes, Y. Özbakir, S. B. Barim, H. Yousefzadeh, S. E. Bozbag and C. Erkey, A remarkable class of nanocomposites: aerogel supported bimetallic nanoparticles, *Front. Mater.*, 2020, **7**, 18.
84. A. Bouarioua and M. Zerdaoui, Photocatalytic activities of TiO_2 layers immobilized on glass substrates by dip-coating technique toward the decolorization of methyl orange as a model organic pollutant, *J. Environ. Chem. Eng.*, 2017, **5**(2), 1565–1574.
85. J. Shao, W. Sheng and M. Wang, *et al.*, In situ synthesis of carbon-doped TiO_2 single-crystal nanorods with a remarkably photocatalytic efficiency, *Appl. Catal., B*, 2017, **209**, 311–319.

86. S. Varnagiris, A. Medvids and M. Lelis, *et al.*, Black carbon-doped TiO_2 films: synthesis, characterization and photocatalysis, *J. Photochem. Photobiol., A*, 2019, **382**, 111941.
87. M. S. Adly, S. M. El-Dafrawy and S. A. El-Hakam, Application of nanostructured graphene oxide/titanium dioxide composites for photocatalytic degradation of rhodamine B and acid green 25 dyes, *J. Mater. Res. Technol.*, 2019, **8**, 5610–5622.
88. R. T. Bento, O. V. Correa and M. F. Pillis, Photocatalytic activity of undoped and sulfur-doped TiO_2 films grown by MOCVD for water treatment under visible light, *J. Eur. Ceram. Soc.*, 2019, **39**, 3498–3504.
89. M. S. Azami, K. Ismail and M. A. M. Ishak, *et al.*, Formation of an amorphous carbon nitride/titania composite for photocatalytic degradation of RR4 dye, *J. Water Process Eng.*, 2020, **35**, 101209.
90. E. Bailón-García, A. Elmouwahidi and M. A. Álvarez, *et al.*, New carbon xerogel-TiO_2 composites with high performance as visible-light photocatalysts for dye mineralization, *Appl. Catal., B*, 2017, **201**, 29–40.
91. P. Muthirulan, C. N. Devi and M. M. Sundaram, Synchronous role of coupled adsorption and photocatalytic degradation on CAC–TiO_2 composite generating excellent mineralization of alizarin cyanine green dye in aqueous solution, *Arabian J. Chem.*, 2017, **10**, S1477–S1483.
92. T. Fang, X. Chen and M. Wang, *et al.*, Cuprous oxide/titanium dioxide composite photocatalytic decolorization of reactive brilliant red X-3B dyes wastewater under visible light, *Res. Chem. Intermed.*, 2020, **46**, 5459–5477.
93. H. Liu, W. Guo and Y. Li, *et al.*, Photocatalytic degradation of sixteen organic dyes by TiO_2/WO_3-coated magnetic nanoparticles under simulated visible light and solar light, *J. Environ. Chem. Eng.*, 2018, **6**, 59–67.
94. G. K. Upadhyay, J. K. Rajput and T. K. Pathak, *et al.*, Synthesis of ZnO:-TiO_2 nanocomposites for photocatalyst application in visible light, *Vacuum*, 2019, **160**, 154–163.
95. A. S. Kutuzova and T. A. Dontsova, Characterization and properties of TiO_2–SnO_2 nanocomposites, obtained by hydrolysis method, *Appl. Nanosci.*, 2019, **9**, 873–880.
96. M. Scarisoreanu, C. Fleaca and I. Morjan, *et al.*, High photoactive TiO_2/SnO_2 nanocomposites prepared by laser pyrolysis, *Appl. Surf. Sci.*, 2017, **418**, 491–498.
97. O. Ibukun, P. E. Evans, P. A. Dowben and H. K. Jeong, Titanium dioxide-molybdenum disulfide for photocatalytic degradation of methylene blue, *Chem. Phys.*, 2019, **525**, 110419.
98. L. Karimi, Combination of mesoporous titanium dioxide with MoS_2 nanosheets for high photocatalytic activity, *Pol. J. Chem. Technol.*, 2017, **19**, 56–60.
99. B. G. T. Keerthana and P. Murugakoothan, Synthesis and characterization of CdS/TiO_2 nanocomposite: methylene blue adsorption and enhanced photocatalytic activities, *Vacuum*, 2019, **159**, 476–481.

100. L. K. Krehula, J. Stjepanović and M. Perlog, *et al.*, Conducting polymer polypyrrole and titanium dioxide nanocomposites for photocatalysis of RR45 dye under visible light, *Polym. Bull.*, 2019, **76**, 1697–1715.
101. R. Hickman, E. Walker and S. Chowdhury, TiO_2-PDMS composite sponge for adsorption and solar mediated photodegradation of dye pollutants, *J. Water Process Eng.*, 2018, **24**, 74–82.
102. U. M. Garusinghe, V. S. Raghuwanshi, W. Batchelor and G. Garnier, Water resistant cellulose—titanium dioxide composites for photocatalysis, *Sci. Rep.*, 2018, **8**, 2306.
103. V. Wongso, C. J. Chen and A. Razzaq, *et al.*, Hybrid kaolin/TiO_2 composite: effect of urea addition towards an efficient photocatalyst for dye abatement under visible light irradiation, *Appl. Clay Sci.*, 2019, **180**, 105158.
104. Y. Li, J. Zhou and Y. Fan, *et al.*, Preparation of environment-friendly 3D eggshell membrane-supported anatase TiO_2 as a reusable photocatalyst for degradation of organic dyes, *Chem. Phys. Lett.*, 2017, **689**, 142–147.
105. Z. Liu, R. Liu and Y. Yi, *et al.*, Photocatalytic degradation of dyes over a xylan/PVA/TiO_2 composite under visible light irradiation, *Carbohydr. Polym.*, 2019, **223**, 115081.
106. S. Zhou, L. Xia, K. Zhang, Z. Fu, Y. Wang, Q. Zhang, L. Zhai, Y. Mao and W. Xu, Titanium dioxide decorated natural cellulosic Juncus effusus fiber for highly efficient photodegradation towards dyes, *Carbohydr. Polym.*, 2020, **232**, 115830.
107. V. S. Raghuwanshi, U. M. Garusinghe, W. Batchelor and G. Garnier, Polyamide-amine-epichlorohydrin (PAE) induced TiO_2 nanoparticles assembly in cellulose network, *J. Colloid Interface Sci.*, 2020, **575**, 317–325.
108. P. Makal and D. Das, Superior photocatalytic dye degradation under visible light by reduced graphene oxide laminated TiO_2-B nanowire composite, *J. Environ. Chem. Eng.*, 2019, **7**, 103358.
109. D. Rogala-Wielgus, B. Majkowska-Marzec, A. Zieliński, K. Roszek and M. Liszewska, Evaluation of adhesion strength, corrosion, and biological properties of the MWCNT/TiO_2 coating intended for medical applications, *RSC Adv.*, 2023, **13**(43), 30108–30117.
110. M. A. Kanjwal, K. K. S. Lo and W. W.-F. Leung, Graphene composite nanofibers as a high-performance photocatalyst for environmental remediation, *Sep. Purif. Technol.*, 2019, **215**, 602–611.
111. S. Singh, H. Mahalingam and P. K. Singh, Polymer-supported titanium dioxide photocatalysts for environmental remediation: A review, *Appl. Catal., A*, 2013, **462**, 178–195.
112. R. Wang, K. Shi, D. Huang, J. Zhang and S. An, Synthesis and degradation kinetics of TiO_2/GO composites with highly efficient activity for adsorption and photocatalytic degradation of MB, *Sci. Rep.*, 2019, **9**(1), 1–9.
113. A. Romero-Moran, J. L. Sanchez-Salas and J. Molina-Reyes, Influence of selected reactive oxygen species on the photocatalytic activity of TiO_2/SiO_2 composite coatings processed at low temperature, *Appl. Catal., B*, 2021, **291**, 119685.

114. J. Krýsa, M. Baudys, X. Vislocka and M. Neumann-Spallart, Composite photocatalysts based on TiO_2–carbon for air pollutant removal: Aspects of adsorption, *Catal. Today*, 2020, **340**, 34–39.
115. R. Ata, O. Sacco, V. Vaiano, L. Rizzo, G. Y. Tore and D. Sannino, Visible light active N-doped TiO_2 immobilized on polystyrene as efficient system for wastewater treatment, *J. Photochem. Photobiol., A*, 2017, **348**, 255–262.
116. B. Srikanth, R. Goutham, R. B. Narayan, A. Ramprasath, K. P. Gopinath and A. R. Sankaranarayanan, Recent advancements in supporting materials for immobilised photocatalytic applications in waste water treatment, *J. Environ. Manage.*, 2017, **200**, 60–78.
117. A. Joseph and A. Vijayanandan, Review on Support Materials Used for Immobilization of Nano-Photocatalysts for Water Treatment Applications, *Inorg. Chim. Acta*, 2022, 121284.
118. I. Grčić, N. Koprivanac and G. L. Puma, Modeling the photocatalytic oxidation of carboxylic acids on aqueous TiO_2 suspensions and on immobilized TiO_2-chitosan thin films in different reactor geometries irradiated by UVA or UVC light sources, *Chem. Eng. J.*, 2021, **422**, 130104.
119. H. M. Ali, F. A. Roghabadi and V. Ahmadi, Solid-supported photocatalysts for wastewater treatment: Supports contribution in the photocatalysis process, *Sol. Energy*, 2023, **255**, 99–125.
120. D. L. Cunha, A. Kuznetsov, C. A. Achete, A. E. da H. Machado and M. Marques, Immobilized TiO_2 on glass spheres applied to heterogeneous photocatalysis: Photoactivity, leaching and regeneration process, *PeerJ*, 2018, **6**, e4464.
121. I. M. Sosnin, S. Vlassov and L. M. Dorogin, Application of polydimethylsiloxane in photocatalyst composite materials: A review, *React. Funct. Polym.*, 2021, **158**, 104781.
122. M. J. López-Muñoz, R. van Grieken, J. Aguado and J. Marugán, Role of the support on the activity of silica-supported TiO_2 photocatalysts: structure of the TiO_2/SBA-15 photocatalysts, *Catal. Today*, 2005, **101**(3–4), 307–314.
123. Z. Long, Q. Li, T. Wei, G. Zhang and Z. Ren, Historical development and prospects of photocatalysts for pollutant removal in water, *J. Hazard. Mater.*, 2020, **395**, 122599.
124. B. Srikanth, R. Goutham, R. B. Narayan, A. Ramprasath, K. P. Gopinath and A. R. Sankaranarayanan, Recent advancements in supporting materials for immobilised photocatalytic applications in waste water treatment, *J. Environ. Manage.*, 2017, **200**, 60–78.
125. M. H. Alhaji, K. Sanaullah, A. Khan, A. Hamza, A. Muhammad, M. S. Ishola, A. R. H. Rigit and S. A. Bhawani, Recent developments in immobilizing titanium dioxide on supports for degradation of organic pollutants in wastewater-A review, *Int. J. Environ. Sci. Technol.*, 2017, **14**, 2039–2052.
126. P. A. Atmianlu, R. Badpa, V. Aghabalaei and M. Baghdadi, A review on the various beds used for immobilization of nanoparticles: Overcoming

the barrier to nanoparticle applications in water and wastewater treatment, *J. Environ. Chem. Eng.*, 2021, **9**(6), 106514.
127. P. N. Hishimone, H. Nagai and M. Sato, Methods of fabricating thin films for energy materials and devices, in *Lithium-ion Batteries-Thin Film for Energy Materials and Devices*, IntechOpen, 2020.
128. V. H. Nguyen, Q. B. Tran, X. C. Nguyen, T. T. T. Ho, M. Shokouhimehr, D. V. N. Vo, S. S. Lam, H. P. Nguyen, C. T. Hoang, Q. V. Ly, W. Peng, S. Y. Kim, T. V. Tung and Q. Van Le, Submerged photocatalytic membrane reactor with suspended and immobilized N-doped TiO_2 under visible irradiation for diclofenac removal from wastewater, *Process Saf. Environ. Prot.*, 2020, **142**, 229–237.
129. R. Zouzelka, M. Remzova, J. Plsek, L. Brabec and J. Rathousky, Immobilized rGO/TiO_2 photocatalyst for decontamination of water, *Catalysts*, 2019, **9**(9), 708.
130. Z. Xin, X. Zhao, H. Ji, T. Ma, H. Li, S. Zhong and Z. Shen, Amorphous carbon-linked TiO_2/carbon nanotube film composite with enhanced photocatalytic performance: The effect of interface contact and hydrophilicity, *Chin. Chem. Lett.*, 2021, **32**(7), 2151–2154.
131. M. Behpour, P. Shirazi and M. Rahbar, Immobilization of the Fe_2O_3/TiO_2 photocatalyst on carbon fiber cloth for the degradation of a textile dye under visible light irradiation, *React. Kinet., Mech. Catal.*, 2019, **127**, 1073–1085.
132. X. Shen, L. Song, L. Luo, Y. Zhang, B. Zhu, J. Liu, Z. Chen and L. Zhang, Preparation of TiO_2/C_3N_4 heterojunctions on carbon-fiber cloth as efficient filter-membrane-shaped photocatalyst for removing various pollutants from the flowing wastewater, *J. Colloid Interface Sci.*, 2018, **532**, 798–807.
133. S. Sun, R. Zhao, Y. Xie and Y. Liu, Photocatalytic degradation of aflatoxin B1 by activated carbon supported TiO_2 catalyst, *Food Control*, 2019, **100**, 183–188.
134. L. Rimoldi, D. Meroni, G. Cappelletti and S. Ardizzone, Green and low cost tetracycline degradation processes by nanometric and immobilized TiO_2 systems, *Catal. Today*, 2017, **281**, 38–44.
135. N. M. Mahmoodi, M. Arami and J. Zhang, Preparation and photocatalytic activity of immobilized composite photocatalyst (titania nanoparticle/activated carbon), *J. Alloys Compd.*, 2011, **509**(14), 4754–4764.
136. S. Das and H. Mahalingam, Novel immobilized ternary photocatalytic polymer film based airlift reactor for efficient degradation of complex phthalocyanine dye wastewater, *J. Hazard. Mater.*, 2020, **383**, 121219.
137. R. Ata, O. Sacco, V. Vaiano, L. Rizzo, G. Y. Tore and D. Sannino, Visible light active N-doped TiO_2 immobilized on polystyrene as efficient system for wastewater treatment, *J. Photochem. Photobiol., A*, 2017, **348**, 255–262.
138. X. Zheng, Z. P. Shen, L. Shi, R. Cheng and D. H. Yuan, Photocatalytic membrane reactors (PMRs) in water treatment: configurations and influencing factors, *Catalysts*, 2017, **7**(8), 224.

139. M. R. Al-Mamun, S. Kader, M. S. Islam and M. Z. H. Khan, Photocatalytic activity improvement and application of UV-TiO_2 photocatalysis in textile wastewater treatment: A review, *J. Environ. Chem. Eng.*, 2019, **7**(5), 103248.
140. L. Liu, X. B. Luo, L. Ding and S. L. Luo, Application of nanotechnology in the removal of heavy metal from water, in *Nanomaterials for the Removal of Pollutants and Resource Reutilization*, Elsevier, 2019, pp. 83–147.
141. N. Gholami, B. Ghasemi, B. Anvaripour and S. Jorfi, Enhanced photocatalytic degradation of furfural and a real wastewater using UVC/TiO_2 nanoparticles immobilized on white concrete in a fixed-bed reactor, *J. Ind. Eng. Chem.*, 2018, **62**, 291–301.
142. M. Kete, U. L. Štangar and M. Tasbihi, Alternative support materials for titania photocatalyst towards degradation of organic pollutants, *J. Water Process Eng.*, 2018, **23**, 142–150.
143. T. E. Berger, C. Regmi, A. I. Schäfer and B. S. Richards, Photocatalytic degradation of organic dye via atomic layer deposited TiO_2 on ceramic membranes in single-pass flow-through operation, *J. Membr. Sci.*, 2020, **604**, 118015.
144. R. Lyubimenko, D. Busko, B. S. Richards, A. I. Schäfer and A. Turshatov, Efficient photocatalytic removal of methylene blue using a metalloporphyrin–poly (vinylidene fluoride) hybrid membrane in a flow-through reactor, *ACS Appl. Mater. Interfaces*, 2019, **11**(35), 31763–31776.
145. R. M. Castellanos, J. P. Bassin, M. Dezotti, R. A. Boaventura and V. J. Vilar, Tube-in-tube membrane reactor for heterogeneous TiO_2 photocatalysis with radial addition of H_2O_2, *Chem. Eng. J.*, 2020, **395**, 124998.
146. H. S. Zakria, M. H. D. Othman, R. Kamaludin, S. H. S. A. Kadir, T. A. Kurniawan and A. Jilani, Immobilization techniques of a photocatalyst into and onto a polymer membrane for photocatalytic activity, *RSC Adv.*, 2021, **11**(12), 6985–7014.
147. H. S. Zakria, M. H. D. Othman, R. Kamaludin, S. H. S. A. Kadir, T. A. Kurniawan and A. Jilani, Immobilization techniques of a photocatalyst into and onto a polymer membrane for photocatalytic activity, *RSC Adv.*, 2021, **11**(12), 6985–7014.
148. N. H. Shafudah, H. Nagai, Y. Suwazono, R. Ozawa, Y. Kudoh, T. Takahashi, T. Onuma and M. Sato, Hydrophilic titania thin films from a molecular precursor film formed via electrospray deposition on a quartz glass substrate precoated with carbon nanotubes, *Coatings*, 2020, **10**(11), 1050.
149. J. C. Colmenares, R. S. Varma and V. Nair, Selective photocatalysis of lignin-inspired chemicals by integrating hybrid nanocatalysis in microfluidic reactors, *Chem. Soc. Rev.*, 2017, **46**(22), 6675–6686.
150. Y. Wang, R. Zhang, J. Li, L. Li and S. Lin, First-principles study on transition metal-doped anatase TiO_2, *Nanoscale Res. Lett.*, 2014, **9**, 1–8.
151. J. J. Li, S. C. Cai, E. Q. Yu, B. Weng, X. Chen, J. Chen, H. Jia and Y. J. Xu, Efficient infrared light promoted degradation of volatile organic compounds over photo-thermal responsive Pt-rGO-TiO_2 composites, *Appl. Catal., B*, 2018, **233**, 260–271.

152. N. Liu, Q. Zhu, N. Zhang, C. Zhang, N. Kawazoe, G. Chen, N. Negishi and Y. Yang, Superior disinfection effect of Escherichia coli by hydrothermal synthesized TiO_2-based composite photocatalyst under LED irradiation: Influence of environmental factors and disinfection mechanism, *Environ. Pollut.*, 2019, **247**, 847–856.
153. A. Rafiq, M. Ikram, S. Ali, F. Niaz, M. Khan, Q. Khan and M. Maqbool, Photocatalytic degradation of dyes using semiconductor photocatalysts to clean industrial water pollution, *J. Ind. Eng. Chem.*, 2021, **97**, 111–128.
154. J. F. Pesqueira, M. F. R. Pereira and A. M. Silva, A life cycle assessment of solar-based treatments (H_2O_2, TiO_2 photocatalysis, circumneutral photo-Fenton) for the removal of organic micropollutants, *Sci. Total Environ.*, 2021, **761**, 143258.
155. M. Jahdi, S. B. Mishra, E. N. Nxumalo, S. D. Mhlanga and A. K. Mishra, Mechanistic pathways for the degradation of SMX drug and floatation of degraded products using F–Pt co-doped TiO_2 photocatalysts, *RSC Adv.*, 2020, **10**(46), 27662–27675.
156. D. B. Silva, G. Buttiglieri and S. Babić, State-of-the-art and current challenges for TiO_2/UV-LED photocatalytic degradation of emerging organic micropollutants, *Environ. Sci. Pollut. Res.*, 2021, **28**, 103–120.
157. J. Chen, S. Loeb and J. H. Kim, LED revolution: fundamentals and prospects for UV disinfection applications, *Environ. Sci.: Water Res. Technol.*, 2017, **3**(2), 188–202.
158. A. Gnanaprakasam, V. M. Sivakumar, P. L. Sivayogavalli and M. Thirumarimurugan, Characterization of TiO_2 and ZnO nanoparticles and their applications in photocatalytic degradation of azodyes, *Ecotoxicol. Environ. Saf.*, 2015, **121**, 121–125.
159. K. S. Varma, R. J. Tayade, K. J. Shah, P. A. Joshi, A. D. Shukla and V. G. Gandhi, Photocatalytic degradation of pharmaceutical and pesticide compounds (PPCs) using doped TiO_2 nanomaterials: A review, *Water-Energy Nexus*, 2020, **3**, 46–61.
160. S. A. H. Deba, B. A. Wols, D. R. Yntema and R. G. Lammertink, Photocatalytic ceramic membrane: Effect of the illumination intensity and distribution, *J. Photochem. Photobiol., A*, 2023, **437**, 114469.
161. G. C. Collazzo, E. L. Foletto, S. L. Jahn and M. A. Villetti, Degradation of Direct Black 38 dye under visible light and sunlight irradiation by N-doped anatase TiO_2 as photocatalyst, *J. Environ. Manage.*, 2012, **98**, 107–111.
162. A. Kumar and G. Pandey, A review on the factors affecting the photocatalytic degradation of hazardous materials, *Mater. Sci. Eng. Int. J.*, 2017, **1**(3), 1–10.
163. N. Davari, M. Farhadian, A. R. S. Nazar and M. Homayoonfal, Degradation of diphenhydramine by the photocatalysts of ZnO/Fe_2O_3 and TiO_2/Fe_2O_3 based on clinoptilolite: structural and operational comparison, *J. Environ. Chem. Eng.*, 2017, **5**(6), 5707–5720.
164. M. R. D. Khaki, M. S. Shafeeyan, A. A. A. Raman and W. M. A. W. Daud, Application of doped photocatalysts for organic pollutant degradation-A review, *J. Environ. Manage.*, 2017, **198**, 78–94.

165. B. Sharma, P. K. Boruah, A. Yadav and M. R. Das, TiO_2–Fe_2O_3 nanocomposite heterojunction for superior charge separation and the photocatalytic inactivation of pathogenic bacteria in water under direct sunlight irradiation, *J. Environ. Chem. Eng.*, 2018, **6**(1), 134–145.
166. F. Ahmedchekkat, M. S. Medjram, M. Chiha and A. M. A. Al-Bsoul, Sonophotocatalytic degradation of Rhodamine B using a novel reactor geometry: Effect of operating conditions, *Chem. Eng. J.*, 2011, **178**, 244–251.
167. M. Mansouri, M. Nademi, M. E. Olya and H. Lotfi, Study of methyl tert-butyl ether (MTBE) photocatalytic degradation with UV/TiO_2-ZnO-CuO nanoparticles, *J. Chem. Health Risks*, 2017, **7**(1), 19–32.
168. H. Anwer, A. Mahmood, J. Lee, K. H. Kim, J. W. Park and A. C. Yip, Photocatalysts for degradation of dyes in industrial effluents: Opportunities and challenges, *Nano Res.*, 2019, **12**, 955–972.
169. A. A. Zewde, L. Zhang, Z. Li and E. A. Odey, A review of the application of sonophotocatalytic process based on advanced oxidation process for degrading organic dye, *Rev. Environ. Health*, 2019, **34**(4), 365–375.
170. Y. Wu, H. Pang, Y. Liu, X. Wang, S. Yu, D. Fu, J. Chen and X. Wang, Environmental remediation of heavy metal ions by novel-nanomaterials: a review, *Environ. Pollut.*, 2019, **246**, 608–620.
171. N. Jallouli, K. Elghniji, H. Trabelsi and M. Ksibi, Photocatalytic degradation of paracetamol on TiO_2 nanoparticles and TiO_2/cellulosic fiber under UV and sunlight irradiation, *Arabian J. Chem.*, 2017, **10**, S3640–S3645.
172. A. Bouarioua and M. Zerdaoui, Photocatalytic activities of TiO_2 layers immobilized on glass substrates by dip-coating technique toward the decolorization of methyl orange as a model organic pollutant, *J. Environ. Chem. Eng.*, 2017, **5**(2), 1565–1574.
173. A. L. Boreen, W. A. Arnold and K. McNeill, Photochemical fate of sulfa drugs in the aquatic environment: sulfa drugs containing five-membered heterocyclic groups, *Environ. Sci. Technol.*, 2004, **38**(14), 3933–3940.
174. S. Kader, M. R. Al-Mamun, M. B. K. Suhan, S. B. Shuchi and M. S. Islam, Enhanced photodegradation of methyl orange dye under UV irradiation using MoO_3 and Ag doped TiO_2 photocatalysts, *Environ. Technol. Innovation*, 2022, **27**, 102476.
175. M. M. Islam and S. Basu, Effect of morphology and pH on (photo) electrochemical degradation of methyl orange using TiO_2/Ti mesh photocathode under visible light, *J. Environ. Chem. Eng.*, 2015, **3**(4), 2323–2330.
176. R. Yuan, Y. Zhu, B. Zhou and J. Hu, Photocatalytic oxidation of sulfamethoxazole in the presence of TiO_2: Effect of matrix in aqueous solution on decomposition mechanisms, *Chem. Eng. J.*, 2019, **359**, 1527–1536.
177. M. Długosz, P. Żmudzki, A. Kwiecień, K. Szczubiałka, J. Krzek and M. Nowakowska, Photocatalytic degradation of sulfamethoxazole in aqueous solution using a floating TiO_2-expanded perlite photocatalyst, *J. Hazard. Mater.*, 2015, **298**, 146–153.
178. S. Kader, M. R. Al-Mamun, M. B. Suhan, S. B. Shuchi and M. S. Islam, Enhanced photodegradation of methyl orange dye under UV irradiation

using MoO_3 and Ag doped TiO_2 photocatalysts, *Environ. Technol. Innovation*, 2022, **27**, 102476.
179. A. S. Gonzalez and S. S. Martinez, Study of the sonophotocatalytic degradation of basic blue 9 industrial textile dye over slurry titanium dioxide and influencing factors, *Ultrason. Sonochem.*, 2008, **15**(6), 1038–1042.
180. M. Dhanasekar, S. Ratha, C. S. Rout and S. V. Bhat, Efficient sono-photocatalytic degradation of methylene blue using nickel molybdate nanosheets under diffused sunlight, *J. Environ. Chem. Eng.*, 2017, **5**(3), 2997–3004.
181. X. J. Wen, C. G. Niu, L. Zhang, C. Liang, H. Guo and G. M. Zeng, Photocatalytic degradation of ciprofloxacin by a novel Z-scheme CeO_2–Ag/AgBr photocatalyst: influencing factors, possible degradation pathways, and mechanism insight, *J. Catal.*, 2018, **358**, 141–154.
182. S. Sugashini, T. Gomathi, R. A. Devi, P. N. Sudha, K. Rambabu and F. Banat, Nanochitosan/carboxymethyl cellulose/TiO_2 biocomposite for visible-light-induced photocatalytic degradation of crystal violet dye, *Environ. Res.*, 2022, **204**, 112047.
183. K. Hashimoto, H. Irie and A. Fujishima, TiO_2 photocatalysis: a historical overview and future prospects, *Jpn. J. Appl. Phys.*, 2005, **44**(12R), 8269.
184. J. Peral, X. Domènech and D. F. Ollis, Heterogeneous photocatalysis for purification, decontamination and deodorization of air, *J. Chem. Technol. Biotechnol.*, 1997, **70**(2), 117–140.
185. G. Mamba, M. A. Mamo, X. Y. Mbianda and A. K. Mishra, Nd, N, S-TiO_2 decorated on reduced graphene oxide for a visible light active photocatalyst for dye degradation: Comparison to its MWCNT/Nd, N, S-TiO_2 analogue, *Ind. Eng. Chem. Res.*, 2014, **53**(37), 14329–14338.
186. L. Karimi, S. Zohoori and M. E. Yazdanshenas, Photocatalytic degradation of azo dyes in aqueous solutions under UV irradiation using nano-strontium titanate as the nanophotocatalyst, *J. Saudi Chem. Soc.*, 2014, **18**(5), 581–588.
187. S. Mozia, A. W. Morawski, M. Toyoda and M. Inagaki, Application of anatase-phase TiO_2 for decomposition of azo dye in a photocatalytic membrane reactor, *Desalination*, 2009, **241**(1–3), 97–105.
188. E. T. Soares, M. A. Lansarin and C. C. Moro, A study of process variables for the photocatalytic degradation of rhodamine B, *Braz. J. Chem. Eng.*, 2007, **24**, 29–36.
189. S. Zhou and A. K. Ray, Kinetic studies for photocatalytic degradation of Eosin B on a thin film of titanium dioxide, *Ind. Eng. Chem. Res.*, 2003, **42**(24), 6020–6033.
190. K. Mehrotra, G. S. Yablonsky and A. K. Ray, Macro kinetic studies for photocatalytic degradation of benzoic acid in immobilized systems, *Chemosphere*, 2005, **60**(10), 1427–1436.
191. Z. Shams-Ghahfarokhi and A. Nezamzadeh-Ejhieh, As-synthesized ZSM-5 zeolite as a suitable support for increasing the photoactivity of semiconductors in a typical photodegradation process, *Mater. Sci. Semicond. Process.*, 2015, **39**, 265–275.

192. K. Hashimoto, H. Irie and A. Fujishima, TiO_2 photocatalysis: a historical overview and future prospects, *Jpn. J. Appl. Phys.*, 2005, **44**(12R), 8269.
193. K. Azad and P. Gajanan, Photodegradation of methyl orange in aqueous solution by the visible light active Co: La: TiO_2 nanocomposite, *Chem. Sci. J*, 2017, **8**(3), 1000164–1000174.
194. J. K. Karlsson, O. J. Woodford, R. Al-Aqar and A. Harriman, Effects of temperature and concentration on the rate of photobleaching of Erythrosine in water, *J. Phys. Chem. A*, 2017, **121**(45), 8569–8576.

CHAPTER 13

Enviro-friendly Nanomaterial Synthesis and Its Utilization for Water Purification

HUDA SHARBINI KAMALUDDIN[a] AND KATABATHINI NARASIMHARAO[a*]

[a]Department of Chemistry, Faculty of Science, King Abdulaziz University, P.O. Box 80203, Jeddah 21589, Saudi Arabia
*E-mail: nkatabathini@kau.edu.sa

13.1 Introduction

Water without coloured elements is essential for all forms of life. Water represents about 71% of the Earth's surface. However, not all of it is suitable for drinking, planting, and harvesting since 97.2% is found in sea water and only 2.8% is fresh water[1] (see Figure 13.1). In recent years, growth in population and the daily growth in demand for water requires an attempt to meet the needs for fresh water. As a result, efforts have been made to use ocean water using desalination by distillation or osmosis. However, these processes are costly and are usually carried out on an industry scale.[2] Industrialization is directly or indirectly threatening to the purity of water resources. Direct pollutants include dumping of waste and harmful by-products in the nearest water source without treatment (*e.g.*, from factories, waste management facilities, refineries). On the other hand, indirect pollution enters the water through groundwater, soil, or the atmosphere, such as *via* acid rain.[3] Modern

Chemistry in the Environment Series No. 12
Novel Materials and Water Purification: Towards a Sustainable Future
Edited by Grigorios L. Kyriakopoulos and Miltiadis G. Zamparas

Published by the Royal Society of Chemistry, www.rsc.org

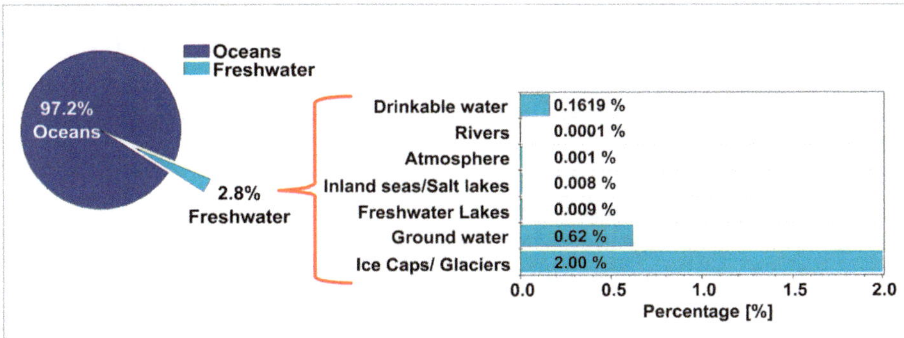

Figure 13.1 The world's water resources.

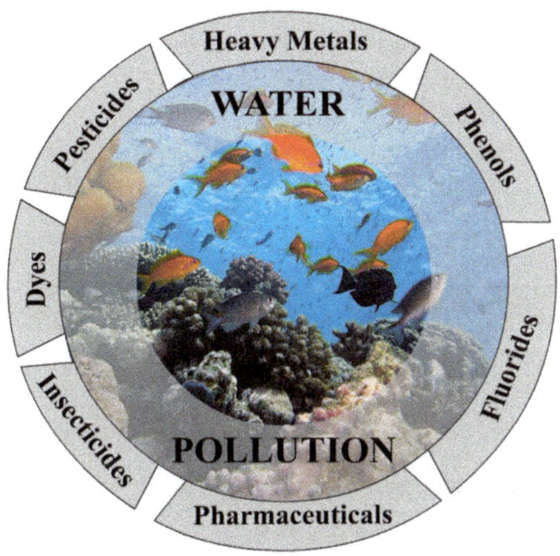

Figure 13.2 The main sources of water pollution.

life, technology, and human activities are also sources of water pollution, such as from waste water and nutrients, municipal, industrial, and agricultural waste, heavy industry, automobiles, and power generation.[4]

Water is known as a "universal solvent" because it can dissolve more substances than other liquids on Earth, which makes it easy to pollute. Water pollution is the dissolution of harmful substances (mainly chemicals and microorganisms) in water bodies, which reduces their quality and threatens humans or the environment. Toxic substances may be heavy metals, pharmaceuticals, insecticides, pesticides, dyes, fluorides, and phenols (see Figure 13.2), and they usually change the chemical or physical properties of the water, such as colour, unpleasant smells, temperature, pH changes (acidity or alkalinity), organic/inorganic solid turbidity, water pressure, *etc.* Water pollution causes disease, ecosystem destruction, and eutrophication due to increased algal growth, which is the formation of a layer on the surface of the water, and hence the reduction of oxygen in the water body. It can also affect the food chain, polluting marine and aquatic animals (fish, shellfish, *etc.*) and poisoning the human body.[3] Pollution affects the environment and humans: every day, 4000 children die of diseases caused by water pollution and unsuitable sanitation; about 1.8 million people (mainly children) are killed yearly by diarrhoea alone.[4]

Consequently, mitigating the negative impacts of water pollution is vital to conserving small freshwater sources. Many efforts have been made to use different techniques. Some treatment techniques include physical, chemical, and biological methods, such as adsorption, electrocoagulation, photocatalytic degradation, membrane separation, and aerobic/anaerobic biodegradation. Adsorption and photocatalytic technologies offer some advantages. Both show simplicity, high removal rates, cost effectiveness, and reduction. Adsorption, or adhesion, is a process of transferring molecules from a gas/liquid to a solid surface (substrates) and is essentially a form of movement of ions or molecules (see Figure 13.3a). This phenomenon occurs due to two types of forces: weak physical forces (van der Waals forces) (known as physisorption) or chemical bond formation and electrostatics (chemisorption).[5,6] This leads to various mechanisms, such as acid–base interactions,

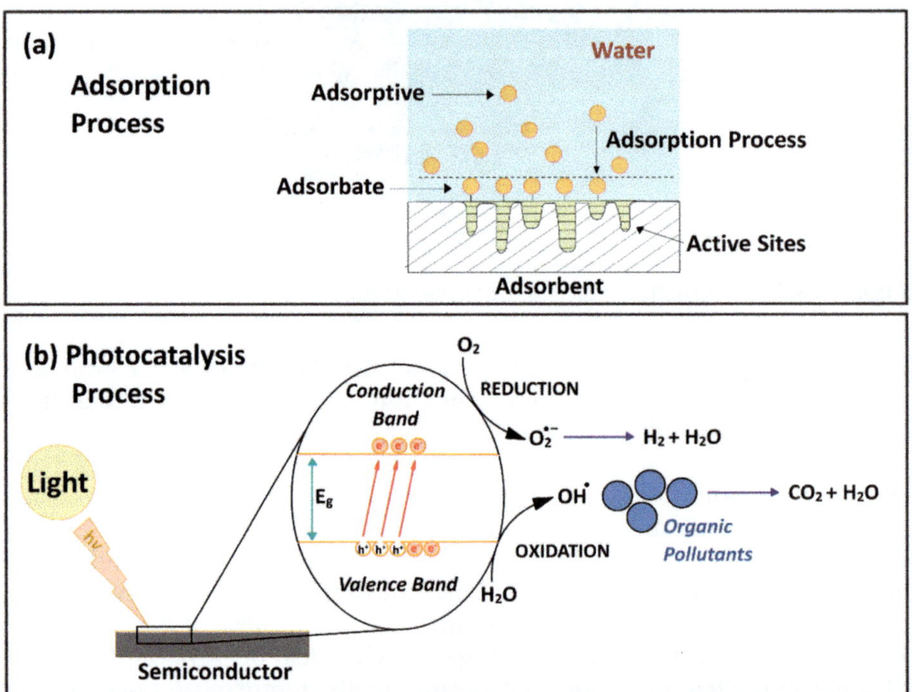

Figure 13.3 Concept of (a) adsorption process and (b) photocatalysis process.

chelation, complexation, hydrogen bonds, electrostatic interactions, diffusion, ion exchange, hole filling, and π interactions.[7] The main characteristics of solid adsorbents are: (1) high selectivity, (2) high adsorption capacity, (3) diameter of pores, (4) distribution of pores, (5) surface active sites, (6) long life, and (7) low cost.[8] Furthermore, the adsorption process for removing contaminants from water depends on the following: (1) the amount of adsorbent, (2) the initial concentration, (3) the pH, and (4) the temperature.[9] The adsorption process has advantages, such as using low-cost materials as absorbents, cost effectiveness, high removal rate, simplicity of operation, and reuse. The photocatalytic activity depends on the materials used,[10] and these materials are often semiconductors, which absorb light (visible or ultraviolet (UV)) depending on the bandgap (E_g), and they transfer electrons from the valence band to the high-energy conduction band, forming active species (free radicals) and forming chemical reactions with molecules (see Figure 13.3b). This process is simple, does not consume significant amounts of energy, does not form harmful by-products, and uses green technologies that reduce inorganic substances or oxidize harmless substances and degrade pollutants into small components, such as water and carbon dioxide.[11]

From a material point of view, substrates and semiconductors have attracted considerable attention in recent decades, especially in their synthesis by green methods using biotechnology. This chapter is intended to provide a platform for researchers to study some efforts using green, safe, and environmental strategies to synthesize substrates and semiconductors, such as carbon materials, metals, metal oxides, and zeolites, using green methods for water treatment, and aims to encourage new ideas on how to pursue comprehensive research of these materials.

13.2 Green Synthesis of Adsorbent and Photocatalyst Materials

Green synthesis methods using plant materials have grown significantly in the last 30 years.[12] In nanotechnology the use of plant-mediated methods is preferred because of the advantages of high efficiency, cost effectiveness, eco-friendliness, speed, and stability. Furthermore, green synthesis is an economical and easy procedure without expensive equipment. In the green process, plants can be used in two ways: (1) using living tissues of biological compounds (such as fruits, bacteria, and other green components), or (2) extracting plants, depending on the biomolecules and secondary metabolites present, usually as reduction or stabilization agents for the formation of metals and oxides.[12–14] When using live plant tissues, many biomaterials are used to clean water, such as cellulose, starch, gelatin, chitin, algae, carbon, chitosan, *etc.* These biomaterials are considered to be more beneficial because they are less expensive, non-toxic, of low energy consumption, biodegradable, biocompatible, and renewable compared to their non-biological counterparts.[15] Plants usually produce phytochemicals, biomolecules, and secondary metabolites, which can be extracted from various parts of the plant, such as roots, leaves, fruits, flowers, seeds, and stems. The extracts obtained from different parts of the plants could act as stabilizing, capping, oxidizing and reducing agents.[16] It is interesting to note that the extract solution includes proteins, enzymes, vitamins, polysaccharides, carbohydrates, organic acids, and amino acids. These phytochemicals behave as stabilizing agents, since external stabilization is restricted.[17]

Consequently, several extraction methods have been reported, such as the use of maceration, Soxhlet extraction, microwaves, supercritical liquids, pulsed electric fields, pressurized liquids, ultrasound, and enzymes. Choosing the most suitable method and solvent to produce biomolecules is very important. Most biomolecule compounds from different plants use a simple maceration method to synthesize adsorbents and photocatalyst materials (see Figure 13.4).[18] In the maceration method, the parts of the plant are dried for several hours and ground into a powder. The obtained powder is mixed with a suitable solvent under constant stirring and finally filtered to obtain an extract solution. The main parameters which we could control in the extraction method are the type of plant material, the solvent (polar or non polar), and the chemical precursors. The mixture of three components is stirred continuously for several hours at room temperature, and eventually, solid nanoparticles are produced (see Figure 13.4).

13.2.1 Carbon-based Materials

Carbon-based/carbonaceous materials are the easiest materials to produce using plant tissues. They include various types, such as activated carbon, biochar, graphene oxide, and carbon nanotubes.[19] These materials are used for water purification as biocarbon and composites with support materials.

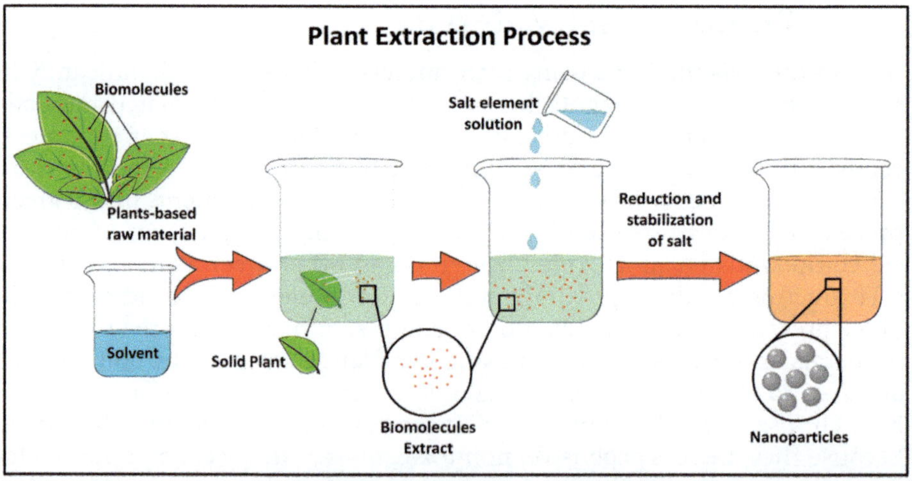

Figure 13.4 A schematic representation of extraction using the maceration method.

13.2.1.1 Biochar Materials

Biochar material is a solid product with a high carbon content, produced from biomass decomposed by thermal chemical processes conditioned by a lack of oxygen,[20] a traditional process called pyrolysis. Biochar is relatively temperature dependent when used in the hydrothermal carbonation process. The higher the processing temperature (between 300 and 800 °C), the lower the biochar yield[21,22] (see Scheme 13.1).

Biochar has many advantages, such as low price, easy-to-obtain, stable and comprehensive sources of raw materials, various processing capabilities, large specific surface area, and structural properties that attract the attention of multiple applications, such as water treatment as an adsorbent.[23,24] Some biochar production from biomass has been reported for water treatment (see Table 13.1). For example, char from grape bagasse is produced by pyrolysis. The adsorption capacity of the adsorbate is 42 mg g^{-1}, and the elimination rate was 96.56%. Similarly, jujube pit biochar (JPB) is a Pb(II) adsorbent obtained from jujube pit through pyrolysis. The JPB texture characteristics are an average diameter of 3387 nm, a pore volume of 0.054 m^3 g^{-1}, and a surface area of 246.9 m^2 g^{-1}. It was found that the kinetic data reached equilibrium in 30 min and that the adsorption capacity was 137.1 mg g^{-1} at pH 6. Interestingly, the JPB still reaches 70% of the original absorption capacity after reuse for five cycles.[25] In another study, nano biochar (NBC) was obtained, a by-product of the gasification of *Gliricidia sepium* from bioenergy waste, and used to remove toxic elements (*e.g.* oxytetracycline (OTC), glyphosate (GL), hexavalent chromium (Cr(VI)), and cadmium (Cd(II))) from water. The remaining biochar was milled on a disc in methanol media, and the colloidal biochar was dispersed in water to obtain NBC after centrifugation. NBC materials have a specific surface area of 28 m^2 g^{-1} and various functional groups, such

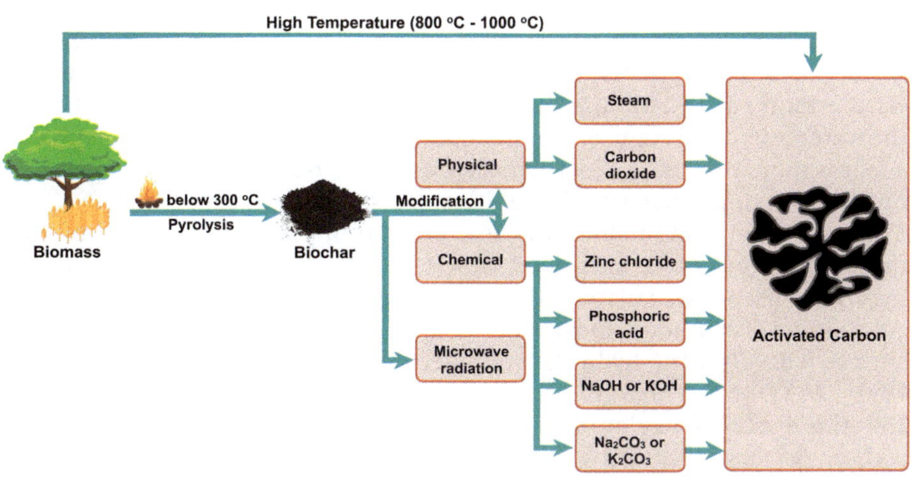

Scheme 13.1 Schematic representation of the formation of biochar materials and activated carbon.

Table 13.1 Different types of green biochar as adsorbate for water purification.

Green adsorbate	Synthesized process	Natural materials	BET ($m^2\ g^{-1}$)	Water contaminant	pH	Adsorption capacity ($mg\ g^{-1}$)
Biochar	Pyrolysis	Grape bagasse	—	Cu(II)		42 [24]
JPB	Pyrolysis	Jujube pit	246.9	Pb(II)		137.1 [25]
NBC	Disk milled in methanol media	Woody biochar as a by-product of *Gliricidia sepium* gasification	28	Oxytetracycline (OTC)	8	15 [26]
				Glyphosate (GL)	7	16.58 [26]
				Hexavalent chromium (Cr(VI))	4	1.2 [26]
				Cadmium (Cd(II))	8	10 [26]
NBC	Planetary ball mill	White pine wood	47.25	Carbamazepine		98[a] [27]

[a]Adsorption capacity (%).

as –OH, C=O, –NH, and $-CH_3$. The adsorption process was conducted at pH = 3–8, and the best adsorption capacity for OTC and Cd(II) was found at pH 8. On the other hand, GL and Cr(VI) showed the best adsorption at pH 7 and 4, respectively (see Table 13.1). It was found that above pH 4, the positive cations of Cd(II) can be bound to the negative charge on the NBC surface. In comparison, Cr(VI) has negative charges that are repelled from the NBC negative surface, which is the reason for the low absorption capacity.[26] According to the same philosophy, white pine wood was used to produce nano biochar using a Planetary ball mill. The optimal grinding parameters were 1.6 hours, 575 rpm, and 4.5 g g^{-1} ball-to-powder ratios at −80 °C. The nano biochar had an average particle size of 60 nm, a specific surface area of 47.25 $m^2\ g^{-1}$, and a pore volume of 0.0335 $cm^3\ g^{-1}$, compared with raw biochar, the specific surface area of which was 3.12 $m^2\ g^{-1}$ with a pore volume of 0.0045 $cm^3\ g^{-1}$. Furthermore, after 24 hours, raw biochar did not adsorb more than 14% of the water micropollutant at room temperature. In comparison, nano biochar adsorbs more than 98% of carbamazepine after 3 hours at room temperature due to the reduced particle size and increased surface area.[27]

Biochar composite material is also used to improve the removal of contamination from waste water (see Table 13.2). For example, orange waste was pyrolyzed (at 650 °C, 2 h) to synthesize biochar and then mixed with a Na_2SiO_3 solution to obtain a composite material of orange waste biochar-magnetic silicate (OBMS) using hydrothermal methods to remove radioactive ions of U(VI). The OBMS compound has a specific surface area of 74.9 $m^2\ g^{-1}$ and contains a stack of agglomerate magnesium silicate layers covered with biocarbon. Because of the active sites of C=C, C–OH, Mg–O–Si and $Mg(OH)_2$ the adsorption capacity was 352.6 mg g^{-1} at 25° C and pH 4.[19] Similarly, multifunctional silver biochar materials (Ag-CMPB) have been produced to adsorb

Table 13.2 Different green reductions for the synthesis of the adsorbent of composite biochar.

Adsorbent	Green reduction	Precursor	Water contaminant	Adsorbent amount ($g L^{-1}$)	Adsorption capacity (%)
OBMS	Orange peel waste	Na_2SiO_3	U(VI)	0.2	352.6 [19]
FABC	Algae of *Aegagropila linnaei* (AL)	$FeSO_4 \cdot 7H_2O$	Bisphenol A (BPA)	200	69.8 [30]
Ag-CMPB	*Citrus maxima* peel (CMP)	$AgNO_3$	Methylene blue (MB)	0.05	95.5[a] [28]
			Rhodamine B (RB)	0.05	69.3[a] [28]
			Methyl orange (MO)	0.05	51.5[a] [28]
MnO_2/PF	Palm frond waste and *Acacia nilotica* seed for extract	$MnCl_2 \cdot 4H_2O$	MO	0.01 to 0.1	91.5 [29]

[a]The adsorption capacity ($mg\ g^{-1}$).

pollutant organic compounds (such as methyl orange (MO), rhodamine B (RB), and methylene blue (MB)). Citrus maxima peel residuals (CMPs) were used to prepare biochar and CMP extracts as a reduction agent for the biosynthesis of Ag-CMPB materials. The CMP extract was mixed with a 1.25 mM $AgNO_3$ solution under light illumination and stirred for 30 minutes at 300 rpm in the dark. Subsequently, 0.53 g of CMP biochar was added to the mixture and stirred for 5 hours under sunlight to obtain the bifunctional material Ag-CMPB. The texture properties of the Ag-CMPB show a specific surface area of 79.2 $m^2\ g^{-1}$, and the diameter was 2.24 nm. The adsorption efficiency for removing organic compounds was in the following sequence: MB > RB > MO dyes, with an absorption capacity of 95.5, 69.3, and 51.5 $mg\ g^{-1}$, respectively.[28] Following the same philosophy, biochar was obtained from palm tree fronds and the extraction of *Acacia nilotica* seeds to synthesize MnO_2 nanoparticles. MnO_2 was loaded onto the surface of the biochar to form a MnO_2/PF composite adsorbent to remove MO dye from a water solution. The slow pyrolysis of palm frond waste at 600 °C and a slow heating rate were used to obtain the biochar. *Acacia nilotica* was washed and powdered to prepare the extract. This was then boiled in distilled water at 100 °C for 15 minutes and then centrifuged to obtain the extract. To produce MnO_2/PF, 2.5 g of biochar and 0.1 M $MnCl_2 \cdot 4H_2O$ with *Acacia nilotica* extract were mixed at room temperature at pH = 11 (1 M KOH) for 30 min. Subsequently, the mixture was filtered, dried at 40 °C, and calcined for 3 hours at 200 °C. The specific surface of the composite was 70.97 $m^2\ g^{-1}$, and the average pore diameter was 0.1363 $m^3\ g^{-1}$. With a pH of 2.5, the removal efficiency of the MO dye was 91.5% because of the presence of functional groups of C–H, C=C, C–O, C=O, and C=C.[29]

13.2.1.2 Bioactivated Carbon Materials

Activated carbon is often used as a sorbent for the separation and purification industry. The most important properties for its activity are porosity, homogeneous structure, high surface area, and radiation stability. However, commercial activated carbon regeneration has many problems and is relatively expensive compared to other materials.[9] Nowadays, researchers are interested in finding cheaper, environmentally sustainable, and nature-rich alternatives by using biomass as raw material, such as cherry stone, tea waste, bamboo, orange peel, rice shell, coconut husk, *Camellia oleifera* shell, *etc.*, to produce bioactivated carbon after biochar modification. In general, there are four ways of obtaining activated carbon from biochar: (1) physical activation, (2) chemical activation, (3) microwave radiation,[31] and (4) directly heating biochar at high temperatures (800–1000 °C) (Scheme 13.1). Physical activation of biochar includes treating with steam and carbon dioxide (CO_2).[32] In the case of chemical activation methods treatment of biochar with different chemical agents is the common practice. However, chemical activation often uses corrosive acids, such as sulfuric acid (H_2SO_4), hydrochloric acid (HCl), and nitric acid (HNO_3),[33] which harm the environment. To minimize the corrosive effect, some efforts have been made to use green methods, such as using (1) potassium or sodium hydroxide (KOH or NaOH), (2) potassium and sodium carbonate (K_2CO_3 or Na_2CO_3), and (3) zinc chloride ($ZnCl_2$).[32] For example, the biochar from jackfruit peel was impregnated with KOH and activated by microwave radiation to form activated carbon.[34] Following the same philosophy, sour cherry (*Prunus cerasus* L.), a waste of the fruit juice industry, was used to synthesize bioactive carbon in the presence of $ZnCl_2$. The researchers pointed out that the optimal conditions for obtaining a high surface area of 1704 $m^2\ g^{-1}$ were an activation temperature of 700 °C and an impregnation ratio of 3 : 1.[35] In addition, palm oil fronds were also used to produce adsorbents of activated biocarbon to remove pesticides such as bentazon, carbofuran, and 2,4-dichlorophenoxyacetic acid (2,4-D) from water. These were synthesized by impregnation methods using KOH; the activation temperature was 750 °C for 2 hours, and the chemical impregnation rate was 2.38. It was observed that bioactive carbon eliminates benzene (8.2%), carbofuran (1.3%), and 2,4-D (9.2%)[36] (see Table 13.3). A novel activated carbon (ACCI) has also been synthesized from the non-living lichen *Cetraria islandica* (L.) Ach (LCI). The aim was to investigate and compare the adsorption efficiency of LCI and ACCI in the removal of malachite green (MG) dye from water. First, the LCI was cleaned and dried in the oven at 105 °C for 24 hours until weight variation in the sample was no longer observed. Then, the dried LCI was froze with liquid nitrogen and sieved (63 μm). Thus, LCI showed a specific surface of 1.103 $m^2\ g^{-1}$,

Table 13.3 Different types of green carbon-based adsorbates for water purification.

Green adsorbate	Activated process	Natural materials	BET (m^2 g^{-1})	Water contaminant	Adsorption capacity (%)
Biochar	Impregnated in KOH, biochar *via* microwave irradiation	Jackfruit peel	1526.4	—	—[34]
Activated biocarbon		Lignite		CN	
		Nut shells			
		Rice hulls			
Activated biocarbon	Impregnated in $ZnCl_2$	Sour cherry stone	1704		—[35]
Activated biocarbon	Impregnated in KOH	Palm oil fronds	1237.13	Pesticides (bentazon, carbofuran, and 2,4-D)	—[36]
ACCI	Impregnated in $ZnCl_2$	Non-living lichen *Cetraria islandica* (L.) Ach.	394.417	Malachite green (MG) dye	93.46[37]

total pore volume of 0.0044 cm^3 g^{-1}, and an average pore diameter of 14.02 nm. ACCI was obtained using the chemical impregnation method in a $ZnCl_2$ solution with a 1/1 (w/w) ratio, dried in a furnace at 105 °C, and activated under a nitrogen atmosphere at 768 °C for 1 h. After activation, the resulting product was refrigerated under a nitrogen atmosphere, left in HCl for 15 minutes, filtered, and then washed. After treatment, the specific surface area was 394.417 m^2 g^{-1}, the total pore volume was 0.1216 cm^3 g^{-1}, and the average diameter was 1.28 nm. It was found that the removal efficiency of ACCI for MG dye was 93.46%, with a short equilibrium time compared to LCI, and an effective removal efficiency of about 84.49%. It was concluded that LCI and ACCI materials are promising sorbents for removing 37 cationic dyes.[37]

13.2.2 Metals

Metal nanomaterials (NMs) and nanoparticles (NPs) have promising properties for water treatment, such as high surface area, polar and non-polar chemistry, controlled and tunable size, easier biodegradation, and reactivity.[38,39] In recent years, many attempts have been made to synthesize NPs in greener protocols. There are two types of synthesized NPs: (1) organic and (2) inorganic. Organic NPs include carbon nanostructures, and inorganic NPs include magnetic, semiconductor, and noble metal NPs.[38] However, inorganic methods, including physical or chemical synthesis, produce toxic and harmful waste, have high costs, and consume energy. On the other hand, organic NP synthesis uses biological or chemical methods that are also known as green protocols. Organic extraction allows biomolecules used as reduction or stabilization agents to be environmentally friendly, efficient, cost effective, and non-toxic. Most biomolecules used are plants, algae, fungi, bacteria, actinomycetes, and yeast.[7] Generally, the literature contains many efforts to produce nanoparticles using alternative green methods for gold (Au NPs) and silver (Ag NPs) nanoparticles in order to clean water of various pollutants, including organic and inorganic compounds, heavy metals, and microorganisms.

13.2.2.1 Gold Nanoparticles

Gold nanoparticles (Au NPs) have gained much attention in many fields due to their unique physicochemical properties, such as attainable sizes, plasmon resonance properties, facial synthesis, and multifunctional capabilities with well-characterized properties.[40] Gold has a polyhedral structure, classified as a face-centred cubic (FCC) structure (Figure 13.5). There are ways to produce Au NPs using various biological resources, such as biosynthesis.

An example of green synthesis of Au NPs is using an unconventional and environmentally friendly cinnamon bark extract as a reduction agent and stabilizer. The cinnamon barks were cleaned, dried, and crushed. The fine cinnamon bark was mixed with distilled water, boiled for 5 minutes, and filtered. Subsequently, the extract was mixed with a solution of (1 mM) gold(III) chloride trihydrate ($HAuCl_4$) and then heated for 15 minutes in a microwave. Transmission electron microscopy (TEM) analysis showed the presence of spherical particles with an average size of 5 to 35 nm, and UV–visible spectroscopy showed a surface plasmon resonance peak of 566 nm. The Au NPs were tested with fluorometese-eosin-Y dye, and the intense eosin peak was suppressed, indicating the activity of the Au NPs.[42] In addition, Au NP biosynthesis was performed using honeybee

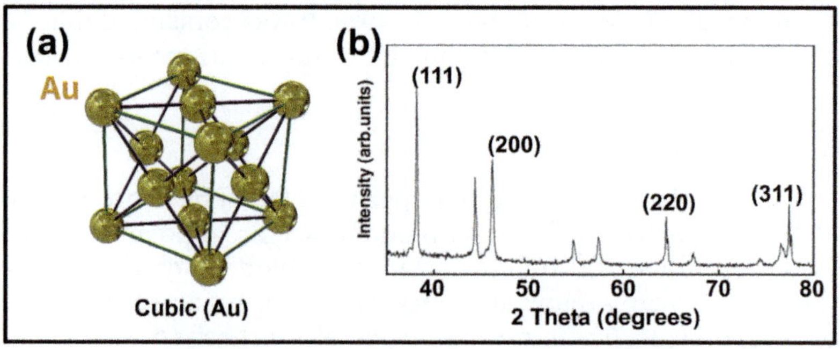

Figure 13.5 (a) Cubic face-centred crystal structure of Au metal nanomaterials with (b) X-ray diffraction (XRD) pattern. Reproduced from ref. 41, https://doi.org/10.1007/s13204-012-0115-7, under the terms of the CC BY 2.0 license https://creativecommons.org/licenses/by/2.0/.

pollen. The pollen extract was produced by mixing 250 grams of dark yellow honeybee pollen with Milli-Q water and homogenizing the mixture into a paste using a ceramic pestle and mortar. The solution was stirred for 15 minutes at ambient temperature and centrifuged to remove unwanted aggregates from the pollen extract. Then, the pollen extract was mixed with (0.5 mM) $AuCl^{-}_4$ solution and incubated at 25 °C for 24 hours. The colour of the solution had changed from light yellow to light magenta, confirming the formation of Au NPs. The biosynthesized Au NPs were shown by TEM to have well-dispersed triangular particle sizes of 7 to 42 nm (see Figure 13.6b). The catalytic activity of biosynthesized Au NPs was studied by reducing 4-nitrophenols (4-NP) to 4-amine phenols (4-AP) in the presence of sodium borohydride ($NaBH_4$) at 23–25 °C and the reduction of 4-NPs occurred for 30 min with a rate constant of $k = 59.178 \times 10^{-3}$ min^{-1}.[43] Another study used *Limnophilia rugosa* leaves for biosynthesis of Au NPs. The leaves were washed and mixed with distilled water for 1 hour at 90 °C in a reflux system. Then, they were re-cooled to room temperature and filtered to obtain an extraction solution. A solution of 1 mM $HAuCl_3$ was mixed with the extract solution at 80 °C for 40 min in the dark. Then the Au NPs were centrifuged for 15 minutes,

Figure 13.6 Representation of Au nanoparticle formation (Au NPs) from various plants with SEM and TEM images. (a) Cinnamon bark and TEM image of Au NPs. TEM reproduced from ref. 42, https://doi.org/10.1007/s11051-020-04983-8, under the terms of the CC BY 4.0 license https://creativecommons.org/licenses/by/4.0/. (b) Honeybee pollen and HRTEM image of Au NPs. HRTEM reproduced from ref. 43 with permission from Elsevier, Copyright 2022. (c) *Limnophila rugosa* leaves and FESEM image of Au NPs. FESEM reproduced from ref. 44, https://doi.org/10.1155/2021/5571663, under the terms of the CC BY 4.0 license https://creativecommons.org/licenses/by/4.0/. (d) *Cistanche deserticola* and HRTEM image of Au NP. HRTEM reproduced from ref. 45 with permission from Elsevier, Copyright 2022.

washed twice with distilled water and once with ethanol, and dried overnight at 90 °C. Scanning electron microscopy (SEM) showed a spherical shape that was not uniformly dispersed and tended to form clusters (see Figure 13.6c). The catalytic activity of 1,4-dinitrobenzene (1,4-DNB) and three isomers of nitrophenol (2-NP, 3-NP, and 4-NP) at ambient temperature were studied within 15 minutes in excess amounts of $NaBH_4$, and the reduction efficiency was 93% after five repeat use.[44] In addition, another plant, *Cistanche deserticola* (CD), was used as a bioreduction agent and stabilized on poly(-styrene-*co*-maleic anhydride) (PSMA) to biosynthesize composite Au@PSMA nanoparticles. CD powder was boiled with distilled water for 1 hour in a flux system, and then cooled to ambient temperature. To obtain the Au@PSMA sample, the PSMA solution was dissolved in a CD extraction at pH 9. Then, the (0.5 mM) HAuCl4 solution was mixed with the CD-PSMA solution at 60 °C for 20 minutes and the resulting suspension was stored at 4–5 °C. High-resolution TEM (HRTEM) showed that Au@PSMA the samples have high sphere crystallinity and an average diameter of 50 nm (Figure 13.6d). The catalytic activity of the Au@PSMA sample showed that 4-NP was degraded to 4-AP within 10 minutes and that the rate constant was 0.127 min^{-1}.[45]

13.2.2.2 Silver Nanoparticles

Silver nanoparticles (Ag NPs) have many applications, including food wrapping, pharmaceutical activities, magnetic, electronic, catalyst materials, wound covering, and water treatment due to their distinctive biological, chemical, physical, and optical properties.[46–48] The silver lattice crystal has a face-centred cubic structure with a bulk coordination number of 12 (see Figure 13.7).

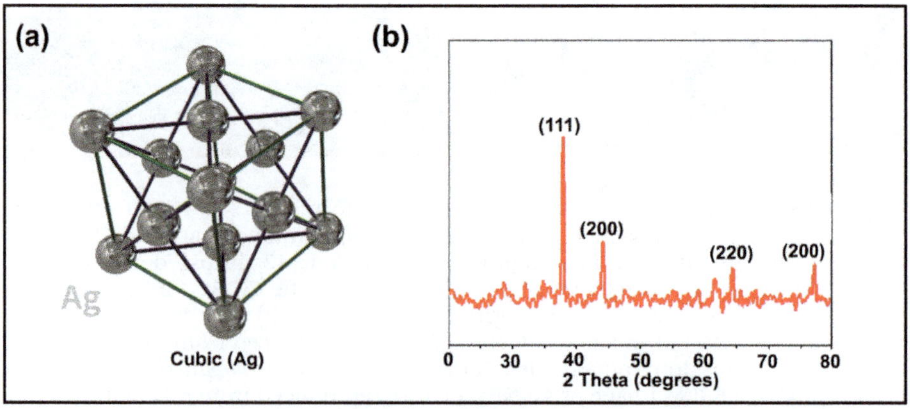

Figure 13.7 (a) Cubic face-centred crystal structure of Ag metal nanomaterials with (b) X-ray diffraction spectrum. XRD reproduced from ref. 49 with permission from Elsevier, Copyright 2023.

Biogenic/green synthetics attract much attention because of their low cost from the use of plant media as reducing and stabilizing agents. For example, Ag nanoparticles were synthesized using *Helichrysum graveolens*. The plant was cleaned, dried at 25 °C, mixed with distilled water, and stirred for 15 min. Subsequently, the mixture was filtered and stored at 2 °C. The plant solution was mixed with (0.1 M) $AgNO_3$ solution and stirred for 6 h. The reduction of $AgNO_3$ was monitored by UV–visible spectrometry (439 nm absorption), and the solution's colour tended to be dark brown. TEM showed that the diameter of the Ag NPs was almost 11 nm (see Figure 13.8a). The photocatalytic activity for MO dye degradation under visible light was approximately 85.39% for 60 min.[48] Likewise, Ag NPs were synthesized using an extract of *Sanvitalia procumbens* (creeping zinnia). The plants were washed and dried in shade at room temperature. Then, the cut plant was heated at 70 °C, stirred for 30 min, cooled down after filtration, and mixed with (0.01 mol L^{-1}) $AgNO_3$ solution added dropwise. The indication of the reduction of Ag ions to Ag NPs was a visual colour change from pale yellow to dark brown. The Ag NPs were separated, washed with distilled water and ethanol, and dried in vacuum ovens. Ag NPs sizes ranged from 10 to 80 nm with a spherical shape (see Figure 13.8b). The degradation/reduction activity of synthesized Ag NPs for Orange G (OG) and Direct blue-15 (DB-15) dyes was successfully observed, and the colour of the dye solution began to disappear after 180 min (Table 13.4).[50] In addition, extracts of jasmine flowers were used to synthesize Ag NPs. The jasmine flowers were washed with tap water, followed by distilled water, and mixed with distilled water to prepare the extract solution, which was heated at 80 °C for 3 h until it became a yellow solution and was then filtered. Then, $AgNO_3$ solution was added drop by drop to the extraction solution, changing the colour from yellow green to red brown. The solution was centrifuged for 5 min, and the precipitates were heated in an oven at 110 °C for 2 h. The silver nanoparticles had a nanofiber shape, and the average crystal size was 22 nm (Figure 13.8c). The synthesized Ag NPs showed a degradation efficiency for MB dyes of about 72% in 2 hours.[51] Similarly, *Thespesia populnea* bark was extracted for the synthesis of Ag NPs. The bark of *Thespesia populnea* was washed with tap water and distilled water. The bark was then added to distilled water, heated for 3 hours at 80 °C and served as a capping agent. The bark extract dissolved in the (1 M) solution of $AgNO_3$ and was stirred for 30 min at room temperature. After 10 minutes of centrifugation for filtration, the sample was dried at 100 °C for 1 hour in hot air. The resulting average crystal size was 19.05 nm, and the SEM showed the sphere shape of the Ag NPs to be 40–50 nm in diameter (Figure 13.8d). The photodegradation efficiency for MB dye was up to 90% under visible light for 120 min.[52] Another study of the biosynthesis Ag NPs used *Trigonella foenum-graecum* L. The leaves were washed with deionized water, dried, and transformed into fine powder. The powder was heated at 60 °C for 45 minutes in deionized water and centrifuged for 15 min. The leaf extract was used to reduce and as a capping agent by mixing it with 1 mM $AgNO_3$ solution and heating it at 60 °C for 5 min. The mixture solution was centrifuged for 15

Figure 13.8 Representation of Ag nanoparticle formation (Ag NPs) from various plants with SEM and TEM images. (a) *Helichrysum graveolens* and TEM image. TEM reproduced from ref. 48 with permission from Springer Nature, Copyright 2020. (b) *Sanvitalia procumbens* (creeping zinnia) and SEM image. SEM reproduced from ref. 50, https://doi.org/10.3390/molecules26206144, under the terms of the CC BY 4.0 license https://creativecommons.org/licenses/by/4.0/. (c) Jasmine flower and SEM image. SEM reproduced from ref. 51 with permission from Elsevier, Copyright 2021. (d) *Thespesia populnea* bark and SEM image. SEM reproduced from ref. 52, https://doi.org/10.1155/2022/7268273, under the terms of the CC BY 4.0 license https://creativecommons.org/licenses/by/4.0/. (e) *Trigonella foenum-graecum* L. and FESEM image. FESEM reproduced from ref. 53, https://doi.org/10.3390/molecules28030951, under the terms of the CC BY 4.0 license https://creativecommons.org/licenses/by/4.0/. (f) *Cassia alata* plant and HRTEM. HRTEM reproduced from ref. 49 with permission from Elsevier, Copyright 2023. (g) *Piper colubrinum* Link and SEM. SEM reproduced from ref. 54, https://doi.org/10.1080/17458080.2021.1970140, under the terms of the CC BY 4.0 license https://creativecommons.org/licenses/by/4.0/. (h) *Amaranthus gangeticus* Linn. leaf and HRTEM image. HRTEM reproduced from ref. 55, https://doi.org/10.1186/s40543-015-0074-1, under the terms of the CC BY 4.0 license https://creativecommons.org/licenses/by/4.0/. (i) Fresh leaves of *Stachys spectabilis* and TEM image. TEM reproduced from ref. 56 with permission from Elsevier, Copyright 2023. (j) *Limnophila rugosa* leaves and FESEM image. FESEM reproduced from ref. 44, https://doi.org/10.1155/2021/5571663, under the terms of the CC BY 4.0 license https://creativecommons.org/licenses/by/4.0/. (k) *Gnetum montanum* (GM) and FESEM image. FESEM reproduced from ref. 57 with permission from Elsevier, Copyright 2023. (l) Leaves of *Phoenix dactylifera* L. and SEM image. SEM reproduced from ref. 58, https://doi.org/10.21203/rs.3.rs-139856/v1, under the terms of the CC BY 4.0 license https://creativecommons.org/licenses/by/4.0/.

min, and the precipitate dried at 3 °C. The shape of the Ag NPs was almost spherical (see Figure 13.8e) with as particle size of 15 to 25 nm, and an average diameter of 21 nm. In the presence of visible light, the catalytic degradation efficiency was measured using several organic dyes: MB dye (96.57%), MO dye (71.45%), and RB dye (92.72%), with a maximum percentage at 20 min.[53] Furthermore, the plant extract of *Cassia alata* (CA) was used to synthesize Ag NPs. The plant CA was collected, rinsed with distilled water, dried under shade for 5 to 7 days, and ground into a fine powder. Then it was mixed with distilled water and refluxed for 30 min and filtered to extract CA. The extract was mixed with (0.1 mM) $AgNO_3$ solution, then (0.5 M) NaOH was added until a pH of 12 was reached. The mixture of CA-mediated Ag NPs was immediately analysed by UV–visible spectroscopy at ambient temperatures. The photocatalytic activity of the CA-mediated Ag NPs was assessed by the decoloration of organic dyes MBs, MO, and rhodamine 6G (R6G) under visible light. The HRTEM of CA-mediated Ag NPs shows a single crystal (50 nm) with a spherical shape (Figure 13.8f). The decoloration rates of the MO, MB and R6G dyes over CA-mediated Ag NPs under visible light were 98.6%, 97.3%, and 98%, respectively.[49]

Another plant used to produce Ag NPs is *Piper colubrinum* Link leaf extract. The leaves were washed several times with distilled water, transformed into a slurry, added to deionized water, passed through cheesecloth, and then filtered and stored at 4 °C. The solution of (1 mM) $AgNO_3$ was heated in a ratio of 9 : 1 and covered with aluminium foil. The leaf extract was added to the silver solution drop by drop with stirring until the colour of the solution changed to dark brown. The obtained suspension was collected, washed several times with ethanol, and analysed by UV–visible spectroscopy. The scanning electron microscopy (SEM) images showed a homogeneous, monodispersed spherical shape (Figure 13.8g), with crystal sizes ranging from 10 to 50 nm. The catalytic activity of the plant-mediated Ag NPs for degradation of MB, bromophenol blue (BPB) and Coomassie brilliant blue (CBB) showed absorbed peaks at 668 nm, 570 nm, and 595 nm, respectively, with a decrease in intensity after 120 min.[54] Similarly, *Amaranthus gangeticus* Linn. leaves were used for the biosynthesis of Ag NPs. The fresh leaves were cleaned with distilled water and cut into small pieces. The small leaves were packed into a compact bag and extracted from solvents (methanol : water = 1 : 3). The solution of (10^{-3} M) $AgNO_3$ was heated for about 5 min at 80 °C. Subsequently, a few drops of 0.05 M NaOH solution were added to the plant leaves extract. The solution was heated for 20 min at 80 °C until its colour turned brownish yellow, indicating the formation of silver nanoparticles (Ag NPs). The HRTEM of the biosynthesized Ag NPs showed a polycrystalline morphology (Figure 13.8h), with average particle sizes of 11 to 15 nm. The catalytic activity of Congo Red dye (CR) occurred within 15 minutes in the presence of $NaBH_4$ and ethanol, and the red colour transformed to colourless.[55] Another study used extract of fresh leaves of *Stachys spectabilis* for the biosynthesis of Ag NPs. The leaves were dried, then crushed into fine powder. The powder was mixed with deionized water, heated at 45 °C for 2.5 min, and filtered, extracted, cooled, and combined with (1.0 mM) $AgNO_3$ solution. Then, the samples were heated at 55 °C for 1.5 hours, centrifuged, and lyophilized to obtain the product. The structure of the spherical crystal, by TEM, is shown in Figure 13.8i, with a crystal size of 10.8 nm. The degradation efficiency for MB dyes was tested in sunlight, and blue dyes changed to colourless after 240 min.[56] In addition, biosynthesis of Ag NPs was carried out using orange peel extract (OPE). Fresh orange peels were washed in distilled water, cut into small pieces, and dried in an oven at 93 °C for 60 min. The peels become a fine powder, which was mixed with distilled water at 60 °C for 60 min. The peel extract was cooled, filtered, and stored at 4 °C. Then (10 mM) $AgNO_3$ stock solution was mixed with the OPE solution, and pH adjustment was made at ambient temperature by mixing with the (1.0 mM) KOH under bright day light. The optimal biosynthesis conditions were 0.0075 g dry peel per mL for OPE concentrations, pH = 11, and (1.5 mM) $AgNO_3$ solution. The crystal size of the OPE Ag NPs was 12.9 nm. Catalytic degradation of MB dye with the Ag NPs occurred in 1 hour.[59] Extract of *Limnophila rugosa* leaves were also used for the biosynthesis of Ag NPs. The leaves were washed with distilled water and mashed, then heated at 90 °C with a 1 hour reflux. The mixture was cooled to ambient temperature, mixed with a solution of (1 mM) $AgNO_3$ under vigorous stirring for 60 min, and then heated at 80 °C in the dark to avoid the photochemical reaction. The suspension was washed several times and dried overnight at 90 °C. The SEM image shows a non-uniform dispersed spherical shape that tends to form clusters (see Figure 13.8j). The biosynthesis catalytic performance was based on reducing 1,4-DNB, 2-NPh, 3-NPh, and 4-NPh in the presence of $NaBH_4$. The efficiency was 93% for 15 min after five repeated cycles.[44] In addition, Ag NPs were synthesized using *Gnetum montanum* (GM) extracts. The dried GM was mixed with distilled water and boiled with flux at 90 °C for 60 minutes. The extracts were mixed with a solution of $AgNO_3$ and stirred at room temperature in the dark until their colour became pale yellow, and the formation of Ag NPs was monitored by UV–visible spectrometry. The SEM images showed uniform dispersed spherical particles with an average size of about 20 nm (Figure 13.8k), while TEM showed monodispersed particles without crystal aggregation. The catalytic activity to reduce 3-NPh and 4-NPh with $NaBH_4$ was completed within 15 min (see Table 13.4).[57]

In addition, leaf extracts were also used for the biosynthesis of composite materials, such as silver and silver oxide nanoparticles (Ag/Ag_2O) NPs, which were extracted from *Phoenix dactylifera* L. leaves. The plant extracts were prepared by washing *Phoenix dactylifera* L. leaves in distilled water, drying them for 12 days in the dark at room temperature, and crushing them into a fine powder. Afterwards, the powder was mixed with distilled water, stirred at ambient temperature for a day, filtered, and the extract was stored at 4 °C. The volume ratio (v/v) of plant extractor/precursor was 1/30, 1/40, and 1/50 with (1 mM) $AgNO_3$ solution and was stirred at ambient temperature for 2 hours. The solution confirmed the transformation of Ag^+ to Ag^0 after 5 min of browning. The Ag/Ag_2O nanoparticles have a spherical

Table 13.4 Different green reductions for synthesizing semiconductors of Ag NPs.

Materials	Green reduction	Ag precursor	Water contaminant	Treatment time (min)	Efficiency of degradation (%)
Ag NPs	*Helichrysum graveolens*	$AgNO_3$	MO dye	60	85.39 [48]
Ag NPs	*Sanvitalia procumbens*	$AgNO_3$	Orang G (OG) dye	180	70.61 [50]
			Direct blue-15 (DB-15)	180	70 [50]
Ag NPs	Jasmine flower	$AgNO_3$	MB dye	120	72 [51]
Ag NPs	*Thespesia populnea* bark	$AgNO_3$	MB dye	120	90 [52]
Ag NPs	*Trigonella foenum-graecum* L.	$AgNO_3$	MB dye	20	96.57 [53]
			MO dye	20	71.45 [53]
			Rhodamine B (RB) dye	20	92.72 [53]
CA-mediated Ag NPs	*Cassia alata* plant	$AgNO_3$	MO dye	120	98.6 [49]
			MB dye	50	97.3 [49]
			Rhodamine 6G (Rh6G) dye	240	98 [49]
Ag NPs	*Piper colubrinum* Link.	$AgNO_3$	MB dye	120	— [54]
			Bromophenol blue (BPB) dye	120	— [54]
			Coomassie brilliant blue dye	120	— [54]
Ag NPs	*Amaranthus gangeticus* Linn. leaf	$AgNO_3$	Congo red (CR) dye	15	— [55]
ss-Ag NPs	*Stachys spectabilis* leaves	$AgNO_3$	MB dye	240	52 [56]
Ag NPs	Orange peel extract (OPE)	$AgNO_3$	MB dye	60	— [59]
Ag NPs	*Limnophila rugosa* leaves	$AgNO_3$	1,4-DNB, 2-NP, 3-NP and 4-NP by $NaBH_4$	15	93 [44]
Ag NPs	*Gnetum montanum* (GM)	$AgNO_3$	3-Nitrophenols and 4-nitrophenol	15	— [57]
Ag/Ag_2O	Leaves of *Phoenix dactylifera* L.	$AgNO_3$	CR dye	60	— [58]
			MB dye	50	— [58]

shape (see Figure 13.8l), and the degradation of catalytic activity of CR and MB dyes was studied. When $NaBH_4$ was present in the aqueous medium with Ag/Ag_2O nanoparticles, the CR and MB dyes were degraded. The red dye was degraded after 60 min, and the blue one was decomposed after 50 min.[58]

13.2.3 Metal Oxides

Metal oxide nanoparticles (MONPs) are metal cores that connect to oxygen using oxo (M–O–M) or hydroxo (M–OH–M) links and form metal-oxo or metal-hydroxo polymers.[60] MONPs are often used in water purification because they have unique characteristics, such as high surface area and activity, and they can replace expensive metal nanoparticles.[61] The most promising MONPs for targeting water contaminants are iron, magnesium, titanium, manganese, and aluminium oxides. In particular, the catalytic activity of nanoparticles is known to depend on the surface and the size of pores.[62]

13.2.3.1 Iron Oxides

Iron oxide nanoparticles (FeO NPs) have attracted much attention in various potential applications, such as magnetic materials, sensors, pigments, adsorbents, and photocatalysts.[63] For example, FeO has four crystalline structures: hematite (α-Fe_2O_3), maghemite (γ-Fe_2O_3), magnetite (Fe_3O_4), and wustite (FeO) (see Figure 13.9).[63–65] The hematite material is an effective iron oxide used as a photocatalyst. On the other hand, maghemite, magnetite, and wustite are mainly used in the areas of nanomagnetism, adsorbate, data storage, energy storage, and drug delivery.[63,66]

When iron oxide nanoparticles (NPs) or magnetic nanoparticles (MNPs) are used as adsorbents or catalysts, their magnetic properties are unique and beneficial, allowing them to be isolated easily when exposed to magnetic fields during separation.[63] It has been reported that many efforts have been made to synthesize an iron oxide adsorbent using green approaches based on plants. A simple maceration technique was used to extract biomolecules to synthesize iron oxides/magnetic nanoparticles as water treatment adsorbents. Subsequently, adsorbents using plant phytochemicals that acted as reduction and capping agents were synthesized and mixed with (0.1 M) $FeCl_3 \cdot 6H_2O$ and (0.2 M) $FeCl_2 \cdot 4H_2O$ [1:1]. The biosynthesized FeO has a spherical shape, with a few cubic shapes, and an average size of 160 nm. All MNPs synthesized showed effective removal efficiency for MB dye and 4-nitroaniline (4-NA) from water over a long period. It was observed that the MNP adsorbed dyes in a short time, above 80% under a wide ranging of conditions, and up to 94% under optimal conditions. Interestingly, MNPs show greater effectiveness in reducing 4-NA in high pH pollutant solutions and at temperatures ranging from 10 to 30 °C.[18] In another report, the extract of the leaf of *Centaurea solstitialis* (Cs) and iron chloride(III) ($FeCl_3$) was used to synthesize adsorbent iron oxide nanoparticles (FeO NPs), which were used in

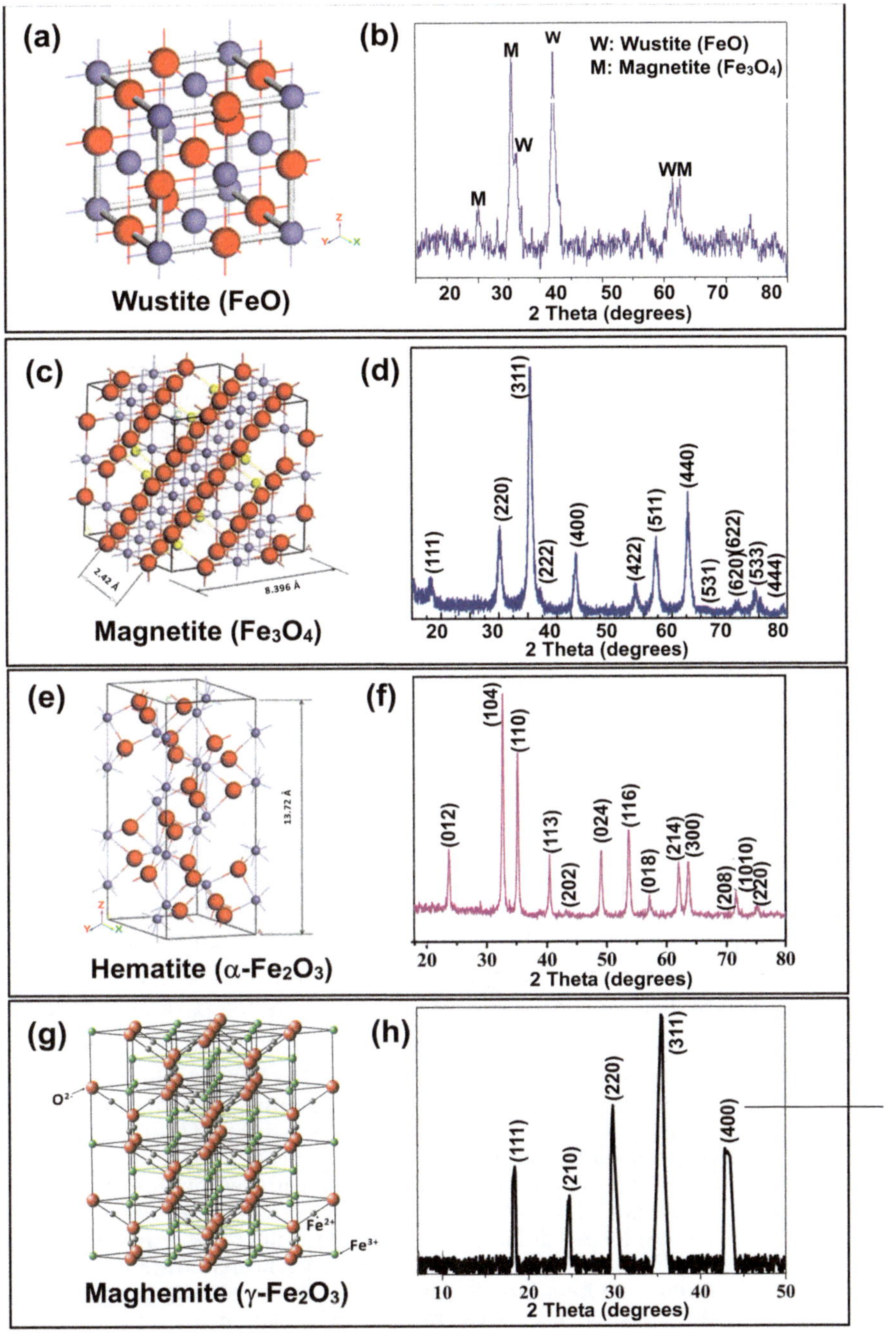

Figure 13.9 Crystal structure and XRD of different FeO nanomaterials. (a) Crystal structure of wustite. Reproduced from ref. 65, https://doi.org/10.5772/66211, under the terms of the CC BY 3.0 license https://creativecommons.org/licenses/by/3.0/. (b) X-ray diffraction of wustite.
(continued)

the Fenton reaction to eliminate Red 180 (RR180) and Basic Red (BR18) dyes. The Cs powder was added to deionized water to form a Cs solution and mixed with 1 M $FeCl_3$ solution. The mixture was incubated in an oven at 70 °C for 24 hours and then separated. The SEM image shows an inhomogeneous spherical morphology (see Figure 13.10a). It was observed that the synthetic adsorbents eliminate dyes under optimal conditions of pH values of 2.5 for RR180 and 3 BR18 dyes. The amount of adsorbent was 0.25 g L^{-1}, and the adsorption capacity exceeded 98%[71] (see Table 13.5). In addition, leaf extracts from *Canthium coromandelicum* (CC) were used to synthesize iron oxide nanoparticles (CC-IONP) by reducing iron(II) chloride ($FeCl_2$). They CC-IONPs were obtained as a catalyst for the degradation of Janus Green B (JG) dye. Using Fourier-transform infrared (FTIR) analysis it was observed that polyphenols and organic acids present in the CC extraction were responsible for capping and stabilization of the CC-IONPs. The obtained material was not uniform in size but was spherical (Figure 13.10b), and the degradation efficiency for JG dye was 97.23% for 180 min with sunlight irradiation.[72] Fresh leaves of neem (*Azardica indica*) were used to synthesize magnetic nanoparticles (MNPs) capable of adsorbing multiple organic pollutants, including MB and 4-NA. Leaves of the plant were dried for 10 days and mixed with 100 ml distilled water for 1 hour at 80 °C. The (1 : 1) ratio between 0.1 M $Fe(Cl)_3{\cdot}H_2O$ solution and 0.2 M $Fe(Cl)_2{\cdot}4H_2O$ solution was prepared in a flask, and the plant extract was added by drop by drop until the solution turned black. The mixture was vigorously stirred for 6 hours at 60 °C, then dried as a suspension and used for adsorption. The XRD data showed the shape (crystal size) of the adsorbent as spherical and cubic (at 277 nm) magnetite (29 nm), and magnetite (45 nm). The adsorption efficiency of the nanoparticles for MB and 4-NA dyes was 94% in 20 min.[18] Further efforts were made using extracts of tangerine peels to synthesize iron oxide to prevent accumulation and reduce the particle diameter. The tangerine peels were cleaned and dried, then milled, sieved into powder, mixed with distilled water, and heated to 80 °C for 15 minutes. FeO NPs were prepared by a co-precipitation method; the 0.1 M $FeCl_3$ solution and 0.1 M $FeCl_2{\cdot}4H_2O$ solution were obtained separately in different

Figure 13.9 *(continued)* Reproduced from ref. 67 with permission from Springer Nature, Copyright 2019. (c) Crystal structure of magnetite. Reproduced from ref. 65, https://doi.org/10.5772/66211, under the terms of the CC BY 3.0 license https://creativecommons.org/licenses/by/3.0/. (d) X-ray diffraction of magnetite. Reproduced from ref. 68 with permission from Elsevier, Copyright 2008. (e) Crystal structure of hematite. Reproduced from ref. 65, https://doi.org/10.5772/66211, under the terms of the CC BY 3.0 license https://creativecommons.org/licenses/by/3.0/, (f) X-ray diffraction of hematite. Reproduced from ref. 69 with permission from the Royal Society of Chemistry. (g) Crystal structure of maghemite. Reproduced from ref. 64, https://doi.org/10.5772/intechopen.101741, under the terms of the CC BY 3.0 license https://creativecommons.org/licenses/by/3.0/. (h) X-ray diffraction of maghemite. Reproduced from ref. 70 with permission from Springer Nature, Copyright 2014.

Figure 13.10 Representation of the formation of iron oxide nanoparticles (Fe_xO_x NPs) from various plant extractions with TEM and SEM images. (a) Leaves of *Centaurea solstitialis* and SEM image. SEM reproduced from ref. 71 with permission from Elsevier, Copyright 2023. (b) *Canthium coromandelicum* and TEM image. TEM reproduced from ref. 72 with permission from Elsevier, Copyright 2021. (c) Tangerine peel and SEM image. SEM reproduced from ref. 73, https://doi.org/10.1186/s40201-015-0237-4, under the terms of the CC BY 4.0 license https://creativecommons.org/licenses/by/4.0/. (d) *Azadirachta indica* leaves and SEM image. SEM reproduced from ref. 74, https://doi.org/10.4236/gsc.2020.104009, under the terms of the CC BY 4.0 license https://creativecommons.org/licenses/by/4.0/. (e) *Cucurbita moschata* leaves and (f) *Beta vulgaris* stalks and TEM images. TEM (e) and (f) reproduced from ref. 75 with permission from Elsevier, Copyright 2020.

Table 13.5 Different green reductions for synthesizing iron oxide.

Adsorbent	Green reduction	Fe precursor	Water contaminant	Adsorbent amount ($g L^{-1}$)	Adsorption capacity (%)
MNPs	Plant phyto-chemicals (fresh leaves of Neem)	$FeCl_3·6H_2O$ and $FeCl_2·4H_2O$	MB dye and 4-nitroaniline (4-NA)	5	—[18]
FeO NPs	*Centaurea solstitialis* leaves	$FeCl_3$	Red 180 (RR180) and Basic Red (BR18)	0.25	98[71]
FeOA NPs	*Cucurbita moschata* leaves	$Fe(NO_3)_3·9H_2O$	Tartrazine and Bordeaux red dyes	0.04	—[75]
FeOB NPs	*Beta vulgaris* stalks	$Fe(NO_3)·_9H_2O$	Tartrazine	0.12	59.79[75]
			Bordeaux red	0.12	59.98[75]
CC-IONPs	*Canthium coromandelicum*	$FeCl_3$	Janus green B dye (JG)	0.02	97.23[72]
FeO NPs	Tangerine peel	$FeCl_3$ and $FeCl_2·4H_2O$	Cadmium ions	0.4	90.00[73]
MNPs	*Azadirachta indica* leaves	$FeCl_3·6H_2O$ and $FeSO_4·7H_2O$	Arsenate (As(v))	3	69.89[a] [74]

[a]The adsorption capacity ($mg g^{-1}$).

concentrations of extract peel solution, which acted as a surfactant and stabilizer. The two solutions were mixed at 80 °C and then 25% NH_4OH was dropped into the mixture with stirring for 20 minutes until the pH exceeded 9. Brown precipitates were collected and washed with distilled water. The SEM images showed an average particle size of about 200 nm when the extraction concentrations were at 4%. For 6% extracting solution, the size of the particles reduces to 50 nm, while at 10% extracting solution the size is 1000 nm indicating that severe aggregation leads to increased size of the nanoparticles. For an adsorption efficiency of FeO NPs at 6%, the removal of cadmium ions (90%) occurs at pH 4.[73] Another study used the *Azadirachta indica* leaves extraction solution for the biosynthesis of magnetic nanoparticles (MNPs). Fresh neem leaves were cleaned and dried in the shade for 7 days, powdered, and mixed with distilled water. The extraction solution was added to the ferric (Fe^{3+}) and ferrous (Fe^{2+}) mixture in a 2:1 molar ratio under a nitrogen atmosphere by the co-precipitation method. Then NH_4OH was added until pH values reached 10–11. Then, the precipitates were collected and washed several times and dried at 60 °C for 12 hours under a vacuum. The XRD pattern showed that the obtained sample has an FCC crystal structure with a crystal size of 8.4 nm, and the SEM image shows the presence of uniform nanometer-sized spherical particles (see Figure 13.10d). The adsorption efficacy for removing arsenic (As(V)) was 69.89 mg g^{-1} at pH 2.[74]

An additional plant was reported for the preparation of two adsorption materials of iron oxide nanoparticles (FeO NPs). Green methods from two agricultural residues were used: *Cucurbita moschata* leaves (FeOA NPs) and *Beta vulgaris* stem (FeOB NPs), and each solution was mixed with 0.1 M iron(III) nitrate solution for the removal of tetrazine and Bordeaux red dyes. The synthetic materials that contained goethite (from FeOB NPs) and maghemite (from FeOA NPs) have diameters of 2 nm (irregular shape) and 20 nm (more round shape), respectively (see Figure 13.10 (e and f)). The adsorption kinetics of FeOA and FeOB NPs were examined and it was found that equilibrium with for FeOA NPs was reached after 1200 min for tartrazine and after 240 min for Bordeaux red, and at 180 min for tartrazine and 120 min for Bordeaux red with FeOB NPs. Furthermore, the highest adsorption capacity of FeOB NPs was 59.79 mg g^{-1} for tartrazine and 59.98 mg g^{-1} for Bordeaux red[75] (see Table 13.5).

For iron oxides as photocatalysts, green technologies have been reported to be very promising as cheap raw materials. Nanoscale iron oxide particles (α-Fe_2O_3) have been synthesized by using papaya (*Carica papaya*) under atmospheric conditions. The study aimed to remove remazol yellow (RR) dye by photocatalysis. The leaves of papaya plants were washed, dried in an oven for 1 hour, and ground into fine powders. The fine powders were boiled for 30 min in deionized water at 80 °C, then extracted by filtration, and stored at 40 °C. The extracted solution was mixed with the (0.1 M) $FeCl_3{\cdot}6H_2O$ solution in a 1:1 ratio at ambient temperature. After that, 1 M NaOH solution was added to the mixture until the pH reached 11. The mixture was stirred for 30 min, and the intense black solution colour appeared, indicating the formation of FeO NPs. The NPs were separated, washed with ethanol and water 2–3 times, and stirred in hot air at 80 °C for 3 hours. The morphology of the NPs was not uniform, and the particles agglomerated, as shown in (Figure 13.11a); the average diameter was 21.59 nm. It was observed that increase in the amount of iron oxide nanoparticles improved dye decolorization, and after 6 hours at pH 2, at a catalyst dose of 0.8 g L^{-1}, the maximum dye degradation rate was 76.6%[76] (see Table 13.6). Another eco-friendly approach to the synthesis of iron oxide nanoparticles (FeO NPs) is to mix leaf extracts from *Psidium guavaja-Moringa oleifera* (PMV) with $FeCl_3$. The fresh PMV leaves were cleaned and dried at ambient temperature (up to 32 °C) for 72 hours in air; the dried leaves were blended to form liquefied powders in a flat-bottom conical flask. The solution was stirred at 80 °C for 1 hour to increase the release of phytochemicals. The mixture solution was cooled to ambient temperature, filtered to extract PMV, and stored at 4 °C. The PMV extract was added to a solution of 0.5 M $FeCl_3$, in a volume ratio of 8:2 (v/v), with vigorous stirring, and the colour changed from orange to dark brown, indicating the formation of NPs. The homogeneous solution was dried in a hot air oven at 100 °C for 24 hours. The dark brown solid was washed, pulverized, and re-dried in a vacuum oven at 60 °C for 4 hours. The TEM image shows that

Figure 13.11 A representation of the formation of iron oxide nanoparticles (Fe_xO_x NPs) from various plant extracts with TEM and SEM images. (a) Carica papaya and FESEM image. FE-SEM reproduced from ref. 76, https://doi.org/10.1016/j.heliyon.2020.e04603, under the terms of the CC BY 4.0 license https://creativecommons.org/licenses/by/4.0/. (b) *Psidium guajava-Moringa oleifera* composite and TEM image. TEM reproduced from ref. 77 with permission from Elsevier, Copyright 2019. (c) *Cynometra ramiflora* and SEM image. SEM reproduced from ref. 62 with permission from Elsevier, Copyright 2018. (d) Amla seed and TEM image. TEM reproduced from ref. 78 with permission from Elsevier, Copyright 2023. (e) *Chlorophytum comosum* leaf and TEM image. TEM reproduced from ref. 79 with permission from Elsevier, Copyright 2021. (f) *Canthium coromandelicum* leaf and TEM image. TEM reproduced from ref. 72 with permission from Elsevier, Copyright 2021. (g) Pomegranate (*Punica granatum*) and SEM image. SEM reproduced from ref. 80 with permission from Elsevier, Copyright 2019.

Table 13.6 Different green reductions for synthesizing iron oxide.

Materials	Green reduction	Fe precursor	Water contaminant	Treatment time (min)	Efficiency degradation (%)
α-Fe_2O_3	*Carica papaya*	$FeCl_3 \cdot 6H_2O$	Remazol yellow (RR) dye	250	76.6 [76]
FeO NPs	*Psidium guajava-Moringa oleifera*	$FeCl_3$	MB dye	60	— [77]
MIO NPs	*Cynometra ramiflora*	$FeCl_2$ and $FeCl_3$	MB dye	110	— [62]
IO NPs	Amla seed	$FeCl_3$	MB dye	90	80.0 [78]
INPs	*Chlorophytum comosum* leaf	$FeCl_3 \cdot 6H_2O$	MO dye	360	77 [79]
CC-IONPs	Leaf *Canthium coromandelicum*	$FeCl_2 \cdot 4H_2O$	Janus green B (JG)	180	97.23 [72]
Fe_2O_3 NPs	Pomegranate seeds (*Punica granatum*)	$FeCl_3$	Reactive blue 4 dye	56	95.08 [80]
FABC	Algae of *Aegagropila linnaei* (AL)	$FeSO_4 \cdot 7H_2O$	Bisphenol A (BPA)	200	69.80 [30]

the PMV-FeO NP clusters have a spherical rod-like shape (see Figure 13.11b). This material demonstrated photocatalytic degradation of MB dyes under visible light. It was observed that after 30 min of sunlight, the colour of the MB changed to light blue, and after 1 hour to light green. The blue to light green colour change could be induced by interactions between PMV-FeO NPs coated with MB and longer exposure to give the high percentage of degradation.[77] The same environmentally friendly methodology was used to synthesize magnetic iron oxide nanoparticles (MIONPs) using *Cynometra ramiflora* (CR), and Fe salts [$FeCl_2$ & FeC13] and the catalytic activity was compared with commercial Fe_2O_3. The CR fruit was washed with water, dried in air, then the fruit flesh was cut, distilled water added, and the mixture boiled for 1 hour. The obtained solution was filtered and stored at 4 °C. The $FeCl_3$ and $FeCl_2$ solutions were mixed in 1:2 molar ratios and added to the extract solutions at ambient temperature. Subsequently, (1 N) NaOH was added until a pH of 11 was reached. The solution became an intense black colour, indicating the formation of nanoparticles. The products were filtrated, washed, and dried in a hot air oven at 60 °C for 6 h. The morphology of biosynthesized MIONP material showed a spherical shape (see Figure 13.11c), with different sizes ranging from 58.50 nm to 78.13 nm. The surface area (BET) and the pore volume were 107.97 $m^2\ g^{-1}$ and 0.1235 $cm^3\ g^{-1}$, respectively. The pores size was 4.58 nm, compared to commercial Fe_2O_3; its surface area was 7.7 $m^2\ g^{-1}$. They observed that the MB dye gradually decreased under sunlight irradiation and eventually disappeared after 110 min. The surface area of MIONP improves the formation of hydroxyl radicals and helps the degradation of MB dyes under sunlight irradiation.[62] In other studies, some biomaterials act as bioreduction and stabilization agents, such as aqueous seed extract of amla seeds (*Phyllanthus emblica*) to synthesize iron oxide nanoparticles (IONPs). The $FeCl_3$ and amla seed extract were mixed to synthesize IONPs. The seeds were washed without removing the kernel, heated with distilled water and extraction solution, and filtered. The (0.01 M) $FeCl_3$ solution was stirred for 15 min. Then the solution was boiled for 10 min and after that the extract solution was added dropwise to the Fe solution. The solution becomes a deep brown colour, and then the precipitates separated from the solution, were washed with acetone, dried at room temperature, and heated at 600 °C for 3 h. The synthesized material has a spherical shape of 2–5 nm size (see Figure 13.11d). They observed that when sodium borohydride ($NaBH_4$) is present, IONPs take about 90 min for 80% of the MB dye to degrade under UV light. However, it took a long time in the absence of $NaBH_4$.[78] The environmentally friendly material of *Chlorophytum comosum* leaf extract has also been used. *Chlorophytum comosum* extract solution was mixed with (0.1 M) ferric chloride ($FeCl_3{\cdot}6H_2O$), and the prepared iron nanoparticles (INPs) were used to eliminate MO dyes. The dried leaves were mixed with deionized water at 80 °C for 1 hour, and the extract was cooled, filtered, and stored at −20 °C. The leaf extract was mixed with the Fe (0.1 M) solution in a round bottle in an inert atmosphere (N_2) and vigorously stirred at ambient temperature for 24 hours. The black suspension was separated, washed with deionized water, dried at 50 °C for 12 h in a vacuum oven, and stored in a tightly sealed container filled with inert gas (N_2). The INPs have particle sizes below 100 nm with some aggregated particles (see Figure 13.11e). The MO dye was mixed with (1 ml) H_2O_2, and the dye colour vanished almost completely with an efficiency is 77% after 6 hours.[79] In addition, the leaf extract *Canthium coromandelicum* (CC) was used to reduce Fe^{3+} in $FeCl_3$ and synthesize iron oxide nanoparticles (CC-IONPs). The CC leaves were collected, washed, dried, and turned into fine powders. The fine powder was dissolved in distilled water, boiled for 30 min, and then filtered. The CC leaf extract was stored at 4 °C. The biological synthesis of IONPs was performed by preparing (0.1 M) Fe solution and stirring it at 80 °C, then adding aqueous extract with stirring for 30 min. The change in colour from light to dark brown-red was due to the appearance of precipitate. The biogenically synthesized CC-IONPs have a spherical shape but not uniform sizes (see Figure 13.11f). CC-IONP exhibited a 97.23% degradation efficiency for Janus Green B (JG) requiring 180 min under sunlight irradiation. It was reported that the size and shape of the synthesized NPs increased the catalytic efficiency.[72] In another report pomegranate seed extract (*Punica granatum*) was used to synthesize Fe_2O_3 NPs. The seeds were cleaned with ultrapure water, ground in, and mixed with water for 2 hours. The extract was filtered and stirred at 4 °C. The extract and the iron chloride solution (1 M) were mixed in a ratio of 1:2 and heated at 70 °C for 15 min until the pale yellow colour changed to brownish black. The mixture was separated, and the precipitate collected and washed with water and ethanol (3–4 times). The collected powder was dried at 60 °C for 3 hours.

The SEM image shows a semi-spherical particle of 25–55 nm distributed uniformly (see Figure 13.11g). The Fe_2O_3 NPs had excellent photocatalytic activity against Reactive blue 4 dyes studied in UV light; the reactivity showed a degradation of 95.08% after 56 min[80] (see Table 13.6).

Researchers have recently attempted to synthesize green FeO-based nanocomposites, which are considered to be better adsorbents. For example, cheap raw materials from the *Aegagropila linnaei* (AL) algae were used for the synthesis of novel biochar/iron oxide (FABC) composites to remove bisphenol A (BPA) under ultraviolet light. The AL was mixed with ferrous sulfate ($FeSO_4{\cdot}7H_2O$) to ensure effective conjugation of Fe_3O_4 nanoparticles to the surface of acidic biochar using hydrothermal methods. It was observed that FABC has a small particle size of 100–150 nm, a surface area of 144.62 $m^2\ g^{-1}$, and a pore volume of 0.259 $cm^3\ g^{-1}$. Furthermore, about 69.8% of BPA was removed after 200 min under acidic conditions (pH 3) suggesting that the activity is adsorption-induced and promotes the absorption of hydrogen ions in Fe_3O_4. At the same time, the active species (OH and O_2) produced by nano iron oxides on the surface of the composites played an essential role in the removal of BPA.[30]

13.2.3.2 Zinc Oxides (ZnO)

Zinc oxide nanoparticles (ZnO NPs) are known as semiconductors with a wide bandgap between 3.10 and 3.37 eV,[81] and are essential as photocatalysts[82,83] and as sorbent for the adsorption process[84] for wastewater remediation. In addition, they are a safe material for food packaging and preservation applications.[85] ZnO NPs have biological activities, such as immobilization of the enzyme glucose oxidase,[86] in anti-inflammatory and antioxidant assays, such as for aspirin and vitamin C,[83,87] as doxorubicin drug carrier against the MDA-MB-231 (TNBC) cell line,[88] antimicrobial,[83,89] antibacterial,[90,91] anticancer,[92] antidiabetic,[90,91] antiparasitic,[91] Iarvicidal,[91] and antibiofilm[89] activity.[89] ZnO has three crystal structures: rock salt, hexagonal wurtzite, and zinc blende (see Figure 13.12); in many cases, the wurtzite crystal structures are used as photocatalysts and substrate materials.

Several efforts have been reported in the literature on the use of green synthesis of ZnO using plants[16,85] (see Table 13.7). Rambabu *et al.*,[82] synthesized DPW-mediated ZnO NPs using *Phoenix dactylifera* waste (date pulp waste, DPW) as a bioreducer, using DP-ZnO NPs to degrade dyes from MB (cationic) and eosin yellow (EY) (anionic) dyes in waste water. They observed, by SEM, that DP-ZnO NPs show a spherical particle of size 31.6 nm (Figure 13.13a), and the bandgap was estimated at 3.25 eV. They found that DP-ZnO NPs showed rapid decomposition of MB and EY of 90.5% and 90.6%, respectively, and a rapid half-life decomposition of 15 min.[82] Manjunatha *et al.*[96] used the extract of the leaves *Artocarpus heterophyllus* to synthesize zinc oxide nanoparticles (ZnO NPs) to remove Congo Red dye from solution (Direct red 28). They calcined the ZnO NPs at 400, 600, and 800 °C for 1 hour. The bandgap energy was calculated and found to be 3.42, 3.38, and 3.35 eV for 400,

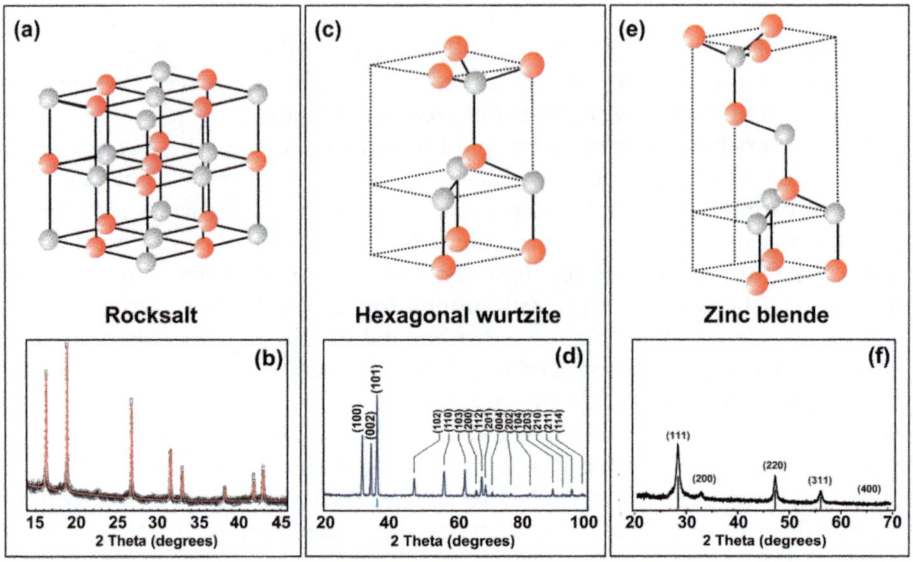

Figure 13.12 Basic crystal structure and XRD spectra of different ZnO nanomaterials: Zn atoms in red and oxygen atoms in white. (a and b) Rock salt. (a) Reproduced from ref. 85 with permission from American Chemical Society, Copyright 2022. (b) Reproduced from ref. 93 with permission from Elsevier, Copyright 2014. (c and d) Hexagonal wurtzite. (c) Reproduced from ref. 85 with permission from American Chemical Society, Copyright 2022. (d) Reproduced from ref. 94 with permission from Elsevier, Copyright 2011. (e and f) Zinc blende. (e) Reproduced from ref. 85 with permission from American Chemical Society, Copyright 2022. (f) Reproduced from ref. 95 with permission from the Royal Society of Chemistry.

Table 13.7 Different green reductions for synthesizing zinc oxide.

ZnO NP Structure	Green reduction	Zn precursor	Water contaminant	Treatment time (min)	Efficiency degradation (%)
Wurtzite	*Phoenix dactylifera* waste	$Zn(NO_3)_2{\cdot}6H_2O$	MB dye	180	90.5 [82]
			Eosin yellow (EY)	180	90.6 [82]
Wurtzite	*Artocarpus heterophyllus* leaves	$Zn(NO_3)_2{\cdot}6H_2O$	CR dye (direct red 28)	60	90 [96]
Wurtzite	Areca nut (*Areca catechu*)	$Zn(NO_3)_2{\cdot}6H_2O$	Rhodamine B (RhB)	120	95 [97]
			MB	90	95 [97]
			Nigrosin	90	96 [97]
Wurtzite	*Myristica fragrans* fruit	$Zn(NO_3)_2{\cdot}6H_2O$	MB	140	88 [91]

Figure 13.13 SEM images of the formation of zinc oxide nanoparticles (ZnO NPs) from various plant extracts. (a) *Phoenix dactylifera* waste and SEM image. SEM reproduced from ref. 82 with permission from Elsevier, Copyright 2021. (b) Areca nut and SEM image. SEM reproduced from ref. 97 with permission from Springer Nature, Copyright 2022. (c) Aloe vera and (d) Cassava starch and SEM images. SEM (c) and (d) reproduced from ref. 98, https://doi.org/10.3389/fchem.2020.571790, under the terms of the CC BY 4.0 license https://creativecommons.org/licenses/by/4.0/.

600, and 800 °C, respectively. They observed that calcined ZnO NPs at 400 °C showed an effective photocatalytic degradation of 90% for 20 ppm CR dye solutions at 1 hour at pH 9.[96] Another plant used was areca nut extract (*Areca catechu*), which was used to synthesize zinc oxide nanoparticles for the degradation of rhodamine B (RhB), MB, and nigrosin under direct sunlight. The ZnO NPs have hexagonal and spherical structures and an average size of 5–30 nm (Figure 13.13b). They found that the RhB solution changed colour after 120 min, resulting in 95% degradation, while the MB dye solution changed colour after 90 min, resulting in 95% degradation; the nigrosin showed 96% degradation, becoming colourless after 90 min.[97] Similarly, the fruit extract of *Myristica fragrans* was used to synthesize ZnO nanoparticles. The *Myristica fragrans* was transformed into a fine powder, soaked with distilled water, and heated for 20 minutes at 150 °C. The powder was kept in an incubator at 37 °C overnight for further extraction. The extracts were filtrated and stored at 4 °C. Then, 4 M $Zn(NO)_3{\cdot}2H_2O$ solution was mixed with the extract under stirring at 60 °C for 2 hours, cooled, centrifuged for 10 minutes, washed, filtered, and dried at 500 °C for 2 hours. The XRD showed particles with pure ZnO material, and the SEM showed particles with semi-spherical shapes and very high aggregation. The bandgap was 2.5 eV, and the photocatalytic activity was tested. It was found that the ZnO NP material gave 88% degradation of MB dye in 140 min under UV light.[91]

Table 13.8 Different green reductions for synthesizing adsorbents of zinc oxide.

Adsorbent Structure	Green reduction	Zn precursor	Water contaminant	Adsorbent amount ($g L^{-1}$)	Adsorption capacity ($mg g^{-1}$)
ZnO-AL Wurtzite	*Aloe vera*	$Zn(NO_3)_2 \cdot 6H_2O$	Cu(II)	250	20.43 [98]
ZnO-ST Wurtzite	Cassava starch				10.96 [98]
ZnO	Lemon peels	$ZnSO_4$	Cu(II) Fe(III)	—	89.9[a] [99] 82.0[a] [99]

[a]The adsorption capacity (%).

ZnO has also been reported as an adsorbent produced by means of environmentally friendly plant-based procedures (see Table 13.8). The synthesis of adsorbent zinc oxide nanoparticles uses two simple green routes to remove copper from waste water: (1) *Aloe vera* (green synthesis) and (2) Cassava starch (gelatinization). For route 1, *Aloe vera* gel (AL) with a concentration of 90% was extracted and mixed with (100 ml) distilled water. Subsequently, 9.40 g zinc nitrate was dissolved in the *Aloe vera* solution, stirred, and rested for 12 hours. The suspension was calcined at 750 °C for 1 hour and denoted as ZnO-AL. For route 2 (gelatinization method), 100 g of natural *Cassava* starch produced by mechanical stirring was mixed with 300 ml water, the suspension was calcined at 750 °C for 1 hour, and the code was ZnO-ST. The XRD and Raman spectra indicated that both routes led to a single phase of ZnO. The SEM images showed that ZnO-AL comprised pseudospheres and hexagonal particles that are not uniform, while the ZnO-ST showed uniform spherical particles (see Figure 13.13c and d). The removal efficiency of for 40 ppm Cu(II) for both routes was the same, but at 80 ppm concentration, the removal efficiency of route 1 was higher than that of the route 2.[98] In other work, lemon peel extracts were used for synthesizing ZnO NPs for treating water contaminated with Cu(II) and Fe(III). The lemon peels were cleaned and dried at 60 °C for 48 hours. Then, the mixture was crushed to a suitable size and immersed in water at a concentration of 250 $g L^{-1}$. The lemon extract was mixed with a $ZnSO_4$ solution at 70 °C for 1 hour until the colour was converted to a white solution because it contained hesperidin flavanol. NH_4OH was added to the solution until a pH of 6.95. The precipitates were washed, centrifuged, and heated at 500 °C for 3 hours. It was found that ZnO was pure, with no carbon, by XRD and this was confirmed by UV–visible spectroscopy at 358 nm. The Cu(II) adsorption efficiency at pH 3–6 was 89.9%, and the Fe(III) adsorption capacity at pH 3.3 was 82.2%.[99]

13.2.3.3 Nickel Oxide (NiO)

Additionally, green methods extend to the formation of nickel oxide nanoparticles (NiO NPs), with a unique cubic crystal structure, as shown in Figure 13.14. The NiO has specific physical, chemical, and optical properties[12]

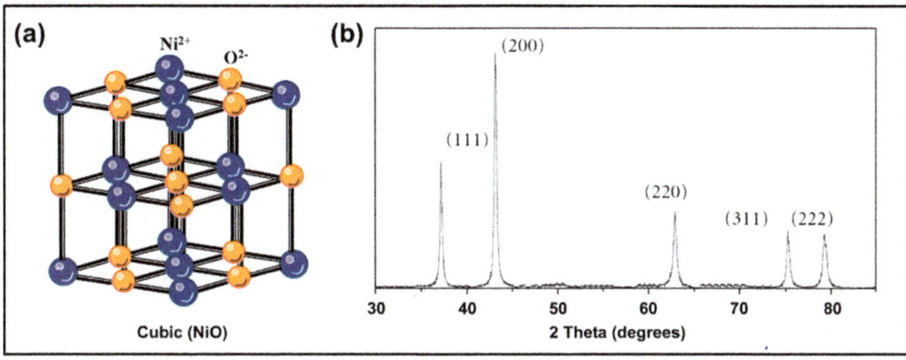

Figure 13.14 (a) Cubic crystal structure of NiO nanomaterials. Reproduced from ref. 108, https://doi.org/10.3390/nano10040636, under the terms of the CC BY 4.0 license https://creativecommons.org/licenses/by/4.0/. (b) X-ray diffraction. Reproduced from ref. 109, https://doi.org/10.1155/2009/795928, under the terms of the CC BY 3.0 license https://creativecommons.org/licenses/by/3.0/.

Table 13.9 Different green reductions for synthesizing adsorbents of nickel oxide.

Adsorbent	Green reduction	Ni precursor	Water contaminant	Adsorbent amount ($g\ L^{-1}$)	Adsorption capacity (%)
NiO NPs	Olive tree leaves	$Ni(NO_3)_2{\cdot}6H_2O$	MO	6	88[110]
NiO NPs	*Syzygium cumini* plant	$NiCl_2{\cdot}6H_2O$	MB	0.02	69[100]
			CR	0.02	99[100]

and is suitable for a variety of applications, such as for adsorbents, photocatalysts, antioxidants,[100] cytotoxicity,[101–103] antibacterial,[101,104,105] antioxidant,[103] anti-oesophagal carcinoma,[103]*in vitro* anticancer activity,[106] and electrochemical.[107]

For example, many adsorbents have been reported (as shown in Table 13.9) with use of the extract of olive leaves as a reduction agent and D-sorbitol as a capping agent, and with $Ni(NO_3)_2$ as a precursor. The leaves of the olive trees were dried, cut into small pieces, and mixed for 3 hours at 65 °C with distilled water. Subsequently, the mixture of extract was merged with 0.001 M $Ni(NO_3)_2{\cdot}6H_2O$ and 0.01 M sorbitol solution added, stirred, and heated at 80 °C until it became dark grey. The SEM image (Figure 13.15a) indicates that the NiO NPs have a highly crystalline spherical shape and there was agglomeration, with a size of 30 to 65 nm. Removal of MB dye was achieved at 96%, under optimum conditions of contact time of 100 min, pH 10, and adsorbent dosage of 6 g L^{-1}. Removal of MO dyes was the most effective, with the maximum removal of 88% with an optimal contact time of 160 min, pH 2, and adsorbent dose 6 g L^{-1}.[110] Another report used a green synthesis method for nickel oxide nanoparticles (NiO NPs) using *Syzygium cumini* leaf extract.

Figure 13.15 A representation of the formation of nickel oxide nanoparticles (NiO NPs) from various extracts with SEM and TEM images: (a) olive tree leaves and SEM image of NiO formation. Reproduced from ref. 110, http://doi.org/10.17576/jsm-2022-5102-17, under the terms of the CC BY 4.0 license https://creativecommons.org/licenses/by/4.0/. (b) Syzygium cumini plant and SEM image of NiO formation. SEM reproduced from ref. 100 with permission from Elsevier, Copyright 2022. (c) *Salvia macrosiphon* Boiss plant and SEM image of NiO formation. SEM reproduced from ref. 111 with permission from Springer Nature, Copyright 2019. (d) *Solanum trilobatum* leaf and SEM image of NiO formation. SEM reproduced from ref. 101 with permission from Elsevier, Copyright 2020. (e) Fresh tea leaves (*Camellia sinensis* plant) and SEM image of NiO formation. SEM reproduced from ref. 112 with permission from Elsevier, Copyright 2021. (f) Leaves of *Evolvulus alsinoides* and HR-TEM image of NiO formation. HR-TEM reproduced from ref. 106 with permission from Elsevier, Copyright 2023. (g) Onion (*Allium cepa* peels) and SEM image of NiO formation. SEM reproduced from ref. 113, https://doi.org/10.2166/wst.2021.237, under the terms of the CC BY 4.0 license https://creativecommons.org/licenses/by/4.0/. (h) Avocado fresh fruits (*Persea americana*) and HRTEM image of NiO formation. HRTEM reproduced from ref. 114 with permission from Elsevier, Copyright 2019. (i) Algae *Sargassum wightii* and SEM image. SEM reproduced from ref. 115 with permission from Elsevier, Copyright 2022. (j) Clitoria ternatea flower and HRTEM. HRTEM reproduced from ref. 105 with permission from Elsevier, Copyright 2022.

Secondary metabolites in *Syzygium cumini* extract were used to reduce nickel ions by stabilizing and capping. The extracts of *Syzygium cumini* plants were dried and converted to a fine powder. The plant powder was mixed with distilled water and stirred at 80–85 °C for 50 min. Then, the mixture was filtered and the plant extract kept at 5 °C. A 2 M $NiCl_2{\cdot}6H_2O$ solution was mixed with the extract solution and then sonicated for 30 min at 75 °C until the colour became greenish brown, and the moisture content was dried at 70 °C. The SEM image shows the formation of spherical NiO NPs, with an estimated XRD crystal size of 10.4 nm (see Figure 13.15b). It was observed that 5 ppm of MB and CR dyes adsorbed on the NiO NPs had a high removal efficiency. The maximum removal for MB was 69% over 150 min at 45 °C with a pH value of 2, while the total removal of CR was 99% over 120 min at 55 °C (pH 8)[100] (see Table 13.9).

Furthermore, NiO NPs are a p-type semiconductor with an energy bandgap between 3.6 eV and 4.0 eV.[104] Green synthesis of NiO NPs by plants was also carried out using extract of *Salvia macrosiphon* Boiss. plants as a capping and stabilizing agent to prevent aggregation of nanoparticles. The extract was mixed with nickel nitrate hexahydrate $Ni(NO_3)_2{\cdot}6H_2O$ and the nickel oxide nanoparticles (NiO NPs) were synthesized to degrade MB dyes under UV light. The *Salvia macrosiphon* was transformed into a powder and dissolved in water by stirring for 3 hours at 60 °C, and the final product was stored at 4 °C. The 0.5 M $Ni(NO_3)_2{\cdot}6H_2O$ was mixed with the extract solution at a temperature of 80 °C for 30 min. The green gels and slurry were observed and dried at 180 °C for 2 hours and calcined at different temperatures of 300, 400, 500, and 600 °C for 2 hours. The nanoparticles were homogeneous and uniformly distributed and spherical in shape (see Figure 13.15c). At 400 °C, the bandgap of NiO NPs was about 2.9 to 3.9 eV. The results showed that the photocatalytic activity of the NiO NPs synthesized at 400 °C degraded MB by about 80% after 5 hours at pH 11.[111] New efforts to synthesize NiO NPs have included extracting the leaves of *Solanum trilobatum* used as fuel, capping, and reduction agent, with $Ni(NO_3)_2{\cdot}6H_2O$ as the Ni precursor. The *Solanum trilobatum* leaves were cleaned, sliced into fine pieces, ground, boiled in deionized water and ethanol, and incubated for 1 day at room temperature. The leaf extract solution was mixed with 0.0001 M $Ni(NO_3)_2{\cdot}6H_2O$ for 1 h, then the homogeneous solution was heated at 250 °C for 15 min, and the brown powder appeared and was dried at 110 °C. The SEM image of the NiO NPs shows a cylindrical and rod-like surface morphology, with a crystal size of 23.21 nm (see Figure 13.15d). The optical properties using reflection spectroscopy (DRS) and photoluminescence spectroscopy (PL) show a bandgap of 3.50 eV. The NiO NPs were an effective material for the degradation of 4-chlorophenol (4-CP), by about 92% under UV light for 210 min. The slow and gradual degradation shows that the conversion to CO_2 and H_2O takes longer because a series of long-lived intermediates occur that depend on pH.[101] Other works have also aimed to synthesize nickel oxide nanoparticles (NiO NPs) using fresh tea leaf extracts (*Camellia sinensis* plant) as reduction agents. The leaves were cleaned, crushed, immersed in deionized water, and boiled at 70–80 °C for 1 hour. The mixture was cooled, mixed with a solution of

$Ni(NO_3)_2{\cdot}6H_2O$, and stirred for 120 min at 70 °C. The mixture becomes colloidal, and the colour changes after 1.5 hours. The solution was centrifuged, washed several times, heated in a vacuum oven at 80 °C, and calcined at 400 °C for 6 hours. The NiO NPs formed a 3–5 nm spherical nanoparticle that agglomerates (see Figure 13.15e). Nanoparticles were studied at different pH (4, 5, 6, 7, 8, and 9) for the degradation of MO, MB, and the combined dyes (MDs) under UV light. It was found that when the pH was 8, the MB degradation was 96.3%, and MO degradation was higher when the pH was 4 (see Table 13.10). It was assumed that high pH values for MB increased the OH^-, leading to strong electrostatic interactions between the cationic dye and the surface of the NiO NPs, whereas MO, with anionic charge, prevented the interaction with the surface of the NiO NPs by creating repulsion. Degradation of the combined dyes (MD) had a maximum percentage of 96.8% when pH was 9, but was 33.24% when the pH was 4.[112] Another study, used extract of the leaves of *Evolvulus alsinoides* for the biosynthesis of nickel oxide nanoparticles for the degradation of rhodamine B (RhB) under visible light.

Table 13.10 Different green reductions for synthesizing nickel oxide.

Materials	Green reduction	Ni precursor	Water contaminant	Treatment time (min)	Dye degradation (%)
NiO NPs	*Salvia macrosiphon* Boiss. plant	$Ni(NO_3)_2{\cdot}6H_2O$	MB dye	300	80.0 [111]
NiO NPs	Onions (*Allium cepa* peels)	$Ni(NO_3)_2{\cdot}6H_2O$	CR direct dye	90	90.0 [113]
NiO NPs	Algae *Sargassum wightii*	$C_4H_6NiO_4$	MB dye	180	85.0 [115]
NiO NPs	*Solanum trilobatum* leaf	$Ni(NO_3)_2{\cdot}6H_2O$	4-Chlorophenol (4-CP)	210	92.0 [101]
NiO NPs	Fresh tea leaves (*Camellia sinensis* plant)	$Ni(NO_3)_2{\cdot}6H_2O$	MB dye MO dye The combined dyes (MD)	24 60 30	96.3 [112] 96.3 [112] 96.8 [112]
NiO NPs	*Clitoria ternatea* flower	$C_4H_6NiO_4$	Fast Green dye (FG) Rose Bengal dye (RB)	150 150	89.0 [105] 77.0 [105]
NiO	*Evolvulus alsinoides* leaves	$Ni(NO_3)_2{\cdot}6H_2O$	Rhodamine B dye (RhB)	—	85.0 [106]
NiO NPs	Fresh *Persea americana* fruits	$Ni(NO_3)_2{\cdot}6H_2O$	Free cyanide (FCN)	30	84.0 [114]
CuNiO NPs	Okra plant	$Ni(NO_3)_2{\cdot}6H_2O$	MB dye	105	78.0 [102]
NiO@ ZnHCC	*Sapindus mukorossi*	$NiCl_2$	Cr(VI)	—	92.0 [116]

The leaves of *Evolvulus alsinoides* were dried and transformed to powder, mixed with water, and stirred at 80 °C for 2 hours. Subsequently, the extraction solution reached pH 7.2 and was stored at 4 °C. The nickel nitrate solution was mixed with the extract solution, stirred for 1 hour until the solution became homogeneous, and then heated at 250 °C for 15 min. The solution was boiled, dehydrated, and decomposed, and gas evolved leaving a dark brown powder. The size of the nanoparticles was about 12 nm. The HRTEM image shows small clusters of spherical and cubic shapes, and the small cluster has a size of 11.6 nm (see Figure 13.15f). The activity against RhB dyes exhibited 85% degradation under visible light.[106] Another green method used onion (*Allium cepa*) peels with a high concentration of quercetin. This polyphenolic compound was used to extract metals from a water solution as a ligand. A dry powder was made and mixed with deionized water. The container was heated at 70 °C for 20 min. Afterwards, the solution was centrifuged and filtered. The extracts were stored in the refrigerator at 4 °C. The 0.1 M $Ni(NO_3)_2 \cdot 6H_2O$ solution was mixed with various concentrations of onion peel extract (10, 20, 30, and 40%) and heated at 70 °C for 40 min, with constant stirring for 3 days. The grey precipitate of nickel oxide nanoparticles was collected by filtration, dried, and stored. The NiO NPs have a spherical and cubic shape, as shown by SEM (see Figure 13.15g). The degradation activity towards Congo Red dye was tested under optimal conditions of 0.02% dye concentration, 0.003 g L^{-1} catalyst dose, pH 6, and temperature at 50 °C, and the dye was decolorized at 90% under UV light.[113] Another plant, fresh *Persea americana* fruit extract, was used to synthesize NiO NPs as photocatalysts for the photodegradation of free cyanide (FCN). The plant was cleaned and heated at 60 °C for 24 hours until the mixture changed to dark brown and changed in colour, indicating the decomposition of the bioactive elements of the seeds in water; the solution was filtered and stored in the refrigerator. The extraction solution was mixed with $Ni(NO_3)_2 \cdot 6H_2O$ solution with constant stirring at 50 °C for 3 hours. The precipitate was collected, washed, and dried at 80 °C for 24 hours, then calcined at 500 °C under open air for 2 hours. The XRD shows that the cube-type structure and morphology has a rhombohedral and spherical shape with slight agglomeration, with a particle size of 11 nm (see Figure 13.15h). The energy bandgap for the synthesized material was 3.75 eV. The photocatalyst was studied under UV light during FCN degradation, and the maximum photocatalytic oxidation efficiency was 84% after 30 min.[114] Further efforts have been made to synthesize green nanoparticles of nickel oxide using extract of the marine brown algae *Sargassum wightii* to break down the hazardous organic dye MB dye in water. The brown algae were transformed into a fine powder and mixed with water for 60 min, then filtered and pure material extracted. The extracted solution was mixed with 0.5 M nickel acetate tetrahydrate ($C_4H_6NiO_4$) and stirred for 2 hours. Then NaOH solution was added until a pH value of 12, whereupon the paste-like solution became brown and was heated for 60 min, and then calcined at 400 °C for 2 hours. The SEM image reveals that the nanoparticles have uniform spherical shapes and aggregate to larger grains of uneven structures (Figure 13.15i) with grain sizes between 100 and 120 nm. Degradation reactivity for dyes in water occurred under 180 min of sunlight, UV light, and LED light. It was observed that under sunlight the degradation of MB showed a high percentage (85%) in 180 min.[115] Furthermore, green synthesis with the medicinal flower extract of *Clitoria ternatea* was also used to synthesize nickel oxide nanoparticles. The flower was dried in the open air, boiled at 60 °C for 1 hour, filtered, kept in an air-tight glass container, and stored at 4 °C. The solution of 0.1 M $C_4H_6NiO_4$ was mixed with the flower extracts, which acts as a reduction and capping agent. The reaction solution changed from violet to blue green, indicating the formation of NiO NPs. The XRD showed formation of NiO NPs, which had a highly crystalline nature with an average size of 9.6 nm, and the TEM showed a hexagonal shape and non-agglomeration with a particle size of 13 nm (see Figure 13.15j). Degradation activity for photocatalytic dyes Fast Green (FG) and Rose Bengal (RB) was carried out under sunlight. It was found that the NiO NPs degraded both FG and RB dyes, by 89% and 77%, respectively.[105]

Recently, researchers have tried to synthesize composite NiO-based metal nanoparticles using green methods for photocatalytic and decontamination procedures. Ghazal *et al.* synthesized copper-doped nickel oxide nanoparticles (Cu-doped NiO NPs) using extract of okra plants by the sol–gel method. The okra plants were chopped into small pieces and washed with distilled water. Then, they were stirred for 2 hours at 60 °C, filtered, and the extract saved. The $Cu_xNi_{(1-x)}O$ mixture was prepared by mixing 0.5 M $Ni(NO_3)_2{\cdot}6H_2O$ with different Cu(II) sulfates of 1% Cu and 3% Cu under stirring. Then, the extract was added to the mixture and then heated at 70 °C for 12 hours. The material obtained was calcined at 400 °C for 2 hours, forming grey Cu-doped NiO NPs (see Figure 13.16). NiO NPs with 3% Cu have the best photocatalytic activity, with 78% decomposition of MB under UV light for 105 min.[102] Nanocomposites that use *Sapindus mukorossi* extract to synthesize a novel nickel

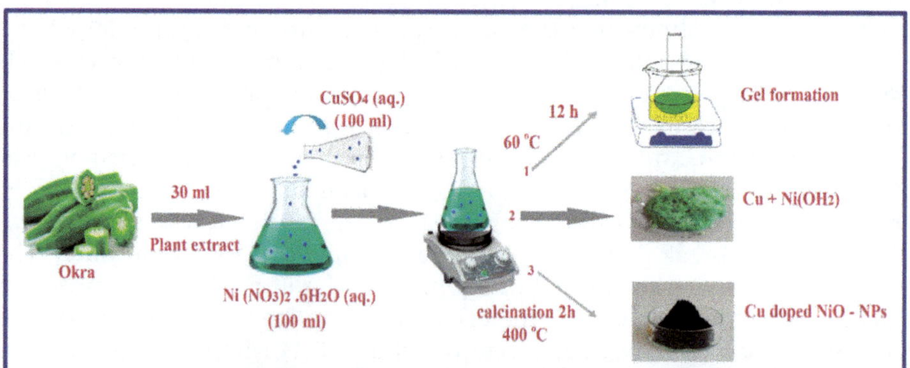

Figure 13.16 Schematic of the green method to form Cu-doped nickel oxide nanoparticles (Cu-doped NiO NPs). Adapted from ref. 102 with permission from Elsevier, Copyright 2021.

oxide-decorated zinc hexacyanocobaltate catalyst (NiO@ZnHCC) to remove toxic Cr(VI) have also been reported. The extract of *Sapindus mukorossi* was mixed with 0.01 M $NiCl_2$ and stirred at 50 °C for 1 hour. After adding 0.01 M NaOH, the precipitate of colourless $Ni(OH)_2$ was filtered and dried at 60 °C for 4 hours. Subsequently, 0.05 M potassium hexacyanocobaltate solution and 0.1 M zinc chloride solution were mixed at 50 °C for 2 hours to form the ZnHCC material. The other process was to combine ZnHCC with NiO by ultrasonication, then water was added dropwise from the extraction solution to a suspended solution of NiO. The obtained product was cleaned with deionized water at 70 °C. The XRD showed a crystalline structure with a cubic form of size 50–100 nm. Using NiO@ZnHCC, 92% removal of Cr(VI) was achieved due to the improved surface area.[116]

13.2.3.4 Silicon Dioxide (SiO_2)

Silicon dioxide (SiO_2), also known as silica, has recently gained attention for a wide range of applications, such as in food, medical industries, coating material,[117] and water treatment.[118] SiO_2 has a linear structure with a unique characteristics of variable and tunable pore sizes, custom interfaces, superior mechanical properties, and large surface area. Silica nanoparticles (Si NPs) have various applications. Therefore, green technologies have been sought to develop the biosynthesis of SiO_2 NP materials; for example, the green synthesis of silica nanoparticles formed from the leaves the weed *Saccharum ravennae* to remove heavy metals of lead (Pb^{2+}) and copper (Cu^{2+}) from water. The leaves of *Saccharum ravennae* (SRL), *Saccharum officinarum* (SOL) and *Oryza sativa* (OSL) were cleaned with distilled water, dried in sunlight, ground and sieved at 150 mm, and stored in an oven at 50 °C. Afterwards, the SRL, SOL, and OSL were heated for 3 hours in a muffle furnace at 700 °C. The powder was treated with 6 N HCl at 70 °C for 12 hours and ash removed by stirring. The suspension was filtered, and the solid form was washed with distilled water. The solid form of each was treated with 2.5 N NaOH to remove impurities and stirred at 70 °C for 12 hours to produce sodium silicate. This was followed by adding, drop by drop, concentrated H_2SO_4 to reach a pH of 2.5, filtering the slurry, washing with distilled water, and drying at 60 °C for 6 hours. Then the dried mass was heated in a muffle furnace at 450 °C for 2 hours for the pyrolysis process, according to the following equations.

$$\text{Leaf biomass ash} + \text{HCl} \rightarrow \text{Leached ash}(\text{SiO}_2) \quad (13.1)$$

$$2\text{NaOH}+\text{SiO}_2 \rightarrow \text{Na}_2\text{SiO}_3+\text{H}_2\text{O} \quad (13.2)$$

$$\text{Na}_2\text{SiO}_3+\text{H}_2\text{SO}_3 \rightarrow \text{SiO}_2+\text{Na}_2\text{SO}_4+\text{H}_2\text{O} \quad (13.3)$$

As a result, a white powder was formed. The morphology of the biosynthesized SNP from the SRL, SOL, and OSL biomass was shown by field-emission SEM (FESEM) to be spherical aggregates of SRL-SiO NPs and OSL-SiO NPs, with an average particle size of 3.56 nm and 39.47 nm, respectively. On the other hand, the SOL-SiO NPs had a hexagonal shape and an average crystal size of 29.13 nm (see Figure 13.17(a–c)). The adsorption capacity for heavy metals (Pb^{2+} and Cu^{2+}) for SRL-SiO NPs (140 mg g^{-1} and 149.25 mg g^{-1}), OSL-SiO NPs (334.7 mg g^{-1} and 274.02 mg g^{-1}), and SOL-SiO NPs (38.55 mg g^{-1} and179.45 mg g^{-1})[119] are shown in Table 13.11. Another precursor of Si used was bamboo leaves for the biosynthesis of SiO_2 NPs to remove toxic solutions (Cu^{2+} and NO^-_3). First, the bamboo leaves were soaked in distilled water for 12 hours, washed and dried under sunlight, and crushed into a fine powder. The bamboo powder was soaked with nitric acid in a conical cup for 24 hours, and then filtered. Subsequently, the mixture was washed with deionized water until a pH of 7.8 was reached an then dried for 12 hours at 90–110 °C. The dried powder was refluxed with 1 N NaOH at 110 °C for 1 hour, forming a sodium silicate solution (SSS) according to the following equation.

$$SiO_2\left(\text{bambooleaves}\right)+NaOH \rightarrow Na_2SiO_3+H_2O \tag{13.4}$$

The SSS solution was mixed with different ratios (3 : 1, 2 : 1, 1 : 1, 1 : 2, and 1 : 3) of ammonia and ethanol, and the diluted SSS was added to the solution dropwise and aged for 1 hour. The precipitates were collected, washed with deionized water, and dried for 3 hours at 50 °C, forming SiO NPs. The SiO NPs were treated with an amine solution and labelled as A-SiO NP. The SEM shows that the SiO NPs had a uniform-sized, monodispersed, and spherical shape (Figure 13.17d), with an average size of 80 nm for SiO_2 NPs and of 85 nm for A-SiO_2 NPs. The biosynthesized A-SiO_2 NP adsorption capacity was better

Figure 13.17 A representation of the formation of silicon dioxide nanoparticles (SiO_2 NPs) from various plant extractions with SEM and TEM images. (a) *Saccharum ravennae* and SEM image of SiO_2 formation. (b) *Saccharum officinarum* and SEM image of SiO_2 formation. (c) Oryza sativa leaf and SEM image of SiO_2 formation. SEM (a), (b) and (c) reproduced from ref. 119 with permission from Elsevier, Copyright 2021. (d) Bamboo leaves and SEM image of SiO_2 formation. SEM reproduced from ref. 120 with permission from Springer Nature, Copyright 2023. (e–g) Rice husk for SiO_2 formation with (e) SEM of SiO_2-AA, (f) SEM of SiO_2-OA, and (g) SEM of SiO_2-HCl. SEM (e), (f), and (g) reproduced from ref. 121, https://doi.org/10.1088/1742-6596/2238/1/012005, under the terms of the CC BY 3.0 license https://creativecommons.org/licenses/by/3.0/.

Table 13.11 Different green reductions for synthesizing semiconductors of silicon oxide.

Materials	Si precursor	Water contaminant	Adsorbent amount (g L^{-1})	Adsorption capacity (mg g^{-1})
SRL-SiO NPs	Leaves of *Saccharum ravennae*	Lead (Pb^{2+}) ion	1	140 [119]
		Copper ion	1	149.25 [119]
OSL-SiO NPs	Leaves of *Saccharum officinarum*	Lead (Pb^{2+}) ion	1	334.7 [119]
		Copper ion	1	274.02 [119]
SOL-SiO NPs	Leaves of *Oryza sativa*	Lead ion	1	338.55 [119]
		Copper ion	1	179.45 [119]
SiO NPs	Bamboo leaves	Copper ion	0.25	121.95 [120]
		Nitrate (NO^-_3) ion	0.25	9.46 [120]
A-SiO NPs	Bamboo leaves treated with amino acid	Copper ion	0.25	196.07 [120]
		Nitrate (NO^-_3) ion	0.25	23.36 [120]
SiO_2-AA	Rice husk (RH)	Triclosan (TCS)	8	2.74 [121]
		Caffeine (CAF)	24	0.75 [121]

than that for SiO NP[120] (see Table 13.11). In addition, rice husk was used as a Si precursor for the biosynthesis of SiO NPs to remove triclosan (TCS) and caffeine (CAF) from aqueous media. The rice husk (RH) was pretreated using a variety of methods: (1) with HCl, (2) with acetate (AA), and (3) with oxalic acid (OA). The solutions of 10 w/w % HCl and 5 w/w % acetic acid were each mixed in a ratio of 1:10 M with the rice husk solution, stirred for 2 hours at 95 °C, and the results labelled as RH-HCl and RH-AA. On the other hand, 0.5 M oxalic acid solution was used as an acid leach for RH-OA in a high-pressure reactor for 2 hours at 155 °C. The SEM images shows that all the produced materials were composed of small-size nanoparticles that agglomerate to form large particles after calcination at 700 °C; the crystal sizes were 5.4 nm for SiO_2-AA (Figure 13.17e), 9.6 nm for SiO_2-OA (Figure 13.17f), and 4.6 nm for SiO_2-HCl (Figure 13.17g). The surface area for all samples was 200 m^2 g^{-1} The adsorption capacity of the nano silica samples from acetic acid (SiO_2-AA) was the best[121] (see Table 13.11).

13.2.3.5 Zeolites

Zeolite materials are micropores with a pore size of 2 nm containing crystalline tetrahedral aluminosilicate with various topologies. The zeolite material has long been recognized for its unique physical, chemical, and adsorption properties for water purification. Among various types of zeolitic structure materials, natural zeolitic clinoptilolite is known for its high surface area for water purification, but its cation exchange capacity is limited.

Most industrial applications use synthesized zeolites. The synthesis process depends on various parameters, such as precursors of aluminium and silicon sources, organic structure direction agents (OSDAs) as templates, water volume during hydrothermal flow, and temperature. In recent years, several attempts have been made to synthesize zeolite using green pathways to minimize environmental pollution and reduce the cost of synthesis, such as using solvent-free methods during hydrothermal flow, sustainable precursors, free templates, mother liquor recycling, *etc.*[122] The seed-directed method is an alternative to using an organic template, which behaves in two ways: (1) as an inorganic nucleus that can crystallize the surrounding species, forming different topologies than the seeds, or (2) as an essential building block unit for the direct crystallization of different seed topologies.[123] In the case of renewable precursors, two precursors should be mentioned. Bio-silica sources were reported using coal fly ash (industrial waste),[124–130] iron ore tailing (IOT),[131] rice husk ash,[132–137] sugarcane bagasse ash,[138] bauxite,[139] and bamboo leaf biomass.[140] For bio-aluminium precursors red mud was reported as being used as a source for synthesizing magnetic zeolite.[140] In addition, bio-precursors could be alumina and silica precursors, such as natural kaolin[141–149] and perlite.[150] Furthermore, these alternative methods should be solvent-free to reduce/eliminate the mother liquor that causes harmful solvents to the environment. Interestingly, the main development in the synthesis of zeolites using environmental approaches has focused on three parameters: the precursor, the seed assistance of free OSDA, and being solvent-free. For example, rice husk was used as a silica source and did not use any organic template (free OSDA) for the green synthesis of Zeolite L to adsorb MB dye. The rice husk was washed with HCl until pH became neutral, dried in air, and combusted at 700 °C for 6 hours to obtain rice husk ash. The rice husk ash was dissolved in distilled water, and the pH was lowered to 1 by using 1 N HCl and stirring for 2 hours. The solution was filtered, and the residual water was washed with distilled water, redissolved in 1 N NaOH, and filtered. The obtained material was rinsed in hot water, cooled to room temperature, and neutralized with HCl to pH 7. The solution was stored in a polypropylene bottle for 18 hours until a gel formed. After that, distilled water was added to the solution to wash the gel, and it was centrifuged for 15 minutes, excess material removed, the washing process was repeated, and it was dried at 80 °C until the gel formed. Subsequently, silica was produced by calcinating for 1 hour at 700 °C. In the presence of KOH, silica was mixed with $Al(OH)_3$, and the molar ratio was 10 SiO_2 : 1 Al_2O_3 : 4 K_2O : 100 H_2O. The solution was heated in an autoclave at 170 °C for 24 hours. The precipitates were washed with demineralized water and heated for 12 hours at 80 °C. The Zeolite L showed a cylindrical morphology with a diameter of 0.049–0.123 μm by SEM. The adsorption capacity (*K*) of green Zeolite L is 1406 mg g^{-1}[134] (see Table 13.12). On the other hand, raw kaolin was used as an aluminium source to produce Zeolite A to remove Cr(III) from tannery waste water. The kaolin was calcinated for 3 hours at 600 °C. The metakaolin forms were treated with 3 M NaOH and stirred at 50 °C for 1 hour until the gel formed. The gel was aged

Table 13.12 Green methods for synthesizing zeolite material.

Materials	Si precursor	Al precursor	Water contaminant	Adsorbent amount ($g L^{-1}$)	Adsorption capacity ($mg g^{-1}$)
Zeolite L	Rice husk	$Al(OH)_3$	MB dye	—	1406 [134]
Zeolite A	Kaolin	Kaolin	Cr(III)	100	200 [141]
Zeolite A	Kaolin R and W	Kaolin R and W	MG dye	0.05	55 [144]

at room temperature for 24 h. Then, the crystallization was carried out in a water bath for 3 hours at 100 °C, and the molar composition was as follows: 1 Na_2O : 1 Al_2O_3 : 2 SiO_2 : 37 H_2O. The bottle was cooled, the material was washed in distilled water until the pH reached less than 10, and dried at 80 °C overnight. Then, 3 M NaOH was added without ageing under hydrothermal flow and crystallization produced Zeolite A. The morphology of Zeolite A has a cube-shaped crystal with an average particle size of 5 µm.[141,142] The adsorption capacity of Zeolite A for the removal of Cr(III) from tannery waste water was 200 mg g^{-1} [141] (see Table 13.12). Similarities were obtained when white and red kaolin were used to synthesize green Zeolite A for adsorption of cationic dyes of malachite green. The red and white kaolins were calcined at 600 °C for 12 hours. Metakaolin was mixed with 5 M NaOH with an 8 : 1 ratio of NaOH/metakaolin. The mixture was treated with hydrothermal treatment and maintained under stirring at 80 °C for 24 hours. The formed zeolite was washed with distilled water, dried at 110 °C, and coded as Zeo-R and Zeo-W. SEM showed that for the obtained materials, Zeo-W and Zeo-R, a cubic crystal structure was formed more for Zeo-R, due to Zeo-R being a more regular shape, than for Zeo-W, and that aggregation of Zeo-W was more pronounced. The adsorption capacity for MG dyes on Zeo-W and Zeo-R was 55 mg g^{-1}.[144]

13.3 Conclusion and Prospects

This chapter focuses on the latest research on biosynthesis of nanosized adsorbents and semiconductors using renewable biomass materials. Nanosized adsorbents and semiconductors are useful for environmental remediation of contaminants, such as dyes, emerging pollutants, and heavy metals from water through degradation, photocatalysis, and adsorption. Several advantages of green methods over conventional methods, such as safety, being less hazardous, cost effectiveness, and durability, have been summarized. The plant-based extracts and tissues produce different biomolecules and secondary metabolites, depending on the type of plant, whereby these biomolecules and secondary metabolites usually act as reducing and/or capping agents and/or stabilizing agents for the formation of carbon-based adsorbents (*e.g.*, biochar and activated carbon), metal nanoparticles (*e.g.*, gold and silver), metal oxide nanomaterials (*e.g.*, iron oxides, zinc oxides, nickel oxides, and silicon dioxide), and zeolite materials for water purification. Based on several research reports, biosynthesis is a promising methodology for synthesizing the required nanomaterials. It is an environmentally sustainable system, and biomass waste, which would otherwise be deposited in landfill sites and waste the energy, could be used for the manufacturing of nanomaterials.

In general, the green synthesis method can produce nanosized materials. However, there is concern about the applicability of green synthesis methods for synthesizing such materials, as the components formed from plant-based materials are still under development, and there is still a lack of information and control on the properties of the synthetic nanomaterials. Further investigations are therefore crucial to understanding the formation mechanism and to improving performance and properties. Biomass is abundant everywhere at a low cost; however, scale-up of this eco-friendly method from laboratory to pilot-scale production is a prerequisite. In addition, multicycle reusable methodologies are essential and will be a new challenge for researchers.

List of Abbreviations

AC	Activated carbon
Ag NPs	Silver nanoparticles
AFM	Atomic force microscopy
Au NPs	Gold nanoparticles
BPB	Bromophenol blue dye
CBB	Coomassie brilliant blue dye
CEC	Cation exchange capacity
CMP	Citrus maxima peel
FCC	Face-centred cubic
GL	Glyphosate
HRTEM	High-resolution transmission electron microscopy
MNPs	Magnetic nanoparticles
MB	Methylene blue dye
MG	Malachite green dye
MO	Methyl orange dye
MONPs	Metal oxide nanoparticles
NBC	Nano biochar
OPE	Orange peel extract
OTC	Oxytetracycline
OBMS	Biochar and magnesium silicate from orange waste
PSMA	Poly(stryrene-comaleic anhydride)
RB	Rhodamine B
SEM	Scanning electron microscopy
SiO_2 NPs	Silica/silicon dioxide nanoparticles
SPR	Surface plasmon resonance
TEM	Transmission electron microscopy
XRD	X-ray diffraction
1,4-DNB	1,4-Dinitrobenzene
2,4-D	2,4-Dichlorophenoxyacetic

4-AP	4-Amino phenol
4-NP	4-Nitro phenol

Acknowledgements

The authors (H.S.K. and K.N.) gratefully acknowledge the Research and Development Office (RDO) of King Abdulaziz University, Jeddah, Saudi Arabia for the Postdoctoral Fellowship to H.S.K.

References

1. Water Facts - Worldwide Water Supply, https://www.usbr.gov/mp/arwec/water-facts-ww-water-sup.html, accessed 15 November 2022.
2. K. Beck, Uses of Seawater, https://sciencing.com/facts-5179246-uses-seawater.html, accessed 15 November 2022.
3. Water and Water Pollution, https://www.toppr.com/guides/biology/natural-resources/water-and-water-pollution/, accessed 16 November 2022.
4. Freshwater, https://www.unep.org/beatpollution/forms-pollution/freshwater, accessed 15 November 2022.
5. Y. Artioli, *Encyclopedia of Ecology*, ed. S. E. Jørgensen, and B. D. Fath, Academic Press, Oxford, 2008, pp. 60–65.
6. A. Ameri, S. Tamjidi, F. Dehghankhalili, A. Farhadi and M. A. Saati, Application of algae as low cost and effective bio-adsorbent for removal of heavy metals from wastewater: a review study, *Environ. Technol. Rev.*, 2020, **9**, 85–110.
7. G. A. V Magalhães-Ghiotto, A. M. de Oliveira, J. P. S. Natal, R. Bergamasco and R. G. Gomes, Green nanoparticles in water treatment: A review of research trends, applications, environmental aspects and large-scale production, *Environ. Nanotechnol., Monit. Manage.*, 2021, **16**, 100526.
8. P. Pourhakkak, M. Taghizadeh, A. Taghizadeh and M. Ghaedi, in *Adsorption: Fundamental Processes and Applications*, ed. M. Ghaedi, Elsevier, 2021, vol. 33, pp. 71–210.
9. M. Ajmal, R. A. K. Rao, R. Ahmad and J. Ahmad, Adsorption studies on Citrus reticulata (fruit peel of orange): removal and recovery of Ni(II) from electroplating wastewater, *J. Hazard. Mater.*, 2000, **79**, 117–131.
10. X. Yang and D. Wang, Photocatalysis: From Fundamental Principles to Materials and Applications, *ACS Appl. Energy Mater.*, 2018, **1**, 6657–6693.
11. G. Ren, H. Han, Y. Wang, S. Liu, J. Zhao, X. Meng and Z. Li, Recent Advances of Photocatalytic Application in Water Treatment: A Review, *Nanomaterials*, 2021, **11**, 1804.
12. W. Ahmad, S. Chandra Bhatt, M. Verma, V. Kumar and H. Kim, A review on current trends in the green synthesis of nickel oxide nanoparticles, characterizations, and their applications, *Environ. Nanotechnol., Monit. Manage.*, 2022, **18**, 100674.
13. R. Britto-Hurtado and M. Cortez-Valadez, in *Micro and Nano Technologies*, ed. U. Shanker, C. M. Hussain, and M. Rani, Elsevier, 2022, pp. 83–127.

14. T. Khan, H. Jalal, K. Karam and M. A. Khan, in *Nanobiotechnology for Plant Protection*, ed. K. A. Abd-Elsalam, Elsevier, 2022, pp. 189–217.
15. M. Zubair, M. Arshad and A. Ullah, in *Nanocellulose: A Sustainable and Renewable Material for Water and Wastewater Treatment*, ed. S. Kalia, Elsevier, 2021, pp. 93–109.
16. A. Rahman, M. H. Harunsani, A. L. Tan and M. M. Khan, Zinc oxide and zinc oxide-based nanostructures: biogenic and phytogenic synthesis, properties and applications, *Bioprocess Biosyst. Eng.*, 2021, **44**, 1333–1372.
17. K. Shivaji, E. S. Monica, A. Devadoss, D. D. Kirubakaran, C. R. Dhas, S. M. Jain and S. Pitchaimuthu, in *Green Photocatalysts*, ed. M. Naushad, S. Rajendran, and E. Lichtfouse, Springer International Publishing, Cham, 2020, pp. 25–46.
18. Y. W. Getahun, J. Gardea-Torresdey, F. S. Manciu, X. Li and A. A. El-Gendy, Green synthesized superparamagnetic iron oxide nanoparticles for water treatment with alternative recyclability, *J. Mol. Liq.*, 2022, **356**, 118983.
19. K. Narasimharao, G. K. R. Angaru, Z. H. Momin, S. Al-Thabaiti, M. Mokhtar, A. Alsheshri, S. Y. Alfaifi, J. R. Koduru and Y.-Y. Chang, Orange waste Biochar-Magnesium Silicate (OBMS) composite for enhanced removal of U(VI) ions from aqueous solutions, *J. Water Process Eng.*, 2023, **51**, 103359.
20. X. Wang, Z. Guo, Z. Hu and J. Zhang, Recent advances in biochar application for water and wastewater treatment: a review, *PeerJ*, 2020, **8**, e9164.
21. M. Ghorbani, E. Amirahmadi, R. W. Neugschwandtner, P. Konvalina, M. Kopecký, J. Moudrý, K. Perná and Y. T. Murindangabo, *Sustainability*, 2022, **14**(22), 14722.
22. R. Li, Y. Wu, X. Lou, H. Li, J. Cheng, B. Shen and L. Qin, *Water*, 2023, **15**(3), 395.
23. M. Gupta, N. Savla, C. Pandit, S. Pandit, P. K. Gupta, M. Pant, S. Khilari, Y. Kumar, D. Agarwal, R. R. Nair, D. Thomas and V. K. Thakur, Use of biomass-derived biochar in wastewater treatment and power production: A promising solution for a sustainable environment, *Sci. Total Environ.*, 2022, **825**, 153892.
24. C. M. S. da Silva, K. da Boit Martinello, S. F. Lütke, M. Godinho, D. Perondi, L. F. O. Silva and G. L. Dotto, Pyrolysis of grape bagasse to produce char for Cu(II) adsorption: a circular economy perspective, *Biomass Convers. Biorefin.*, 2022, 2190–6823.
25. J. Gao, Y. Liu, X. Li, M. Yang, J. Wang and Y. Chen, A promising and cost-effective biochar adsorbent derived from jujube pit for the removal of Pb(II) from aqueous solution, *Sci. Rep.*, 2020, **10**, 7473.
26. S. Ramanayaka, D. C. W. Tsang, D. Hou, Y. S. Ok and M. Vithanage, Green synthesis of graphitic nanobiochar for the removal of emerging contaminants in aqueous media, *Sci. Total Environ.*, 2020, **706**, 135725.

27. M. Naghdi, M. Taheran, S. K. Brar, T. Rouissi, M. Verma, R. Y. Surampalli and J. R. Valero, A green method for production of nanobiochar by ball milling- optimization and characterization, *J. Cleaner Prod.*, 2017, **164**, 1394–1405.
28. N. D. N. Tran, T. H. Bui, A. P. Nguyen, T.-T. Nguyen, V. M. Nguyen, N. L. Duong and T. Nguyen, The ability of silver-biochar green-synthesized from Citrus maxima peel to adsorb pollutant organic compounds and antibacterial activity, *Green Chem. Lett. Rev.*, 2022, **15**, 18–27.
29. A. Taha and S. Daffalla, Biochar Derived from Palm Waste Supported Greenly Synthesized MnO_2 Nanoparticles as a Novel Adsorbent for Wastewater Treatment, *Catalysts*, 2023, **13**(2), 451.
30. C. Yu, J. Tang, H. Su, J. Huang, F. Liu, L. Wang and H. Sun, Development of a novel biochar/iron oxide composite from green algae for bisphenol-A removal: Adsorption and Fenton-like reaction, *Environ. Technol. Innovation*, 2022, **28**, 102647.
31. B. Thomas and S. George, Production of Activated Carbon from Natural Sources, *Trends Green Chem.*, 2015, **1**, 1–5.
32. M. Gęca, M. Wiśniewska and P. Nowicki, Biochars and activated carbons as adsorbents of inorganic and organic compounds from multicomponent systems – A review, *Adv. Colloid Interface Sci.*, 2022, **305**, 102687.
33. H.-C. Tao, H.-R. Zhang, J.-B. Li and W.-Y. Ding, Biomass based activated carbon obtained from sludge and sugarcane bagasse for removing lead ion from wastewater, *Bioresour. Technol.*, 2015, **192**, 611–617.
34. B. Bhushan, A. Nayak and S. Kotnala, Green synthesis of highly porous activated carbon from jackfruit peel: Effect of operating factors on its physico-chemical characteristics, *Mater. Today: Proc.*, 2021, **44**, 187–191.
35. D. Angin, Production and characterization of activated carbon from sour cherry stones by zinc chloride, *Fuel*, 2014, **115**, 804–811.
36. J. M. Salman, Optimization of preparation conditions for activated carbon from palm oil fronds using response surface methodology on removal of pesticides from aqueous solution, *Arabian J. Chem.*, 2014, **7**, 101–108.
37. H. Koyuncu and A. R. Kul, Synthesis and characterization of a novel activated carbon using nonliving lichen *cetraria islandica* (L.) ach. and its application in water remediation: Equilibrium, kinetic and thermodynamic studies of malachite green removal from aqueous media, *Surf. Interfaces*, 2020, **21**, 100653.
38. H. Barabadi, B. Tajani, M. Moradi, K. Damavandi Kamali, R. Meena, S. Honary, M. A. Mahjoub and M. Saravanan, Penicillium Family as Emerging Nanofactory for Biosynthesis of Green Nanomaterials: A Journey into the World of Microorganisms, *J. Cluster Sci.*, 2019, **30**, 843–856.
39. R. Baby, M. Z. Hussein, A. H. Abdullah and Z. Zainal, Nanomaterials for the Treatment of Heavy Metal Contaminated Water, *Polymers*, 2022, **14**, 583.
40. P. B. Santhosh, J. Genova and H. Chamati, *Chemistry*, 2022, **4**, 345–369.

41. M. Singh, R. Kalaivani, S. Manikandan, N. Sangeetha and A. K. Kumaraguru, Facile green synthesis of variable metallic gold nanoparticle using Padina gymnospora, a brown marine macroalga, *Appl. Nanosci.*, 2013, **3**, 145–151.
42. O. S. ElMitwalli, O. A. Barakat, R. M. Daoud, S. Akhtar and F. Z. Henari, Green synthesis of gold nanoparticles using cinnamon bark extract, characterization, and fluorescence activity in Au/eosin Y assemblies, *J. Nanopart. Res.*, 2020, **22**, 309.
43. B. Kumar, K. Smita, Y. Angulo, A. Debut and L. Cumbal, Honeybee pollen assisted biosynthesis of nanogold and its application as catalyst in reduction of 4-nitrophenol, *Heliyon*, 2022, **8**, e10191.
44. V. T. Le, N. N. Q. Ngu, T. P. Chau, T. D. Nguyen, V. T. Nguyen, T. L. H. Nguyen, X. T. Cao and V.-D. Doan, Silver and Gold Nanoparticles from *Limnophila rugosa* Leaves: Biosynthesis, Characterization, and Catalytic Activity in Reduction of Nitrophenols, *J. Nanomater.*, 2021, **2021**, 5571663.
45. T. H. A. Nguyen, T. T. V. Le, B. A. Huynh, N. V. Nguyen, V. T. Le, V.-D. Doan, V. A. Tran, A.-T. Nguyen, X. T. Cao and Y. Vasseghian, Novel biogenic gold nanoparticles stabilized on poly(styrene-co-maleic anhydride) as an effective material for reduction of nitrophenols and colorimetric detection of Pb(II), *Environ. Res.*, 2022, **212**, 113281.
46. A. Jain, S. Ranjan, N. Dasgupta and C. Ramalingam, Nanomaterials in food and agriculture: An overview on their safety concerns and regulatory issues, *Crit. Rev. Food Sci. Nutr.*, 2018, **58**, 297–317.
47. P. Mathur, S. Jha, S. Ramteke and N. K. Jain, Pharmaceutical aspects of silver nanoparticles, *Artif. Cells, Nanomed., Biotechnol.*, 2018, **46**, 115–126.
48. M. E. Taghavizadeh Yazdi, M. S. Amiri, S. Akbari, M. Sharifalhoseini, F. Nourbakhsh, M. Mashreghi, E. Yousefi, M. R. Abbasi, M. Modarres and A. Es-haghi, Green Synthesis of Silver Nanoparticles Using *Helichrysum graveolens* for Biomedical Applications and Wastewater Treatment, *BioNanoScience*, 2020, **10**, 1121–1127.
49. K. Nagaraj, P. Thangamuniyandi, S. Kamalesu, M. Dixitkumar, A. K. Saini, S. K. Sharma, J. Naman, J. Priyanshi, C. Uthra, S. Lokhandwala, N. M. Parekh, S. Radha, S. Sakthinathan, T.-W. Chiu and C. Karuppiah, Silver nanoparticles using *Cassia Alata* and its catalytic reduction activities of Rhodamine6G, Methyl orange and methylene blue dyes, *Inorg. Chem. Commun.*, 2023, **155**, 110985.
50. M. Aslam, F. Fozia, A. Gul, I. Ahmad, R. Ullah, A. Bari, R. A. Mothana and H. Hussain, *Molecules*, 2021, **26**(20), 6144.
51. M. Aravind, A. Ahmad, I. Ahmad, M. Amalanathan, K. Naseem, S. M. M. Mary, C. Parvathiraja, S. Hussain, T. S. Algarni, M. Pervaiz and M. Zuber, Critical green routing synthesis of silver NPs using jasmine flower extract for biological activities and photocatalytical degradation of methylene blue, *J. Environ. Chem. Eng.*, 2021, **9**, 104877.

52. M. Yahya Tahir, A. Ahmad, A. A. Alothman, M. S. S. Mushab and S. Ali, Green Synthesis of Silver Nanoparticles Using *Thespesia populnea* Bark Extract for Efficient Removal of Methylene Blue (MB) Degradation *via* Photocatalysis with Antimicrobial Activity and for Anticancer Activity, *Bioinorg. Chem. Appl.*, 2022, **2022**, 7268273.
53. M. Moond, S. Singh, S. Sangwan, P. Devi, A. Beniwal, J. Rani, A. Kumari and S. Rani, *Molecules*, 2023, **28**(3), 951.
54. R. Santhoshkumar, A. H. Parvathy and E. V Soniya, Phytosynthesis of silver nanoparticles from aqueous leaf extracts of *Piper colubrinum*: characterisation and catalytic activity, *J. Exp. Nanosci.*, 2021, **16**, 294–308.
55. H. Kolya, P. Maiti, A. Pandey and T. Tripathy, Green synthesis of silver nanoparticles with antimicrobial and azo dye (Congo red) degradation properties using *Amaranthus gangeticus* Linn leaf extract, *J. Anal. Sci. Technol.*, 2015, **6**, 33.
56. R. Erenler, E. N. Gecer, I. Hosaflioglu and L. Behcet, Green synthesis of silver nanoparticles using *Stachys spectabilis*: Identification, catalytic degradation, and antioxidant activity, *Biochem. Biophys. Res. Commun.*, 2023, **659**, 91–95.
57. V.-D. Doan, V. T. Le, D. L. Tran, T. L. H. Nguyen, D. C. Nguyen, A.-T. Nguyen and V. T. Le, Catalytic reduction of nitrophenols using *Gnetum montanum* extract capped silver nanoparticles, *Mol. Catal.*, 2023, **534**, 112804.
58. S. Laouini, A. Bouafia and M. Tedjani, Catalytic Activity for Dye Degradation and Characterization of Silver/Silver Oxide Nanoparticles Green Synthesized by Aqueous Leaves Extract of *Phoenix Dactylifera* L., *Res. Sq.*, 2021, 2–42.
59. C. Simatupang, V. K. Jindal and R. Jindal, Biosynthesis of Silver Nanoparticles Using Orange Peel Extract for Application in Catalytic Degradation of Methylene Blue Dye: 10.32526/ennrj/19/202100088, *Environ. Nat. Resour. J.*, 2021, **19**, 468–480.
60. G. Pal, P. Rai and A. Pandey, Characterization and Applications of Nanoparticles, in *Micro and Nano Technologies*, ed. A. K. Shukla, and S. B. T.-G. S. Iravani, Elsevier, 2019, pp. 1–26.
61. U. Kamran, H. N. Bhatti, M. Iqbal and A. Nazir, Green Synthesis of Metal Nanoparticles and their Applications in Different Fields: A Review, *Z. Phys. Chem.*, 2019, **233**, 1325–1349.
62. S. Bishnoi, A. Kumar and R. Selvaraj, Facile synthesis of magnetic iron oxide nanoparticles using inedible *Cynometra ramiflora* fruit extract waste and their photocatalytic degradation of methylene blue dye, *Mater. Res. Bull.*, 2018, **97**, 121–127.
63. S. D. Roy, K. C. Das and S. S. Dhar, Conventional to green synthesis of magnetic iron oxide nanoparticles; its application as catalyst, photocatalyst and toxicity: A short review, *Inorg. Chem. Commun.*, 2021, **134**, 109050.
64. K. C. Verma and N. Goyal, in *Iron Oxide Nanoparticles*, ed. X.-L. Huang, IntechOpen, Rijeka, 2022, ch. 7.

65. X. Yu and J. Zhou, in *Study of Grain Boundary Character*, ed. T. Tanski, and W. Borek, IntechOpen, Rijeka, 2017, ch. 4.
66. S. K. Maji, N. Mukherjee, A. Mondal and B. Adhikary, Synthesis, characterization and photocatalytic activity of α-Fe_2O_3 nanoparticles, *Polyhedron*, 2012, **33**, 145–149.
67. N. Rosdi, R. S. Azis, M. S. Mustaffa, N. H. Abdullah, S. Sulaiman and T. T. Ling, Synthesis and characterization of Mg–Ti substituted barium hexaferrite ($BaMg_{0.6}Ti_{0.6}Fe_{108}O_{19}$) derived from millscale waste for microwave application, *J. Mater. Sci.: Mater. Electron.*, 2019, **30**, 8636–8644.
68. M. E. Compeán-Jasso, F. Ruiz, J. R. Martínez and A. Herrera-Gómez, Magnetic properties of magnetite nanoparticles synthesized by forced hydrolysis, *Mater. Lett.*, 2008, **62**, 4248–4250.
69. R. Wahab, F. Khan and A. A. Al-Khedhairy, Hematite iron oxide nanoparticles: apoptosis of myoblast cancer cells and their arithmetical assessment, *RSC Adv.*, 2018, **8**, 24750–24759.
70. Z. Muneeb Ur Rahman, M. Usman, S. Zia-ur-Rahman, A. Ullah and I. Won Kim, Static magnetic properties of Maghemite nanoparticles, *J. Korean Phys. Soc.*, 2014, **65**, 1925–1929.
71. Z. Isik, R. Bouchareb, H. Arslan, S. Özdemir, S. Gonca, N. Dizge, D. Balakrishnan and S. V. S. Prasad, Green synthesis of iron oxide nanoparticles derived from water and methanol extract of *Centaurea solstitialis* leaves and tested for antimicrobial activity and dye decolorization capability, *Environ. Res.*, 2023, **219**, 115072.
72. C. Sudhakar, M. Poonkothai, T. Selvankmuar, K. Selvam, G. Rajivgandhi, M. Z. Siddiqi, N. S. Alharbi, S. Kadaikunnan and N. Vijayakumar, Biomimetic synthesis of iron oxide nanoparticles using *Canthium coromandelicum* leaf extract and its antibacterial and catalytic degradation of Janus green, *Inorg. Chem. Commun.*, 2021, **133**, 108977.
73. M. H. Ehrampoush, M. Miria, M. H. Salmani and A. H. Mahvi, Cadmium removal from aqueous solution by green synthesis iron oxide nanoparticles with tangerine peel extract, *J. Environ. Health Sci. Eng.*, 2015, **13**, 84.
74. K. Parajuli, A. K. Sah and H. Paudyal, Green Synthesis of Magnetite Nanoparticles Using Aqueous Leaves Extracts of *Azadirachta indica* and Its Application for the Removal of As(v) from Water, *Green Sustainable Chem.*, 2020, **10**, 1–16.
75. A. C. de Lima Barizão, M. F. Silva, M. Andrade, F. C. Brito, R. G. Gomes and R. Bergamasco, Green synthesis of iron oxide nanoparticles for tartrazine and bordeaux red dye removal, *J. Environ. Chem. Eng.*, 2020, **8**, 103618.
76. M. S. H. Bhuiyan, M. Y. Miah, S. C. Paul, T. Das Aka, O. Saha, M. M. Rahaman, M. J. I. Sharif, O. Habiba and M. Ashaduzzaman, Green synthesis of iron oxide nanoparticle using Carica papaya leaf extract: application for photocatalytic degradation of remazol yellow RR dye and antibacterial activity, *Heliyon*, 2020, **6**, e04603.

77. N. Madubuonu, S. O. Aisida, A. Ali, I. Ahmad, T. Zhao, S. Botha, M. Maaza and F. I. Ezema, Biosynthesis of iron oxide nanoparticles *via* a composite of Psidium guavaja-Moringa oleifera and their antibacterial and photocatalytic study, *J. Photochem. Photobiol., B*, 2019, **199**, 111601.
78. I. Ashraf, N. B. Singh and A. Agarwal, Green synthesis of iron oxide nanoparticles using Amla seed for methylene blue dye removal from water, *Mater. Today: Proc.*, 2023, **72**, 311–316.
79. L. Shaker Ardakani, V. Alimardani, A. M. Tamaddon, A. M. Amani and S. Taghizadeh, Green synthesis of iron-based nanoparticles using *Chlorophytum comosum* leaf extract: methyl orange dye degradation and antimicrobial properties, *Heliyon*, 2021, **7**, e06159.
80. I. Bibi, N. Nazar, S. Ata, M. Sultan, A. Ali, A. Abbas, K. Jilani, S. Kamal, F. M. Sarim, M. I. Khan, F. Jalal and M. Iqbal, Green synthesis of iron oxide nanoparticles using pomegranate seeds extract and photocatalytic activity evaluation for the degradation of textile dye, *J. Mater. Res. Technol.*, 2019, **8**, 6115–6124.
81. K. Davis, R. Yarbrough, M. Froeschle, J. White and H. Rathnayake, Band gap engineered zinc oxide nanostructures *via* a sol–gel synthesis of solvent driven shape-controlled crystal growth, *RSC Adv.*, 2019, **9**, 14638–14648.
82. K. Rambabu, G. Bharath, F. Banat and P. L. Show, Green synthesis of zinc oxide nanoparticles using *Phoenix dactylifera* waste as bioreductant for effective dye degradation and antibacterial performance in wastewater treatment, *J. Hazard. Mater.*, 2021, **402**, 123560.
83. F. Rahman, M. A. Majed Patwary, M. A. Bakar Siddique, M. S. Bashar, M. A. Haque, B. Akter, R. Rashid, M. A. Haque and A. K. M. Royhan Uddin, Green synthesis of zinc oxide nanoparticles using *Cocos nucifera* leaf extract: characterization, antimicrobial, antioxidant and photocatalytic activity, *R. Soc. Open Sci.*, 2022, **9**, 220858.
84. W. S. Al-Arjan, Zinc Oxide Nanoparticles and Their Application in Adsorption of Toxic Dye from Aqueous Solution, *Polymers*, 2022, **14**, 3086.
85. M. Zare, K. Namratha, S. Ilyas, A. Sultana, A. Hezam, L. Sunil, M. A. Surmeneva, R. A. Surmenev, M. B. Nayan, S. Ramakrishna, S. Mathur and K. Byrappa, Emerging Trends for ZnO Nanoparticles and Their Applications in Food Packaging, *ACS Food Sci. Technol.*, 2022, **2**, 763–781.
86. A. Jayachandran, T. R. Aswathy and A. S. Nair, Green synthesis and characterization of zinc oxide nanoparticles using Cayratia pedata leaf extract, *Biochem. Biophys. Rep.*, 2021, **26**, 100995.
87. N. Vijayakumar, V. K. Bhuvaneshwari, G. K. Ayyadurai, R. Jayaprakash, K. Gopinath, M. Nicoletti, S. Alarifi and M. Govindarajan, Green synthesis of zinc oxide nanoparticles using Anoectochilus elatus, and their biomedical applications, *Saudi J. Biol. Sci.*, 2022, **29**, 2270–2279.

88. M. Batool, S. Khurshid, W. M. Daoush, S. A. Siddique and T. Nadeem, Green Synthesis and Biomedical Applications of ZnO Nanoparticles: Role of PEGylated-ZnO Nanoparticles as Doxorubicin Drug Carrier against MDA-MB-231(TNBC) Cells Line, *Crystals*, 2021, **11**, 344.
89. S. Ş. Doğan and A. Kocabaş, Green synthesis of ZnO nanoparticles with Veronica multifida and their antibiofilm activity, *Hum. Exp. Toxicol.*, 2020, **39**, 319–327.
90. N. Bala, S. Saha, M. Chakraborty, M. Maiti, S. Das, R. Basu and P. Nandy, Green synthesis of zinc oxide nanoparticles using *Hibiscus subdariffa* leaf extract: effect of temperature on synthesis, anti-bacterial activity and anti-diabetic activity, *RSC Adv.*, 2015, **5**, 4993–5003.
91. S. Faisal, H. Jan, S. A. Shah, S. Shah, A. Khan, M. T. Akbar, M. Rizwan, F. Jan, W. Noreen Akhtar, A. Khattak and S. Syed, Green Synthesis of Zinc Oxide (ZnO) Nanoparticles Using Aqueous Fruit Extracts of Myristica fragrans: Their Characterizations and Biological and Environmental Applications, *ACS Omega*, 2021, **6**, 9709–9722.
92. A. Umamaheswari, S. L. Prabu, S. A. John and A. Puratchikody, Green synthesis of zinc oxide nanoparticles using leaf extracts of Raphanus sativus var. Longipinnatus and evaluation of their anticancer property in A549 cell lines, *Biotechnol. Rep.*, 2021, **29**, e00595.
93. P. S. Sokolov, A. N. Baranov, A. M. T. Bell and V. L. Solozhenko, Low-temperature thermal expansion of rock-salt ZnO, *Solid State Commun.*, 2014, **177**, 65–67.
94. A. K. Zak, W. H. A. Majid, M. Darroudi and R. Yousefi, Synthesis and characterization of ZnO nanoparticles prepared in gelatin media, *Mater. Lett.*, 2011, **65**, 70–73.
95. X. Liang, P. Guo, G. Wang, R. Deng, D. Pan and X. Wei, Dilute magnetic semiconductor Cu_2MnSnS_4 nanocrystals with a novel zincblende and wurtzite structure, *RSC Adv.*, 2012, **2**, 5044–5046.
96. C. Vidya, C. Manjunatha, M. N. Chandraprabha, R. Megha and M. A. L. Antony Raj, Hazard free green synthesis of ZnO nano-photo-catalyst using Artocarpus Heterophyllus leaf extract for the degradation of Congo red dye in water treatment applications, *J. Environ. Chem. Eng.*, 2017, **5**, 3172–3180.
97. V. B. Raghavendra, S. Shankar, M. Govindappa, A. Pugazhendhi, M. Sharma and S. C. Nayaka, Green Synthesis of Zinc Oxide Nanoparticles (ZnO NPs) for Effective Degradation of Dye, Polyethylene and Antibacterial Performance in Waste Water Treatment, *J. Inorg. Organomet. Polym. Mater.*, 2022, **32**, 614–630.
98. J. de O. Primo, C. Bittencourt, S. Acosta, A. Sierra-Castillo, J.-F. Colomer, S. Jaerger, V. C. Teixeira and F. J. Anaissi, Synthesis of Zinc Oxide Nanoparticles by Ecofriendly Routes: Adsorbent for Copper Removal From Wastewater, *Front. Chem.*, 2020, **8**, 571790.
99. D. Eissa, R. H. Hegab, A. Abou-Shady and Y. H. Kotp, Green synthesis of ZnO, MgO and SiO2 nanoparticles and its effect on irrigation water, soil properties, and Origanum majorana productivity, *Sci. Rep.*, 2022, **12**, 5780.

100. T. Riaz, A. Munnwar, T. Shahzadi, M. Zaib, S. Shahid, M. Javed, S. Iqbal, K. Rizwan, M. Waqas, B. Khalid, N. S. Awwad, H. A. Ibrahium and M. A. Bajaber, Phyto-mediated synthesis of nickel oxide (NiO) nanoparticles using leaves' extract of Syzygium cumini for antioxidant and dyes removal studies from wastewater, *Inorg. Chem. Commun.*, 2022, **142**, 109656.
101. A. A. Ezhilarasi, J. J. Vijaya, K. Kaviyarasu, X. Zhang and L. J. Kennedy, Green synthesis of nickel oxide nanoparticles using *Solanum trilobatum* extract for cytotoxicity, antibacterial and photocatalytic studies, *Surf. Interfaces*, 2020, **20**, 100553.
102. S. Ghazal, N. Khandannasab, H. A. Hosseini, Z. Sabouri, A. Rangrazi and M. Darroudi, Green synthesis of copper-doped nickel oxide nanoparticles using okra plant extract for the evaluation of their cytotoxicity and photocatalytic properties, *Ceram. Int.*, 2021, **47**, 27165–27176.
103. Y. Zhang, B. Mahdavi, M. Mohammadhosseini, E. Rezaei-Seresht, S. Paydarfard, M. Qorbani, M. Karimian, N. Abbasi, H. Ghaneialvar and E. Karimi, Green synthesis of NiO nanoparticles using Calendula officinalis extract: Chemical characterization, antioxidant, cytotoxicity, and anti-esophageal carcinoma properties, *Arabian J. Chem.*, 2021, **14**, 103105.
104. S. O. Aisida, K. Ugwu, P. A. Akpa, A. C. Nwanya, U. Nwankwo, S. S. Botha, P. M. Ejikeme, I. Ahmad, M. Maaza and F. I. Ezema, Biosynthesis of silver nanoparticles using bitter leave (Veronica amygdalina) for antibacterial activities, *Surf. Interfaces*, 2019, **17**, 100359.
105. S. Prabhu, T. Daniel Thangadurai, P. Vijai Bharathy and P. Kalugasalam, Synthesis and characterization of nickel oxide nanoparticles using Clitoria ternatea flower extract: Photocatalytic dye degradation under sunlight and antibacterial activity applications, *Results Chem.*, 2022, **4**, 100285.
106. N. Sundaresan, S. Ravichandran and I. kaliappan, Biosynthesis of nickel oxide nanoparticles using *Evolvulus alsinoides* extract and their potential photocatalytic and invitro anticancer activity, *Inorg. Chem. Commun.*, 2023, **150**, 110489.
107. P. Karpagavinayagam, A. Emi Princess Prasanna and C. Vedhi, Eco-friendly synthesis of nickel oxide nanoparticles using Avicennia Marina leaf extract: Morphological characterization and electrochemical application, *Mater. Today: Proc.*, 2022, **48**, 136–142.
108. C.-C. Diao, C.-Y. Huang, C.-F. Yang and C.-C. Wu, *Nanomaterials*, 2020, **10**(4), 636.
109. H. Qiao, Z. Wei, H. Yang, L. Zhu and X. Yan, Preparation and Characterization of NiO Nanoparticles by Anodic Arc Plasma Method, *J. Nanomater.*, 2009, **2009**, 795928.
110. I. M. Rashid, S. D. Salman, A. K. Mohammed and Y. S. Mahdi, Green Synthesis of Nickle Oxide Nanoparticles for Adsorption of Dyes, *Sains Malays.*, 2022, **2**, 533–546.
111. Z. Sabouri, N. Fereydouni, A. Akbari, H. A. Hosseini, A. Hashemzadeh, M. S. Amiri, R. Kazemi Oskuee and M. Darroudi, Plant-based synthesis

of NiO nanoparticles using *salvia macrosiphon* Boiss extract and examination of their water treatment, *Rare Met.*, 2020, **39**, 1134–1144.
112. C. Kalita and P. Saikia, Magnetically separable tea leaf mediated nickel oxide nanoparticles for excellent photocatalytic activity, *J. Indian Chem. Soc.*, 2021, **98**, 100213.
113. M. A. Rafique, S. Kiran, S. Javed, I. Ahmad, S. Yousaf, N. Iqbal, G. Afzal and F. Rani, Green synthesis of nickel oxide nanoparticles using *Allium cepa* peels for degradation of Congo red direct dye: an environmental remedial approach, *Water Sci. Technol.*, 2021, **84**, 2793–2804.
114. A. K. H. Bashir, L. C. Razanamahandry, A. C. Nwanya, K. Kaviyarasu, W. Saban, H. E. A. Mohamed, S. K. O. Ntwampe, F. I. Ezema and M. Maaza, Biosynthesis of NiO nanoparticles for photodegradation of free cyanide solutions under ultraviolet light, *J. Phys. Chem. Solids*, 2019, **134**, 133–140.
115. P. Ramesh and A. Rajendran, Green synthesis of nickel oxide nanoparticles for photodegradation analysis, *Mater. Today: Proc.*, 2022, **68**, 367–372.
116. M. Rani, J. Yadav and U. Shanker, Green synthesis, kinetics and photoactivity of novel nickel oxide-decorated zinc hexacyanocobaltate catalyst for efficient removal of toxic Cr(VI), *J. Environ. Chem. Eng.*, 2021, **9**, 105073.
117. E. Akhayere, D. Kavaz and A. Vaseashta, Efficacy Studies of Silica Nanoparticles Synthesized Using Agricultural Waste for Mitigating Waterborne Contaminants, *Appl. Sci.*, 2022, **12**(18), 9279.
118. L. A. September, N. Kheswa, N. S. Seroka and L. Khotseng, Green synthesis of silica and silicon from agricultural residue sugarcane bagasse ash – a mini review, *RSC Adv.*, 2023, **13**, 1370–1380.
119. D. Sachan, A. Ramesh and G. Das, Green synthesis of silica nanoparticles from leaf biomass and its application to remove heavy metals from synthetic wastewater: A comparative analysis, *Environ. Nanotechnol., Monit. Manage.*, 2021, **16**, 100467.
120. P. Sharma, J. Kherb, J. Prakash and R. Kaushal, A novel and facile green synthesis of SiO_2 nanoparticles for removal of toxic water pollutants, *Appl. Nanosci.*, 2023, **13**, 735–747.
121. J. Tejedor, V. H. Guerrero, K. Vizuete and A. Debut, Environmentally friendly synthesis of silicon dioxide nanoparticles and their application for the removal of emerging contaminants in aqueous media, *J. Phys.: Conf. Ser.*, 2022, **2238**, 12005.
122. A. Maghfirah, M. M. Ilmi, A. T. N. Fajar and G. T. M. Kadja, A review on the green synthesis of hierarchically porous zeolite, *Mater. Today Chem.*, 2020, **17**, 100348.
123. G. T. M. Kadja, I. R. Kadir, A. T. N. Fajar, V. Suendo and R. R. Mukti, Revisiting the seed-assisted synthesis of zeolites without organic structure-directing agents: insights from the CHA case, *RSC Adv.*, 2020, **10**, 5304–5315.

124. P. Solanki, V. Gupta and R. Kulshrestha, Synthesis of Zeolite from Fly Ash and Removal of Heavy Metal Ions from Newly Synthesized Zeolite, *E-J. Chem.*, 2010, **7**, 356150.
125. N. Murayama, H. Yamamoto and J. Shibata, Mechanism of zeolite synthesis from coal fly ash by alkali hydrothermal reaction, *Int. J. Miner. Process.*, 2002, **64**, 1–17.
126. J. L. X. Hong, T. Maneerung, S. N. Koh, S. Kawi and C.-H. Wang, Conversion of Coal Fly Ash into Zeolite Materials: Synthesis and Characterizations, Process Design, and Its Cost-Benefit Analysis, *Ind. Eng. Chem. Res.*, 2017, **56**, 11565–11574.
127. D. Längauer, V. Čablík, S. Hredzák, A. Zubrik, M. Matik and Z. Danková, Preparation of Synthetic Zeolites from Coal Fly Ash by Hydrothermal Synthesis, *Materials*, 2021, **14**(5), 1267.
128. X. Ren, R. Qu, S. Liu, H. Zhao, W. Wu, H. Song, C. Zheng, X. Wu and X. Gao, Synthesis of Zeolites from Coal Fly Ash for Removal of Harmful Gaseous Pollutants: A Review, *Aerosol Air Qual. Res.*, 2020, **20**, 1127–1144.
129. K. Ojha, N. C. Pradhan and A. N. Samanta, Zeolite from fly ash: synthesis and characterization, *Bull. Mater. Sci.*, 2004, **27**, 555–564.
130. X. Ren, L. Xiao, R. Qu, S. Liu, D. Ye, H. Song, W. Wu, C. Zheng, X. Wu and X. Gao, Synthesis and characterization of a single phase zeolite A using coal fly ash, *RSC Adv.*, 2018, **8**, 42200–42209.
131. P. Zhang, S. Li, P. Guo and C. Zhang, Seed-Assisted, OSDA-Free, Solvent-Free Synthesis of ZSM-5 Zeolite from Iron Ore Tailings, *Waste Biomass Valorization*, 2020, **11**, 4381–4391.
132. R. M. Mohamed, I. A. Mkhalid and M. A. Barakat, Rice husk ash as a renewable source for the production of zeolite NaY and its characterization, *Arabian J. Chem.*, 2015, **8**, 48–53.
133. C. G. Flores, H. Schneider, J. S. Dornelles, L. B. Gomes, N. R. Marcilio and P. J. Melo, Synthesis of potassium zeolite from rice husk ash as a silicon source, *Clean. Eng. Technol.*, 2021, **4**, 100201.
134. C. Sundari, S. Setiadji, Y. Rohmatullah, S. Sanusi, D. F. Nurbaeti, I. Novianti, I. Farida, A. Nurohmah and A. L. Ivansyah, Synthesis of zeolite L using rice husk ash silica for adsorption of methylene blue: kinetic and adsorption isotherm, *MATEC Web Conf.*, 2018.
135. C. Zhang, S. Li and S. Bao, Sustainable Synthesis of ZSM-5 Zeolite from Rice Husk Ash Without Addition of Solvents, *Waste Biomass Valorization*, 2019, **10**, 2825–2835.
136. J. Madhu, A. Santhanam, M. Natarajan and D. Velauthapillai, CO2 adsorption performance of template free zeolite A and X synthesized from rice husk ash as silicon source, *RSC Adv.*, 2022, **12**, 23221–23239.
137. D. Prasetyoko, Z. Ramli, S. Endud, H. Hamdan and B. Sulikowski, Conversion of rice husk ash to zeolite beta, *Waste Manag.*, 2006, **26**, 1173–1179.
138. T. Sriatun and L. Suyati, Synthesis of Zeolite from Sugarcane Bagasse Ash Using Cetyltrimethylammonium Bromide as Structure Directing Agent, *Indones. J. Chem.*, 2018, **18**, 159–165.

139. C. Belviso, M. Mancinelli and A. Lettino, A green process for zeolite synthesis: low-temperature vapor phase treatment of natural bauxites, *J. Mater. Sci.*, 2022, **57**, 16619–16631.
140. E.-P. Ng, J.-H. Chow, R. R. Mukti, O. Muraza, T. C. Ling and K.-L. Wong, Hydrothermal synthesis of zeolite a from bamboo leaf biomass and its catalytic activity in cyanoethylation of methanol under autogenic pressure and air conditions, *Mater. Chem. Phys.*, 2017, **201**, 78–85.
141. L. Ayele, E. Pérez, Á. Mayoral, Y. Chebude and I. Díaz, Synthesis of zeolite A using raw kaolin from Ethiopia and its application in removal of Cr(III) from tannery wastewater, *J. Chem. Technol. Biotechnol.*, 2018, **93**, 146–154.
142. L. Ayele, J. Pérez-Pariente, Y. Chebude and I. Díaz, Conventional *versus* alkali fusion synthesis of zeolite A from low grade kaolin, *Appl. Clay Sci.*, 2016, **132–133**, 485–490.
143. N. Sazali, Z. Harun, T. Abdullahi, N. H. Kamarudin, N. Sazali, M. R. Jamalludin, S. K. Hubadillah and S. S. Alias, The Route of Hydrothermal Synthesis Zeolite-A from the Low-Grade Perak kaolin, Malaysia, *Silicon*, 2022, **14**, 7257–7273.
144. P. M. Pereira, B. F. Ferreira, N. P. Oliveira, E. J. Nassar, K. J. Ciuffi, M. A. Vicente, R. Trujillano, V. Rives, A. Gil, S. Korili and E. H. De Faria, *Appl. Sci.*, 2018, **8**(4), 608.
145. T. S. Jamil, H. S. Ibrahim, I. H. Abd El-Maksoud and S. T. El-Wakeel, Application of zeolite prepared from Egyptian kaolin for removal of heavy metals: I. Optimum conditions, *Desalination*, 2010, **258**, 34–40.
146. T. S. Jamil, H. H. Abdel Ghafar, H. S. Ibrahim and I. H. Abd El-Maksoud, Removal of methylene blue by two zeolites prepared from naturally occurring Egyptian kaolin as cost effective technique, *Solid State Sci.*, 2011, **13**, 1844–1851.
147. M. M. Selim and I. H. Abd El-Maksoud, Hydrogenation of edible oil over zeolite prepared from local kaolin, *Microporous Mesoporous Mater.*, 2004, **74**, 79–85.
148. Y. He, S. Tang, S. Yin and S. Li, Research progress on green synthesis of various high-purity zeolites from natural material-kaolin, *J. Cleaner Prod.*, 2021, **306**, 127248.
149. H. Didik Prasetyoko, M. Santoso, I. Qoniah, W. L. Leaw, P. B. D. Firda and H. Nur, A review on synthesis of kaolin-based zeolite and the effect of impurities, *J. Chin. Chem. Soc.*, 2020, **67**, 911–936.
150. P. F. Corregidor, D. E. Acosta and H. A. Destéfanis, Kinetic Study of Seed-Assisted Crystallization of ZSM-5 Zeolite in an OSDA-Free System Using a Natural Aluminosilicate as Starting Source, *Ind. Eng. Chem. Res.*, 2018, **57**, 13713–13720.

Subject Index

activated carbon thin film (ACTF), 104, 106, 115
adsorption studies, 209, 211–212, 300
  batch adsorption studies, 212
    competitive adsorption study, 213–214, 222
    exploratory study, 216
    exploratory study, RGY and SB, 212
    isotherm, 213, 220–221
    kinetics, 212–213, 218–220
    regeneration and reuse studies, 213, 221–222
    RGY solution pH, 212, 216–218
  fixed-bed column studies, 214, 222–226
  mechanism, 2
  thermodynamics, 228–231
advanced oxidation processes (AOPs), 258
  dye degradation, solution, 258
aerogels, 36–37
Ag@MXene, 56
antifouling, 50, 112–117
auto-self-healing property, 117–118
  self-healing predicted mechanism, 118–120

bacteria, 33–34
bioactivated carbon materials, 306–307
bioadsorbent, 210
  and dye, 210
biochar materials, 303–306
bioremediation, water, 33
  aerogels, 36–37
  bacteria, 33–34
  carbonaceous materials, 37
  green nanoparticles, environmental remediation, 38
  hydrogels, 36–37
  protein nanofibrils, 35–36
  seaweeds, 36
bisphenol A, 184
Bohart–Adams model, 224

carbon composites, 69, 71
carbon nanotubes, 69, 71
catalytic membranes, 166
  biocatalytic membranes, 171
    enzyme immobilization approaches, 173–174
    principles, 171–173
  photocatalytic membranes, 167–168
    approaches, 169–171
    photocatalytic principle, 168–169
cellular dysfunction, 2
cellulose, 103, 104
  macro and micro sources of, 107
cellulose nanocrystals (CNCs), 107–108
cellulose nanofibrils (CNFs), 108
charcoal, 26

chemically modified graphene, 188
and fabrication methods, 188–189
chemical reduction, 85
chitin, 2
chitosan, 1, 2, 71
circular (bio)economy, 79
competitive adsorption study, 213–214
Congo red (CR) dye separation, 54
contamination, water, 25
conventional membranes, 155–156
conventional methods, wastewater treatment, 235–237

desalination, 2, 40, 54–59, 105
double-walled carbon nanotubes (DWCNTs), 69
dye
aquatic pollutant, 256–257
degradation, solution, 258

electrochemical reduction, 131
electro-responsive membranes (ERMs), 161–162, 175
anodic potential stimuli, 163–164
cathodic potential stimuli, 164–166
electrochemical stimulus, 162–163
electrospinning, 105
environmental pollutants, 173

Fenton process, 88–90

gold nanoparticles (Au NPs), 308–310
graphene, 68–69, 127, 184
binary composites, 134–135
derivatives, 127–128
preparation of, 128–130
preparation of, 128–130
tertiary composites, 135–136
graphene-based 3D nanostructures, 189
and fabrication methods, 189–190
graphene-based composites, 68–69
graphene binary composites, 134–135
graphene tertiary composites, 135–136
sorption, metal ions, 136–139
synthesis methods, 133–134
graphene-based macrostructures (GBMs), 184, 185, 203
adsorbates removal, 199
anionic contaminants, uptake, 195–196
biomass-based GBMs, 192
cationic contaminants, uptake, 193–195
emerging contaminants, 201–202
inorganic GBM hybrids, 191–192
MOF-based GBMs, 192–193
for multicomponent removal, 200–201
non-ionic pollutants removal, 196–199
for oily contaminants, 199–200
organic/inorganic pollutants, adsorption, 193
polymer-based GBMs, 190–191
pristine graphene, 186
and synthetic methods, 185
graphene-based membranes, 139
separation mechanism and properties, 142–146
transportation mechanisms, 146
for water filtration, 139–142
graphene composites, 132
photodegradation, organic contaminants, 132–133
graphene oxide (GO), 128, 133, 203
and fabrication methods, 186–187
reduction, 130–132

graphite, 127
green synthesis methods, 301
adsorbent and photocatalyst materials, 301–302
carbon-based materials, 302
bioactivated carbon materials, 306–307
biochar materials, 303–306
gold nanoparticles (Au NPs), 308–310
iron oxide nanoparticles (FeO NPs), 316–325
metals, 307–308
nickel oxide (NiO), 328–335
silicon dioxide (SiO_2) (silica), 335–337
silver nanoparticles (Ag NPs), 310–316
zeolites, 337–339
zinc oxide nanoparticles (ZnO NPs), 325–328

heavy metal ions, 183

industrial and agricultural wastewater sources, 235
iron oxide nanoparticles (FeO NPs), 238, 316–325
BET surface area analysis, 248
biosynthesis, 238–241
contaminants removal, 241–244
electron microscopy, 245–246
FTIR spectroscopy, 244
nitrogen adsorption/desorption isotherm, 247–248
preparation methods, 238
thermogravimetric analysis, 246–247
XRD, 244–245
isothermal titration calorimetry (ITC), 214–215

laser ablation, 84

maghemite (γ-Fe_2O_3), 86
magnetic materials
Fenton and photo-Fenton process, 88–90
photoactivity, 90–91
magnetic materials synthesis, 81
maghemite (γ-Fe_2O_3), 86
magnetite (Fe_3O_4), 83–85
nanozerovalent iron, 81–83
magnetic NPs (MNPs), 241, 242
magnetite (Fe_3O_4), 83–85
membrane technology, 49, 105
limitations of, 50
microfibrillated cellulose (MFC), 109
microwave-assisted synthesis, 85
microwave reduction, 131
multi-walled carbon nanotubes (MWCNTs), 69
municipal wastewater sources, 235
MXene, 50, 51, 59, 60
membranes preparation, methods, 51–54

nanocellulose, 107, 120
auto-self-healing property, 117–118
self-healing predicted mechanism, 118–120
cellulose nanocrystals (CNCs), 107–108
cellulose nanofibrils (CNFs), 108
membranes, antifouling, 112–117
microfibrillated cellulose (MFC), 109
using in membranes, advantages, 111–112
nanotechnology, 103, 237–238, 248
nanozerovalent iron, 81–83
next-generation materials, 156–157
nickel oxide (NiO), 328–335

organic pollutants, 86
catalytic reduction, NaBH4, 86–88

photocatalysis, 26–28, 41, 81, 281, 282
photocatalyst performance, 259
fundamental principles, 259–262
photocatalytic advanced oxidation (PAO) technologies, 26, 27
photocatalytic oxidation, pollutants, 90
bare magnetic materials, photoactivity, 90–91
heterojunction-based magnetic photocatalysts, 91–92
insulator role, 92–94
photo-Fenton process, 88–90
physisorption, 300
polymeric membranes, 50
characterization, 2–3
elemental compositional analysis, 3
FT-IR spectroscopy studies, 3–4
scanning electron microscopy investigation, 3
X-ray diffraction (XRD) spectra, 4–5
chitosan-based nanomaterials and, 8–12
chitosan-based polymers, performance comparison, 14
chitosan–metal oxide-based, 12–13
initial dye concentration, 13–14
water contaminated with pharmaceuticals, treatment, 14
zero-, one-, two-, and three-dimensional nanomaterials for, 5–8
polymeric responsive membranes
direct stimulation, 157
electro-responsive membranes (ERMs), 161–162
anodic potential stimuli, 163–164
cathodic potential stimuli, 164–166
electrochemical stimulus, 162–163
field-responsive polymeric membranes, 160–161
hygro-responsive membranes, 159
indirect stimulation, 160
ionic strength-responsive membranes, 159
pH-responsive membranes, 157–159
temperature-responsive membranes, 160
pristine graphene, 186

quaternary ammonium anion exchanger/quaternized bioadsorbent (STEA), 209, 224–226, 231
adsorption studies, 211–212
characterization of, 211
chemical and structural characterization, 215–216
RGY adsorption enthalpy, 226–228
synthesis of, 210–211

reactive oxygen species (ROS), 79
reduced graphene oxide (rGO), 128, 130–133, 203
and fabrication methods, 187–188
remazol golden yellow (RGY) RNL dye, 210, 231
remediation, water, 25
innovative membrane separation techniques, 32
photocatalysis, 26–28

polymer membranes,
enhanced filtration, 32–33
water purification nanocomposites, advances, 28–30
ZnO, biocompatible
semiconductor, 30–32
responsive materials, 156–157

self-healing predicted mechanism,
118–120
silicon dioxide (SiO_2) (silica),
335–337
silver nanoparticles (Ag NPs),
310–316
single-walled carbon nanotubes
(SWCNTs), 69
sol–gel route, 85
sputtering, 84
sugarcane bagasse (SB), 209
characterization of, 211
chemical and structural
characterization, 215–216
Sustainable Development Goals
(SDGs), 102
sustainable materials, 38, 40
SWOT (strengths, weaknesses,
opportunities, threats) analysis,
80

techno-economic feasibility
analysis, 74
technologies, nanocellulose
membranes, 109
bottom-up methods, 109
top-down methods, 110–111
temperature-responsive
membranes, 175
thermal decomposition, 84
thermal reduction, 131
TiO_2-based composites, 262, 267,
281, 282
catalyst design, factors,
269–270
supportive substrate,
270–274
drawbacks, literature,
268–269
initial dye concentration,
277–278
light intensity and irradiation
time, 274–277
search design exclusion
criteria, 265
grouping, synthesis,
265
inclusion criteria,
264–265
search strategy, PRISMA 2020
checklist, 265–266
solution pH and water matrix,
278–279
specific research questions,
263–264
study selection and data
extraction, 266–267
temperature, influence,
279–281
water decolorization,
267–268
photocatalytic material,
262–263
titanium dioxide (TiO_2)/titania,
259, 262
toxic substances, 299

universal solvent, 299

zeolites, 337–339
zerovalent iron (nZVI)
nanoparticles, 243, 244
zinc oxide nanoparticles (ZnO NPs),
325–328